# Achieve Your Potential

The author, Mark Dugopolski, has developed specific content in MyMathLab® to ensure you have many resources to help you achieve success in mathematics—and beyond! The MyMathLab features described here will help you:

- **Review** math skills and concepts you may have forgotten
- **Retain** new concepts as you move through your math course
- **Develop** skills that will help with your transition from college to your profession

## Adaptive Study Plan

The Study Plan will help you study more efficiently and effectively.

Your performance and activity are assessed continually in real time, providing a personalized experience based on your individual needs.

**Study Plan**

Study Plan

You have earned 0 of 687 mastery points (MP).

Practice these objectives and then take a Quiz Me to prove mastery and earn more points.

View progress

What to work on next

**GR.1 Real Number System**

Identify types of numbers. | Practice | Quiz Me | 0 of 1 MP

More Objectives to practice and master | View all chapters

**GR.1 Real Number System**

Graph numbers on a number line. | Practice | Quiz Me | 0 of 1 MP
Find absolute values. | Practice | Quiz Me | 0 of 1 MP
Calculate with real numbers. | Practice | Quiz Me | 0 of 1 MP
Use the properties of real numbers. | Practice | Quiz Me | 0 of 1 MP

MyMathLab®
with KNEWTON Adaptive Learning

## Skills for Success

The Skills for Success Module supports your continued success in college.

This module provides tutorials and guidance on a variety of topics, including transitioning to college, online learning, time management and more.

Additional content is provided to help with the development of professional skills such as resume writing or creation and interview preparation.

## Getting Ready

Are you frustrated when you know you learned a math concept in the past, but you can't quite remember the skill when it's time to use it? Don't worry!

The author has included Getting Ready material so you can brush up on forgotten material efficiently by taking a quick skill review quiz to pinpoint the areas where you need help.

Then, a personalized homework assignment provides additional practice on those forgotten concepts, right when you need it.

## Review Exercises

As you work through your math course, these MyMathLab® exercises support ongoing review to help you maintain essential skills.

The ability to recall important math concepts as you continually acquire new mathematical skills will help you be successful in this math course and in your future math courses.

Scan the code or visit **www.mymathlab.com/success-stories** to learn how students across the country have used MyMathLab to achieve their potential and be successful in their math courses.

# Trigonometry

**Mark Dugopolski**

*Southeastern Louisiana University*

**PEARSON**

Boston   Columbus   Indianapolis   New York   San Francisco   Upper Saddle River
Amsterdam   Cape Town   Dubai   London   Madrid   Milan   Munich   Paris   Montréal   Toronto
Delhi   Mexico City   São Paulo   Sydney   Hong Kong   Seoul   Singapore   Taipei   Tokyo

**Editor in Chief:** Anne Kelly
**Acquisitions Editor:** Kathryn O'Connor
**Content Editor:** Kerri Consalvo
**Editorial Assistant:** Judith Garber
**Senior Managing Editor:** Karen Wernholm
**Production Project Manager:** Mary Sanger
**Associate Director of Design, USHE North and West:** Andrea Nix
**Cover Designer:** Barbara T. Atkinson
**Cover Image:** NorSob/Shutterstock
**Marketing Manager:** Peggy Lucas
**Marketing Assistant:** Justine Goulart
**Senior Author Support/Technology Specialist:** Joe Vetere
**Procurement Specialist:** Debbie Rossi
**Associate Media Producer:** Nicholas Sweeny
**MathXL Project Supervisor:** Kristina Evans
**Test Gen Senior Content Developer:** John Flanagan
**Rights and Permissions Advisor:** Dana Weightman
**Image Manager:** Rachel Youdelman
**Photo Research:** Integra
**Design, Production Coordination, Technical Illustrations, and Composition:** Cenveo® Publisher Services

Photo credits appear on page 354, which constitutes a continuation of the copyright page.

Many of the designations used by manufacturers and sellers to distinguish their products are claimed as trademarks. Where those designations appear in this book, and Pearson was aware of a trademark claim, the designations have been printed in initial caps or all caps.

**Library of Congress Cataloging-in-Publication Data**

Dugopolski, Mark.
  Trigonometry / Mark Dugopolski, Southeastern Louisiana University. -- 4th edition.
    pages cm
  ISBN 0-321-92348-0
1. Trigonometry--Textbooks.  I. Title.
  QA531.D764 2014
  516.24--dc23

2013023105

7  17

**PEARSON**

http://www.pearsonhighered.com

ISBN-10:  0-321-92348-0
ISBN-13: 978-0-321-92348-6

# CONTENTS

Preface vii
Resources for Success x

## P Algebraic Prerequisites 1

P.1 The Cartesian Coordinate System 2
P.2 Functions 13
P.3 Families of Functions, Transformations, and Symmetry 20
P.4 Compositions and Inverses 30
Chapter P Highlights 40
Chapter P Review Exercises 42
Chapter P Test 43

## 1 Angles and the Trigonometric Functions 44

1.1 Angles and Degree Measure 45
1.2 Radian Measure, Arc Length, and Area 54
1.3 Angular and Linear Velocity 64
1.4 The Trigonometric Functions 71
1.5 Right Triangle Trigonometry 81
1.6 The Fundamental Identity and Reference Angles 97
Chapter 1 Highlights 104
Chapter 1 Review Exercises 106
Chapter 1 Test 108

## 2 Graphs of the Trigonometric Functions 109

2.1 The Unit Circle and Graphing 110
2.2 The General Sine Wave 122
2.3 Graphs of the Secant and Cosecant Functions 133
2.4 Graphs of the Tangent and Cotangent Functions 140
2.5 Combining Functions 148
Chapter 2 Highlights 155
Chapter 2 Review Exercises 156
Chapter 2 Test 157
Tying It All Together 158

# 3 Trigonometric Identities  160

**3.1**   Basic Identities  161
**3.2**   Verifying Identities  170
**3.3**   Sum and Difference Identities for Cosine  177
**3.4**   Sum and Difference Identities for Sine and Tangent  184
**3.5**   Double-Angle and Half-Angle Identities  189
**3.6**   Product and Sum Identities  198
        Chapter 3 Highlights  208
        Chapter 3 Review Exercises  209
        Chapter 3 Test  211
        Tying It All Together  211

# 4 Solving Conditional Trigonometric Equations  213

**4.1**   The Inverse Trigonometric Functions  214
**4.2**   Basic Sine, Cosine, and Tangent Equations  225
**4.3**   Equations Involving Compositions  237
**4.4**   Trigonometric Equations of Quadratic Type  245
        Chapter 4 Highlights  253
        Chapter 4 Review Exercises  253
        Chapter 4 Test  255
        Tying It All Together  256

# 5 Applications of Trigonometry  257

**5.1**   The Law of Sines  258
**5.2**   The Law of Cosines  268
**5.3**   Area of a Triangle  278
**5.4**   Vectors  285
**5.5**   Applications of Vectors  295
        Chapter 5 Highlights  304
        Chapter 5 Review Exercises  306
        Chapter 5 Test  308
        Tying It All Together  309

# 6 Complex Numbers, Polar Coordinates, and Parametric Equations 310

6.1 Complex Numbers 311
6.2 Trigonometric Form of Complex Numbers 319
6.3 Powers and Roots of Complex Numbers 326
6.4 Polar Equations 332
6.5 Parametric Equations 343
Chapter 6 Highlights 350
Chapter 6 Review Exercises 351
Chapter 6 Test 352
Tying It All Together 353

**Online Appendix A: Solutions to Try This Exercises**
**Online Appendix B: More Outside the Box**
**Photo Credits** 354
**Answers to Exercises** A-1
**Index of Applications** I-0
**Index** I-2
**Function Gallery** I-6

# PREFACE

The fourth edition of *Trigonometry* is designed for a variety of students with different mathematical needs. For those students who will take additional mathematics, this text will provide the proper foundation of skills, understanding, and insights for success in subsequent courses. For those students who will not further pursue mathematics, the extensive emphasis on applications and modeling will demonstrate the usefulness and applicability of trigonometry in the modern world. I am always on the lookout for real-life applications of mathematics, and I have included many problems that people actually encounter on the job. With an emphasis on problem solving, this text provides students an excellent opportunity to sharpen their reasoning and thinking skills. With increased problem-solving capabilities, students will have confidence to tackle problems that they encounter outside the mathematics classroom.

## New to This Edition

For this edition of *Trigonometry*, I have updated explanations, examples, exercises, and art in response to users of the third edition. Here are the major changes in this edition:

- The number of fill-in-the-blank exercises has been increased to help students master the concepts.
- The exercise sets now contain new headings that group the exercises by type: Concepts, Skills, and Modeling.
- Exercise sets now contain QR codes that link the exercises to videos on the Web.
- A new *Video Notebook* gives students a structured place to take notes as they watch the videos. Definitions, examples, and important concepts are highlighted, and helpful hints are pointed out along the way. The notebook is available in MyMathLab®.
- Many new Outside the Box exercises have been added so that there are now two of them at the end of every section.
- A review of square roots and the rules for simplifying them is now included in Section P.1 along with the Pythagorean theorem.
- Chapter 4, Solving Conditional Trigonometric Equations, has been revamped for better flow and organization. It now contains 139 new exercises
- Suggested homework problems are indicated in the Annotated Instructor's Edition. Each section contains approximately 20 underlined exercise numbers that can be used for an initial assignment of homework exercises.

## Continuing Features

With each new edition, all of the features are reviewed to make sure they are providing a positive impact on student success. The continuing features of the text are listed here.

### Strategies for Success

- **Chapter Opener** Each chapter begins with chapter opener text that discusses a real-world situation in which the mathematics of the chapter is used. Examples and exercises that relate back to the chapter opener are included in the chapter.
- **Try This** Occurring immediately after every example in the text is an exercise that is very similar to the example. These problems give students the opportunity

to immediately check their understanding of the example. Solutions to all *Try This* exercises are in Online Appendix A, which can be found in MyMathLab or in the Instructor Resource Center.

- **Summaries** of important concepts are included to help students clarify ideas that have multiple parts.
- **Strategies** contain general guidelines for solving certain types of problems. They are designed to help students sharpen their problem-solving skills.
- **Procedures** are similar to *Strategies*, but are more specific and more algorithmic. *Procedures* are designed to give students a step-by-step approach for solving a specific type of problem.
- **Function Galleries** Located throughout the text, these function summaries are also gathered together at the end of the text. These graphical summaries are designed to help students link the visual aspects of various families of functions to the properties of the functions.
- **Hints** Selected applications include hints that are designed to encourage students to start thinking about the problem at hand. A *Hint* logo HINT is used where a hint is given.
- **Graphing Calculator Discussions** Optional graphing calculator discussions have been included in the text. They are clearly marked by graphing calculator icons so that they can be easily skipped if desired. Although the graphing calculator discussions are optional, all students will benefit from reading them. In this text, the graphing calculator is used as a tool to support and enhance conclusions, not to make conclusions.

### Section Exercises and Review

- **For Thought** Each exercise set is preceded by a set of ten true/false questions that review the basic concepts in the section, help check student understanding, and offer opportunities for writing and discussion. The answers to all *For Thought* exercises are included in the back of the *Student Edition*.
- **Exercise Sets** The exercise sets range from easy to challenging and are arranged in order of increasing difficulty. Those exercises that require a graphing calculator are optional and are marked with an icon.
- **Writing/Discussion and Cooperative Learning Exercises** These exercises deepen students' understanding by giving them the opportunity to express mathematical ideas both in writing and to their classmates during small group or team discussions.
- **Outside the Box** Found throughout the text, these problems are designed to get students and instructors to do some mathematics just for fun. I enjoyed solving these problems and hope that you will also. The problems can be used for individual or group work. They may or may not have anything to do with the sections in which they are located and might not even require any techniques discussed in the text. So be creative and try thinking *Outside the Box*. The answers are given in the *Annotated Instructor's Edition* only, and complete solutions can be found in the *Instructor's Solutions Manual*. Online Appendix B contains 34 extra Outside the Box problems.
- **Pop Quizzes** Included at the end of every section of the text, the *Pop Quizzes* give instructors and students convenient quizzes that can be used in the classroom to check understanding of the basics. The answers appear in the *Annotated Instructor's Edition* only.
- **Linking Concepts** This feature is located at the end of nearly every exercise set. It is a multipart exercise or exploration that can be used for individual or group work. The idea of this feature is to use concepts from the current section along with concepts from preceding sections or chapters to solve problems that illustrate the links among various ideas. Some parts of these questions are open-ended and require somewhat more thought than standard skill-building exercises. Answers are given in the *Annotated Instructor's Edition* only, and full solutions can be found in the *Instructor's Solutions Manual*.

**Chapter Review**

- **Highlights** This end-of-chapter feature contains an overview of all of the concepts presented in the chapter, along with brief examples to illustrate the concepts.
- **Chapter Review Exercises** These exercises are designed to give students a comprehensive review of the chapter without reference to individual sections and to prepare students for a chapter test.
- **Chapter Test** The problems in the *Chapter Test* are designed to measure the student's readiness for a typical one-hour classroom test. Instructors may also use them as a model for their own end-of-chapter tests. Students should be aware that their in-class test could vary from the *Chapter Test* due to different emphasis placed on the topics by individual instructors.
- **Tying It All Together** Found at the end of most chapters in the text, these exercises help students review selected concepts from the present and prior chapters and require students to integrate multiple concepts and skills.
- **Index of Applications** The many applications contained within the text are listed in the *Index of Applications* that appears at the end of the text. The applications are page referenced and grouped by subject matter.

# RESOURCES FOR SUCCESS

## MyMathLab® Online Course (access code required)

MyMathLab from Pearson is the world's leading online resource in mathematics, integrating interactive homework, assessment, and media in a flexible, easy-to-use format. It provides **engaging experiences** that personalize, stimulate, and measure learning for each student. And it comes from an **experienced partner** with educational expertise and an eye on the future.

To learn more about how MyMathLab combines proven learning applications with powerful assessment, visit **www.mymathlab.com** or contact your Pearson representative.

MyMathLab helps prepare students and gets them thinking more conceptually and visually through the following features:

### Getting Ready ⟶

Students refresh prerequisite topics through assignable skill review quizzes and personalized homework integrated in MyMathLab.

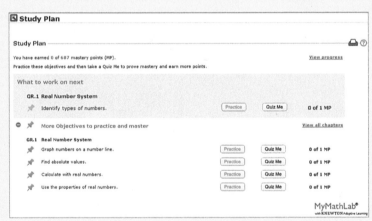

### Adaptive Study Plan

The Study Plan makes studying more efficient and effective for every student. Performance and activity are assessed continually in real time. The data and analytics are used to provide personalized content, reinforcing concepts that target each student's strengths and weaknesses.

### Skills for Success Modules

Integrated within MyMathLab, these modules help students succeed in collegiate courses and prepare for future professions.

### Review Your Knowledge

Review Exercises in the text are now assignable in MyMathLab and require students to recall previously learned content and skills. By helping students maintain essential skills throughout the course, they retain information that prepares them for future math courses.

### Enhanced Graphing Functionality

New functionality within the graphing utility allows graphing of 3-point quadratic functions, 4-point cubic graphs, and transformations in exercises.

## Video Assessment

Video assessment is tied to the video lecture for each section of the book to check students' understanding of important math concepts. Instructors can assign these questions as a prerequisite to homework assignments.

## Instructor Resources

Additional resources can be downloaded from www.pearsonhighered.com or hardcopy resources can be ordered from your sales representative.

### Ready to Go Courses

Now it is even easier to get started with MyMathLab. The Ready to Go MyMathLab course option includes author-chosen preassigned homework, integrated review, and more.

### TestGen®

TestGen® (www.pearsoned.com/testgen) enables instructors to build, edit, print, and administer tests using a computerized bank of questions developed to cover all the objectives of the text.

### PowerPoint® Lecture Slides

Classroom presentation slides are geared specifically to sequence the text. Available in MyMathLab.

### Annotated Instructor's Edition

Provides answers beside the text exercises for most exercises and in an answer section at the back of the book for all others. Groups of exercises are keyed back to corresponding examples from the section.

### Instructor's Solutions Manual

Provides complete solutions to all text exercises, including the *For Thought* and *Linking Concepts* exercises.

## Student Resources

Additional resources are available to help students succeed.

### Lecture Video

Example and content videos provide comprehensive coverage of each section and topic in the text in an engaging format that stresses student interaction. Includes optional subtitles in English and Spanish. All videos are assignable within MyMathLab.

### Student's Solutions Manual

Provides detailed solutions to odd-numbered exercises in the text.

### NEW Video Notebook

This printed guide gives students a structured place to take notes as they watch the videos. Definitions, examples, and important concepts are highlighted, and helpful hints are pointed out along the way. The notebook is available in MyMathLab.

### Graphing Calculator Manual

Provides instructions and keystroke operations for the TI-83/84 Plus and TI-89. Contains worked-out examples taken directly from the text. Available in MyMathLab.

# ACKNOWLEDGMENTS

Thanks to the many professors and students who have used this text in previous editions. I am always glad to hear from users of my texts. You can e-mail me at bookinit@charter.net. Thanks to the following reviewers, whose comments and suggestions were invaluable in preparing this fourth edition:

Douglas B. Aichele, *Oklahoma State University*

Richard Araiza, *Albuquerque Technical Vocational Institute*

Angela Beltran-Aguilar, *Mira Costa College*

David C. Bramlett, *Jackson State University*

Thomas R. Caplinger, *University of Memphis*

Marte M. Carter, *Central Missouri State University*

Frederick Chapple, *Baltimore City Community College*

Julane B. Crabtree, *Johnson County Community College*

Diane Dexter Ellis, *Mississippi State University*

Thomas L. Fitzkee, *Francis Marion University*

Ken Harrelson, *Oklahoma City Community College*

Janet Harter, *Amarillo College*

Brian Hickey, *East Central College*

Timothy D. Hudson, *East Carolina University*

Evan Innerst, *Canada College*

Christine Ingrid Keller, *Southwest Texas State University*

Kimberly Johnson, *Mesa Community College*

Carol Miller, *Ivy Technical Community College–Muncie*

John Nichols, *Oklahoma Baptist University*

Robert Parr, *Blinn College–Bryan*

Pat Simms, *Ouachita Technical College*

Donald P. Skow, *University of Texas, Pan American*

Edward Swenson, *Forsyth Technical Community College*

Alexandra Tauson, *Community College of Allegheny County*

Kirk C. Trigsted, *University of Idaho*

John Vogt, *Newman University*

Thanks to Edgar N. Reyes, Southeastern Louisiana University, for working all of the exercises and writing the *Solutions Manuals*; Rebecca W. Muller, Southeastern Louisiana University, for writing the *Instructor's Testing Manual*; Darryl Nester, Bluffton College, for writing the *Graphing Calculator Manual*; and Dylan Faullin, for writing the *Video Notebook*. I wish to express my thanks to Deana Richmond and Helen Medley for accuracy checking. A special thanks to Cenveo, the compositor, for the superb work they did on this book.

Finally, it has been another wonderful experience working with the talented and dedicated Pearson Education team. I wish to thank them for their assistance, encouragement, and direction on this project: Greg Tobin, President, for his superior guidance; Katie O'Connor, Acquisitions Editor, whose ideas were central in making this revision a success; Peggy Lucas, Marketing Manager, for creatively tying all the ideas together to get the message out; Kerri Consalvo, Content Editor, for keeping it all on track with poise; Mary Sanger, Production Project Manager, whose knowledge was integral to the production of this text; Barbara Atkinson, Senior Designer, whose eye for design helped improve the book's look dramatically; Nick Sweeny, Associate Media Producer, who put together a top-notch media package; Judith Garber, Editorial Assistant, and Justine Goulart, Marketing Assistant, who ran the behind-the-scenes work; and Joseph Vetere, Senior Author Support/Technical Specialist, for his expertise in the production of this book.

As always, thanks to my wife, Cheryl, whose love, encouragement, understanding, support, and patience are invaluable.

**Mark Dugopolski**
*Ponchatoula, Louisiana*

# P

# Algebraic Prerequisites

**P.1** The Cartesian Coordinate System

**P.2** Functions

**P.3** Families of Functions, Transformations, and Symmetry

**P.4** Compositions and Inverses

In the 1995 America's Cup race the defending *Young America* kept up a good face, but from the start it was clear that the New Zealand entry, *Black Magic,* was sailing higher and faster in the 11-knot breeze and rolling swells. For two-and-a-half years the crew had done everything they could to prepare for that moment.

Gone are the days when raw sailing ability and stamina won races like the America's Cup. New Zealand's *Black Magic* sported a mathematically created design that reduced its drag and optimized its stability and speed.

**WHAT YOU WILL LEARN**

**Preparation is just as important in trigonometry as it is in sailing. In this chapter we review some of the basics of algebra that are necessary for success in trigonometry. Throughout this chapter, you will see that even basic concepts of algebra have applications in business, science, engineering, and even sailing.**

# P.1 The Cartesian Coordinate System

**Table P.1**

| Toppings $x$ | Cost $y$ |
|:---:|:---:|
| 0 | $ 5 |
| 1 | 7 |
| 2 | 9 |
| 3 | 11 |
| 4 | 13 |

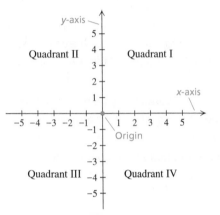

**Figure P.1**

In this chapter we study pairs of variables and how they are related. For example, $p$ might represent the price of gasoline and $n$ the number of gallons that you consume in a month; $x$ might be the number of toppings on a medium pizza and $y$ the cost of that pizza; or $h$ might be the height of a two-year-old child and $w$ the weight. To picture relationships between pairs of variables we use a two-dimensional coordinate system.

## A Two-Dimensional Coordinate System

Suppose that the cost of a medium pizza is $5 plus $2 per topping. Table P.1 shows a chart in which $x$ is the number of toppings and $y$ is the cost. To indicate that $x = 3$ and $y = 11$ go together, we can use the **ordered pair** $(3, 11)$. The ordered pair $(2, 9)$ indicates that a 2-topping pizza costs $9. The **first coordinate** of the ordered pair represents the number of toppings and the **second coordinate** represents the cost. The assignment of what the coordinates represent is arbitrary, but once made it is kept fixed. For example, in the present context the ordered pair $(11, 3)$ indicates that an 11-topping pizza costs $3, and it is not the same as $(3, 11)$. This is the reason they are called "ordered pairs."

Note that parentheses are used to indicate ordered pairs and also to indicate intervals of real numbers. For example, $(3, 7)$ could be an ordered pair or the interval of real numbers between 3 and 7. However, the meaning of this notation should always be clear from the discussion.

Ordered pairs are pictured on a plane in the **rectangular coordinate system** or **Cartesian coordinate system,** named after the French mathematician René Descartes (1596–1650). The Cartesian coordinate system consists of two number lines drawn perpendicular to one another, intersecting at zero on each number line as in Fig. P.1. The point at which they intersect is called the **origin.** The horizontal number line is the $x$-**axis** and the vertical number line is the $y$-**axis.** The two number lines divide the plane into four regions called **quadrants,** numbered as in Fig. P.1. The quadrants do not include any points on the axes.

We call a plane with a rectangular coordinate system the **coordinate plane** or the $xy$-**plane.** Every ordered pair of real numbers $(a, b)$ corresponds to a point $P$ in the $xy$-plane. For this reason, ordered pairs of numbers are often called **points.** The numbers $a$ and $b$ are called the **coordinates** of point $P$. Locating the point $P$ that corresponds to $(a, b)$ in the $xy$-plane is referred to as **plotting** or **graphing** the point, and $P$ is called the **graph** of $(a, b)$.

### EXAMPLE 1 Plotting points

Plot the points $(3, 5)$, $(4, -5)$, $(-3, 4)$, $(-2, -5)$, and $(0, 2)$ in the $xy$-plane.

**Solution**

The point $(3, 5)$ is located 3 units to the right of the origin and 5 units above the $x$-axis as shown in Fig. P.2. The point $(4, -5)$ is located 4 units to the right of the origin and 5 units below the $x$-axis. The point $(-3, 4)$ is located 3 units to the left of the origin and 4 units above the $x$-axis. The point $(-2, -5)$ is located 2 units to the left of the origin and 5 units below the $x$-axis. The point $(0, 2)$ is on the $y$-axis because its first coordinate is zero.

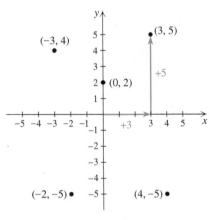

**Figure P.2**

**TRY THIS.** Plot $(-3, -2)$, $(-1, 3)$, $(5/2, 0)$, and $(2, -3)$. ∎

Note that for points in quadrant I, both coordinates are positive. In quadrant II the first coordinate is negative and the second is positive, whereas in quadrant III, both coordinates are negative. In quadrant IV the first coordinate is positive and the second is negative.

# The Pythagorean Theorem

To find a formula for the distance between two points in the Cartesian coordinate system we need the Pythagorean theorem. The Pythagorean theorem gives a relationship between the legs and hypotenuse of a right triangle.

**The Pythagorean Theorem**

In a right triangle the sum of the squares of the lengths of the legs is equal to the square of the length of the hypotenuse.

$$a^2 + b^2 = c^2$$

The Pythagorean theorem is probably the most widely known theorem in mathematics. It was known to the Babylonians 4000 years ago and has been proven by such notables as Euclid, Leonardo da Vinci, and James Garfield (the 20th U.S. president). You can find many proofs on the Internet. The first site that this author found contained 54 proofs.

If any two sides of a right triangle are known, the Pythagorean theorem can be used to find the third side, as shown in the next example.

**EXAMPLE 2**   **Applying the Pythagorean theorem**

Find the unknown side in each given triangle.

**a.**                      **b.**

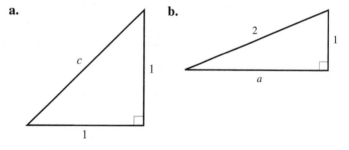

**Solution**

**a.** The lengths of the legs are each 1 unit and the hypotenuse is unknown. Use $c^2 = a^2 + b^2$ with $a = 1$ and $b = 1$:

$$c^2 = 1^2 + 1^2$$
$$c^2 = 2$$
$$c = \pm\sqrt{2}$$

Since the length of the hypotenuse is positive, $c = \sqrt{2}$.

**b.** The hypotenuse is 2, one leg is 1, and the other leg is unknown. Use $a^2 + b^2 = c^2$ with $c = 2$ and $b = 1$:

$$a^2 + 1^2 = 2^2$$
$$a^2 = 3$$
$$a = \pm\sqrt{3}$$

Since the length of a leg of a triangle is positive, $a = \sqrt{3}$.

**TRY THIS.** The legs of a right triangle are 2 feet and 3 feet. Find the length of the hypotenuse.

## Simplifying Square Roots

The exact lengths of the missing sides of the triangles in Example 2 are $\sqrt{2}$ and $\sqrt{3}$. Values for these roots, such as 1.414 and 1.732, that you get from a calculator are approximate. Since these numbers are irrational, they can't be expressed exactly in decimal form. In trigonometry we will often encounter square roots in their exact form. There are two rules to remember when working with square roots:

**Rule: Product and Quotient**

> For any real numbers $a$ and $b$
>
> **1.** $\sqrt{a} \cdot \sqrt{b} = \sqrt{ab}$      Product rule for square roots
>
> **2.** $\dfrac{\sqrt{a}}{\sqrt{b}} = \sqrt{\dfrac{a}{b}}$   $(b \neq 0)$      Quotient rule for square roots
>
> provided that all of the roots are real numbers.

When using square roots to express exact results we use the product and quotient rules to write them in *simplified form*. (Note: In $\sqrt{a}$, the **radicand** is $a$.)

**Definition: Simplified Form for Square Roots**

> An expression involving a square root is in **simplified form** if it has
>
> **1.** *no* perfect squares as factors of the radicand,
> **2.** *no* fractions in the radicand, and
> **3.** *no* square roots in a denominator.

Removing a square root from a denominator is called **rationalizing the denominator.** The next example shows how the product and quotient rules are used to rationalize denominators and get radical expressions into simplified form.

**EXAMPLE 3**   Simplifying square roots

Write each expression in simplified form.

**a.** $\sqrt{20}$   **b.** $\sqrt{\dfrac{5}{9}}$   **c.** $\sqrt{\dfrac{32}{5}}$

**Solution**

**a.** $\sqrt{20} = \sqrt{4 \cdot 5}$      Identify 4 as a perfect square factor.

$\qquad\quad = \sqrt{4} \cdot \sqrt{5}$      Product rule for square roots

$\qquad\quad = 2\sqrt{5}$      Simplify $\sqrt{4}$.

**b.** $\sqrt{\dfrac{5}{9}} = \dfrac{\sqrt{5}}{\sqrt{9}}$      Quotient rule for square roots

$\qquad\quad = \dfrac{\sqrt{5}}{3}$      Simplify $\sqrt{9}$.

**c.** $\sqrt{\dfrac{32}{5}} = \dfrac{\sqrt{32}}{\sqrt{5}}$      Quotient rule for square roots

$\qquad\quad = \dfrac{\sqrt{16} \cdot \sqrt{2}}{\sqrt{5}}$      Product rule for square roots

$\qquad\quad = \dfrac{4 \cdot \sqrt{2}}{\sqrt{5}}$      Simplify.

$\qquad\quad = \dfrac{4 \cdot \sqrt{2} \cdot \sqrt{5}}{\sqrt{5} \cdot \sqrt{5}}$      Rationalize the denominator.

$\qquad\quad = \dfrac{4\sqrt{10}}{5}$      Simplified form

**Figure P.3**

**TRY THIS.** Simplify $\sqrt{50}$, $\sqrt{\dfrac{3}{16}}$, and $\sqrt{\dfrac{12}{7}}$.   ■

## The Distance Formula

Consider the points $A(x_1, y_1)$ and $B(x_2, y_2)$ shown in Fig. P.3. Let $d$ represent the length of the line segment $\overline{AB}$. Now $\overline{AB}$ is the hypotenuse of the right triangle in Fig. P.3. The distance between $A$ and $C$ is $|x_2 - x_1|$ and the distance between $B$ and $C$ is $|y_2 - y_1|$. Using the Pythagorean theorem we have

$$d^2 = |x_2 - x_1|^2 + |y_2 - y_1|^2.$$

The absolute value symbols can be replaced by parentheses, because $|a - b|^2 = (a - b)^2$ for any real numbers $a$ and $b$. Since the distance between two points is nonnegative, we have the following formula.

**The Distance Formula**

> The distance $d$ between the points $(x_1, y_1)$ and $(x_2, y_2)$ is given by the formula
> $$d = \sqrt{(x_2 - x_1)^2 + (y_2 - y_1)^2}.$$

**EXAMPLE 4**   **Applying the distance formula**

Find the distance between $(5, -3)$ and $(-1, -6)$.

**Solution**

Let $(x_1, y_1) = (5, -3)$ and $(x_2, y_2) = (-1, -6)$. See Fig. P.4. The distance is the same regardless of which point is chosen as $(x_1, y_1)$, or $(x_2, y_2)$. Substitute these values into the distance formula:

$$\begin{aligned} d &= \sqrt{(-1 - 5)^2 + (-6 - (-3))^2} \\ &= \sqrt{(-6)^2 + (-3)^2} \\ &= \sqrt{36 + 9} = \sqrt{45} = \sqrt{9}\sqrt{5} = 3\sqrt{5} \end{aligned}$$

The exact distance between the points is $3\sqrt{5}$. Note that $\sqrt{36 + 9} \neq \sqrt{36} + \sqrt{9}$, but $\sqrt{9 \cdot 5} = \sqrt{9} \cdot \sqrt{5}$.

**TRY THIS.** Find the distance between $(-3, -2)$ and $(-1, 4)$.   ■

## The Midpoint Formula

When you average two test scores (by finding their sum and dividing by 2) you are finding a number midway between the two scores. Likewise, the midpoint of the line segment with endpoints $(x_1, y_1)$ and $(x_2, y_2)$ is found by dividing the sum of the $x$-coordinates by 2 and the sum of the $y$-coordinates by 2. See Fig. P.5.

**Figure P.4**

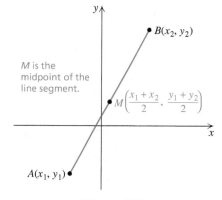

**Figure P.5**

**The Midpoint Formula**

The midpoint of the line segment with endpoints $(x_1, y_1)$ and $(x_2, y_2)$ is

$$\left(\frac{x_1 + x_2}{2}, \frac{y_1 + y_2}{2}\right).$$

The midpoint formula is illustrated in the next example. Note the strange-looking coordinates in part (b). In trigonometry we will often perform arithmetic with the number $\pi$.

**EXAMPLE 5**    **Applying the midpoint formula**

Find the midpoint of the line segment with the given endpoints.

**a.** $(1, -3), (5, 4)$   **b.** $\left(\frac{\pi}{2}, 0\right), (\pi, 0)$

**Solution**

**a.** To find the midpoint add the coordinates and divide by 2:

$$\left(\frac{1 + 5}{2}, \frac{-3 + 4}{2}\right) = \left(3, \frac{1}{2}\right)$$

**b.** To find the midpoint add the coordinates and divide by 2:

$$\left(\frac{\frac{\pi}{2} + \pi}{2}, \frac{0 + 0}{2}\right) = \left(\frac{\frac{\pi}{2} + \frac{2\pi}{2}}{2}, 0\right) = \left(\frac{\frac{3\pi}{2}}{2}, 0\right) = \left(\frac{3\pi}{4}, 0\right)$$

**TRY THIS.** Find the midpoint of the line segment with endpoints $(3, -2)$ and $(5, 6)$. ∎

## The Circle

A **circle** is the set of all points in a plane that lie a fixed distance from a given point in the plane. The fixed distance is the **radius** and the given point is the **center.** The distance formula can be used to write an equation for a circle with center $(h, k)$ and radius $r$ $(r > 0)$. A point $(x, y)$ is on the circle shown in Fig. P.6 if and only if it satisfies the equation

$$\sqrt{(x - h)^2 + (y - k)^2} = r.$$

Using the definition of square root, we can write the following equation.

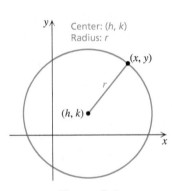

Center: $(h, k)$
Radius: $r$

$(x, y)$

$r$

$(h, k)$

**Figure P.6**

**Theorem: Equation for a Circle**

The equation for a circle with center $(h, k)$ and radius $r$ for $r > 0$ is

$$(x - h)^2 + (y - k)^2 = r^2.$$

The form $(x - h)^2 + (y - k)^2 = r^2$ is the **standard form** for the equation of a circle. If $(h, k)$ is $(0, 0)$, then the circle is centered at the origin and its equation is of the form $x^2 + y^2 = r^2$. If $r = 1$, the circle is called a **unit circle.**

**EXAMPLE 6**    **Graphing a circle**

Sketch the graph of each circle.

**a.** $x^2 + y^2 = 1$   **b.** $(x - 1)^2 + (y + 2)^2 = 9$

**Figure P.7**

**Figure P.8**

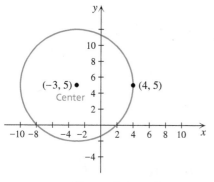

**Figure P.10**

### Solution

**a.** The circle has center $(0, 0)$ and radius 1. You can draw the circle as shown in Fig. P.7 with a compass. To draw the circle by hand locate the points that lie 1 unit above, below, right, and left of the center as shown in Fig. P.7, then sketch a circle through these points.

**b.** This circle has center $(1, -2)$ and radius 3. Plot the center and the points that lie 3 units above, below, left, and right of the center as shown in Fig. P.8. Then sketch a circle through these points.

To support the conclusions of part (b) with a graphing calculator you must first solve the equation for $y$:

$$(x - 1)^2 + (y + 2)^2 = 9$$
$$(y + 2)^2 = 9 - (x - 1)^2$$
$$y + 2 = \pm\sqrt{9 - (x - 1)^2}$$
$$y = -2 \pm \sqrt{9 - (x - 1)^2}$$

Now enter $y_1$ and $y_2$ as shown in Fig. P.9(a). Set the viewing window as shown in Fig. P.9(b). The graph in Fig. P.9(c) supports our previous conclusions. A circle will look round only if the same unit distance is used on both axes.

(a)                    (b)                    (c)

**Figure P.9**

**TRY THIS.** Sketch the graph of $(x + 2)^2 + (y - 4)^2 = 25$.  ■

In the next example we write the equation for a circle from a description of the circle.

**EXAMPLE 7** Writing the equation of a circle

Write the standard equation for a circle with center $(-3, 5)$ and passing through $(4, 5)$ as shown in Fig. P.10.

### Solution

Since the distance between the center $(-3, 5)$ and $(4, 5)$ is 7 units, the radius of the circle is 7. Use $h = -3$, $k = 5$, and $r = 7$ in the standard equation for a circle:

$$(x - (-3))^2 + (y - 5)^2 = 7^2$$

So the equation of the circle is $(x + 3)^2 + (y - 5)^2 = 49$.

**TRY THIS.** Write the standard equation for the circle with center $(2, -1)$ and passing through $(3, 6)$.  ■

## The Line

Any circle in the coordinate plane is the solution set to an equation of the form $(x - h)^2 + (y - k)^2 = r^2$. Likewise, any straight line in the coordinate plane is the solution set to an equation of another form.

**Definition: Linear Equation in Two Variables (Standard Form)**

A **linear equation** in two variables $x$ and $y$ is an equation of the form

$$Ax + By = C$$

where $A$, $B$, and $C$ are real numbers and $A$ and $B$ are not both zero.

The graph of any linear equation is a straight line. A linear equation can be written in many different forms. The equations

$$x = 5 - y, \qquad y = \frac{1}{2}x - 9, \qquad x = 4, \qquad \text{and} \qquad y = 5$$

are all linear equations because they could all be written in the form $Ax + By = C$.

There is only one line containing any two distinct points in the $xy$-plane. So to graph a linear equation we need locate only two points that satisfy the equation and draw a line through them. But which points do we use? Since all lines look alike, what distinguishes one line from another is its location. The best way to show the location is to show the points where the line crosses the $x$- and $y$-axes. These points are called the **$x$-intercept** and **$y$-intercept.**

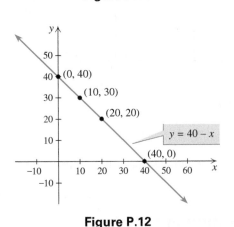

**Figure P.11**

### EXAMPLE 8  Graphing lines and showing the intercepts

Graph each equation. Find and show the intercepts.

**a.** $2x - 3y = 9$  **b.** $y = 40 - x$

**Solution**

**a.** Since the $y$-coordinate of the $x$-intercept is 0, we replace $y$ with 0 in the equation:

$$2x - 3(0) = 9$$
$$2x = 9$$
$$x = \frac{9}{2}$$

To find the $y$-intercept, we replace $x$ with 0 in the equation:

$$2(0) - 3y = 9$$
$$-3y = 9$$
$$y = -3$$

**Figure P.12**

The $x$-intercept is $(9/2, 0)$ and the $y$-intercept is $(0, -3)$. Locate the intercepts and draw the line as shown in Fig. P.11. To check, locate a point such as $(3, -1)$, which also satisfies the equation, and see if the line goes through it.

**b.** If $x = 0$, then $y = 40 - 0 = 40$ and the $y$-intercept is $(0, 40)$. If $y = 0$, then $0 = 40 - x$ or $x = 40$. The $x$-intercept is $(40, 0)$. Draw a line through these points as shown in Fig. P.12. Check that $(10, 30)$ and $(20, 20)$ also satisfy $y = 40 - x$ and the line goes through these points.

 The calculator graph shown in Fig. P.13 is consistent with the graph in Fig. P.12. Note that the viewing window is set to show the intercepts.

**Figure P.13**

**TRY THIS.** Graph $2x + 5y = 10$ and determine the intercepts. ∎

### EXAMPLE 9  Graphing horizontal and vertical lines

Graph each equation in the rectangular coordinate system.

**a.** $y = 3$  **b.** $x = 4$

**Solution**

**a.** The equation $y = 3$ is equivalent to $0 \cdot x + y = 3$. Because $x$ is multiplied by 0, we can choose any value for $x$ as long as we choose 3 for $y$. So ordered pairs such as $(-3, 3)$, $(-2, 3)$, and $(4, 3)$ satisfy the equation $y = 3$. The graph of $y = 3$ is the horizontal line shown in Fig. P.14.

**b.** The equation $x = 4$ is equivalent to $x + 0 \cdot y = 4$. Because $y$ is multiplied by 0, we can choose any value for $y$ as long as we choose 4 for $x$. So ordered pairs such as $(4, -3)$, $(4, 2)$, and $(4, 5)$ satisfy the equation $x = 4$. The graph of $x = 4$ is the vertical line shown in Fig. P.15.

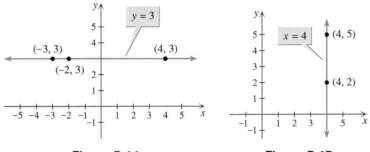

| Figure P.14 | Figure P.15 |

**TRY THIS.** Graph $y = 5$ in the rectangular coordinate system. ■

**FOR THOUGHT...** True or False? Explain.

**1.** The point $(2, -3)$ is in quadrant II.

**2.** The point $(4, 0)$ is in quadrant I.

**3.** The distance between $(a, b)$ and $(c, d)$ is $\sqrt{(a - b)^2 + (c - d)^2}$.

**4.** The equation $3x^2 + y = 5$ is a linear equation.

**5.** The graph of $x = 5$ is a vertical line.

**6.** $\sqrt{7^2 + 9^2} = 7 + 9$.

**7.** The origin lies midway between $(1, 3)$ and $(-1, -3)$.

**8.** The distance between $(3, -7)$ and $(3, 3)$ is 10.

**9.** The $x$-intercept for the graph of $3x - 2y = 7$ is $(7/3, 0)$.

**10.** The graph of $x^2 + y^2 = 9$ is a circle centered at $(0, 0)$ with radius 9.

Scan this code to link to section lecture videos.

## P.1 EXERCISES

**CONCEPTS**

*Fill in the blank.*

**1.** If $x$ and $y$ are real numbers, then $(x, y)$ is a(n) _____ pair of real numbers.

**2.** Ordered pairs are graphed in the rectangular coordinate system or the _____ coordinate system.

**3.** The horizontal number line in the rectangular coordinate system is the _____.

**4.** The intersection of the $x$-axis and the $y$-axis is the _____.

**5.** According to the _____, the sum of the squares of the legs in a right triangle is equal to the square of the hypotenuse.

**6.** The set of all points in a plane that lie a fixed distance from a given point in the plane is a(n) _____.

**7.** An equation of the form $Ax + By = C$ is a(n) _____ in two variables.

**8.** The point where a line crosses the $y$-axis is the _____.

## SKILLS

The figure for Exercises 9–18 shows 10 points in the xy-plane. For each point write the corresponding ordered pair and name the quadrant in which it lies or the axis on which it lies.

**9.** A        **10.** B        **11.** C        **12.** D

**13.** E        **14.** F        **15.** G        **16.** H

**17.** I        **18.** J

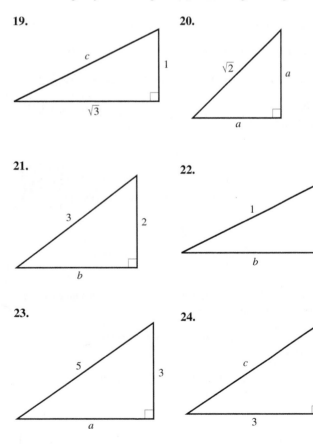

**Figure for Exercises 9–18**

Find the length of the missing side(s) in each right triangle.

**19.**                                    **20.**

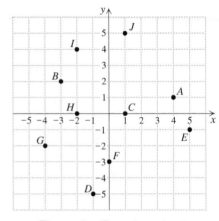

**21.**                                    **22.**

**23.**                                    **24.**

Simplify each expression involving square roots.

**25.** $\sqrt{28}$     **26.** $\sqrt{50}$     **27.** $\sqrt{\dfrac{5}{9}}$     **28.** $\sqrt{\dfrac{3}{16}}$

**29.** $\sqrt{\dfrac{2}{3}}$     **30.** $\sqrt{\dfrac{3}{5}}$     **31.** $\sqrt{\dfrac{12}{5}}$     **32.** $\sqrt{\dfrac{25}{3}}$

**33.** $\dfrac{1}{\sqrt{3}}$     **34.** $\dfrac{3}{\sqrt{2}}$     **35.** $\dfrac{\sqrt{2}}{\sqrt{3}}$     **36.** $\dfrac{\sqrt{5}}{\sqrt{2}}$

For each pair of points find the distance between them and the midpoint of the line segment joining them.

**37.** $(1, 3), (4, 7)$        **38.** $(-3, -2), (9, 3)$

**39.** $(-1, -2), (1, 0)$        **40.** $(-1, 0), (1, 2)$

**41.** $(0, 0), \left(\sqrt{2}/2, \sqrt{2}/2\right)$        **42.** $(0, 0), \left(\sqrt{3}, 1\right)$

**43.** $\left(\sqrt{18}, \sqrt{12}\right), \left(\sqrt{8}, \sqrt{27}\right)$

**44.** $\left(\sqrt{50}, \sqrt{20}\right), \left(\sqrt{72}, \sqrt{45}\right)$

**45.** $(1.2, 4.8), (-3.8, -2.2)$        **46.** $(-2.3, 1.5), (4.7, -7.5)$

**47.** $(\pi, 0), (\pi/2, 1)$        **48.** $(0, 0), (\pi/2, 1)$

**49.** $(\pi, 0), (2\pi, 0)$        **50.** $(\pi, 1), (\pi/2, 1)$

Determine the center and radius of each circle and sketch the graph.

**51.** $x^2 + y^2 = 16$        **52.** $x^2 + y^2 = 1$

**53.** $(x + 6)^2 + y^2 = 36$        **54.** $x^2 = 9 - (y - 3)^2$

**55.** $(x - 2)^2 = 8 - (y + 2)^2$

**56.** $(y + 2)^2 = 20 - (x - 4)^2$

Write the standard equation for each circle.

**57.** Center at $(0, 0)$ with radius $\sqrt{7}$

**58.** Center at $(0, 0)$ with radius $2\sqrt{3}$

**59.** Center at $(-2, 5)$ with radius $1/2$

**60.** Center at $(-1, -6)$ with radius $1/3$

**61.** Center at $(3, 5)$ and passing through the origin

**62.** Center at $(-3, 9)$ and passing through the origin

**63.** Center at $(0, 0)$ and passing through $\left(\sqrt{2}/2, \sqrt{2}/2\right)$

**64.** Center at $(0, 0)$ and passing through $\left(\sqrt{3}/2, 1/2\right)$

*Write the standard equation for each of the following circles.*

**65.**          **66.**

**67.**          **68.**

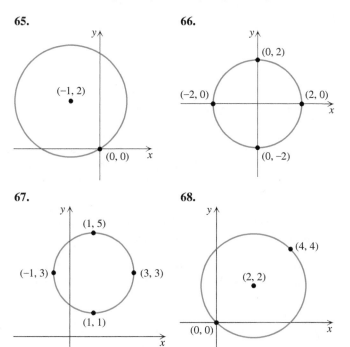

*Find all real numbers a such that the given point is on the circle* $x^2 + y^2 = 1$.

**69.** $(a, 3/5)$          **70.** $(a, -1/2)$

**71.** $(-2/5, a)$          **72.** $(2/3, a)$

*Sketch the graph of each linear equation. Be sure to find and show the x- and y-intercepts.*

**73.** $y = 3x - 4$          **74.** $y = 5x - 5$

**75.** $3x - y = 6$          **76.** $5x - 2y = 10$

**77.** $x + y = 80$          **78.** $2x + y = -100$

**79.** $x = 3y - 90$          **80.** $x = 80 - 2y$

**81.** $\dfrac{1}{2}x - \dfrac{1}{3}y = 600$          **82.** $\dfrac{2}{3}y - \dfrac{1}{2}x = 400$

**83.** $2x + 4y = 0.01$          **84.** $3x - 5y = 1.5$

*Graph each equation in the rectangular coordinate system.*

**85.** $x = 5$     **86.** $y = -2$     **87.** $y = 4$

**88.** $x = -3$     **89.** $x = -4$     **90.** $y = 5$

**91.** $y - 1 = 0$     **92.** $5 - x = 4$

*Graph each of the following equations on a graphing calculator. Choose a viewing window that shows both the x- and y-intercepts, then draw the graph on paper with the x- and y-axes labeled appropriately.*

**93.** $y = x - 20$          **94.** $y = 999x - 100$

**95.** $500x + y = 3000$          **96.** $200x - 300y = 1$

## MODELING

*Solve each problem.*

**97.** A right triangle has legs with lengths 6 ft and 8 ft. What is the length of the hypotenuse?

**98.** A right triangle has a hypotenuse of 10 ft. If one leg is 4 ft, then what is the length of the other leg?

**99.** *Full Conduit* A conduit with an inside diameter of 4 cm can accommodate two round wires each having a diameter of 2 cm, as shown in the accompanying figure. Two smaller wires will also fit on the sides.
   **a.** What is the diameter of each smaller wire?

   **b.** Find the equations of the two smallest circles in the accompanying figure.

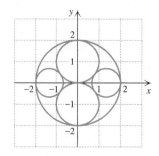

**Figure for Exercise 99**

**100.** *Perpendicular Tangents* A circle of radius 5 intersects a circle of radius 12 so that the tangent lines at the points of intersection are perpendicular. What is the distance between the centers of the circles?

**101.** *More Wires* A round conduit with an inside diameter of 4 cm can accommodate two circular wires each having a diameter of 2 cm, as shown in the accompanying figure. The smallest circle in the figure is tangent to the circle of radius 2, a circle of radius 1, and the x-axis. Find its equation.

**102.** *Three More* Three more circles could be positioned in the accompanying figure so that they are tangent to the circle of radius 2, a circle of radius 1, and the x-axis. Find their equations.

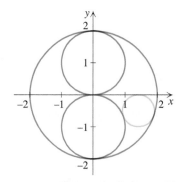

**Figure for Exercises 101 and 102**

**103.** *First Marriage* The median age at first marriage for women went from 20.8 in 1970 to 26.1 in 2012 as shown in the figure (Census Bureau, www.census.gov). Find the midpoint of the line segment in the figure and interpret your result.

**Figure for Exercise 103**

**104.** *Unmarried Couples* The number of unmarried-couple households, *h* (in millions), can be approximated using the equation

$$h = 0.229n + 2.913,$$

where *n* is the number of years since 1990 (Census Bureau, www.census.gov).
**a.** Find and interpret the *n*-intercept for the line.

**b.** Find and interpret the *h*-intercept for the line.

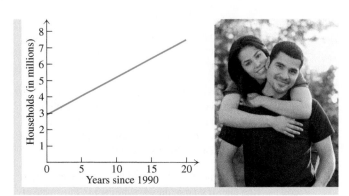

**Figure for Exercise 104**

**WRITING/DISCUSSION**

**105.** *Finding Points* Can you find two points such that their coordinates are integers and the distance between them is 10? $\sqrt{10}$? $\sqrt{19}$? Explain.

**106.** *Equidistant Midpoint* A right triangle has vertices $(0, 0)$, $(1, 0)$, and $\left(1, \sqrt{3}\right)$. Find the midpoint of the hypotenuse. Find the distance from the midpoint of the hypotenuse to each vertex.

**107.** *Cooperative Learning* Working in small groups, plot the points $(-1, 3)$ and $(4, 1)$ on a sheet of graph paper. Assuming that these two points are adjacent vertices of a square, find the other two vertices. Now select your own pair of adjacent vertices and "complete the square." Next, start with the points $(x_1, y_1)$ and $(x_2, y_2)$ as adjacent vertices of a square and write expressions for the coordinates of the other two vertices.

**108.** *Cooperative Learning* Repeat the previous exercise, assuming that the first two points are opposite vertices of a square.

**OUTSIDE THE BOX**

**109.** *Paying Up* A king agreed to pay his gardener one dollar's worth of titanium per day for seven days of work on the castle grounds. The king has a seven-dollar bar of titanium that is segmented so that it can be broken into seven one-dollar pieces, but it is bad luck to break a seven-dollar bar of titanium more than twice. How can the king make two breaks in the bar and pay the gardener exactly one dollar's worth of titanium per day for seven days?

**Figure for Exercise 109**

**110.** *Midpoints* Find the exact area of the triangle in which $(3, -1)$, $(4, 4)$ and $(1, 2)$ are the midpoints of its three sides.

**P.1 POP QUIZ**

**1.** Find the distance between $(-1, 3)$ and $(3, 5)$.

**2.** Find the center and radius for the circle $(x - 3)^2 + (y + 5)^2 = 81$.

**3.** Find the center and radius for the circle $x^2 + 4x + y^2 - 10y = -28$.

**4.** Find the equation of the circle that passes through the origin and has center at $(3, 4)$.

**5.** Find all intercepts for $2x - 3y = 12$.

**6.** Which point is on both of the lines $x = 5$ and $y = -1$?

 Functions

In Section P.1 we studied ordered pairs and graphed sets of ordered pairs. In this section we continue studying sets of ordered pairs, particularly those in which the second coordinate is determined by the first coordinate.

## The Function Concept

If you spend $10 on gasoline, then the price per gallon determines the number of gallons that you get. The number of hours that you sleep before a test might have an influence on but does not determine your grade on the test. If the value of a variable $y$ is determined by the value of another variable $x$, then $y$ **is a function of** $x$. The phrase "is a function of" means "is determined by." *If there is more than one value for $y$ corresponding to a particular $x$-value, then $y$ is not determined by $x$, and $y$ is not a function of $x$.*

The value of one variable might be determined by the values of two or more other variables. For example, the volume of a right circular cylinder is a function of its radius and height ($V = \pi r^2 h$). The volume of a rectangular box is a function of its length, width, and height ($V = LWH$). The basic trigonometric functions that we will be studying in this text are all functions of one variable, but on occasion we will encounter functions of more than one variable.

**EXAMPLE 1** Applying the function concept

Decide whether $a$ is a function of $b$, $b$ is a function of $a$, or neither.

**a.** Let $a$ represent a positive integer smaller than 100 and $b$ represent the number of divisors of $a$.
**b.** Let $a$ represent the age of a U.S. citizen and $b$ represent the number of days since his/her birth.
**c.** Let $a$ represent the age of a U.S. citizen and $b$ represent his/her annual income.
**d.** Let $b$ represent any real number and $a$ represent its square.

**Solution**

**a.** We can determine the number of divisors of any positive integer smaller than 100. So $b$ is a function of $a$. We cannot determine the integer knowing the number of divisors, because different integers have the same number of divisors. So $a$ is not a function of $b$.
**b.** The number of days since a person's birth certainly determines the age of the person in the usual way. So $a$ is a function of $b$. However, you cannot determine the number of days since a person's birth from the age. You need more information. For example, the number of days since the birth for two one-year-olds could be 370 or 380 days. So $b$ is not a function of $a$.
**c.** We cannot determine the income from the age or the age from the income. We would need more information. Even though age and income are related, the relationship is not strong enough to say that either one is a function of the other.
**d.** Since every number has a unique square, the square is determined by the number. So $a$ is a function of $b$. You cannot determine what the number is from knowing its square. For example, if you know the square is 4, then the number is either 2 or $-2$. So $b$ is not a function of $a$.

**TRY THIS.** Let $p$ be the price of a grocery item and $t$ be the amount of sales tax at 5% on that item. Determine whether $p$ is a function of $t$, $t$ is a function of $p$, or neither. ■

The functions that we study will usually be defined by formulas. For example, if $y$ is the square of $x$, then $y$ is a function of $x$. We can write the formula or equation

$y = x^2$ and we say that $y = x^2$ is a function. Because $y$ is determined by $x$, we say that $y$ is the **dependent variable** and $x$ is the **independent variable.** When we use the variables $x$ and $y$, we always use $x$ for the independent variable and $y$ for the dependent variable.

A function provides a rule for pairing values of the independent variable with values of the dependent variable. The function $y = x^2$ pairs 3 with 9, 4 with 16, and so on. We write this pairing as $(3, 9)$, $(4, 16)$, and so on. Of course, there are infinitely many ordered pairs that satisfy $y = x^2$. This set of ordered pairs can be thought of as the function.

**Definition: Function**

A **function** is a set of ordered pairs in which no two pairs have the same first coordinate and different second coordinates.

For example, the set $\{(1, 1), (2, 4), (-2, 4)\}$ is a function because no two pairs have the same first coordinate and different second coordinates. The set $\{(2, 2), (9, 3), (9, -3)\}$ is not a function because $(9, 3)$ and $(9, -3)$ have the same first coordinate and have different second coordinates. Note that we do not allow the same first coordinate with different second coordinates, because we want the second coordinate to be determined by the first coordinate.

A calculator is a function machine. Built-in functions on a calculator are marked with symbols such as $\sqrt{x}$, $x^2$, $x!$, $10^x$, $e^x$, $\ln(x)$, $\sin(x)$, $\cos(x)$, etc. When you provide the first coordinate and use one of these functions, the calculator finds the appropriate second coordinate. The $y$-coordinate is certainly determined by the $x$-coordinate, because the calculator will not produce two different $y$-coordinates corresponding to a given $x$-coordinate.

## Constructing Functions

The ordered pairs in a function can be specified by a verbal description, list, table, formula, or graph. The most common and efficient way to specify the ordered pairs of a function is to give a formula that will produce them. For example, the function $A = \pi r^2$ determines the area of a circle when the radius is given and the function $y = \sqrt{x}$ determines the nonnegative square root of a nonnegative number. In the next example we find a formula for, or construct, a function.

**EXAMPLE 2** Constructing a function

Given that a square has diagonal of length $d$ and side of length $s$, write the area $A$ as a function of the length of the diagonal.

**Solution**

The area of a square is given by $A = s^2$. The diagonal is the hypotenuse of a right triangle as shown in Fig. P.16. Apply the Pythagorean theorem to the right triangle in the figure:

$$s^2 + s^2 = d^2$$
$$2s^2 = d^2$$
$$s^2 = \frac{d^2}{2}$$

Since $A = s^2$ and $s^2 = \frac{d^2}{2}$ we have $A = \frac{d^2}{2}$, which expresses the area of a square as a function of the length of its diagonal.

**TRY THIS.** A square has perimeter $P$ and sides of length $s$. Write the side as a function of the perimeter. ∎

**Figure P.16**

# Graphs of Functions

The graph of a function is a picture of all points whose ordered pairs make up the function. We can graph a function by calculating enough ordered pairs to determine the shape of the graph. When you graph functions, try to anticipate what the graph will look like, and after the graph is drawn, pause to reflect on the shape of the graph and the type of function that produced it. If you have a graphing calculator, use it to help you graph the equations in the following example. Remember that a graphing calculator shows only finitely many points and a graph consists of infinitely many points. After looking at the display of a graphing calculator, you must still decide what the entire graph looks like.

The **domain** of a function is the set of all first coordinates of the ordered pairs. The **range** of a function is the set of all second coordinates of the ordered pairs. If the domain is not specified, it is assumed to be all real numbers that can be used in the expression defining the function. The corresponding second coordinates must also be real numbers.

**EXAMPLE 3**  **Graphing by plotting ordered pairs**

Graph each function and state the domain and range.

**a.** $y = x^2$  **b.** $y = |x| - 40$

**Solution**

**a.** Make a table of ordered pairs that satisfy $y = x^2$:

| $x$ | $-2$ | $-1$ | 0 | 1 | 2 |
|---|---|---|---|---|---|
| $y = x^2$ | 4 | 1 | 0 | 1 | 4 |

These ordered pairs indicate a graph in the shape shown in Fig. P.17. This curve is called a **parabola**. The domain is $(-\infty, \infty)$ because any real number can be used for $x$ in $y = x^2$. Since all $y$-coordinates are nonnegative, the graph is on or above the $x$-axis. The range is $[0, \infty)$.

The calculator graph in Fig. P.18 supports these conclusions. The calculator plots many more points than we are willing to plot by hand. □

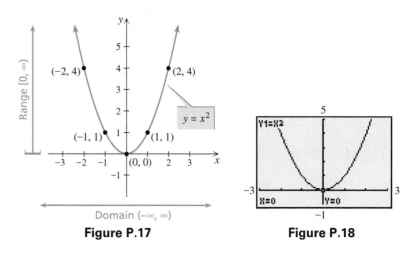

**Figure P.17**                    **Figure P.18**

**b.** Make a table of ordered pairs that satisfy $y = |x| - 40$:

| $x$ | $-40$ | $-20$ | 0 | 20 | 40 |
|---|---|---|---|---|---|
| $y = |x| - 40$ | 0 | $-20$ | $-40$ | $-20$ | 0 |

Plotting these ordered pairs suggests the V-shaped graph of Fig. P.19 on the next page. From the graph we see that the domain is $(-\infty, \infty)$ and the range is $[-40, \infty)$.

The calculator graph in Fig. P.20 supports these conclusions.

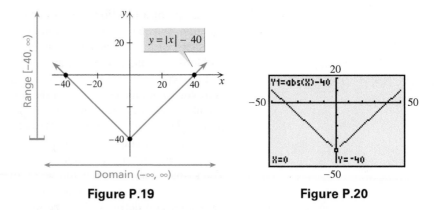

**Figure P.19**                                        **Figure P.20**

**TRY THIS.**  Graph $y = |x| + 2$ and state the domain and range.  ∎

The next example shows two functions whose domain is not the entire set of real numbers.

**EXAMPLE 4**    **Graphing by plotting ordered pairs**

Graph each function and state the domain and range.

**a.** $y = \sqrt{x - 20}$   **b.** $y = \dfrac{1}{x}$

**Solution**

**a.** Make a table of ordered pairs that satisfy $y = \sqrt{x - 20}$.

| $x$ | 20 | 21 | 24 | 29 | 36 |
|---|---|---|---|---|---|
| $y = \sqrt{x - 20}$ | 0 | 1 | 2 | 3 | 4 |

Plot these ordered pairs and sketch a curve through them as shown in Fig. P.21. This curve is half of a parabola. If $\sqrt{x - 20}$ is a real number, then $x - 20 \geq 0$, or $x \geq 20$. So the domain is $[20, \infty)$. On the graph we see that the $y$-coordinates range from 0 upward. So the range is the nonnegative real numbers, $[0, \infty)$.

The calculator graph in Fig. P.22 supports these conclusions.  □

**Figure P.21**                                        **Figure P.22**

**b.** Make a table of ordered pairs that satisfy $y = 1/x$:

| $x$ | −2 | −1 | −1/2 | 1/2 | 1 | 2 |
|---|---|---|---|---|---|---|
| $y = 1/x$ | −1/2 | −1 | −2 | 2 | 1 | 1/2 |

Note that as $x$ gets larger, $y$ gets smaller, and vice versa. Plotting these ordered pairs suggests the graph shown in Fig. P.23. The domain is all real numbers except 0, $(-\infty, 0) \cup (0, \infty)$, and the range is also $(-\infty, 0) \cup (0, \infty)$.

The calculator graph in Fig. P.24 supports these conclusions.

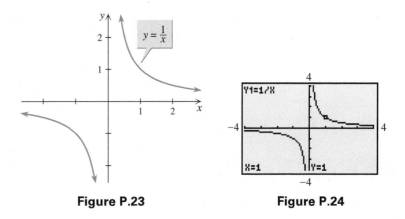

**Figure P.23**        **Figure P.24**

**TRY THIS.** Graph $y = \sqrt{x + 3}$ and state the domain and range. ■

## Function Notation

If $y$ is a function of $x$, we may use the expression $f(x)$, rather than $y$, for the dependent variable. The notation $f(x)$ is read "the value of $f$ at $x$" or simply "$f$ of $x$." So $y$ and $f(x)$ are both symbols for the second coordinate when the first coordinate is $x$, and we may write $y = f(x)$. For example, we could write $f(x) = \sqrt{x} - 40$ for the function $y = \sqrt{x} - 40$. The notation $f(x)$ is called **function notation.**

**EXAMPLE 5**    **Using function notation**

Find each of the following for the function $f(x) = \sqrt{2x + 9}$.

**a.** $f(20)$    **b.** $x$, if $f(x) = 0$    **c.** $f(-6)$    **d.** $f(-x)$

**Solution**

**a.** The expression $f(20)$ is the second coordinate when the first coordinate is 20. To find the value of $f(20)$ replace $x$ by 20 in the formula $f(x) = \sqrt{2x + 9}$:

$$f(20) = \sqrt{2 \cdot 20 + 9} = \sqrt{49} = 7$$

**b.** To find a value of $x$ so that the second coordinate is 0, we must solve $\sqrt{2x + 9} = 0$:

$$\sqrt{2x + 9} = 0$$
$$2x + 9 = 0$$
$$2x = -9$$
$$x = -\frac{9}{2}$$

So $x = -9/2$.

**c.** The expression $f(-6)$ is undefined, because the domain of $f(x) = \sqrt{2x + 9}$ is the interval $[-9/2, \infty)$.

**d.** The expression $f(-x)$ is the second coordinate when the first coordinate is $-x$. So replace $x$ by $-x$ in the formula $f(x) = \sqrt{2x + 9}$:

$$f(-x) = \sqrt{2(-x) + 9} = \sqrt{-2x + 9}$$

**TRY THIS.** Let $f(x) = \sqrt{x + 1}$    **a.** Find $f(8)$.    **b.** Find $x$ if $f(x) = 5$. ■

We may use letters other than $f$ in function notation. For example, we could have $h(x) = \sqrt{x}$ and $g(x) = x^2$. If a function describes some real application, then

a letter that fits the situation is usually used. For example, if watermelons are \$3.00 apiece, then the cost of $n$ watermelons, in dollars, is given by the function $C(n) = 3n$. The cost of 5 watermelons is $C(5) = 3 \cdot 5 = \$15$. In trigonometry the abbreviations sin, cos, and tan are used rather than a single letter to name the trigonometric functions. The dependent variables are written as $\sin(x)$, $\cos(x)$, and $\tan(x)$.

## FOR THOUGHT...   True or False? Explain.

1. The number of gallons of gas that can be purchased for \$20 is a function of the price per gallon.

2. Each student's exam grade is a function of the student's IQ.

3. Any set of ordered pairs is a function.

4. If $y = x^2$, then $y$ is a function of $x$.

5. If $x$ is the independent variable, then $f(x)$ is the dependent variable.

6. The domain of $f(x) = 1/x$ is $(-\infty, 0) \cup (0, \infty)$.

7. The domain of $g(x) = |x - 3|$ is $[3, \infty)$.

8. The range of $y = 8 - x^2$ is $(-\infty, 8]$.

9. If $f(t) = \dfrac{t - 2}{t + 2}$, then $f(0) = -1$.

10. If $f(x) = x - 5$ and $f(a) = 0$, then $a = 5$.

## P.2 EXERCISES

Scan this code to link to section lecture videos.

### CONCEPTS

*Fill in the blank.*

1. A set of ordered pairs in which no two have the same first coordinate and different second coordinates is a(n) _____

2. For a set of ordered pairs, the variable corresponding to the first coordinate is the _____ variable and the variable corresponding to the second coordinate is the _____ variable.

3. For a set of ordered pairs, the set of all first coordinates is the _____ and the set of all second coordinates is the _____.

4. The graph of $y = x^2$ is a(n) _____.

5. If the value of $y$ is determined by the value of $x$, then $y$ is a(n) _____ of $x$.

6. We write $y = x^2$ in _____ notation as $f(x) = x^2$.

### SKILLS

*For each pair of variables determine whether a is a function of b, b is a function of a, or neither.*

7. $a$ is the radius of any U.S. coin and $b$ is its circumference.

8. $a$ is the length of any rectangle with width 5 in. and $b$ is its perimeter.

9. $a$ is the length of any piece of U.S. paper currency and $b$ is its denomination.

10. $a$ is the diameter of any U.S. coin and $b$ is its value.

11. $a$ is the universal product code for any item at Walmart and $b$ is its price.

12. $a$ is the final exam score for any student in your class and $b$ is the student's semester grade.

13. $a$ is the time spent studying for the final exam for any student in your class and $b$ is the student's final exam score.

14. $a$ is the age of any adult male and $b$ is his shoe size.

15. $a$ is the height of any car in inches and $b$ is its height in centimeters.

16. $a$ is the cost of mailing any first-class letter and $b$ is its weight.

17. $a$ is any real number and $b$ is the cube of that number.

18. $a$ is any real number and $b$ is the fourth power of that number.

19. $a$ is any real number and $b$ is the absolute value of that number.

20. $a$ is any nonnegative real number and $b$ is a square root of that number.

*Consider a square with side of length s, diagonal of length d, perimeter P, and area A. Make a sketch.*

21. Write $A$ as a function of $s$.    22. Write $s$ as a function of $A$.

23. Write $s$ as a function of $d$.    24. Write $d$ as a function of $s$.

25. Write $P$ as a function of $s$.    26. Write $s$ as a function of $P$.

27. Write $A$ as a function of $P$.    28. Write $d$ as a function of $A$.

*Graph each function by plotting points and state the domain and range. If you have a graphing calculator, use it to check your results.*

**29.** $y = 2x - 1$            **30.** $y = -x + 3$

**31.** $y = 5$                **32.** $y = -4$

**33.** $y = x^2 - 20$          **34.** $y = x^2 + 50$

**35.** $y = 40 - x^2$          **36.** $y = -10 - x^2$

**37.** $y = x^3$              **38.** $y = -x^3$

**39.** $y = \sqrt{x - 10}$       **40.** $y = \sqrt{x + 30}$

**41.** $y = \sqrt{x} + 30$      **42.** $y = \sqrt{x} - 50$

**43.** $y = |x| - 40$         **44.** $y = 2|x|$

**45.** $y = |x - 20|$         **46.** $y = |x + 30|$

*Let $f(x) = 3x^2 - x$, $g(x) = 4x - 2$, and $k(x) = |x + 3|$. Find the following.*

**47.** $f(2)$                **48.** $f(-4)$

**49.** $g(-1)$               **50.** $g(-2)$

**51.** $k(5)$                **52.** $k(-4)$

**53.** $f(0.56)$              **54.** $g(-0.5)$

**55.** $f(-1) + g(-1)$        **56.** $f(3) \cdot k(3)$

**57.** $f(-5) - k(-5)$        **58.** $\dfrac{f(1)}{g(1)}$

**59.** $f(a)$                **60.** $g(b)$

**61.** $f(-x)$               **62.** $g(-x)$

**63.** $x$, if $f(x) = 0$      **64.** $x$, if $g(x) = 3$

**65.** $a$, if $k(a) = 4$      **66.** $t$, if $f(t) = 10$

## MODELING

*Solve each problem.*

**67.** *Aerobics* If a woman in an aerobic dance class burns 353 calories per hour, express the number of calories burned, $C$, as a function of the number of hours danced, $n$.

**68.** *Paper Consumption* If the average American uses 580 lb of paper annually, express the total annual paper consumption of the United States, $P$, as a function of the size of the population, $n$.

**69.** *Cost of Window Cleaning* If a window cleaner charges $50 per visit plus $35 per hour, express the total charge $C$ as a function of the number of hours worked, $n$.

**70.** *Cost of a Sundae* At Doubledipski's ice cream parlor, the cost of a sundae is $2.50 plus $0.50 for each topping. Express the total cost of a sundae, $C$, as a function of the number of toppings, $n$.

**71.** *Capsize Control* The capsize screening value $C$ is an indicator of a sailboat's suitability for extended offshore sailing. $C$ is determined by the function $C = 4D^{-1/3}B$, where $D$ is the displacement of the boat in pounds and $B$ is its beam (or width) in feet. Sketch the graph of this function for $B$ ranging from 0 to 20 ft, assuming that $D$ is fixed at 22,800 lb. Find $C$ for the Island Packet 40, which has a displacement of 22,800 lb and a beam of 12 ft 11 in. (Island Packet Yachts, www.ipy.com).

12 ft 11 in.      $D = 22,800$ lb

**Figure for Exercises 71 and 72**

**72.** *Limiting the Beam* The International Offshore Rules require that the capsize screening value $C$ (from the previous exercise) be less than or equal to 2 for safety. What is the maximum allowable beam (to the nearest inch) for a boat with a displacement of 22,800 lb? For a fixed displacement, is a boat more or less likely to capsize as its beam gets larger?

**73.** *Bore and Stroke* The displacement $D$ of an engine is given by the function

$$D = \frac{\pi}{4}B^2 \cdot S \cdot N,$$

where $B$ is the bore (the diameter of a cylinder), $S$ is the stroke (distance the piston moves), and $N$ is the number of cylinders. The two-cylinder Harley-Davidson Evo engine has a bore of 3.498 in. and a stroke of 4.250 in. Find its displacement.

**74.** *Rebuilt Engine* If the bore of the Evo engine from the last exercise is increased by 0.020 in., then what is the increase in displacement?

**75.** *Finding the Bore* Find a formula that expresses the bore $B$ as a function of the displacement $D$, stroke $S$, and number of cylinders $N$. See Exercise 73.

**76.** *Compression Ratio* The compression ratio $CR$ for an engine is a function of the bore $B$, the stroke $S$, and the clearance volume $V$, where

$$CR = 1 + \frac{\pi B^2 \cdot S}{4V}.$$

The clearance volume is the volume of air in the cylinder when the piston is at top dead center. Find a formula that expresses the clearance volume $V$ as a function of $CR$, $B$, and $S$.

**REVIEW**

**77.** According to the _____ theorem, the sum of the squares of the _____ of a right triangle is equal to the square of the _____ .

**78.** The equation $(x - 3)^2 + (x + 2)^2 = 36$ is the equation of a(n) _____ with _____ 6 and _____ $(3, -2)$.

**79.** Find the distance between the points $(2, -4)$ and $(-3, -6)$.

**80.** Find the midpoint of the line segment whose endpoints are $(4, -8)$ and $(-6, 16)$.

**81.** Find the $x$- and $y$-intercepts for the line $4x - 6y = 40$.

**82.** The sides of a rectangle are 3 feet and 7 feet. Find the length of the diagonal.

**OUTSIDE THE BOX**

**83.** *Powers of Three* What is $x$ if

$$\frac{1}{27} \cdot 3^{100} \cdot \frac{1}{81} \cdot 9^x = \frac{1}{3} \cdot 3^{x}?$$

**84.** *Sum of Cubes* If $a + b = 3$ and $a^2 + b^2 = 89$, then what is the value of $a^3 + b^3$?

---

## P.2 POP QUIZ

**1.** Is the radius of a circle a function of its area?

**2.** Express the side of a square $s$ as a function of its area $A$.

**3.** Let $a$ be a positive real number and $b$ be a real number such that $b^2 = a$. Is $b$ a function of $a$?

**4.** What is the domain of $y = \sqrt{x - 1}$?

**5.** What is the range of $y = x^2 + 2$?

**6.** What is $f(3)$ if $f(x) = 3x + 6$?

**7.** Find $a$ if $f(a) = 10$ and $f(x) = 2x - 4$.

---

## P.3 Families of Functions, Transformations, and Symmetry

We have already seen different functions that have similar graphs. In this section we see how those similar graphs are related to each other. If $a$, $h$, and $k$ are real numbers with $a \neq 0$, then $y = af(x - h) + k$ is a **transformation** of the function $y = f(x)$. All of the transformations of a function form a **family of functions.** For example, any function of the form $y = a\sqrt{x - h} + k$ is in the square-root family, any function of the form $y = a|x - h| + k$ is in the absolute-value family, and any function of the form $y = a(x - h)^2 + k$ is in the square or quadratic family. All of the functions in a family of functions have similar graphs. The graph of any function in the square family is called a **parabola.** We will now see what effect each of the numbers $a$, $h$, and $k$ has on the graph of the original function $y = f(x)$.

### Horizontal Translation

If you can move a graph to the left or right (without changing its shape) so that it coincides with another graph, then the graphs are *horizontal translations* of each other. This geometric idea is made more precise using algebra in the following definition.

**Definition: Translation to the Right or Left**

If $h > 0$, then the graph of $y = f(x - h)$ is a **translation of $h$ units to the right** of the graph of $y = f(x)$. If $h < 0$, then the graph of $y = f(x - h)$ is a **translation of $|h|$ units to the left** of the graph of $y = f(x)$.

According to the order of operations, the first operation to perform in the formula $y = af(x - h) + k$ is to subtract $h$ from $x$. Then $f(x - h)$ is multiplied by $a$, and finally $k$ is added on. The order is important here, and we will study the effects of these numbers in the order $h$, $a$, and $k$.

### EXAMPLE 1   Translations to the right or left

Graph $f(x) = \sqrt{x}$, $g(x) = \sqrt{x - 3}$, and $h(x) = \sqrt{x + 5}$ on the same coordinate plane.

**Solution**

First sketch $f(x) = \sqrt{x}$ through $(0, 0)$ $(1, 1)$, and $(4, 2)$ as shown in Fig. P.25. Since the first operation of $g$ is to subtract 3, we get the corresponding points by adding 3 to each $x$-coordinate. So $g$ goes through $(3, 0)$, $(4, 1)$, and $(7, 2)$. Since the first operation of $h$ is to add 5, we get corresponding points by subtracting 5 from the $x$-coordinates. So $h$ goes through $(-5, 0)$, $(-4, 1)$, and $(-1, 2)$.

The calculator graphs of $f$, $g$, and $h$ are shown in Fig. P.26. Be sure to note the difference between $y = \sqrt{\phantom{x}} (x) - 3$ and $y = \sqrt{\phantom{x}} (x - 3)$ on a calculator.

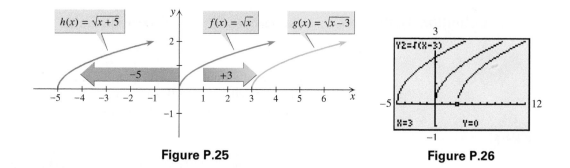

Figure P.25                    Figure P.26

**TRY THIS.** Graph $f(x) = \sqrt{x}$, $g(x) = \sqrt{x - 2}$, and $h(x) = \sqrt{x + 1}$ on the same coordinate plane.  ∎

Notice that $y = \sqrt{x - 3}$ lies 3 units to the *right* and $y = \sqrt{x + 5}$ lies 5 units to the *left* of $y = \sqrt{x}$. The next example shows two more horizontal translations.

### EXAMPLE 2   Horizontal translations

Sketch the graph of each function.

**a.** $f(x) = |x - 1|$
**b.** $f(x) = (x + 3)^2$

**Solution**

**a.** The function $f(x) = |x - 1|$ is in the absolute-value family and its graph is a translation one unit to the right of $g(x) = |x|$. Calculate a few ordered pairs to get an accurate graph. The points $(0, 1)$, $(1, 0)$, and $(2, 1)$ are on the graph of $f(x) = |x - 1|$ shown in Fig. P.27 on the next page.

**b.** The function $f(x) = (x + 3)^2$ is in the square family and its graph is a translation three units to the left of the graph of $g(x) = x^2$. Calculate a few ordered pairs to get an accurate graph. The points $(-3, 0)$, $(-2, 1)$, and $(-4, 1)$ are on the graph shown in Fig. P.28 on the next page.

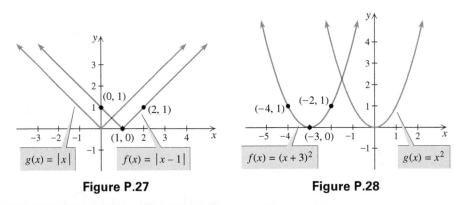

**Figure P.27**          **Figure P.28**

**TRY THIS.** Graph $f(x) = (x - 2)^2$.

## Reflection

If a graph is a mirror image of another, then the graphs are *reflections* of each other. As with horizontal translation, this idea is also made precise by using algebra.

**Definition: Reflection**

The graph of $y = -f(x)$ is a **reflection** in the *x*-axis of the graph of $y = f(x)$.

In the general formula $y = af(x - h) + k$, a reflection is accomplished by choosing $a = -1$ and $h = k = 0$. All points on the graph of $y = -f(x)$ can be obtained by simply changing the signs of all of the *y*-coordinates of the points on the graph of $y = f(x)$.

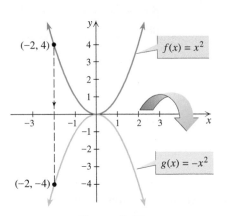

**Figure P.29**

### EXAMPLE 3   Graphing using reflection

Graph each pair of functions on the same coordinate plane.

**a.** $f(x) = x^2$, $g(x) = -x^2$   **b.** $f(x) = x^3$, $g(x) = -x^3$
**c.** $f(x) = |x|$, $g(x) = -|x|$

**Solution**

**a.** The graph of $f(x) = x^2$ goes through $(0, 0)$, $(\pm 1, 1)$, and $(\pm 2, 4)$. The graph of $g(x) = -x^2$ goes through $(0, 0)$, $(\pm 1, -1)$, and $(\pm 2, -4)$ as shown in Fig. P.29.

**b.** Make a table of ordered pairs for $f$ as follows:

| $x$ | $-2$ | $-1$ | $0$ | $1$ | $2$ |
|---|---|---|---|---|---|
| $f(x) = x^3$ | $-8$ | $-1$ | $0$ | $1$ | $8$ |

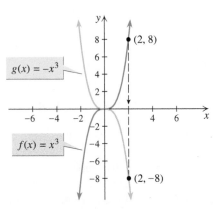

**Figure P.30**

Sketch the graph of $f$ through these ordered pairs as shown in Fig. P.30. Since $g(x) = -f(x)$, the graph of $g$ can be obtained by reflecting the graph of $f$ in the *x*-axis. Each point on the graph of $f$ corresponds to a point on the graph of $g$ with the opposite *y*-coordinate. For example, $(2, 8)$ on the graph of $f$ corresponds to $(2, -8)$ on the graph of $g$. Both graphs are shown in Fig. P.30.

**c.** The graph of $f$ is the familiar V-shaped graph of the absolute value function as shown in Fig. P.31. Since $g(x) = -f(x)$, the graph of $g$ can be obtained by reflecting the graph of $f$ in the *x*-axis. Each point on the graph of $f$ corresponds to a point on the graph of $g$ with the opposite *y*-coordinate. For example, $(2, 2)$ on $f$ corresponds to $(2, -2)$ on $g$. Both graphs are shown in Fig. P.31.

**TRY THIS.** Graph $f(x) = \sqrt{x}$ and $g(x) = -\sqrt{x}$ on the same coordinate plane.

We do not include reflections in the $y$-axis in our function families, but it is interesting to note that you get a reflection in the $y$-axis simply by replacing $x$ with $-x$ in the formula for the function. For example, graph $y_1 = \sqrt{x}$ and $y_2 = \sqrt{-x}$ on your graphing calculator.

## Stretching and Shrinking

Two graphs that have similar shapes may be related by a vertical *stretching* or *shrinking* of one into the other. This geometric idea is defined precisely using algebra as follows.

**Definitions:**
**Stretching and Shrinking**

The graph of $y = af(x)$ is obtained from the graph of $y = f(x)$ by

1. **stretching** the graph of $y = f(x)$ by $a$ when $a > 1$, or
2. **shrinking** the graph of $y = f(x)$ by $a$ when $0 < a < 1$.

Note that stretching and shrinking are defined for positive values of $a$. If $a$ is negative, then reflection occurs along with stretching or shrinking.

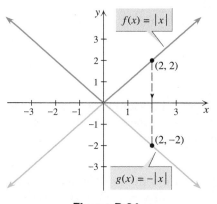

**Figure P.31**

**EXAMPLE 4** **Graphing using stretching and shrinking**

In each case graph the three functions on the same coordinate plane.

**a.** $f(x) = \sqrt{x}, g(x) = 2\sqrt{x}, h(x) = \dfrac{1}{2}\sqrt{x}$

**b.** $f(x) = x^2, g(x) = 2x^2, h(x) = \dfrac{1}{2}x^2$

**Solution**

**a.** The graph of $f(x) = \sqrt{x}$ goes through $(0, 0)$ $(1, 1)$, and $(4, 2)$ as shown in Fig. P.32. The graph of $g$ is obtained by stretching the graph of $f$ by a factor of 2. So $g$ goes through $(0, 0), (1, 2)$, and $(4, 4)$. The graph of $h$ is obtained by shrinking the graph of $f$ by a factor of $\frac{1}{2}$. So $h$ goes through $(0, 0), \left(1, \frac{1}{2}\right)$, and $(4, 1)$.  The functions $f$, $g$, and $h$ are shown on a graphing calculator in Fig. P.33. Note how the viewing window affects the shape of the graph. The curves do not appear as separated on the calculator as they do in Fig. P.32. □

**b.** The graph of $f(x) = x^2$ is the familiar parabola shown in Fig. P.34. We stretch it by a factor of 2 to get the graph of $g$ and shrink it by a factor of $\frac{1}{2}$ to get the graph of $h$.

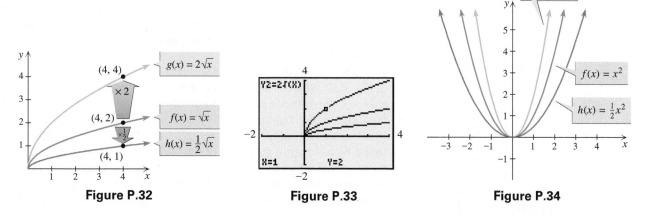

**Figure P.32**       **Figure P.33**       **Figure P.34**

**TRY THIS.** Graph $f(x) = |x|$, $g(x) = 3|x|$, and $h(x) = \frac{1}{3}|x|$ on the same coordinate plane. ■

Note that the graph of $y = 2x^2$ has exactly the same shape as the graph of $y = x^2$ if we simply change the scale on the $y$-axis as in Fig. P.35.

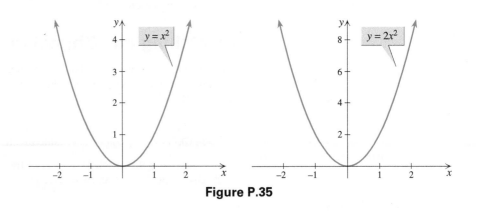

**Figure P.35**

## Vertical Translation

In the formula $y = af(x - h) + k$, the last operation is addition of $k$. If we add the constant $k$ to all $y$-coordinates on the graph of $y = f(x)$, the graph will be translated up or down depending on whether $k$ is positive or negative.

**Definition: Translation Upward or Downward**

If $k > 0$, then the graph of $y = f(x) + k$ is a **translation of $k$ units upward** of the graph of $y = f(x)$. If $k < 0$, then the graph of $y = f(x) + k$ is a **translation of $|k|$ units downward** of the graph of $y = f(x)$.

**Figure P.36**

**Figure P.37**

**EXAMPLE 5**  **Translations upward or downward**

Graph the three given functions on the same coordinate plane.

**a.** $f(x) = \sqrt{x}, g(x) = \sqrt{x} + 3, h(x) = \sqrt{x} - 5$
**b.** $f(x) = x^2, g(x) = x^2 + 2, h(x) = x^2 - 3$

**Solution**

**a.** First sketch $f(x) = \sqrt{x}$ through $(0, 0)$, $(1, 1)$, and $(4, 2)$ as shown in Fig. P.36. Since $g(x) = \sqrt{x} + 3$ we can add 3 to the $y$-coordinate of each point to get $(0, 3)$, $(1, 4)$, and $(4, 5)$. Sketch $g$ through these points. Every point on $f$ can be moved up 3 units to obtain a corresponding point on $g$. We now subtract 5 from the $y$-coordinates on $f$ to obtain points on $h$. So $h$ goes through $(0, -5)$, $(1, -4)$, and $(4, -3)$.

   The relationship between $f$, $g$, and $h$ can be seen with a graphing calculator in Fig. P.37. You should experiment with your graphing calculator to see how a change in the formula changes the graph. □

**b.** First sketch the familiar graph of $f(x) = x^2$ through $(\pm 2, 4)$, $(\pm 1, 1)$, and $(0, 0)$ as shown in Fig. P.38. Since $g(x) = f(x) + 2$, the graph of $g$ can be obtained by translating the graph of $f$ upward two units. Since $h(x) = f(x) - 3$, the graph of $h$ can be obtained by translating the graph of $f$ downward three units. For example, the point $(-1, 1)$ on the graph of $f$ moves up to $(-1, 3)$ on the graph of $g$ and down to $(-1, -2)$ on the graph of $h$ as shown in Fig. P.38.

**TRY THIS.** Graph $f(x) = \sqrt{x}, g(x) = \sqrt{x} + 1$, and $h(x) = \sqrt{x} - 2$ on the same coordinate plane. ■

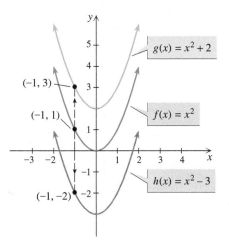

**Figure P.38**

# Multiple Transformations

Any combination of stretching, shrinking, reflecting, horizontal translation, and vertical translation transforms one function into a new function. If a transformation does not change the shape of a graph, then it is a **rigid** transformation. If the shape changes, then it is **nonrigid.** Stretching and shrinking are nonrigid transformations. Translating (horizontally or vertically) and reflection are rigid transformations.

When graphing a function involving more than one transformation, $y = af(x - h) + k$, apply the transformations in the order that we discussed them: $h$-$a$-$k$. Remember that this order is simply the order of operations.

> **PROCEDURE**
>
> ## Multiple Transformations
>
> To graph $y = af(x - h) + k$ apply the transformations in the following order:
>
> **1.** Horizontal translation ($h$).
> **2.** Reflecting/stretching/shrinking ($a$).
> **3.** Vertical translation ($k$).

Note that the order in which you reflect and stretch or reflect and shrink does not matter. It does matter that you do vertical translation last. For example, if $y = x^2$ is reflected in the $x$-axis and then translated up one unit, the equation for the graph is $y = -x^2 + 1$. If $y = x^2$ is translated up one unit and then reflected in the $x$-axis, the equation for the graph in the final position is $y = -(x^2 + 1)$ or $y = -x^2 - 1$. Changing the order has resulted in different functions.

**EXAMPLE 6** Graphing using several transformations

Use transformations to graph each function.

**a.** $y = -2(x - 3)^2 + 4$
**b.** $y = 4 - 2\sqrt{x + 1}$

**Solution**

**a.** This function is in the square family. So the graph of $y = x^2$ is translated 3 units to the right, reflected and stretched by a factor of 2, and finally translated 4 units upward. See Fig. P.39.

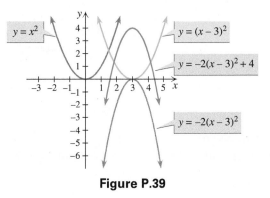

**Figure P.39**

**b.** Rewrite the function as $y = -2\sqrt{x + 1} + 4$ and recognize that it is in the square-root family. So we start with the graph of $y = \sqrt{x}$. The graph of $y = \sqrt{x + 1}$ is a horizontal translation one unit to the left of $y = \sqrt{x}$. Stretch

by a factor of 2 to get $y = 2\sqrt{x + 1}$. Reflect in the $x$-axis to get $y = -2\sqrt{x + 1}$. Finally, translate 4 units upward to get the graph of $y = -2\sqrt{x + 1} + 4$. All of these graphs are shown in Fig. P.40.

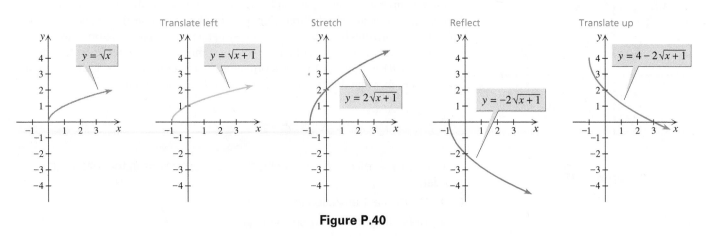

**Figure P.40**

**TRY THIS.**  Graph $y = 4 - 2\,|x + 1|$.  ∎

## Symmetry

The graph of $g(x) = -x^2$ is a reflection in the $x$-axis of the graph of $f(x) = x^2$. If the paper were folded along the $x$-axis, the graphs would coincide. See Fig. P.41. The symmetry that we call reflection occurs between two functions, but the graph of $f(x) = x^2$ has a symmetry within itself. Points such as $(2, 4)$ and $(-2, 4)$ are on the graph and are equidistant from the $y$-axis. Folding the paper along the $y$-axis brings all such pairs of points together. See Fig. P.42. The reason for this symmetry about the $y$-axis is the fact that $f(-x) = f(x)$ for any real number $x$. We get the same $y$-coordinate whether we evaluate the function at a number or at its opposite.

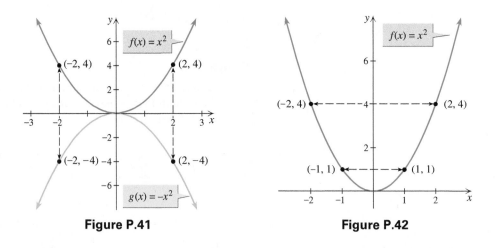

**Figure P.41**               **Figure P.42**

**Definition: Symmetric about the $y$-Axis**

If $f(-x) = f(x)$ for every value of $x$ in the domain of the function $f$, then $f$ is an **even function** and its graph is **symmetric about the $y$-axis.**

The graph of $y = f(-x)$ is a reflection in the $y$-axis of the graph of $y = f(x)$. If the function is even, these graphs are identical. For an example, graph $y = x^2$ and $y = (-x)^2$. If the function is not even, the graphs are not identical, but each is still a reflection in the $y$-axis of the other. For an example, graph $y = x^3$ and $y = (-x)^3$.

Consider the graph of $f(x) = x^3$ shown in Fig. P.43. It is not symmetric about the $y$-axis like the graph of $f(x) = x^2$, but it has another kind of symmetry. On the graph of $f(x) = x^3$ we find pairs of points such as $(2, 8)$ and $(-2, -8)$. These points are equidistant from the origin and on opposite sides of the origin. So the symmetry of this graph is about the origin. In this case, $f(x)$ and $f(-x)$ are not equal, but $f(-x) = -f(x)$.

**Definition: Symmetric about the Origin**

> If $f(-x) = -f(x)$ for every value of $x$ in the domain of the function $f$, then $f$ is an **odd function** and its graph is **symmetric about the origin.**

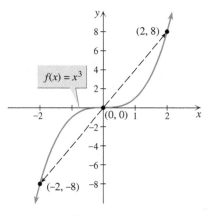

$f(x) = x^3$

(2, 8)

(0, 0)

(-2, -8)

**Figure P.43**

We can look at a graph and see if it is symmetric about the $y$-axis or the origin, but it makes graphing easier and more accurate if we can identify symmetry *before* graphing. Using the definitions, we can determine whether a function is even, odd, or neither from the formula defining the function. If we know about symmetry (or the lack of it) in advance, then we know what to expect when we plot points.

**EXAMPLE 7**  **Determining symmetry in a graph**

Discuss the symmetry of the graph of each function.

**a.** $f(x) = 5x^3 - x$  **b.** $f(x) = |x| + 3$  **c.** $f(x) = x^2 - 3x + 6$

**Solution**

**a.** Replace $x$ by $-x$ in the formula for $f(x)$ and simplify:

$$f(-x) = 5(-x)^3 - (-x) = -5x^3 + x$$

Is $f(-x)$ equal to $f(x)$ or the opposite of $f(x)$? Since $-f(x) = -5x^3 + x$, we have $f(-x) = -f(x)$. So $f$ is an odd function and the graph is symmetric about the origin.

**b.** Since $|-x| = |x|$ for any real number $x$, we have $f(-x) = |-x| + 3 = |x| + 3$. Because $f(-x) = f(x)$, the function is even and the graph is symmetric about the $y$-axis.

**c.** In this case, $f(-x) = (-x)^2 - 3(-x) + 6 = x^2 + 3x + 6$. So $f(-x) \neq f(x)$, and $f(-x) \neq -f(x)$. This function is neither odd nor even, and its graph has neither type of symmetry.

**TRY THIS.** Discuss the symmetry of the graph of $f(x) = -2x^2 + 5$.  ∎

## FOR THOUGHT...  True or False? Explain.

**1.** The graph of $f(x) = (-x)^4$ is a reflection in the $x$-axis of the graph of $g(x) = x^4$.

**2.** The graph of $f(x) = x^2 - 4$ lies 4 units to the right of the graph of $g(x) = x^2$.

**3.** The graph of $y = |x + 2| + 2$ is a translation 2 units to the right and 2 units upward of the graph of $y = |x|$.

**4.** The graph of $f(x) = -3$ is a reflection in the $x$-axis of the graph of $g(x) = 3$.

**5.** The functions $y = x^2 + 4x + 1$ and $y = (x + 2)^2 - 3$ have the same graph.

**6.** The graph of $y = -(x - 3)^2 - 4$ can be obtained by moving $y = x^2$ a distance of 3 units to the right and down 4 units, and then reflecting in the $x$-axis.

**7.** If $f(x) = -x^3 + 2x^2 - 3x + 5$, then $f(-x) = x^3 + 2x^2 + 3x + 5$.

**8.** The graphs of $f(x) = -\sqrt{x}$ and $g(x) = \sqrt{-x}$ are identical.

**9.** If $f(x) = x^3 - x$, then $f(-x) = -f(x)$.

**10.** The graph of $y = x^3$ is symmetric with respect to the origin.

## P.3 EXERCISES

### CONCEPTS

*Fill in the blank.*

**1.** Translating and reflecting are _____ transformations.

**2.** Stretching and shrinking are _____ transformations.

**3.** The graph of $y = -f(x)$ is a(n) _____ of the graph of $y = f(x)$.

**4.** The graph of $y = f(x) + k$ is a(n) _____ of the graph of $y = f(x)$ if $k > 0$ or a(n) _____ if $k < 0$.

**5.** The graph of $y = f(x - h)$ is a translation to the _____ of the graph of $y = f(x)$ if $h > 0$ or a translation to the _____ if $h < 0$.

**6.** The graph of $y = af(x)$ is obtained by _____ the graph of $y = f(x)$ if $a > 1$ or _____ the graph of $y = f(x)$ if $0 < a < 1$.

**7.** If $f(-x) = -f(x)$ for every $x$ in the domain of $f$, then $f$ is a(n) _____ function.

**8.** If $f(-x) = f(x)$ for every $x$ in the domain of $f$, then $f$ is a(n) _____ function.

**9.** The graph of $y = af(x - h) + k$ is a(n) _____ of the graph of $y = f(x)$.

**10.** All transformations of a function form a(n) _____ of functions.

### SKILLS

*Sketch the graphs of each pair of functions on the same coordinate plane.*

**11.** $f(x) = \sqrt{x}$, $g(x) = -\sqrt{x}$

**12.** $f(x) = x^2 + 1$, $g(x) = -x^2 - 1$

**13.** $y = x$, $y = -x$

**14.** $y = \sqrt{4 - x^2}$, $y = -\sqrt{4 - x^2}$

**15.** $f(x) = |x|$, $g(x) = |x| - 4$

**16.** $f(x) = \sqrt{x}$, $g(x) = \sqrt{x} + 3$

**17.** $f(x) = x$, $g(x) = x + 3$     **18.** $f(x) = x^2$, $g(x) = x^2 - 5$

**19.** $y = x^2$, $y = (x - 3)^2$     **20.** $y = |x|$, $y = |x + 2|$

**21.** $f(x) = x^3$, $g(x) = (x + 1)^3$

**22.** $f(x) = \sqrt{x}$, $g(x) = \sqrt{x - 3}$

**23.** $y = \sqrt{x}$, $y = 3\sqrt{x}$     **24.** $y = |x|$, $y = \dfrac{1}{3}|x|$

**25.** $y = x^2$, $y = \dfrac{1}{4}x^2$     **26.** $y = x^2$, $y = 4 - x^2$

*The figure for Exercises 27–34 shows eight parabolas. Match each function with its graph (a)–(h).*

**27.** $y = x^2$     **28.** $y = (x - 4)^2 + 2$

**29.** $y = (x + 4)^2 - 2$     **30.** $y = -2(x - 2)^2$

**31.** $y = -2(x + 2)^2$     **32.** $y = -\dfrac{1}{2}x^2 - 4$

**33.** $y = \dfrac{1}{2}(x + 4)^2 + 2$     **34.** $y = -2(x - 4)^2 - 2$

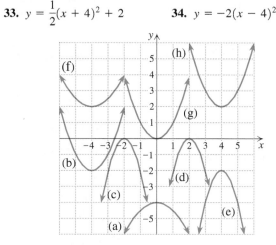

**Figure for Exercises 27–34**

*Write the equation of each graph in its final position.*

**35.** The graph of $y = x^2$ is translated 10 units to the right and 4 units upward.

**36.** The graph of $y = \sqrt{x}$ is translated 5 units to the left and 12 units downward.

**37.** The graph of $y = |x|$ is reflected in the $x$-axis, stretched by a factor of 3, then translated 7 units to the right and 9 units upward.

**38.** The graph of $y = x$ is stretched by a factor of 2, reflected in the $x$-axis, then translated 8 units downward and 6 units to the left.

**39.** The graph of $y = \sqrt{x}$ is stretched by a factor of 3, translated 5 units upward, then reflected in the $x$-axis.

**40.** The graph of $y = x^2$ is translated 13 units to the right and 6 units downward, then reflected in the $x$-axis.

*Use transformations to graph each function.*

**41.** $y = \sqrt{x - 1} + 2$     **42.** $y = \sqrt{x + 5} - 4$

**43.** $y = |x - 1| + 3$     **44.** $y = |x + 3| - 4$

**45.** $y = 3x - 40$

**46.** $y = -4x + 200$

**47.** $y = \frac{1}{2}x - 20$

**48.** $y = -\frac{1}{2}x + 40$

**49.** $y = -\frac{1}{2}|x| + 40$

**50.** $y = 3|x| - 200$

**51.** $y = -\frac{1}{2}|x + 4|$

**52.** $y = 3|x - 2|$

**53.** $y = -(x - 3)^2 + 1$

**54.** $y = -(x + 2)^2 - 4$

**55.** $y = -2(x + 3)^2 - 4$

**56.** $y = 3(x + 1)^2 - 5$

**57.** $y = -2\sqrt{x + 3} + 2$

**58.** $y = -\frac{1}{2}\sqrt{x + 2} + 4$

*Determine whether the graph of each function is symmetric about the y-axis or the origin. Indicate whether the function is even, odd, or neither.*

**59.** $f(x) = x^4$

**60.** $f(x) = x^4 - 2x^2$

**61.** $f(x) = x^4 - x^3$

**62.** $f(x) = x^3 - x$

**63.** $f(x) = (x + 3)^2$

**64.** $f(x) = (x - 1)^2$

**65.** $f(x) = \sqrt{x}$

**66.** $f(x) = |x| - 9$

**67.** $f(x) = x$

**68.** $f(x) = -x$

**69.** $f(x) = 3x + 2$

**70.** $f(x) = x - 3$

**71.** $f(x) = x^3 - 5x + 1$

**72.** $f(x) = x^6 - x^4 + x^2$

**73.** $f(x) = |x - 2|$

**74.** $f(x) = (x^2 - 2)^3$

**75.** $f(x) = 1 + \frac{1}{x^2}$

**76.** $f(x) = \sqrt{9 - x^2}$

**77.** $f(x) = |x^2 - 3|$

**78.** $f(x) = \sqrt{x^2 + 3}$

*Match each function with its graph (a)–(h).*

**79.** $y = 2 + \sqrt{x}$

**80.** $y = \sqrt{2 + x}$

**81.** $y = \sqrt{x^2}$

**82.** $y = \sqrt{\frac{x}{2}}$

**83.** $y = \frac{1}{2}\sqrt{x}$

**84.** $y = 2 - \sqrt{x - 2}$

**85.** $y = -2\sqrt{x}$

**86.** $y = -\sqrt{-x}$

(a)   (b)

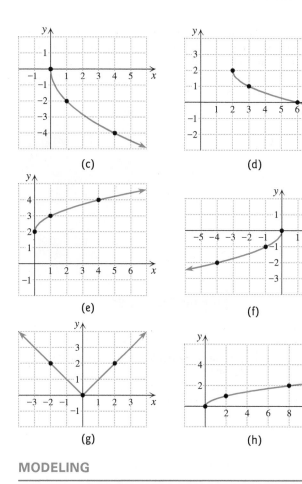

(c)   (d)   (e)   (f)   (g)   (h)

## MODELING

*Solve each problem.*

**87.** *Across-the-Board Raise* Each teacher at C. F. Gauss Elementary School is given an across-the-board raise of $2000. Write a function that *transforms* each old salary $x$ into a new salary $N(x)$.

**88.** *Cost-of-Living Raise* Each registered nurse at Blue Hills Memorial Hospital is first given a 5% cost-of-living raise and then a $3000 merit raise. Write a function that *transforms* each old salary $x$ into a new salary $N(x)$. Does it make any difference in which order these raises are given? Explain.

**89.** *Unemployment Versus Inflation* The Phillips curve shows the relationship between the unemployment rate $x$ and the inflation rate $y$. If the equation of the curve is $y = 1 - \sqrt{x}$ for a certain third world country, then for what value of $x$ is the inflation rate 50%?

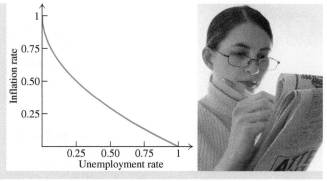

**Figure for Exercise 89**

**90.** *Production Function* The production function shows the relationship between inputs and outputs. A manufacturer of custom windows produces $y$ windows per week using $x$ hours of labor per week, where $y = 1.75\sqrt{x}$. How many hours of labor are required to keep production at 28 windows per week?

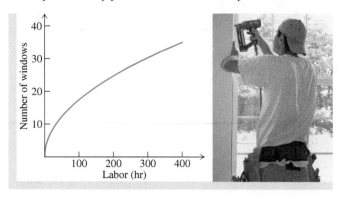

**Figure for Exercise 90**

**WRITING/DISCUSSION**

**91.** Graph each pair of functions (without simplifying the second function) on the same screen of a graphing calculator and explain what each exercise illustrates.

  **a.** $y = x^4 - x^2, y = (-x)^4 - (-x)^2$

  **b.** $y = x^3 - x, y = (-x)^3 - (-x)$

  **c.** $y = x^4 - x^2, y = (x + 1)^4 - (x + 1)^2$

  **d.** $y = x^3 - x, y = (x - 2)^3 - (x - 2) + 3$

**92.** Graph $y = x^3 + 6x^2 + 12x + 8$ on a graphing calculator. The graph is a transformation of the graph of a simpler function. What is the transformation?

**REVIEW**

**93.** Find the equation of a circle with center $(0, 0)$ and radius 1.

**94.** Find the equation of a horizontal line through $(0, 5)$.

**95.** Find the equation of a vertical line through $(4, 0)$.

**96.** What equation expresses the area of a square as a function of its side?

**97.** Find $f(6)$ and $f(-x)$ if $f(x) = 2x^2 - 3x$.

**98.** Find $a$ if $f(a) = 9$ and $f(x) = 2x^2 + 1$.

**OUTSIDE THE BOX**

**99.** *Door Prizes* One thousand tickets numbered 1 through 1000 are placed in a bowl. The tickets are drawn for door prizes one at a time without replacement. How many tickets must be drawn from the bowl to be certain that one of the numbers on a selected ticket is twice as large as another selected number?

**100.** *Completing the Rectangle* A rectangle has opposite vertices at $(1, -1)$ and $(3, 5)$. The other two vertices lie on the line $y = 2$. Find their coordinates.

**P.3 POP QUIZ**

**1.** What is the equation of the curve $y = \sqrt{x}$ after it is translated 8 units upward?

**2.** What is the equation of the curve $y = x^2$ after it is translated 9 units to the right?

**3.** What is the equation of the curve $y = x^3$ after it is reflected in the $x$-axis?

**4.** Find the domain and range for $y = -2\sqrt{x - 1} + 5$.

**5.** If the curve $y = x^2$ is translated 6 units to the right, stretched by a factor of 3, reflected in the $x$-axis, and translated 4 units upward, then what is the equation of the curve in its final position?

**6.** Is $y = \sqrt{4 - x^2}$ even, odd, or neither?

# P.4 Compositions and Inverses

In this section we study compositions and inverses. In the composition of two functions, the output of one function is the input of another function. Inverse functions are functions that undo each other.

## Composition of Functions

If $w$ is a function of $t$ and $z$ is a function of $w$, then $z$ is a function of $t$. The function in which $z$ is determined from $t$ is the *composition* of the other two functions.

EXAMPLE **1** **Composition of functions given as formulas**

Express the area of a circle as a function of its diameter.

**Solution**

The equation $r = d/2$ expresses the radius as a function of the diameter. The equation $A = \pi r^2$ expresses the area as a function of the radius. By substituting $d/2$ for $r$ we get

$$A = \pi \frac{d^2}{4},$$

which expresses the area as a function of the diameter.

**TRY THIS.** The diameter of a circle is a function of the radius $(d = 2r)$, and the radius is a function of the circumference $(r = C/(2\pi))$. Express $d$ as a function of $C$. ∎

In Example 1 it was merely a matter of substitution to find the composition of two functions that are given as formulas. The composition of two functions given in function notation is a function that is defined as follows.

**Definition: Composition of Functions**

If $f$ and $g$ are two functions, the **composition** of $f$ and $g$, written $f \circ g$, is a function that is defined by the equation

$$(f \circ g)(x) = f(g(x)),$$

provided that $g(x)$ is in the domain of $f$. The composition of $g$ and $f$ is written $g \circ f$.

Note that to find $(f \circ g)(x)$ using function notation, we substitute $g(x)$ for $x$ in the function $f$.

EXAMPLE **2** **Evaluating compositions defined by function notation**

Let $f(x) = \sqrt{x}$, $g(x) = 2x - 1$, and $h(x) = x^2$. Evaluate each expression.

**a.** $(f \circ g)(5)$    **b.** $(g \circ h)(x)$    **c.** $(h \circ g)(x)$

**Solution**

**a.** The expression $(f \circ g)(5)$ means to apply $f$ to $g(5)$:

$$(f \circ g)(5) = f(g(5)) \quad \text{Definition of composition}$$
$$= f(9) \quad \text{Since } g(5) = 2 \cdot 5 - 1 = 9$$
$$= \sqrt{9}$$
$$= 3$$

**b.** The expression $(g \circ h)(x)$ means to apply $g$ to $h(x)$:

$$(g \circ h)(x) = g(h(x)) \quad \text{Definition of composition}$$
$$= g(x^2) \quad \text{Since } h(x) = x^2$$
$$= 2x^2 - 1 \quad \text{Substitute } x^2 \text{ for } x \text{ in } g(x) = 2x - 1.$$

**c.** The expression $(h \circ g)(x)$ means that we are to apply $h$ to $g(x)$:

$$(h \circ g)(x) = h(g(x)) \qquad \text{Definition of composition}$$
$$= h(2x - 1) \qquad \text{Since } g(x) = 2x - 1$$
$$= (2x - 1)^2 \qquad \text{Substitute } 2x - 1 \text{ for } x \text{ in } h(x) = x^2.$$
$$= 4x^2 - 4x + 1 \qquad \text{Simplify.}$$

**TRY THIS.** Let $h(x) = \sqrt{x - 1}$ and $j(x) = 2x$. Find $(h \circ j)(5)$ and $(j \circ h)(5)$. ■

Note that for the composition $f \circ g$ to be defined at $x$, $g(x)$ must be in the domain of $f$. So the domain of $f \circ g$ is the set of all values of $x$ in the domain of $g$ for which $g(x)$ is in the domain of $f$. The diagram shown in Fig. P.44 will help you to understand the composition of functions.

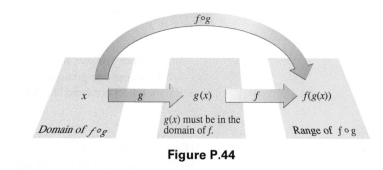

**Figure P.44**

In Example 2 we combined functions to form a composition function. In the next example we start with a function and express it as a composition of simpler functions. Viewing a function as a composition of simpler functions will help you understand inverse functions, which are discussed next.

**EXAMPLE 3** **Expressing a function as a composition**

Express $H(x) = (x + 3)^2$ as a composition of simpler functions.

**Solution**

To find the second coordinate of an ordered pair for the function $H(x) = (x + 3)^2$, we start with $x$, add 3 to $x$, then square the result. These two operations can be accomplished by composition, using $f(x) = x + 3$ followed by $g(x) = x^2$:

$$(g \circ f)(x) = g(f(x)) = g(x + 3) = (x + 3)^2$$

So the function $H$ is the same as the composition of $g$ and $f$, $H = g \circ f$. Notice that $(f \circ g)(x) = x^2 + 3$ and it is not the same as $H(x)$.

**TRY THIS.** Express $K(x) = \sqrt{x - 3}$ as a composition of simpler functions. ■

## One-to-One and Invertible Functions

Consider a medium pizza that costs \$5 plus \$2 per topping. Table P.2 shows the ordered pairs of the function that determines the cost. Note that for each number of toppings there is a unique cost and for each cost there is a unique number of toppings. There is a **one-to-one correspondence** between the domain and range of this function. The function is a *one-to-one function.* For a function that is not one-to-one, consider a Wendy's menu. Every item corresponds to a unique price, but the price \$0.99 corresponds to many different items.

**Table P.2**

| Toppings $x$ | Cost $y$ |
|---|---|
| 0 | \$ 5 |
| 1 | 7 |
| 2 | 9 |
| 3 | 11 |
| 4 | 13 |

**Definition:
One-to-One Function**

If a function has no two ordered pairs with different first coordinates and the same second coordinate, then the function is a **one-to-one** function.

**Table P.3**

| Cost $x$ | Toppings $y$ |
|----------|--------------|
| $ 5 | 0 |
| 7 | 1 |
| 9 | 2 |
| 11 | 3 |
| 13 | 4 |

Because the function in Table P.2 is one-to-one, we can determine the number of toppings when given the cost as shown in Table P.3. Of course, we could just read Table P.2 backward, but we make a new table to emphasize that there is a new function under discussion. Table P.3 is the *inverse function* of the function in Table P.2.

A function is a set of ordered pairs in which no two ordered pairs have the same first coordinates and different second coordinates. If we interchange the $x$- and $y$-coordinates in each ordered pair of a function, as in Tables P.2 and P.3, the resulting set of ordered pairs might or might not be a function. If the original function is one-to-one, then the set obtained by interchanging the coordinates in each ordered pair is a function, the inverse function. If a function is one-to-one, then it has an inverse function or is **invertible.**

**Definition:
Inverse Function**

The **inverse** of a one-to-one function $f$ is the function $f^{-1}$ (read "$f$ inverse"), where the ordered pairs of $f^{-1}$ are obtained by interchanging the coordinates in each ordered pair of $f$.

In this notation, the number $-1$ in $f^{-1}$ does not represent a negative exponent. It is merely a symbol for denoting the inverse function.

If an invertible function is given by a formula, then the inverse function is found by interchanging $x$ and $y$ in the formula and solving for $y$. This method is called the **switch-and-solve** method.

**EXAMPLE 4** The Switch-and-Solve Method

The function $y = 2x + 5$ gives the cost of a pizza where $5 is the basic cost and $x$ is the number of toppings at $2 each. Find the inverse function.

**Solution**

To find the inverse, interchange $x$ and $y$:

$$y = 2x + 5 \quad \text{The function}$$

$$x = 2y + 5 \quad \text{The inverse function}$$

Now solve for $y$:

$$x - 5 = 2y$$

$$\frac{x - 5}{2} = y$$

The inverse function is $y = \frac{x - 5}{2}$, where $x$ is the cost and $y$ is the number of toppings.

**TRY THIS.** The function $y = 40x + 25$ gives the cost, in dollars $y$, for a service call that takes $x$ hours. Find the inverse function. ∎

If we had used function notation in Example 4, we would have $f(x) = 2x + 5$ and $f^{-1}(x) = \frac{x - 5}{2}$. In the next example we find the inverse of a function expressed in function notation.

EXAMPLE **5** The Switch-and-Solve Method with Function Notation

Find the inverse of the function $f(x) = 2x^3 + 1$.

**Solution**

Write the function with $y$ instead of $f(x)$ and then interchange $x$ and $y$:

$$f(x) = 2x^3 + 1 \quad \text{The original function}$$

$$y = 2x^3 + 1 \quad \text{Replace } f(x) \text{ with } y.$$

$$x = 2y^3 + 1 \quad \text{Interchange } x \text{ and } y.$$

Now solve for $y$:

$$x - 1 = 2y^3$$

$$\frac{x - 1}{2} = y^3$$

$$\sqrt[3]{\frac{x - 1}{2}} = y$$

Now replace $y$ by $f^{-1}(x)$ to get $f^{-1}(x) = \sqrt[3]{\frac{x - 1}{2}}$.

**TRY THIS.** Find the inverse of $h(x) = x^3 - 5$. ∎

Example 5 suggests the following steps for finding an inverse function using function notation.

> **PROCEDURE**
>
> ### Finding $f^{-1}(x)$ by the Switch-and-Solve Method
>
> To find the inverse of an invertible function given in function notation:
> 1. Replace $f(x)$ by $y$.
> 2. Interchange $x$ and $y$.
> 3. Solve the equation for $y$.
> 4. Replace $y$ by $f^{-1}(x)$.

Since the coordinates in the ordered pairs are interchanged, the domain of $f^{-1}$ is the range of $f$, and the range of $f^{-1}$ is the domain of $f$ as shown in Fig. P.45. The functions $f$ and $f^{-1}$ are inverses of each other. Because each function undoes what the other does, the composition of a function and its inverse is the identity function. If we start with $x$ and apply the function followed by its inverse function we end up with $x$.

Figure P.45

**EXAMPLE 6**　**Compositions of inverse functions**

Let $f(x) = 2x + 1$ and $f^{-1}(x) = \dfrac{x-1}{2}$. Evaluate each expression.

**a.** $(f^{-1} \circ f)(x)$　**b.** $(f \circ f^{-1})(x)$

**Solution**

**a.** $(f^{-1} \circ f)(x) = f^{-1}(f(x)) = f^{-1}(2x + 1) = \dfrac{2x + 1 - 1}{2} = x$

**b.** $(f \circ f^{-1})(x) = f(f^{-1}(x)) = f\left(\dfrac{x-1}{2}\right) = 2\left(\dfrac{x-1}{2}\right) + 1 = x$

**TRY THIS.** Let $f(x) = x^3 + 7$ and $f^{-1}(x) = \sqrt[3]{x - 7}$. Find $(f^{-1} \circ f)(x)$. ■

Remember that the inverse function undoes the function. In Example 6, $f(3) = 7$ and $f^{-1}(7) = 3$.

The function $f(x) = x^2$ is not invertible because it is not one-to-one: $f(2) = 4$ and $f(-2) = 4$. Even though $f(x) = x^2$ is not invertible, it is convenient to think of square and square root as inverse functions. This problem is easily corrected by restricting the domain of the square function. The function $f(x) = x^2$ for $x \geq 0$ is invertible, and its inverse is $f^{-1}(x) = \sqrt{x}$. The domain of the function must be the range of the inverse function and vice versa.

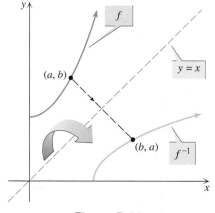

**Figure P.46**

## Graphs of $f$ and $f^{-1}$

If a point $(a, b)$ is on the graph of an invertible function $f$, then $(b, a)$ is on the graph of $f^{-1}$. See Fig. P.46. Since the points $(a, b)$ and $(b, a)$ are symmetric with respect to the line $y = x$, the graph of $f^{-1}$ is a reflection of the graph of $f$ with respect to the line $y = x$.

**EXAMPLE 7**　**Graphing a function and its inverse**

Find the inverse of the function $f(x) = \sqrt{x - 1}$ and graph both $f$ and $f^{-1}$ on the same coordinate axes.

**Solution**

If we interchange $x$ and $y$ in $y = \sqrt{x - 1}$, we get $x = \sqrt{y - 1}$. Now solve for $y$:

$$x = \sqrt{y - 1}$$
$$x^2 = y - 1$$
$$x^2 + 1 = y$$

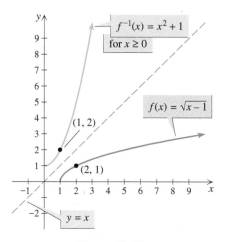

**Figure P.47**

Since $y \geq 0$ in $y = \sqrt{x - 1}$, we must have $x \geq 0$ in $y = x^2 + 1$. So $f^{-1}(x) = x^2 + 1$ for $x \geq 0$. The graphs of $f$ and $f^{-1}$ are shown in Fig. P.47. Notice that the graph of $f^{-1}$ is a reflection of the graph of $f$.

**TRY THIS.** Find the inverse of $f(x) = \sqrt{x + 2}$ and graph $f$ and $f^{-1}$. ■

## Finding Inverses Using Composition

It is no surprise that the inverse of $f(x) = x^2$ for $x \geq 0$ is the function $f^{-1}(x) = \sqrt{x}$. For nonnegative numbers, taking the square root undoes what squaring does. It is also no surprise that the inverse of $f(x) = 3x$ is $f^{-1}(x) = x/3$ or that the inverse of $f(x) = x + 9$ is $f^{-1}(x) = x - 9$. If an invertible function involves a single operation,

it is usually easy to write the inverse function because for each common operation there is an inverse operation. If an invertible function is composed of more than one operation, we find the inverse function by applying the inverse operations in the opposite order from the order in which they appear in the original function. If $f(x) = \sqrt{x} + 2$ (range $[2, \infty)$) then $f^{-1}(x) = (x - 2)^2$ for $x \geq 2$. The domain of the inverse is the range of the function.

**EXAMPLE 8** **Finding inverses using composition**

Find the inverse of each function.

**a.** $f(x) = 2x + 1$ **b.** $g(x) = \dfrac{x^3 + 5}{2}$

**Solution**

**a.** The function $f(x) = 2x + 1$ is a composition of multiplying $x$ by 2 and then adding 1. So the inverse function is a composition of subtracting 1 and then dividing by 2: $f^{-1}(x) = (x - 1)/2$. See Fig. P.48.

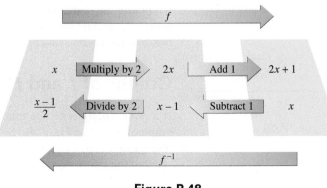

**Figure P.48**

**b.** The function $g(x) = (x^3 + 5)/2$ is a composition of cubing $x$, adding 5, and dividing by 2. The inverse is a composition of multiplying by 2, subtracting 5, and then taking the cube root: $g^{-1}(x) = \sqrt[3]{2x - 5}$. See Fig. P.49.

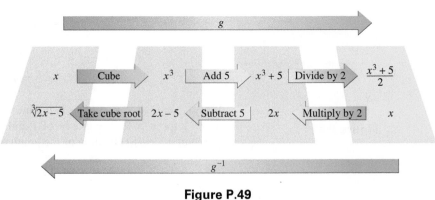

**Figure P.49**

**TRY THIS.** Find the inverse of $f(x) = \frac{2}{3}x + 6$.

## FOR THOUGHT... True or False? Explain.

1. If $s = P/4$ and $A = s^2$, then $A$ is a function of $P$.

2. If $t = x - 1$ and $y = t^2$, then $y = x^2 - 1$.

3. If $f(x) = \sqrt{x}$ and $g(x) = x - 2$, then $(f \circ g)(x) = \sqrt{x} - 2$.

4. If $f(x) = 5x$ and $g(x) = x/5$, then $(f \circ g)(x) = (g \circ f)(x) = x$.

5. If $F(x) = (x - 9)^2$, $g(x) = x^2$, and $h(x) = x - 9$, then $F = h \circ g$.

6. If $f(x) = 2x + 1$, then $f^{-1}(x) = \frac{1}{2}x - 1$.

7. If $f(x) = x^3 - 5$, then $f^{-1}(x) = \sqrt[3]{x + 5}$.

8. If $g(x) = x^2$, then $g^{-1}(x) = \sqrt{x}$.

9. Every function has an inverse function.

10. The only functions that are invertible are the one-to-one functions.

Scan this code to link to section lecture videos.

## P.4 EXERCISES

### CONCEPTS

*Fill in the blank.*

1. For two functions $f$ and $g$, the function $f \circ g$ is the _____ of $f$ and $g$.

2. If a function has no two ordered pairs with different first coordinates and the same second coordinate, then the function is _____ .

3. If a function is one-to-one, then it is _____ .

4. If $f$ is one-to-one, then the function obtained by interchanging the coordinates in each ordered pair of $f$ is the _____ of $f$.

5. The _____ method gives four steps for finding the inverse of a function defined by a formula.

6. The graphs of $f$ and $f^{-1}$ are _____ with respect to the line $y = x$.

### SKILLS

*Use the two given functions to write y as a function of x.*

7. $y = 2a - 3, a = 3x + 1$

8. $y = -4d - 1, d = -3x - 2$

9. $y = w^2 - 2, w = x + 3$

10. $y = 3t^2 - 3, t = x - 1$

11. $y = 3m - 1, m = \dfrac{x + 1}{3}$

12. $y = 2z + 5, z = \dfrac{1}{2}x - \dfrac{5}{2}$

13. $y = 2k^3 - 1, k = \sqrt[3]{\dfrac{x + 1}{2}}$

14. $y = \dfrac{s^3 + 1}{5}, s = \sqrt[3]{5x - 1}$

*Let $f(x) = 3x - 1$, $g(x) = x^2 + 1$, and $h(x) = \dfrac{x + 1}{3}$. Evaluate each expression.*

15. $f(g(-1))$

16. $g(f(-1))$

17. $(f \circ h)(5)$

18. $(h \circ f)(-7)$

19. $(f \circ g)(4)$

20. $(g \circ h)(-9)$

21. $f(g(x))$

22. $g(f(x))$

23. $(g \circ f)(x)$

24. $(f \circ g)(x)$

25. $(h \circ g)(x)$

26. $(g \circ h)(x)$

*Let $f(x) = |x|$, $g(x) = x - 7$, and $h(x) = x^2$. Write each of the following functions as a composition of functions chosen from f, g, and h.*

27. $F(x) = x^2 - 7$

28. $G(x) = |x| - 7$

29. $H(x) = (x - 7)^2$

30. $M(x) = |x - 7|$

31. $N(x) = |x^2|$

32. $R(x) = |x|^2$

*Find the inverse of each function.*

33. $y = 3x - 7$

34. $y = -2x + 5$

35. $y = 2 + \sqrt{x - 3}$

36. $y = \sqrt{3x - 1}$

37. $f(x) = -x - 9$

38. $f(x) = -x + 3$

39. $f(x) = -\dfrac{1}{x}$

40. $f(x) = x$

**41.** $f(x) = \sqrt[3]{x - 9} + 5$

**42.** $f(x) = \sqrt[3]{\dfrac{x}{2}} + 5$

**43.** $f(x) = (x - 2)^2$ for $x \geq 2$

**44.** $f(x) = (x + 3)^2$ for $x \geq -3$

*Show that $(f \circ f^{-1})(x) = x$ and $(f^{-1} \circ f)(x) = x$ for each given pair of functions.*

**45.** $f(x) = 2x - 1, f^{-1}(x) = \dfrac{1}{2}x + \dfrac{1}{2}$

**46.** $f(x) = \dfrac{x + 1}{3}, f^{-1}(x) = 3x - 1$

**47.** $f(x) = 4x + 4, f^{-1}(x) = 0.25x - 1$

**48.** $f(x) = 20 - 5x, f^{-1}(x) = -0.2x + 4$

**49.** $f(x) = \sqrt[3]{4 - 3x}, f^{-1}(x) = \dfrac{4 - x^3}{3}$

**50.** $f(x) = x^3 + 5, f^{-1}(x) = \sqrt[3]{x - 5}$

**51.** $f(x) = \sqrt[3]{x^5 - 1}, f^{-1}(x) = \sqrt[5]{x^3 + 1}$

**52.** $f(x) = x^{3/5} - 3, f^{-1}(x) = (x + 3)^{5/3}$

*Find the inverse of each function and graph both $f$ and $f^{-1}$ on the same coordinate plane.*

**53.** $f(x) = 3x + 2$

**54.** $f(x) = -x - 8$

**55.** $f(x) = x^2 - 4$ for $x \geq 0$

**56.** $f(x) = 1 - x^2$ for $x \geq 0$

**57.** $f(x) = x^3$

**58.** $f(x) = -x^3$

**59.** $f(x) = \sqrt{x} - 3$

**60.** $f(x) = \sqrt{x - 3}$

*Each of the following functions is invertible. Find the inverse using composition.*

**61.** $f(x) = 5x + 2$

**62.** $f(x) = 0.5x - 1$

**63.** $f(x) = 2x - 88$

**64.** $f(x) = 3x + 99$

**65.** $f(x) = -3x + 7$          **66.** $f(x) = -5x + 1$

**67.** $f(x) = 4 - 3x$          **68.** $f(x) = 5 - 2x$

**69.** $f(x) = \dfrac{x}{2} - 9$          **70.** $f(x) = \dfrac{x}{3} + 6$

**71.** $f(x) = \dfrac{1}{x} + 3$          **72.** $f(x) = \dfrac{1}{x + 3}$

**73.** $f(x) = \sqrt[3]{x} - 9$          **74.** $f(x) = \sqrt[3]{x - 9}$

**75.** $f(x) = 2x^3 - 7$          **76.** $f(x) = -x^3 + 4$

## MODELING

*Solve each problem.*

**77.** Write the area $A$ of a square with a side of length $s$ as a function of its diagonal $d$.

**78.** Write the perimeter of a square $P$ as a function of the area $A$.

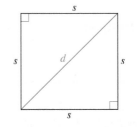

**Figure for Exercises 77 and 78**

**79.** *Withholding* A student sells ice cream bars. Her weekly salary is given by $S(x) = 0.40x + 200$, where $x$ is the number of ice cream bars sold. Her employer determines the amount withheld for taxes using the function $W(x) = 0.20x$, where $x$ is her weekly salary. Write the amount withheld as a function of the number of ice cream bars sold.
    HINT  Use composition.

**80.** *Making Radios* The number of radios shipped by ABC Electronics per week to its factory stores is given by $S(x) = 0.90x$, where $x$ is the number manufactured. The number of radios purchased per week is given by $P(x) = 0.80x$, where $x$ is the number shipped. Write the number purchased per week as a function of the number manufactured.

**81.** *Price of a Car* If the tax on a new car is 8% of the purchase price $P$, express the total cost $C$ as a function of the purchase price. Express the purchase price $P$ as a function of the total cost $C$.

**82.** *Volume of a Cube* Express the volume of a cube $V(x)$ as a function of the length of a side $x$. Express the length of a side of a cube $S(x)$ as a function of the volume $x$.

**83.** *Poiseuille's Law* Under certain conditions, the velocity $V$ of blood in a vessel at distance $r$ from the center of the vessel is given by $V = 500(5.625 \times 10^{-5} - r^2)$ where $0 \leq r \leq 7.5 \times 10^{-3}$. Write $r$ as a function of $V$.

**84.** *Expenditures Versus Income* The amount of household expenditures $E$ (in dollars) is related to income $I$ (in dollars) by the equation $E = 0.4I + 10,000$. Write $I$ as a function of $E$. For what income is $E = I$?

**85.** *Depreciation Rate* A new Lexus sells for about $50,000. The function $r = 1 - \left(\dfrac{V}{50,000}\right)^{1/5}$ expresses the depreciation rate $r$ for a five-year-old car as a function of the value $V$ of a five-year-old car. What is the depreciation rate for a car that is worth $18,000 after five years? Write $V$ as a function of $r$.

**Figure for Exercise 85**

**86.** *Growth Rate* One measurement of the quality of a mutual fund is its average annual growth rate over the last 10 years. The function $r = \left(\dfrac{P}{10,000}\right)^{1/10} - 1$ expresses the average annual growth rate $r$ as a function of the present value $P$ of an investment of $10,000 made 10 years ago. An investment of $10,000 in Fidelity's Contrafund in 2002 was worth $21,931 in 2012 (Fidelity Investments, www.fidelity.com). What was the annual growth rate for that period? Write $P$ as a function of $r$.

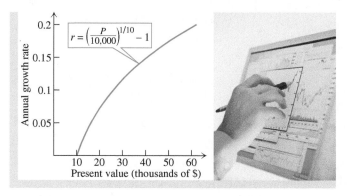

**Figure for Exercise 86**

**WRITING/DISCUSSION**

**87.** If $f(x) = 2x + 1$ and $g(x) = 3x - 5$, verify that $(f \circ g)^{-1} = g^{-1} \circ f^{-1}$.

**88.** Explain why $(f \circ g)^{-1} = g^{-1} \circ f^{-1}$ for any invertible functions $f$ and $g$. Discuss any restrictions on the domains and ranges of $f$ and $g$ for this equation to be correct.

**REVIEW**

**89.** Find the equation of a circle with center $(0, 1)$ and radius 3.

**90.** Find the midpoint of the line segment that has endpoints $(\pi/3, 1)$ and $(\pi/2, 1)$.

**91.** The equation $d = s\sqrt{2}$ expresses the diagonal of a square as a _____ of its side.

**92.** The graph of $f(x) = -x^2$ is a _____ in the $x$-axis of the graph of $f(x) = x^2$.

**93.** Find the domain and range for the function $f(x) = x^2 + 1$.

**94.** Find the domain and range for the function $f(x) = -\sqrt{x - 2} + 3$.

**OUTSIDE THE BOX**

**95.** *Maximizing the Sum* Consider the following sum:

$$\frac{1}{1} + \frac{2}{2} + \frac{3}{3} + \frac{4}{4} + \frac{5}{5} + \cdots + \frac{2015}{2015}$$

You are allowed to rearrange the numerators of the fractions in any way that you choose, but keep the denominators as they are given. What arrangement would give the largest sum for the 2015 fractions?

**96.** *Compositions* Suppose $f$ is a linear function such that $(f \circ f \circ f)(x) = 27x + 26$. Find the $y$-intercept for the graph of $f$.

**P.4** POP QUIZ

**1.** Write the area of a circle as a function of its diameter.

**2.** Suppose $y = 3a - 1$ and $a = 4x - 5$. Write $y$ as a function of $x$.

**3.** If $f(x) = x^2$ and $g(x) = x - 2$, find and simplify $(f \circ g)(5)$.

**4.** Let $f(x) = |x|$, $g(x) = x - 5$, $h(x) = x^3$, and $M(x) = |x|^3 - 5$. Express $M$ as a composition of $f$, $g$, and $h$ (not necessarily in that order).

**5.** If $f(x) = 2x - 1$, then what is $f^{-1}(x)$?

**6.** If $g(x) = \sqrt[3]{x + 1} - 4$, then what is $g^{-1}(x)$?

# Highlights

## P.1 The Cartesian Coordinate System

| | | |
|---|---|---|
| **Pythagorean Theorem** | A triangle is a right triangle if and only if the sum of the squares of the legs is equal to the square of the hypotenuse. | A triangle with sides 3, 4, and 5 is a right triangle because $3^2 + 4^2 = 5^2$. |
| **Square Roots:** | For any real numbers $a$ and $b$, | |
| **Product Rule** | $\sqrt{a} \cdot \sqrt{b} = \sqrt{ab}$ | $\sqrt{2} \cdot \sqrt{3} = \sqrt{6}$ |
| **Quotient Rule** | $\dfrac{\sqrt{a}}{\sqrt{b}} = \sqrt{\dfrac{a}{b}}\ (b \neq 0)$ <br><br> provided that all of the roots are real numbers. | $\dfrac{\sqrt{6}}{\sqrt{3}} = \sqrt{2}$ |
| **Simplified Form** | 1. *no* perfect squares as factors of the radicand, <br><br> 2. *no* fractions in the radicand, and <br><br> 3. *no* square roots in a denominator. | $\sqrt{12} = 2\sqrt{3}$ <br><br> $\sqrt{\dfrac{3}{4}} = \dfrac{\sqrt{3}}{2}$ <br><br> $\dfrac{1}{\sqrt{3}} = \dfrac{1 \cdot \sqrt{3}}{\sqrt{3} \cdot \sqrt{3}} = \dfrac{\sqrt{3}}{3}$ |
| **Distance Formula** | The distance between $(x_1, y_1)$ and $(x_2, y_2)$ is $\sqrt{(x_2 - x_1)^2 + (y_2 - y_1)^2}$. | For $(1, 2)$ and $(4, -2)$, $\sqrt{(4 - 1)^2 + (-2 - 2)^2} = 5$. |
| **Midpoint Formula** | The midpoint of the line segment with endpoints $(x_1, y_1)$ and $(x_2, y_2)$ is $\left(\dfrac{x_1 + x_2}{2}, \dfrac{y_1 + y_2}{2}\right)$. | For $(0, -4)$ and $(6, 2)$, $\left(\dfrac{0 + 6}{2}, \dfrac{-4 + 2}{2}\right) = (3, -1)$. |
| **Equation of a Circle** | The graph of $(x - h)^2 + (y - k)^2 = r^2\,(r > 0)$ is a circle with center $(h, k)$ and radius $r$. | Circle: $(x - 1)^2 + (y + 2)^2 = 9$, center $(1, -2)$, radius 3 |
| **Equation of a Line** | Standard form: $Ax + By = C$ where $A$ and $B$ are not both zero <br> Vertical line: $x = h$ <br> Horizontal line: $y = k$ | Line: $2x + 3y = 6$ <br><br> Vertical line: $x = 5$ <br> Horizontal line: $y = 7$ |

## P.2 Functions

| | | |
|---|---|---|
| **Function** | A set of ordered pairs in which no two ordered pairs have the same first coordinate and different second coordinates | $\{(1, 5), (2, 3), (3, 5)\}$ |
| **Independent Variable** | A variable used to represent the first coordinate of an ordered pair | $\{(x, y)\,|\,y = 2x\}$ <br> Independent variable $x$ |
| **Dependent Variable** | A variable used to represent the second coordinate of an ordered pair | Dependent variable $y$ |

| **Graph of a Function** | A picture in the Cartesian coordinate system of all ordered pairs of a function | 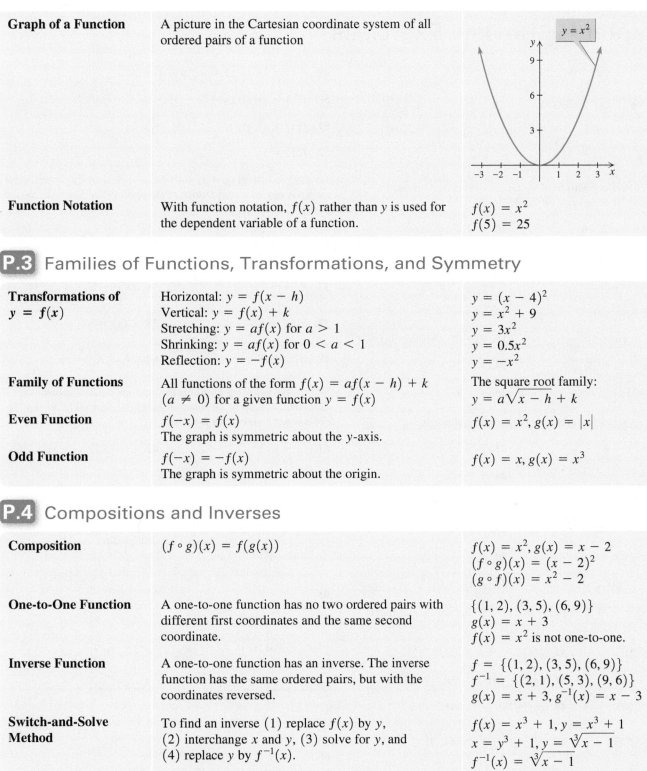 |
| --- | --- | --- |
| **Function Notation** | With function notation, $f(x)$ rather than $y$ is used for the dependent variable of a function. | $f(x) = x^2$ $f(5) = 25$ |

## P.3 Families of Functions, Transformations, and Symmetry

| **Transformations of** $y = f(x)$ | Horizontal: $y = f(x - h)$ Vertical: $y = f(x) + k$ Stretching: $y = af(x)$ for $a > 1$ Shrinking: $y = af(x)$ for $0 < a < 1$ Reflection: $y = -f(x)$ | $y = (x - 4)^2$ $y = x^2 + 9$ $y = 3x^2$ $y = 0.5x^2$ $y = -x^2$ |
| --- | --- | --- |
| **Family of Functions** | All functions of the form $f(x) = af(x - h) + k$ $(a \neq 0)$ for a given function $y = f(x)$ | The square root family: $y = a\sqrt{x - h} + k$ |
| **Even Function** | $f(-x) = f(x)$ The graph is symmetric about the $y$-axis. | $f(x) = x^2, g(x) = |x|$ |
| **Odd Function** | $f(-x) = -f(x)$ The graph is symmetric about the origin. | $f(x) = x, g(x) = x^3$ |

## P.4 Compositions and Inverses

| **Composition** | $(f \circ g)(x) = f(g(x))$ | $f(x) = x^2, g(x) = x - 2$ $(f \circ g)(x) = (x - 2)^2$ $(g \circ f)(x) = x^2 - 2$ |
| --- | --- | --- |
| **One-to-One Function** | A one-to-one function has no two ordered pairs with different first coordinates and the same second coordinate. | $\{(1, 2), (3, 5), (6, 9)\}$ $g(x) = x + 3$ $f(x) = x^2$ is not one-to-one. |
| **Inverse Function** | A one-to-one function has an inverse. The inverse function has the same ordered pairs, but with the coordinates reversed. | $f = \{(1, 2), (3, 5), (6, 9)\}$ $f^{-1} = \{(2, 1), (5, 3), (9, 6)\}$ $g(x) = x + 3, g^{-1}(x) = x - 3$ |
| **Switch-and-Solve Method** | To find an inverse (1) replace $f(x)$ by $y$, (2) interchange $x$ and $y$, (3) solve for $y$, and (4) replace $y$ by $f^{-1}(x)$. | $f(x) = x^3 + 1, y = x^3 + 1$ $x = y^3 + 1, y = \sqrt[3]{x - 1}$ $f^{-1}(x) = \sqrt[3]{x - 1}$ |
| **Graph of** $f^{-1}$ | Reflect the graph of $f$ about the line $y = x$ to get the graph of $f^{-1}$. |  |

# Chapter P Review Exercises

*Simplify each expression involving square roots.*

**1.** $\sqrt{98}$        **2.** $\sqrt{200}$        **3.** $\sqrt{\dfrac{9}{5}}$        **4.** $\sqrt{\dfrac{16}{13}}$

**5.** $\sqrt{\dfrac{5}{6}}$        **6.** $\sqrt{\dfrac{3}{8}}$        **7.** $\dfrac{8}{\sqrt{2}}$        **8.** $\dfrac{9}{\sqrt{3}}$

*For each pair of points, find the distance between them and the midpoint of the line segment joining them.*

**9.** $(-3, 5), (2, -6)$        **10.** $(-1, 1), (-2, -3)$

**11.** $(\pi/2, 1), (\pi, 1)$        **12.** $(\pi/2, 2), (\pi/3, 2)$

*Graph each equation.*

**13.** $x^2 + y^2 = 9$        **14.** $x^2 + y^2 = 25$

**15.** $x^2 + (y - 1)^2 = 1$        **16.** $(x - 2)^2 + (y + 1)^2 = 1$

**17.** $2x - 5y = 20$        **18.** $y = 2x - 50$

**19.** $x = 5$        **20.** $y = 3$

*Graph each function and state the domain and range.*

**21.** $y = x - 4$        **22.** $y = -x + 4$

**23.** $y = 4$        **24.** $y = -2$

**25.** $y = x^2 - 3$        **26.** $y = 6 - x^2$

**27.** $y = x^3 - 1$        **28.** $y = -x^3 + 1$

**29.** $y = \sqrt{x - 1}$        **30.** $y = \sqrt{x + 3}$

**31.** $f(x) = |x| - 4$        **32.** $f(x) = 3|x|$

*Let $f(x) = x^2 + 3$ and $g(x) = 2x - 7$. Find and simplify each expression.*

**33.** $f(-3)$        **34.** $g(3)$        **35.** $g(12)$

**36.** $f(-1)$        **37.** $(g \circ f)(-3)$        **38.** $(f \circ g)(3)$

**39.** $f(g(2))$        **40.** $g(f(-2))$        **41.** $(f \circ g)(x)$

**42.** $(g \circ f)(x)$        **43.** $g\left(\dfrac{x + 7}{2}\right)$        **44.** $f\left(\sqrt{x - 3}\right)$

**45.** $g^{-1}(x)$        **46.** $g^{-1}(-3)$

*Use transformations to graph each pair of functions on the same coordinate plane.*

**47.** $f(x) = \sqrt{x}, g(x) = 2\sqrt{x + 3}$

**48.** $f(x) = \sqrt{x}, g(x) = -2\sqrt{x} + 3$

**49.** $f(x) = |x|, g(x) = -2|x + 2| + 4$

**50.** $f(x) = |x|, g(x) = |x - 1| - 3$

**51.** $f(x) = x^2, g(x) = (x - 2)^2 + 1$

**52.** $f(x) = x^2, g(x) = -2x^2 + 4$

*Let $f(x) = \sqrt[3]{x}, g(x) = x - 4, h(x) = x/3$, and $j(x) = x^2$. Write each function as a composition of the appropriate functions chosen from f, g, h, and j.*

**53.** $N(x) = \dfrac{1}{3}x^2$        **54.** $P(x) = \left(\dfrac{x}{3}\right)^2$

**55.** $R(x) = \dfrac{x - 4}{3}$        **56.** $Q(x) = \dfrac{x}{3} - 4$

**57.** $F(x) = \sqrt[3]{x - 4}$        **58.** $G(x) = -4 + \sqrt[3]{x}$

**59.** $H(x) = \sqrt[3]{x^2 - 4}$        **60.** $M(x) = \left(\dfrac{x - 4}{3}\right)^2$

*Each of the following graphs is the graph of a function involving absolute value. Write an equation for each graph and state the domain and range of the function.*

**61.**

**62.**

**63.**

**64.**

**65.**

**66.**

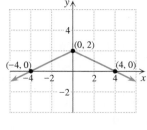

*Discuss the symmetry of the graph of each function.*

**67.** $f(x) = x^4 - 9x^2$

**68.** $f(x) = |x| - 99$

**69.** $f(x) = -x^3 - 5x$

**70.** $f(x) = \dfrac{15}{x}$

**71.** $f(x) = -x + 1$

**72.** $f(x) = |x - 1|$

**73.** $f(x) = x^2 + 1$

**74.** $f(x) = \sqrt{16 - x^2}$

*Graph each pair of functions on the same coordinate plane.*

**75.** $f(x) = \sqrt{x + 3}, g(x) = x^2 - 3$ for $x \geq 0$

**76.** $f(x) = (x - 2)^3, g(x) = \sqrt[3]{x} + 2$

**77.** $f(x) = 2x - 4, g(x) = \dfrac{1}{2}x + 2$

**78.** $f(x) = -\dfrac{1}{2}x + 4, g(x) = -2x + 8$

*Find the inverse of each function.*

**79.** $y = 5x$

**80.** $y = x + 9$

**81.** $y = 3x - 21$

**82.** $y = 2x - 5$

**83.** $f(x) = x^3 - 1$

**84.** $f(x) = \sqrt[3]{x} + 5$

**85.** $f(x) = \dfrac{1}{x - 3}$

**86.** $f(x) = \dfrac{1}{x} + 5$

**87.** $f(x) = x^2$ for $x \geq 0$

**88.** $f(x) = \sqrt{x - 3}$

*Solve each problem.*

**89.** Find the length of the hypotenuse of a right triangle that has legs with lengths $\sqrt{5}$ and 2.

**90.** A right triangle has a hypotenuse of length 3 and one leg with length 1. Find the length of the other leg.

**91.** Find the $x$- and $y$-intercepts for the graph of $3x - 4y = 9$.

**92.** Find the $x$- and $y$-intercepts for the graph of $y = 2x - 3$.

**93.** Write in standard form the equation of the circle that has center $(-3, 5)$ and radius $\sqrt{3}$.

**94.** What is the center and radius for the circle $(x - 3)^2 + (y + 1)^2 = 16$?

**95.** Find the equations of the lines through the origin that are tangent to $(x - 3)^2 + y^2 = 1$.

**96.** Find the equations of the lines through the origin that are tangent to $(x - 2)^2 + (y - 2)^2 = 2$.

**97.** Write the radius of a circle as a function of the circumference.

**98.** Write the diameter of a circle, $d$, as a function of its area, $A$.

## OUTSIDE THE BOX

**99.** *Wrong Division*  Is it possible to divide the integers from 1 through 9 into three sets of any size so that the product of the integers in each of the three sets is less than 72?

**100.** *Area of a Parallelogram*  A parallelogram with an area of 580 has vertices $(0, 0)$, $(5, 30)$, and $(20, a)$. If $a > 0$, then what is the value of $a$?

# Chapter  Test

*State the domain and range of each function.*

**1.** $y = |x|$

**2.** $f(x) = \sqrt{x - 9}$

**3.** $f(x) = 2 - x^2$

*Sketch the graph of each function.*

**4.** $y = 2x - 3$

**5.** $f(x) = \sqrt{x - 5}$

**6.** $y = 2|x| - 4$

**7.** $f(x) = -(x - 2)^2 + 5$

*Let $f(x) = \sqrt{x + 2}$ and $g(x) = 3x - 1$. Find and simplify each of the following expressions.*

**8.** $f(7)$

**9.** $g(2)$

**10.** $(f \circ g)(2)$

**11.** $g^{-1}(x)$

**12.** $g^{-1}(20)$

*Simplify each expression involving square roots.*

**13.** $\sqrt{72}$

**14.** $\sqrt{\dfrac{25}{6}}$

**15.** $\dfrac{20}{\sqrt{8}}$

*Solve each problem.*

**16.** Find the midpoint of the line segment whose endpoints are $(3, 5)$ and $(-3, 6)$.

**17.** Find the distance between the points $(-2, 4)$ and $(3, -1)$.

**18.** Write the equation in standard form for a circle with center $(-4, 1)$ and radius $\sqrt{7}$.

**19.** What is the center and radius of the circle $(x - 4)^2 + (y + 1)^2 = 6$?

**20.** If $f(x) = 3x - 9$ and $f(a) = 6$, then what is $a$?

**21.** Discuss the symmetry of the graph of the function $f(x) = x^4 - 3x^2 + 9$.

**22.** Find the inverse of the function $g(x) = 3 + \sqrt[3]{x - 2}$.

**23.** Write a formula expressing the area of a square $A$ as a function of its perimeter $P$.

# 1

# Angles and the Trigonometric Functions

**1.1** Angles and Degree Measure

**1.2** Radian Measure, Arc Length, and Area

**1.3** Angular and Linear Velocity

**1.4** The Trigonometric Functions

**1.5** Right Triangle Trigonometry

**1.6** The Fundamental Identity and Reference Angles

Since the beginning of human flight, humans have been using the air to spy on their neighbors. In 1960 the U-2 spy plane flew at an altitude of 13 miles to photograph the Soviet Union and Cuba. Today, spy planes have been replaced with unmanned aerial vehicles (UAVs) and satellites. At least 10 types of UAVs have been used in Iraq. Pictures taken from UAVs and satellites from an altitude of 700 miles are used to determine sizes of buildings and troop movements.

Determining the sizes of objects without measuring them physically is one of the principal applications of trigonometry. Long before we dreamed of traveling to the moon, trigonometry was used to calculate the distance to the moon.

**In this chapter we will explore many applications of trigonometric functions, ranging from finding the velocity of a lawnmower blade to estimating the size of a building in an aerial photograph.**

# 1.1 Angles and Degree Measure

Trigonometry was first studied by the Greeks, Egyptians, and Babylonians and used in surveying, navigation, and astronomy. Using trigonometry, they had a powerful tool for finding areas of plots of land, lengths of sides, and measures of angles, without physically measuring them. We begin our study of trigonometry by studying angles and their degree measures.

## Angles

In geometry a **ray** is defined as a point on a line together with all points of the line on one side of that point. Fig. 1.1(a) shows ray $\overrightarrow{AB}$. An **angle** is defined as the union of two rays with a common endpoint, the **vertex.** The angle shown in Fig. 1.1(b) can be named $\angle A$, $\angle BAC$, or $\angle CAB$. (Read the symbol $\angle$ as "angle.") Angles are also named using Greek letters such as $\alpha$ (alpha), $\beta$ (beta), $\gamma$ (gamma), or $\theta$ (theta).

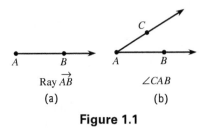

**Figure 1.1**

An angle is often thought of as being formed by rotating one ray away from a fixed ray as indicated by the arrow in Fig. 1.2(a). The fixed ray is the **initial side** and the rotated ray is the **terminal side**. An angle whose vertex is the center of a circle as shown in Fig. 1.2(b) is a **central angle** and the arc of the circle through which the terminal side moves is the **intercepted arc.** An angle in **standard position** is located in a rectangular coordinate system with the vertex at the origin and the initial side on the positive $x$-axis as shown in Fig. 1.2(c).

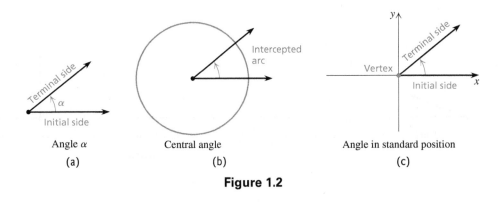

**Figure 1.2**

## Degree Measure of Angles

The measure $m(\alpha)$ of an angle $\alpha$ indicates the amount of rotation to the terminal side from the initial side. It is found by using any circle centered at the vertex. The circle is divided into 360 equal arcs and each arc is one **degree** $(1°)$. No one knows for sure why 360 was chosen to divide up the circle. It is most likely connected to the fact that some ancient calendars used 360 as the number of days in a year and that the sun changes position in the sky by approximately one degree each day.

**Definition:**
**Degree Measure**

The **degree measure of an angle** is the number of degrees in the intercepted arc of a circle centered at the vertex. The degree measure is positive if the rotation is counterclockwise and negative if the rotation is clockwise.

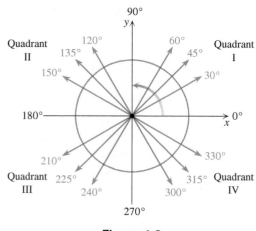

**Figure 1.3**

Figure 1.3 shows the positions of the terminal sides of some angles in standard position with common positive measures between 0° and 360°. An angle with measure between 0° and 90° is an **acute angle.** An angle with measure between 90° and 180° is an **obtuse angle.** An angle of exactly 180° is a **straight angle.** See Fig. 1.4. A 90° angle is a **right angle.** An angle in standard position is said to lie in the quadrant where its terminal side lies. If the terminal side is on an axis, the angle is a **quadrantal angle.** We often think of the degree measure of an angle as the angle itself. For example, we write $m(\alpha) = 60°$ or $\alpha = 60°$, and we say that 60° is an acute angle.

| Acute angle | Obtuse angle | Straight angle | Right angle | Quadrantal angle |

**Figure 1.4**

Coterminal angles

**Figure 1.5**

The initial side of an angle may be rotated in a positive or negative direction to get to the position of the terminal side. For example, if the initial side shown in Fig. 1.5 rotates clockwise for one quarter of a revolution to get to the terminal position, then the measure of the angle is −90°. If the initial side had rotated counterclockwise, then the measure of the angle would be 270°. If the initial side had rotated clockwise for one and a quarter revolutions to get to the terminal position, then the angle would be −450°. The angles −90°, −450°, and 270° are **coterminal angles**—they have the same terminal side. The degree measures of coterminal angles differ by a multiple of 360° (one complete revolution).

**Coterminal Angles**

Angles $\alpha$ and $\beta$ are coterminal if and only if there is an integer $k$ such that

$$m(\beta) = m(\alpha) + k360°.$$

An angle formed by two rays can be thought of as one angle with infinitely many different measures, or infinitely many different coterminal angles.

### EXAMPLE **1** Finding coterminal angles

Find two positive angles and two negative angles that are coterminal with $-50°$.

**Solution**

Since any angle of the form $-50° + k360°$ is coterminal with $-50°$, there are infinitely many possible answers. For simplicity, we choose the positive integers 1 and 2 and the negative integers $-1$ and $-2$ for $k$ to get the following angles:

$$-50° + 1 \cdot 360° = 310°$$

$$-50° + 2 \cdot 360° = 670°$$

$$-50° + (-1)360° = -410°$$

$$-50° + (-2)360° = -770°$$

The angles $310°$, $670°$, $-410°$, and $-770°$ are coterminal with $-50°$.

**TRY THIS.** Find two positive and two negative angles coterminal with $10°$.    ■

In the next example we use the definition of coterminal to determine whether two given angles are coterminal.

### EXAMPLE **2** Determining whether angles are coterminal

Determine whether angles in standard position with the given measures are coterminal.

**a.** $m(\alpha) = 190°, m(\beta) = -170°$
**b.** $m(\alpha) = 150°, m(\beta) = 880°$

**Solution**

**a.** If there is an integer $k$ such that $190 + 360k = -170$, then $\alpha$ and $\beta$ are coterminal.

$$190 + 360k = -170$$

$$360k = -360$$

$$k = -1$$

Since the equation has an integral solution, $\alpha$ and $\beta$ are coterminal.

**b.** If there is an integer $k$ such that $150 + 360k = 880$, then $\alpha$ and $\beta$ are coterminal.

$$150 + 360k = 880$$

$$360k = 730$$

$$k = \frac{73}{36}$$

Since there is no integral solution to the equation, $\alpha$ and $\beta$ are not coterminal.

**TRY THIS.** Determine whether $-690°$ and $390°$ are coterminal.    ■

Quadrantal angles such as 0°, 90°, 180°, 270°, and 360° have terminal sides on an axis and do not lie in any quadrant. Any angle that is not coterminal with a quadrantal angle lies in one of the four quadrants depending on the location of its terminal side. For example, any angle between 0° and 90° lies in quadrant I and any angle between 90° and 180° lies in quadrant II. To determine the quadrant in which an angle lies, add or subtract multiples of 360° (one revolution) to obtain a coterminal angle with a measure between 0° and 360°.

**EXAMPLE 3** **Determining in which quadrant an angle lies**

Name the quadrant in which each angle lies.

**a.** 230°  **b.** −580°  **c.** 1380°

**Solution**

**a.** Since 180° < 230° < 270°, a 230° angle lies in quadrant III.
**b.** We must add 2(360°) to −580°, to get an angle between 0° and 360°:

$$-580° + 2(360°) = 140°$$

So 140° and −580° are coterminal. Since 90° < 140° < 180°, 140° lies in quadrant II and so does −580°.

**c.** From 1380° we must subtract 3(360°) to obtain an angle between 0° and 360°:

$$1380° - 3(360°) = 300°$$

So 1380° and 300° are coterminal. Since 270° < 300° < 360°, 300° lies in quadrant IV and so does 1380°.

**TRY THIS.** Name the quadrant in which −890° lies. ∎

## Minutes and Seconds

Each degree is divided into 60 equal parts called **minutes,** and each minute is divided into 60 equal parts called **seconds.** A minute (min) is 1/60 of a degree (deg), and a second (sec) is 1/60 of a minute or 1/3600 of a degree. An angle with measure 44°12′30″ is an angle with a measure of 44 degrees, 12 minutes, and 30 seconds. Historically, angles were measured by using the degrees-minutes-seconds format, but with calculators it is convenient to have the fractional parts of a degree written as a decimal number such as 7.218°. Some calculators can handle angles in degrees-minutes-seconds format and even convert them to decimal degree format.

**EXAMPLE 4** **Converting degrees-minutes-seconds to decimal degrees**

Convert the measure 44°12′30″ to decimal degrees.

**Solution**

Since 1 degree = 60 minutes and 1 degree = 3600 seconds, we get

$$12 \text{ min} = 12 \text{ min} \cdot \frac{1 \text{ deg}}{60 \text{ min}} = \frac{12}{60} \text{ deg} \qquad \text{and}$$

$$30 \text{ sec} = 30 \text{ sec} \cdot \frac{1 \text{ deg}}{3600 \text{ sec}} = \frac{30}{3600} \text{ deg}.$$

So

$$44°12′30″ = \left( 44 + \frac{12}{60} + \frac{30}{3600} \right)° \approx 44.2083°.$$

**Figure 1.6**

 A graphing calculator can convert to decimal degrees as shown in Fig. 1.6.

**TRY THIS.** Convert 35°15′12″ to decimal degrees.    ■

Note that the conversion of Example 4 was done by *cancellation of units*. Minutes in the numerator "canceled" with minutes in the denominator to give the result in degrees, and seconds in the numerator "canceled" with seconds in the denominator to give the result in degrees.

**EXAMPLE 5** Converting decimal degrees to degrees-minutes-seconds

Convert the measure 44.235° to degrees-minutes-seconds format.

**Solution**

First convert 0.235° to minutes. Since 1 degree = 60 minutes,

$$0.235 \text{ deg} = 0.235 \text{ deg} \cdot \frac{60 \text{ min}}{1 \text{ deg}} = 14.1 \text{ min.}$$

So 44.235° = 44°14.1′. Now convert 0.1′ to seconds. Since 1 minute = 60 seconds,

$$0.1 \text{ min} = 0.1 \text{ min} \cdot \frac{60 \text{ sec}}{1 \text{ min}} = 6 \text{ sec.}$$

So 44.235° = 44°14′6″.

**Figure 1.7**

 A graphing calculator can convert to degrees-minutes-seconds as shown in Fig. 1.7.

**TRY THIS.** Convert 56.321° to degrees-minutes-seconds.    ■

Note again how the original units canceled in Example 5, giving the result in the desired units of measurement.

In the next example, we perform computations with angle measurements in the degrees-minutes-seconds format.

**EXAMPLE 6** Computations with degrees-minutes-seconds

Perform the indicated operations. Express answers in degrees-minutes-seconds format.

**a.** 13°45′33″ + 9°33′39″
**b.** 45° − 6°45′30″
**c.** (21°27′36″)/2

**Solution**

**a.** First add the degrees, minutes, and seconds:

$$\begin{array}{r} 13°45′33″ \\ 9°33′39″ \\ \hline 22°78′72″ \end{array}$$

Since 72″ = 1′ + 12″, we have 22°78′72″ = 22°79′12″. Since 79′ = 1° + 19′, we have 22°79′12″ = 23°19′12″. So

$$13°45′33″ + 9°33′39″ = 23°19′12″.$$

**b.** First note that $45° = 44°60' = 44°59'60''$:

$$44°59'60''$$
$$\underline{-6°45'30''}$$
$$38°14'30''$$

So $45° - 6°45'30'' = 38°14'30''$.

**c.** First divide the degrees, minutes, and seconds by 2.

$$\frac{21°27'36''}{2} = \left(10\frac{1}{2}\right)°\left(13\frac{1}{2}\right)'18''$$

Since $\frac{1}{2}' = 30''$ and $\frac{1}{2}° = 30'$, we have

$$\left(10\frac{1}{2}\right)°\left(13\frac{1}{2}\right)'18'' = \left(10\frac{1}{2}\right)°13'48'' = 10°43'48''.$$

**TRY THIS.** Find the difference $36°5' - 22°33'12''$. ∎

**Figure 1.8**

Calculators can do the computations that we did in Example 6. But you should not depend on a calculator. Depending on the mode, the calculator in Fig. 1.8 gets different results for $45° - 6°45'30''$. □

## FOR THOUGHT... True or False? Explain.

1. An angle is the union of two rays with a common endpoint.

2. The common endpoint of the two rays of an angle is called the vertex.

3. If a circle is divided into 360 equal arcs, then each arc is one degree.

4. The lengths of the rays of $\angle A$ determine the degree measure of $\angle A$.

5. The degree measure of an angle cannot be negative.

6. An angle of $540°$ is a quadrantal angle.

7. Angles of $20°$ and $380°$ are coterminal.

8. Angles of $5°$ and $-365°$ are coterminal.

9. $25°66' = 26°6'$

10. $25°20'40'' = 25.34°$

## 1.1 EXERCISES

### CONCEPTS

*Fill in the blank.*

1. A(n) _____ is a union of two rays with a common endpoint.

2. An angle whose vertex is the center of a circle is a(n) _____ angle.

3. An angle in _____ is located in a rectangular coordinate system with the vertex at the origin and the initial side on the positive *x*-axis.

4. An angle with a measure between $0°$ and $90°$ is a(n) _____ angle.

5. An angle with a measure between $90°$ and $180°$ is a(n) _____ angle.

6. An angle with a measure of $90°$ is a(n) _____ angle.

7. Angles in standard position with the same initial side and the same terminal side are _____ angles.

8. An angle in standard position whose terminal side is on an axis is a(n) _____ angle.

9. A(n) _____ is one sixtieth of a degree.

10. A(n) _____ is one sixtieth of a minute.

## SKILLS

*Find the degree measures of two positive and two negative angles that are coterminal with each given angle.*

**11.** 60°          **12.** 45°

**13.** 30°          **14.** 90°

**15.** 225°         **16.** 300°

**17.** −45°         **18.** −30°

**19.** −90°         **20.** −135°

**21.** −210°        **22.** −315°

*Determine whether the angles in each given pair are coterminal.*

**23.** 40°, −320°          **24.** 20°, 380°

**25.** 4°, −364°           **26.** 8°, −368°

**27.** 155°, 1235°         **28.** 272°, 1712°

**29.** 22°, −1058°         **30.** −128°, 592°

**31.** 312.4°, −227.6°     **32.** −87.3°, 812.7°

*Name the quadrant in which each angle lies.*

**33.** 85°    **34.** 110°    **35.** −125°    **36.** −200°

**37.** −740°  **38.** −1230°  **39.** 933°     **40.** 1568°

*Find the negative angle between 0° and −360° that is coterminal with the given positive angle.*

**41.** 50°    **42.** 40°    **43.** 140°

**44.** 190°   **45.** 270°   **46.** 210°

*Find the degree measure of the smallest positive angle that is coterminal with each angle.*

**47.** 400°           **48.** 540°

**49.** −340°          **50.** −180°

**51.** −1100°         **52.** −840°

**53.** 900.54°        **54.** 1235.6°

*Match the following degree measures with their angles (a)–(h).*

**55.** 30°    **56.** 45°    **57.** 60°    **58.** 120°

**59.** −330°  **60.** −315°  **61.** −60°   **62.** −120°

a.          b.

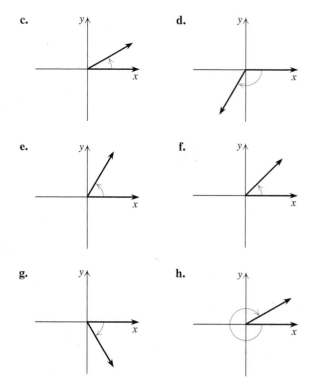

c.          d.

e.          f.

g.          h.

*Convert each angle to decimal degrees. When necessary round to four decimal places.*

**63.** 13°12′     **64.** 45°6′      **65.** −8°30′18″

**66.** −5°45′30″  **67.** 28°5′9″    **68.** 44°19′32″

**69.** 155°34′52″ **70.** 200°44′51″

*Convert each angle to degree-minutes-seconds. Round to the nearest whole number of seconds.*

**71.** 75.5°    **72.** 39.25°   **73.** 39.4°    **74.** 17.8°

**75.** −17.33°  **76.** −9.12°   **77.** 18.123°  **78.** 122.786°

*Perform each computation without a calculator. Express the answer in degrees-minutes-seconds format. Use a calculator to check your answers.*

**79.** 24°15′ + 33°51′          **80.** 99°35′ + 66°48′

**81.** 55°11′ − 23°37′          **82.** 76°6′ − 18°54′

**83.** 16°23′41″ + 44°43′39″    **84.** 7°55′42″ + 8°22′28″

**85.** 90° − 7°44′35″           **86.** 180° − 49°39′45″

**87.** 66°43′6″ − 5°51′53″      **88.** 34°39′12″ − 9°49′18″

**89.** 2(32°36′37″)             **90.** 3(89°41′56″)

**91.** 3(15°53′42″)             **92.** 4(9°36′40″)

**93.** (43°13′8″)/2             **94.** (34°40′15″)/3

**95.** (13°10′9″)/3             **96.** (5°18′40″)/4

*Find the degree measure of the angle α in each figure.*

**97.**
**98.**

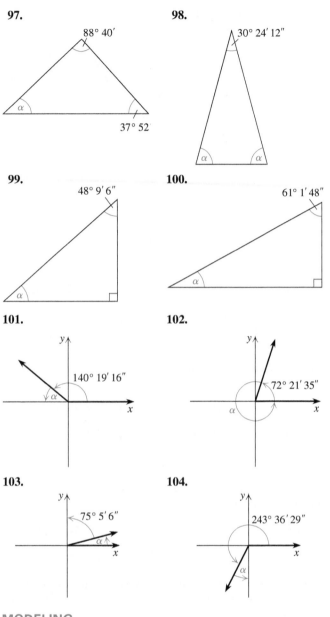

**99.**
**100.**

**101.**
**102.**

**103.**
**104.**

**MODELING**

*Solve each problem.*

**105.** *Angle between Spokes*  The wheel shown in the accompanying figure has 17 spokes. The angle between the spokes is approximately 21.17647°. Write this angle in the degrees-minutes-seconds format, to the nearest tenth of a second.

21.17647°

**Figure for Exercise 105**

**106.** *Angle of Elevation*  The angle of elevation shown in the accompanying figure is 37.243°. Write this angle in the degrees-minutes-seconds format, to the nearest tenth of a second.

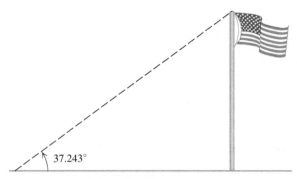

37.243°

**Figure for Exercise 106**

**107.** *Angles on a Bicycle*  The angles between the tubes on a bicycle frame are shown in the accompanying figure. Find the sum of the three angles. Convert each angle to decimal degree format (to four decimal places) and find their sum again.

73° 37′

49° 41′

56° 42′

**Figure for Exercise 107**

**108.** *Angular Window*  The angles between the sides of a triangular window are shown in the accompanying figure. Find the sum of the three angles. Convert each angle to decimal degree format (to four decimal places) and find their sum again.

125° 14′

27° 23′

27° 23′

**Figure for Exercise 108**

**109.** *Property Angles* The angles between the sides of a piece of property are shown in the accompanying figure. Find the sum of the four angles. Convert each angle to decimal degree format (to four decimal places) and find their sum again.

**Figure for Exercise 109**

**110.** *Machine Parts* Find the sum of the four angles on the metal plate shown in the accompanying figure. Convert each angle to decimal degree format (to four decimal places) and find the sum again.

**Figure for Exercise 110**

**111.** *Clock Angle* At 3:20 what is the acute angle between the minute hand and the hour hand on a standard clock?

**Figure for Exercise 111**

**112.** *Clock Angle* At 7:10:18 what is the acute angle between the minute hand and the second hand on a standard clock?

**REVIEW**

**113.** Let $f(x) = x^2 - 1$ and $g(x) = 3x - 2$. Find $f(g(2))$ and $g(f(2))$.

**114.** Find $f^{-1}(x)$, if $f(x) = 5x - 9$.

**115.** Find the distance between the points $(4, 5)$ and $(1, 1)$.

**116.** Find the equation of the circle with center $(4, -8)$ and radius $\sqrt{3}$.

**117.** Find the exact length of the diagonal of a square whose sides are 5 feet each.

**118.** Find the domain and range for the function $f(x) = 5 + \frac{1}{2}\sqrt{x + 3}$.

**OUTSIDE THE BOX**

**119.** *Triangles and Circles* Each of the five circles in the accompanying diagram has radius $r$. The four right triangles are congruent and each hypotenuse has length 1. Each line segment that appears to intersect a circle intersects it at exactly one point (a point of tangency). Find $r$.

**Figure for Exercise 119**

**120.** *An Odd Integer* Find the odd integer between 2000 and 2300 that is divisible by 11 and 13.

**POP QUIZ**

**1.** In which quadrant does $-120°$ lie?

**2.** Find the degree measure of the smallest positive angle that is coterminal with $1267°$.

**3.** What is the degree measure of the smallest positive angle that is coterminal with $-394°$?

**4.** Are $-240°$ and $60°$ coterminal?

**5.** Convert $70°30'36''$ to decimal degrees.

**6.** Convert $32.82°$ to degrees-minutes-seconds format.

**7.** Find the sum of $12°45'34''$ and $39°38'53''$. Express the answer in degrees-minutes-seconds format.

# Radian Measure, Arc Length, and Area

Degree measure of angles is used mostly in applied areas such as surveying, navigation, and engineering. Radian measure of angles is used more in scientific fields and results in simpler formulas in trigonometry and calculus.

## Radian Measure of Angles

For radian measure of angles in standard position we use a **unit circle** (a circle with radius 1) centered at the origin as shown in Fig. 1.9.

**Figure 1.9**

| Definition: | The **radian measure** of the angle $\alpha$ in standard position is the directed length |
| :--- | :--- |
| **Radian Measure** | of the intercepted arc on the unit circle. |

**Directed length** means that the radian measure is positive if the rotation of the terminal side is counterclockwise and negative if the rotation is clockwise. Figure 1.10 shows an angle $\alpha$ with a radian measure of 1 radian and an angle $\beta$ with a radian measure of $-1$ radian. In each case the length of the intercepted arc is 1 unit, but clockwise rotation corresponds to a negative radian measure.

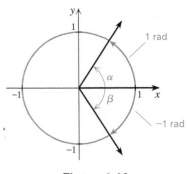

**Figure 1.10**

Note that instead of a unit circle, we could have used a circle of any radius $r$. If the radius is $r$, then the radian measure is the length of the intercepted arc divided by $r$. This definition gives the same result because the length of the intercepted arc is directly proportional to the radius of the circle.

Since the radius of the unit circle is the real number 1 without any dimension (such as feet or inches), the length of an intercepted arc is a real number without any dimension and so the radian measure of an angle is also a real number without any

dimension. One **radian** (abbreviated 1 rad) is the real number 1. If $s$ is the length of the intercepted arc on the unit circle for an angle $\alpha$ as shown in Fig. 1.9, we write $m(\alpha) = s$ radians or simply $m(\alpha) = s$.

Because the circumference of a circle with radius $r$ is $2\pi r$, the circumference of the unit circle is $2\pi$. If the initial side rotates 360° (one complete revolution), then the length of the intercepted arc is $2\pi$. So an angle of 360° has a radian measure of $2\pi$ radians. We express this relationship as $360° = 2\pi$ rad or simply $360° = 2\pi$. Dividing each side by 2 yields $180° = \pi$, which is the basic relationship to remember for conversion between these two measurements of angles.

 Use the MODE key on a graphing calculator to set the calculator to radian or degree mode as shown on the third line in Fig. 1.11. □

**Figure 1.11**

**Degree-Radian Conversion**

> Conversion from degrees to radians or radians to degrees is based on
>
> $$180 \text{ degrees} = \pi \text{ radians}.$$

To convert degrees to radians or radians to degrees we use 180 degrees $= \pi$ radians and cancellation of units. For example,

$$1 \text{ deg} = 1 \text{ deg} \cdot \frac{\pi \text{ rad}}{180 \text{ deg}} = \frac{\pi}{180} \text{rad} \approx 0.0175 \text{ rad}$$

and

$$1 \text{ rad} = 1 \text{ rad} \cdot \frac{180 \text{ deg}}{\pi \text{ rad}} = \frac{180}{\pi} \text{deg} \approx 57.3 \text{ deg}.$$

 If your calculator is in radian mode, as in Fig. 1.12(a), pressing ENTER converts degrees to radians. When the calculator is in degree mode, as in Fig. 1.12(b), pressing ENTER converts radians to degrees. □

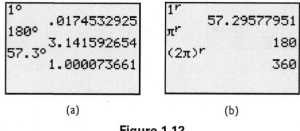

(a)          (b)

**Figure 1.12**

Figure 1.13(a) shows an angle of 1° and Fig. 1.13(b) shows an angle of 1 radian. An angle of 1° intercepts an arc whose length is $\frac{1}{360}$ of the circumference of the circle. An angle of 1 radian intercepts an arc on the unit circle whose length is the radius of the unit circle. In short, a 1 radian angle intercepts a 1 radius arc.

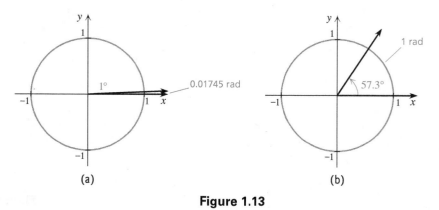

(a)          (b)

**Figure 1.13**

**Figure 1.14**

**EXAMPLE 1** Converting from degrees to radians

Convert the degree measures to radians.

**a.** 270° **b.** −23.6°

**Solution**

**a.** To convert degrees to radians, multiply the degree measure by $\pi$ rad/180 deg:

$$270° = 270 \text{ deg} \cdot \frac{\pi \text{ rad}}{180 \text{ deg}} = \frac{3\pi}{2} \text{ rad}$$

The exact value $3\pi/2$ rad is approximately 4.71 rad; but when a measure in radians is a simple multiple of $\pi$, we usually write the exact value.

Check this result with a calculator in radian mode as shown in Fig. 1.14.

**b.** $-23.6° = -23.6 \text{ deg} \cdot \dfrac{\pi \text{ rad}}{180 \text{ deg}} \approx -0.412 \text{ rad}$

**TRY THIS.** Convert 210° to radians. ∎

**EXAMPLE 2** Converting from radians to degrees

Convert the radian measures to degrees.

**a.** $\dfrac{7\pi}{6}$ **b.** 12.3

**Solution**

Multiply the radian measure by 180 deg/$\pi$ rad:

**Figure 1.15**

**a.** $\dfrac{7\pi}{6} = \dfrac{7\pi}{6} \text{ rad} \cdot \dfrac{180 \text{ deg}}{\pi \text{ rad}} = 210°$

**b.** $12.3 = 12.3 \text{ rad} \cdot \dfrac{180 \text{ deg}}{\pi \text{ rad}} \approx 704.7°$

Check these answers in degree mode as shown in Fig. 1.15.

**TRY THIS.** Convert the radian measure $\frac{5\pi}{3}$ to degrees. ∎

Figure 1.16 shows angles with common measures in standard position. Coterminal angles in standard position have radian measures that differ by an integral multiple of $2\pi$ (their degree measures differ by an integral multiple of 360°).

**Figure 1.16**

**Figure 1.17**

**Figure 1.18**

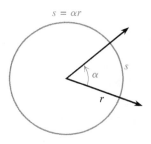

**Figure 1.19**

**EXAMPLE 3** **Finding coterminal angles using radian measure**

Find two positive and two negative angles using radian measure that are coterminal with $\pi/6$.

**Solution**

All angles coterminal with $\pi/6$ have a radian measure of the form $\pi/6 + k(2\pi)$, where $k$ is an integer. To find some angles, let $k$ be 1, 2, $-1$, and $-2$:

$$\frac{\pi}{6} + 1(2\pi) = \frac{\pi}{6} + \frac{12\pi}{6} = \frac{13\pi}{6}$$

$$\frac{\pi}{6} + 2(2\pi) = \frac{\pi}{6} + \frac{24\pi}{6} = \frac{25\pi}{6}$$

$$\frac{\pi}{6} + (-1)(2\pi) = \frac{\pi}{6} - \frac{12\pi}{6} = -\frac{11\pi}{6}$$

$$\frac{\pi}{6} + (-2)(2\pi) = \frac{\pi}{6} - \frac{24\pi}{6} = -\frac{23\pi}{6}$$

The angles $13\pi/6$, $25\pi/6$, $-11\pi/6$, and $-23\pi/6$ are coterminal with $\pi/6$.

**TRY THIS.** Find two positive and two negative angles coterminal with $5\pi/3$. ∎

## Arc Length of a Circle

If a 30° central angle intercepts an arc of length $s$ on a circle of radius $r$, then a 60° central angle intercepts an arc of length $2s$ and a 90° angle intercepts an arc of length $3s$. See Fig. 1.17. In general, a central angle $\alpha$ in a circle of radius $r$ intercepts an arc whose length $s$ is directly proportional to the measure of the central angle. Since a central angle of $2\pi$ (or 360°) intercepts an arc whose length is the circumference of the circle, we have

$$\frac{\text{arc length}}{\text{circumference}} = \frac{m(\alpha) \text{ in radians}}{2\pi \text{ radians}} = \frac{m(\alpha) \text{ in degrees}}{360 \text{ degrees}}.$$

So if $\alpha$ is the radian measure of an angle, $s$ is the arc length, and $2\pi r$ is the circumference, we have $\frac{s}{2\pi r} = \frac{\alpha}{2\pi}$. Multiplying each side of this equation by $2\pi r$ yields the arc-length formula $s = \alpha r$. See Fig. 1.18. (Using radian measure rather than degrees eliminates $\pi$ and results in a simpler formula.)

**Theorem:**
**Arc-Length Formula**

> The length $s$ of an arc intercepted by a central angle of $\alpha$ *radians* on a circle of radius $r$ is given by
>
> $$s = \alpha r.$$

You can solve problems involving central angles and arc length by using the arc-length formula, a simple proportion, or by noting that arc length is a fraction of the circumference of the circle. All three methods are shown in the next example.

**EXAMPLE 4** **Finding arc length**

**a.** A central angle of $\pi/2$ intercepts an arc on the surface of the earth that runs from the equator to the North Pole, as shown in Fig. 1.19. Using 3950 miles as the radius of the earth, find the length of the intercept arc to the nearest mile.

**Figure 1.20**

**b.** The wagon wheel shown in Fig. 1.20 has a diameter of 28 inches and an angle of 30° between the spokes. What is the length of the arc $s$ (to the nearest hundredth of an inch) between two adjacent spokes?

**Solution**

**a.** Using $s = \alpha r$, we have

$$s = \frac{\pi}{2}(3950) \approx 6205 \text{ miles.}$$

We can also solve the problem using a proportion:

$$\frac{s}{2\pi(3950)} = \frac{\pi/2}{2\pi}$$

$$s = 2\pi(3950) \cdot \frac{\pi/2}{2\pi} \approx 6205 \text{ miles}$$

Since $\frac{\pi/2}{2\pi} = \frac{1}{4}$, the length of the arc is $\frac{1}{4}$ of the circumference of the earth:

$$\frac{1}{4} \cdot 2\pi \cdot 3950 \approx 6205 \text{ miles}$$

**b.** To use the arc-length formula, you must convert 30° to radians:

$$30° = 30 \text{ deg} \cdot \frac{\pi \text{ rad}}{180 \text{ deg}} = \frac{\pi}{6} \text{ rad}$$

Now using $s = \alpha r$, we have

$$s = \frac{\pi}{6}(14) \approx 7.33 \text{ inches.}$$

We can also solve the problem using a proportion:

$$\frac{s}{2\pi(14)} = \frac{30°}{360°}$$

$$s = 2\pi(14)\frac{30}{360} \approx 7.33 \text{ inches}$$

Note that $\frac{30}{360} = \frac{1}{12}$ and the arc length is $\frac{1}{12}$ of the circumference of the circle.

**TRY THIS.** Find the length of an arc intercepted by a central angle of 45° on a circle of radius 8 inches.                        ■

If $\alpha$ is negative, the formula $s = \alpha r$ gives a negative number for the length of the arc. So $s$ is a directed length. Note that the formula $s = \alpha r$ applies only if $\alpha$ is in radians.

Since the radian measure of an angle is a dimensionless real number, the product of radians and inches in Example 4 is given in inches.

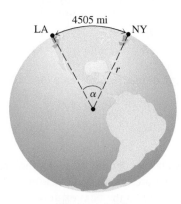

**Figure 1.21**

**EXAMPLE 5**    **Finding the central angle**

The distance between Los Angeles and New York City is 4505 miles. Find the central angle to the nearest tenth of a degree that intercepts an arc of 4505 miles on the surface of the earth (radius 3950 miles) as shown in Fig. 1.21.

**Solution**

From the formula $s = \alpha r$ we get $\alpha = s/r$:

$$\alpha = \frac{4505 \text{ mi}}{3950 \text{ mi}} \approx 1.1405 \text{ rad}$$

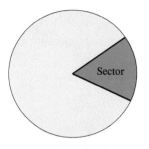

**Figure 1.22**

Multiply by 180 deg/$\pi$ rad to find the angle in degrees:

$$\alpha = 1.1405 \text{ rad} \cdot \frac{180 \text{ deg}}{\pi \text{ rad}} \approx 65.3°$$

**TRY THIS.** Find the central angle in degrees that intercepts an arc of length 3 cm on a circle of radius 50 cm.  ■

## Area of a Sector of a Circle

A **sector** of a circle is the region between two radii of a circle, as shown in Fig. 1.22. The area of a sector of a circle with radius $r$ is directly proportional to the measure of the central angle $\alpha$. Since a central angle of $2\pi$ (or 360°) forms a sector that covers the entire circle, we have

$$\frac{\text{area of sector}}{\text{area of circle}} = \frac{m(\alpha) \text{ in radians}}{2\pi \text{ radians}} = \frac{m(\alpha) \text{ in degrees}}{360 \text{ degrees}}.$$

So if the area of a sector is $A$ and its central angle is $\alpha$ radians, we have $\frac{A}{\pi r^2} = \frac{\alpha}{2\pi}$. Multiplying each side of this equation by $\pi r^2$ yields the formula for the area of a sector. The formula for the area is simplest when the angle is measured in radians because $\pi$ cancels out:

**Theorem: Area of a Sector**

The area $A$ of a sector with central angle $\alpha$ (in radians) in a circle of radius $r$ is given by

$$A = \frac{\alpha r^2}{2}.$$

**Figure 1.23**

You can solve problems involving sectors just like we solved problems involving central angles and arc length. You can use the formula for the area of a sector, a proportion, or note that the area of a sector is a fraction of the area of the circle. All three methods are shown in the next example.

**EXAMPLE 6**  **Finding the area of a sector of a circle**

A center-pivot irrigation system is used to water a circular field with radius 200 feet, as shown in Fig. 1.23. In three hours the system waters a sector with a central angle of $\pi/8$. What area (in square feet) is watered in that time?

**Solution**

If $A$ is the area of the sector, then using the formula for the area of a sector we have

$$A = \frac{(\pi/8)200^2}{2} \approx 7854 \text{ ft}^2.$$

We could also find the area by writing and solving a proportion:

$$\frac{A}{\pi \cdot 200^2} = \frac{\pi/8}{2\pi}$$

$$A = \frac{\pi/8}{2\pi} \cdot \pi \cdot 200^2 \approx 7854 \text{ ft}^2$$

Since $\frac{\pi/8}{2\pi} = \frac{1}{16}$, the area of the sector is 1/16 of the area of the circle:

$$A = \frac{1}{16} \cdot \pi \cdot 200^2 \approx 7854 \text{ ft}^2$$

**TRY THIS.** A sector of a circle has a central angle 36°. If the radius of the circle is 88 feet, then what is the area of the sector?  ■

## FOR THOUGHT... True or False? Explain.

1. The radian measure of an angle cannot be negative.

2. The radius of the unit circle is $\pi$.

3. The circumference of a unit circle is $2\pi$.

4. A central angle of one radian in a unit circle intercepts an arc of length one.

5. To convert radian measure to degree measure multiply the radian measure by $\frac{180}{\pi}$.

6. To convert degree measure to radian measure multiply the degree measure by $\frac{\pi}{360}$.

7. $180° = \pi$ rad

8. $45° = \frac{\pi}{2}$ rad

9. $60° = \frac{\pi}{3}$ rad

10. The length of the arc intercepted by a central angle of $\pi/4$ radians on a circle of radius 4 is $\pi$.

## 1.2 EXERCISES

### CONCEPTS

*Fill in the blank.*

1. A circle with radius 1 is a(n) _____ circle.

2. The _____ measure of an angle in standard position is the directed length of the intercepted arc on the unit circle.

3. The length $s$ of an arc intercepted by a central angle of $\alpha$ radians on a circle of radius $r$ is given by the formula _____.

4. The area $A$ of a sector with central angle $\alpha$ (in radians) in a circle of radius $r$ is given by the formula _____.

### SKILLS

*Find the missing degree or radian measure for each position of the terminal side shown. For Exercise 5, use degrees between 0° and 360° and radians between 0 and $2\pi$. For Exercise 6, use degrees between $-360°$ and 0° and radians between $-2\pi$ and 0. Practice these two exercises until you have memorized the degree and radian measures corresponding to these common angles.*

5.

6.

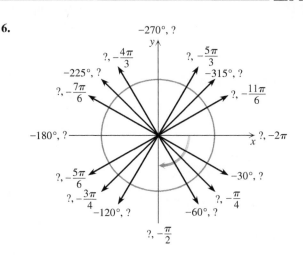

*Convert each degree measure to radian measure. Give exact answers (in terms of $\pi$).*

7. $45°$

8. $30°$

9. $90°$

10. $60°$

11. $120°$

12. $180°$

13. $150°$

14. $225°$

15. $18°$

16. $48°$

*Convert each radian measure to degree measure.*

17. $\frac{\pi}{3}$

18. $\frac{\pi}{6}$

19. $\frac{5\pi}{12}$

20. $\frac{17\pi}{12}$

21. $\frac{3\pi}{4}$

22. $\frac{5\pi}{4}$

23. $-6\pi$

24. $-9\pi$

*Convert each radian measure to degree measure to the nearest thousandth of a degree. Use the value of π found on your calculator.*

**25.** 2.39          **26.** 0.452

**27.** −0.128        **28.** −1.924

*Convert each degree measure to radian measure. Use the value of π found on your calculator and round answers to three decimal places.*

**29.** 37.4°         **30.** 125.3°        **31.** −13°47′

**32.** −99°15′       **33.** 53°37′6″       **34.** 187°49′36″

*Find the radian measure for two positive and two negative angles that are coterminal with the given angle.*

**35.** $\dfrac{\pi}{3}$          **36.** $\dfrac{\pi}{4}$

**37.** $\dfrac{\pi}{2}$          **38.** $\pi$

**39.** $\dfrac{2\pi}{3}$          **40.** $\dfrac{5\pi}{6}$

**41.** 1.2          **42.** 2

*For each given angle name the quadrant in which the terminal side lies.*

**43.** $\dfrac{5\pi}{12}$     **44.** $\dfrac{13\pi}{12}$     **45.** $-\dfrac{6\pi}{7}$

**46.** $-\dfrac{39\pi}{20}$   **47.** $-\dfrac{13\pi}{8}$    **48.** $-\dfrac{11\pi}{8}$

**49.** $\dfrac{17\pi}{3}$     **50.** $\dfrac{19\pi}{4}$     **51.** 3

**52.** 23.1          **53.** −7.3          **54.** −2

*Match the following radian measures with their angles (a)–(h).*

**55.** $\pi/3$       **56.** $\pi/6$       **57.** $2\pi/3$      **58.** $\pi/4$

**59.** $-\pi/3$      **60.** $-4\pi/3$     **61.** $-11\pi/6$    **62.** $-3\pi/4$

**a.**    **b.**

**c.**    **d.**

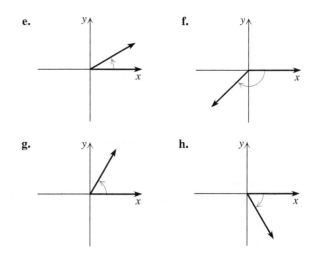

**e.**    **f.**

**g.**    **h.**

*Find the measure in radians of the smallest positive angle that is coterminal with each given angle. For angles given in terms of π, express the answer in terms of π. Otherwise, round to the nearest hundredth.*

**63.** $3\pi$        **64.** $6\pi$        **65.** $-\dfrac{\pi}{2}$

**66.** $-\dfrac{3\pi}{2}$     **67.** $\dfrac{9\pi}{2}$      **68.** $\dfrac{19\pi}{2}$

**69.** $-\dfrac{5\pi}{3}$     **70.** $-\dfrac{7\pi}{6}$     **71.** $-\dfrac{13\pi}{3}$

**72.** $-\dfrac{19\pi}{4}$    **73.** 8.32          **74.** −23.55

*Perform the indicated operation. Express the result in terms of π.*

**75.** $\pi - \dfrac{\pi}{4}$     **76.** $2\pi - \dfrac{\pi}{3}$     **77.** $\dfrac{\pi}{2} + \dfrac{\pi}{3}$

**78.** $\dfrac{\pi}{2} + \dfrac{\pi}{6}$     **79.** $\dfrac{\pi}{2} + \dfrac{\pi}{4}$     **80.** $\dfrac{\pi}{3} + \dfrac{\pi}{4}$

**81.** $\dfrac{\pi}{3} - \pi$     **82.** $\dfrac{7\pi}{6} - 2\pi$

*Find the length of the arc intercepted by the given central angle α in a circle of radius r. Round to the nearest tenth.*

**83.** $\alpha = \pi/4, r = 12$ ft     **84.** $\alpha = 1, r = 4$ cm

**85.** $\alpha = 3°, r = 4000$ mi     **86.** $\alpha = 60°, r = 2$ m

**87.** $\alpha = 1.3, r = 26.1$ m     **88.** $\alpha = \pi/8, r = 30$ yd

*Find the radius of the circle in which the given central angle α intercepts an arc of the given length s. Round to the nearest tenth.*

**89.** $\alpha = 1, s = 1$ mi     **90.** $\alpha = 0.004, s = 99$ km

**91.** $\alpha = 180°, s = 10$ km     **92.** $\alpha = 360°, s = 8$ m

**93.** $\alpha = \pi/6, s = 500$ ft     **94.** $\alpha = \pi/3, s = 7$ in.

*Find the exact area of the sector of the circle with the given radius and central angle.*

**95.** $r = 6, \alpha = 30°$     **96.** $r = 4, \alpha = 45°$

**97.** $r = 12, \alpha = \pi/3$     **98.** $r = 8, \alpha = \pi/12$

## MODELING

*Solve each problem.*

**99.** *Distance to North Pole*  Peshtigo, Wisconsin, is on the 45th parallel. This means that an arc from Peshtigo to the North Pole subtends a central angle of 45° as shown in the figure. If the radius of the earth is 3950 mi, then how far (to the nearest mile) is it from Peshtigo to the North Pole?

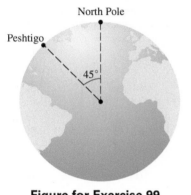

**Figure for Exercise 99**

**100.** *Distance to the Helper*  A surveyor sights her 6-ft 2-in. helper on a nearby hill as shown in the figure. If the angle of sight between the helper's feet and head is 0°37′, then approximately how far (to the nearest foot) is the helper?

**HINT** Assume that the helper is the arc of a circle intercepted by an angle of 0°37′.

**Figure for Exercise 100**

**101.** *Photographing the Earth*  From an altitude of 161 mi, the Orbiting Geophysical Observatory (OGO-1) can photograph a path on the surface of the earth that is approximately 2000 mi wide. Find the central angle, to the nearest tenth of a degree, that intercepts an arc of 2000 mi on the surface of the earth (radius 3950 mi).

**102.** *Margin of Error in a Field Goal*  The kicker aims to kick the ball midway between the uprights. To score a 50-yd field goal, what is the maximum number of degrees that the actual trajectory can deviate from the intended trajectory? What is the maximum number of degrees for a 20-yd field goal? Give answers to the nearest hundredth of a degree.

**HINT** Assume that the distance between the goal posts, 18.5 ft, is the length of an arc on a circle of radius 50 yd as shown in the figure.

**Figure for Exercise 102**

**103.** *Eratosthenes Measures the Earth*  Over 2200 years ago Eratosthenes read in the Alexandria library that at noon on June 21 a vertical stick in Syene cast no shadow. So on June 21 at noon Eratosthenes set out a vertical stick in Alexandria and found an angle of 7° in the position shown in the drawing. Eratosthenes reasoned that since the sun is so far away, sunlight must be arriving at the earth in parallel rays. With this assumption he concluded that the earth is round and the central angle in the drawing must also be 7°. He then paid a man to pace off the distance between Syene and Alexandria and found it to be 800 km. From these facts, calculate the circumference of the earth (to the nearest kilometer) as Eratosthenes did and compare his answer with the circumference calculated by using the currently accepted radius of 6378 km.

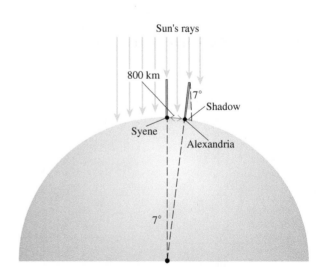

**Figure for Exercise 103**

**104.** *Another Unit* Eratosthenes did not really use kilometers as stated in the previous exercise. The unit he used for distance was a stadion. He used 500 stadia as the distance between the two cities. Find the circumference of the earth in stadia. The exact size of a stadion is unknown.

**105.** *Rotating Cogs* A bicycle has a chain that connects a chain ring of radius 6 in. with a cog of radius 2 in. as shown in the accompanying figure. If the chain ring rotates through an angle of 20°, then through how many degrees does the cog rotate?

**6 in.**

**2 in.**

chain ring

cog

**Figure for Exercise 105**

**106.** *Cadence and Speed* A CatEye bike computer measures cadence and speed. Cadence is the number of revolutions per minute of the pedals. A cyclist has a cadence of 40 rpm on a bike with 700-mm diameter wheels. What is her speed in miles per hour (to the nearest tenth) using the chain ring and cog dimensions from the previous exercise.

700 mm

**Figure for Exercise 106**

**107.** *Area of a Sector* A rope of length 6 feet is arranged in the shape of a sector of a circle with central angle $\theta$ radians, as shown in the accompanying figure. Write the area of the sector $A$ as a function of $\theta$.

**Figure for Exercises 107 and 108**

**108.** *Maximum Sector* Use a graphing calculator to find the angle $\theta$ in radians that maximizes the area of the sector in the previous exercise.

**109.** *Area of Irrigation* A central-pivot irrigation system is watering a circular field with a radius of 150 ft. The system rotates $\pi/6$ radians in one hour. What area (to the nearest square foot) is watered in one hour?

**110.** *Weather Radar* A weather radar system scans a circular area of radius 30 mi. In one second it scans a sector with central angle of 75°. What area (to the nearest square mile) is scanned in that time?

**111.** *Area of a Slice of Pizza* If a 16-in.-diameter pizza is cut into 6 slices of the same size, then what is the area of each slice to the nearest tenth of a square inch?

**112.** *Area of a Slice of Pizza* A slice of pizza with a central angle of $\pi/7$ is cut from a pizza with a radius of 10 in. What is the area of the slice to the nearest tenth of a square inch?

**113.** *Venn Watering Diagram* An irrigation specialist draws three circles on the ground of radius 10 meters so that the center of the second circle is on the first circle and the center of the third circle is an intersection point of the first two circles. At the center of each circle he places a sprinkler that waters in a circular pattern with a 10-meter radius. Find the exact and approximate total area watered by the three sprinklers.

**114.** *More Efficient Design* A smarter irrigation specialist draws a circle of radius 10 meters and finds three equally spaced points on the circle. At each of these three points she places a sprinkler that waters in a circular pattern with a radius of 10 meters.
  **a.** Find the exact and approximate total area watered by the three sprinklers.
  **b.** To water a very large field with a minimum number of these sprinklers, how would you place them without drawing circles?

**115.** *Volume of a Cup* Sunbelt Paper Products makes conical paper cups by cutting a 4-in. circular piece of paper on the radius and overlapping the paper by an angle $\alpha$ as shown in the figure.
  **a.** Find the volume of the cup to the nearest tenth of a cubic inch if $\alpha = 30°$.
  **b.** Write the volume of the cup as a function of $\alpha$.

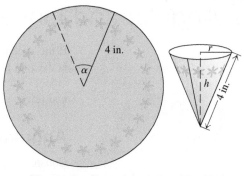

4 in.

$\alpha$

$r$

$h$

4 in.

**Figure for Exercise 115 and 116**

**116.** *Maximizing the Volume* The volume of the cup in Exercise 115 is zero if $\alpha$ is 0° or 360°. Somewhere in between, the volume reaches a maximum.

  **a.** Use the maximum feature of a graphing calculator to find the angle $\alpha$ to the nearest hundredth of a degree that gives the maximum volume for the cup.

  **b.** What is the maximum volume for the cup to the nearest tenth of a cubic inch?

## WRITING/DISCUSSION

**117.** *Discussion* What device is used for measuring angles? Is it marked in degrees or radians? Why?

**118.** *Cooperative Learning* In small groups, discuss the difference between radian measure and degree measure of an angle. Exactly how big is a unit circle? Why is a radian a real number? Could we define radian measure using any size circle?

## REVIEW

**119.** A(n) _____ is a union of two _____ with a common endpoint.

**120.** Convert 48.23° to degrees-minutes-seconds format.

**121.** Find the equation of a vertical line through $(\pi, 0)$.

**122.** What point lies midway between $(\pi/2, 0)$ and $(\pi, 0)$?

**123.** The graph of $f(x) = (x - 3)^2$ lies 3 units to the _____ of the graph of $f(x) = x^2$.

**124.** Find $w$ if $f(w) = 5$ and $f(x) = -3x - 7$.

## OUTSIDE THE BOX

**125.** *The Survivor* There are 13 contestants on a reality television show. They are instructed to each take a seat at a circular table containing 13 chairs that are numbered consecutively with the numbers 1 through 13. The producer then starts at number 1 and tells that contestant that he is a survivor. That contestant then leaves the table. The producer skips a contestant and tells the contestant in chair number 3 that he is a survivor. That contestant then leaves the table. The producer continues around the table skipping a contestant and telling the next contestant that he is a survivor. Each survivor leaves the table. The last person left at the table is *not* a survivor and must leave the show.

  **a.** For $n = 13$ find the unlucky number $k$, for which the person sitting in chair $k$ must leave the show.

  **b.** Find $k$ for $n = 8, 16,$ and 41.

  **c.** Find a formula for $k$.

**126.** *Ordered Triples* How many ordered triples of real numbers satisfy the following conditions. The product of $x$ and $y$ is $z$. The product of $x$ and $z$ is $y$. The product of $y$ and $z$ is $x$.

---

## 1.2 POP QUIZ

**1.** Convert 270° to radian measure.

**2.** Convert $7\pi/4$ to degree measure.

**3.** Are $-3\pi/4$ and $5\pi/4$ coterminal?

**4.** What is the radian measure of the smallest positive angle that is coterminal with $-3\pi/4$?

**5.** Find the exact length of the arc intercepted by a central angle of 60° in a circle with radius 30 feet.

**6.** Find the central angle to the nearest tenth of a degree that intercepts an arc of length 1 foot on a circle of radius 8 feet.

**7.** A sector of a circle has a central angle of 90°. If the radius of the circle is 8 inches, then what is the exact area?

---

## 1.3 Angular and Linear Velocity

**Figure 1.24**

If the speedometer in your car says 50 mph, then your velocity is 50 mph. Velocity is the rate at which the location of an object is changing with respect to time. If an object is in motion on a circle we can discuss two types of velocity, angular velocity and linear velocity.

### Angular Velocity

Consider a point located on a helicopter blade that is rotating at 400 revolutions per minute as shown in Fig. 1.24. As the blade rotates, an angle is formed by the initial

and terminal positions of the blade. In one revolution of the blade, the point rotates through an angle of $2\pi$ radians. The *angular velocity* of the point is the rate at which the angle is changing. Although 400 revolutions per minute could be considered an angular velocity, we will express angular velocity in radians per unit of time. To find the angular velocity we simply convert 400 rev/min to radians/minute using $2\pi$ rad $= 1$ rev and cancellation of units:

$$\frac{400 \text{ rev}}{1 \text{ min}} = \frac{400 \text{ rev}}{1 \text{ min}} \cdot \frac{2\pi \text{ rad}}{1 \text{ rev}} = \frac{800\pi \text{ rad}}{1 \text{ min}} \approx 2513 \text{ rad/min}$$

So the angular velocity of the point is approximately 2513 rad/min. The angular velocity does not depend on the length of the blade (or the radius of the circular path), but only on the number of revolutions per unit of time.

We use the Greek letter $\omega$ (omega) to represent angular velocity, and define it as follows.

**Definition:**
**Angular Velocity**

If a point is in motion on a circle through an angle of $\alpha$ radians in time $t$, then its **angular velocity** $\omega$ is given by

$$\omega = \frac{\alpha}{t}.$$

Angular velocity can be expressed with many different units of time. For example, an angular velocity can be given as radians per hour, radians per week, radians per year, and so on.

**EXAMPLE 1** **Changing the units**

Convert the angular velocity 240 rad/hr to rad/min.

**Solution**

Use the fact that 1 hr $= 60$ min and the cancellation of units method:

$$\omega = \frac{240 \text{ rad}}{1 \text{ hr}} = \frac{240 \text{ rad}}{\text{hr}} \cdot \frac{1 \text{ hr}}{60 \text{ min}} = 4 \text{ rad/min}$$

**TRY THIS.** Convert the angular velocity of 3000 rad/hr to rad/sec.

**EXAMPLE 2** **Finding angular velocity**

A typical 22-inch lawnmower blade rotates at a rate of 2500 revolutions per minute (rpm). What is the angular velocity in radians per second of a point on the tip of the blade?

**Solution**

Use the fact that $2\pi$ radians $= 1$ revolution and 60 seconds $= 1$ minute to convert rev/min into rad/sec using cancellation of units:

$$\omega = \frac{2500 \text{ rev}}{\text{min}} = \frac{2500 \text{ rev}}{\text{min}} \cdot \frac{2\pi \text{ rad}}{1 \text{ rev}} \cdot \frac{1 \text{ min}}{60 \text{ sec}} \approx 261.8 \text{ rad/sec}$$

So the angular velocity of the point is approximately 261.8 rad/sec.

**TRY THIS.** Find the angular velocity in radians per second for a particle that is moving in a circular path at 5 revolutions per second on a circle of radius 10 feet.

## Linear Velocity

Linear velocity is expressed in the same manner whether the motion is on a circle or in a straight line. The *linear velocity* of a point in motion is the rate at which the distance is changing. If a point is in motion on a circle then the distance traveled by the point in a unit of time is the length of an arc on the circle. We use the letter $v$ to represent linear velocity for a point in circular motion and define it as follows.

**Definition: Linear Velocity**

> If a point is in motion on a circle of radius $r$ through an angle of $\alpha$ radians in time $t$, then its **linear velocity** $v$ is given by
>
> $$v = \frac{s}{t},$$
>
> where $s$ is the arc length determined by $s = \alpha r$.

**Figure 1.25**

1.2 m

**EXAMPLE 3** Linear velocity of a propeller

A propeller with a radius of 1.2 meters is rotating at 1400 revolutions per minute as shown in Fig. 1.25. What is the linear velocity in meters per minute for a point on the tip of the propeller?

**Solution**

The point is traveling on a circle with circumference $2\pi(1.2)$ or $2.4\pi$ meters. According to the definition of linear velocity we divide the distance traveled (or arc length) by the amount of time. But this will be accomplished if we simply convert rev/min to m/min by cancellation of units.

$$v = \frac{1400 \text{ rev}}{1 \text{ min}} = \frac{1400 \text{ rev}}{1 \text{ min}} \cdot \frac{2.4\pi \text{ m}}{1 \text{ rev}} = \frac{3360\pi \text{ m}}{1 \text{ min}} \approx 10{,}556 \text{ m/min.}$$

Note that the last fraction is distance over time. So the point has a linear velocity of approximately 10,556 m/min.

**TRY THIS.** Find the linear velocity in meters per second for a particle that is moving in a circular path at 6 revolutions per second on a circle of radius 8 meters. ■

**Figure 1.26**

22 in.

**EXAMPLE 4** Linear velocity of a lawnmower blade

What is the linear velocity in miles per hour of the tip of a 22-inch lawnmower blade that is rotating at 2500 revolutions per minute? See Fig. 1.26.

**Solution**

The tip of the blade is traveling on a circle with a circumference of $2\pi(11)$ or $22\pi$ in. So the tip travels $22\pi$ in. for one revolution of the blade. Now use cancellation of units to convert 2500 rev/min into mi/hr:

$$v = \frac{2500 \text{ rev}}{1 \text{ min}} = \frac{2500 \text{ rev}}{1 \text{ min}} \cdot \frac{22\pi \text{ in.}}{1 \text{ rev}} \cdot \frac{60 \text{ min}}{1 \text{ hr}} \cdot \frac{1 \text{ ft}}{12 \text{ in.}} \cdot \frac{1 \text{ mi}}{5280 \text{ ft}}$$

$$\approx 163.6 \text{ mi/hr}$$

So the linear velocity of the tip of the blade is 163.6 mi/hr. (Keep your hands and feet away from the rotating blade.)

**TRY THIS.** Find the linear velocity in miles per hour for a particle that is moving in a circular path at 1000 revolutions per minute on a circle with a diameter of 14 inches. ■

# Linear Velocity in Terms of Angular Velocity

Linear velocity is arc length over time ($v = s/t$) and angular velocity is angle over time ($\omega = \alpha/t$). Since $s = \alpha r$, we have

$$v = \frac{s}{t} = \frac{\alpha r}{t} = r \cdot \frac{\alpha}{t} = r\omega.$$

We have found a relationship between the two velocities.

**Theorem: Linear Velocity in Terms of Angular Velocity**

If $v$ is the linear velocity of a point on a circle of radius $r$, and $\omega$ is its angular velocity, then

$$v = r\omega.$$

**Figure 1.27**

Any point on the surface of the earth (except at the poles) makes one revolution ($2\pi$ radians) about the axis of the earth in 24 hours. So the angular velocity of a point on the earth is $2\pi/24$ or $\pi/12$ radians per hour. The linear velocity of a point on the surface of the earth depends on its distance from the axis of the earth.

**EXAMPLE 5** Linear velocity on the surface of the earth

What is the linear velocity in miles per hour of a point on the equator? See Fig. 1.27.

**Solution**

A point on the equator has an angular velocity of $\pi/12$ radians per hour on a circle of radius 3950 miles. Using the formula $v = \omega r$, we get

$$v = \frac{\pi}{12} \text{ rad/hr} \cdot 3950 \text{ mi} = \frac{3950\pi}{12} \text{ mi/hr} \approx 1034 \text{ mi/hr}.$$

Note that we write angular velocity as rad/hr, but radians are omitted from the answer in miles per hour because radians are simply real numbers.

**TRY THIS.** Find the linear velocity in feet per second for a point on the equator. ■

## FOR THOUGHT... True or False? Explain.

1. $\frac{240 \text{ rev}}{1 \text{ min}} = 4$ rev/hr

2. $\frac{60\pi \text{ rad}}{1 \text{ hr}} = \pi$ rad/min

3. $\frac{4 \text{ rev}}{1 \text{ sec}} \cdot \frac{2\pi \text{ ft}}{1 \text{ rev}} = 8\pi$ rev/ft

4. $\frac{5\pi \text{ rad}}{1 \text{ hr}} \cdot \frac{60 \text{ min}}{1 \text{ hr}} = 300\pi$ rad/min

5. A velocity of 50 rad/min is a linear velocity.

6. A velocity of 32.2 ft/hr is an angular velocity.

7. The angular velocity of a point on a rotating wheel is 40 in./sec.

8. The linear velocity of a point that is 1 ft from the center of a wheel rotating at 1 rev/sec is $2\pi$ ft/sec.

9. A point that is 1 in. from the center of a rotating wheel on a car has the same angular velocity as a point that is 2 in. from the center.

10. As the earth rotates, Miami and Boston have the same linear velocity.

## 1.3 EXERCISES

### CONCEPTS

*Fill in the blank.*

1. If a point is in motion on a circle through an angle of $\alpha$ radians in time $t$, then its _____ $\omega$ is given by $\omega = \alpha/t$.

2. The _____ of a point in motion is the rate at which the distance is changing.

3. If a point is in motion on a circle of radius $r$ through an angle of $\alpha$ radians in time $t$, then its _____ $v$ is given by $v = s/t$, where $s$ is the arc length determined from $s = \alpha r$.

4. If $v$ is the _____ velocity of a point on a circle of radius $r$ and $\omega$ is its _____ velocity, then $v = r\omega$.

### SKILLS

*Find each product. Be sure to indicate the units for the answer. Round approximate answers to the nearest tenth.*

5. $\dfrac{300 \text{ rad}}{1 \text{ hr}} \cdot \dfrac{1 \text{ hr}}{60 \text{ min}}$

6. $\dfrac{2 \text{ rad}}{3 \text{ sec}} \cdot \dfrac{60 \text{ sec}}{1 \text{ min}}$

7. $\dfrac{4 \text{ rev}}{1 \text{ sec}} \cdot \dfrac{2\pi \text{ rad}}{1 \text{ rev}}$

8. $\dfrac{99.6 \text{ rad}}{2 \text{ sec}} \cdot \dfrac{1 \text{ rev}}{2\pi \text{ rad}}$

9. $\dfrac{55 \text{ rev}}{1 \text{ min}} \cdot \dfrac{6\pi \text{ ft}}{1 \text{ rev}}$

10. $\dfrac{33.3 \text{ rev}}{1 \text{ sec}} \cdot \dfrac{5.6 \text{ ft}}{1 \text{ rev}}$

11. $\dfrac{10 \text{ rad}}{1 \text{ min}} \cdot \dfrac{1 \text{ rev}}{2\pi \text{ rad}} \cdot \dfrac{60 \text{ min}}{1 \text{ hr}}$

12. $\dfrac{4400 \text{ rev}}{45 \text{ sec}} \cdot \dfrac{0.87 \text{ ft}}{1 \text{ rev}} \cdot \dfrac{1 \text{ yd}}{3 \text{ ft}}$

*Perform each conversion. Round approximate answers to the nearest tenth.*

13. 30 rev/min = _____ rad/min

14. 60 rev/sec = _____ rad/sec

15. 120 rev/hr = _____ rev/min

16. 150 rev/sec = _____ rev/hr

17. 180 rev/sec = _____ rad/hr

18. 3000 rad/hr = _____ rev/sec

19. 30 mi/hr = _____ ft/sec

20. 150 ft/sec = _____ mi/hr

*A windmill for generating electricity has a blade that is 30 feet long. Depending on the wind, it rotates at various velocities. In each case, find the angular velocity in rad/sec (to the nearest tenth) for the tip of the blade. Use 30 days/month.*

21. 500 rev/sec

22. 300 rev/sec

23. 433.2 rev/min

24. 11,000 rev/hr

25. 50,000 rev/day

26. 999,000 rev/mo

*A common speed for an electric motor is 3450 revolutions per minute. Saw blades of various diameters can be attached to such a motor. Determine the linear velocity in mi/hr for a point on the edge of a blade with each given diameter.*

27. 6 in.

28. 8 in.

29. 10 in.

30. 12 in.

31. 14 in.

32. 16 in.

### MODELING

*Solve each problem. Round approximate answers to the nearest tenth.*

33. *Velocity of a Record* Find the angular velocity in radians per minute and linear velocity in inches per minute for a point on the edge of a 6.5-in.-diameter record spinning at 45 rev/min.

34. *Velocity of an Album* Find the angular velocity in radians per second and linear velocity in inches per minute for a point on the edge of a 12-in.-diameter record spinning at $33\frac{1}{3}$ rev/min.

35. *London Eye* The London Eye, shown in the accompanying figure, can carry 800 passengers in 32 capsules around a circle measuring 424 meters in circumference. If one revolution takes 30 minutes, then what is the linear velocity of a capsule in feet per second? Round to the nearest tenth.

**Figure for Exercise 35**

36. *First Ferris Wheel* The first Ferris wheel was built for the Chicago Exposition in 1893. It held 2100 people, had a diameter of 250 ft, and took 30 min to make one revolution. Find the linear velocity in ft/sec for a person riding the Ferris wheel.

37. *Lawnmower Blade* What is the linear velocity (in miles per hour) of the tip of a lawnmower blade spinning at 2800 rev/min for a lawnmower that cuts a 20-in.-wide path?

**38.** *Router Bit* A router bit makes 45,000 rev/min. What is the linear velocity (in miles per hour) of the outside edge of a bit that cuts a 1-in.-wide path?

**39.** *Table Saw* The blade on a table saw rotates at 3450 revolutions per minute. How much faster (in ft/sec) does a 12-in.-diameter blade strike a piece of wood than a 10-in.-diameter blade?

**40.** *Automobile Tire* If a car runs over a nail at 55 mi/hr and the nail is lodged in the tire tread 13 in. from the center of the wheel, then what is the angular velocity of the nail in radians per hour?

**41.** *Linear Velocity Near the North Pole* Find the linear velocity for a point on the surface of the earth that is 1 mi from the North Pole.
**HINT** Assume that the point travels around a circle of radius 1 mi.

**42.** *Linear Velocity of Peshtigo* What are the linear and angular velocities for Peshtigo, Wisconsin (on the 45th parallel), with respect to its rotation around the axis of the earth? ( See Exercise 99 in Section 1.2.)

**43.** *Achieving Synchronous Orbit* The space shuttle orbits the earth in about 90 min at an altitude of about 125 mi, but a communication satellite must always remain above a fixed location on the earth, in synchronous orbit with the earth. Since the earth rotates 15° per hour, the angular velocity of a communication satellite must be 15° per hour. To achieve a synchronous orbit with the earth, the radius of the orbit of a satellite must be 6.5 times the radius of the earth (3950 mi). What is the linear velocity in miles per hour of such a satellite?

**44.** *A Small Ferris Wheel* A student is riding a 20-m-diameter Ferris wheel that is making three revolutions per minute.
**a.** What is the student's linear velocity?

**b.** What is the student's angular velocity?

**c.** If there are eight equally spaced seats on the Ferris wheel, then what is the length of the arc between two adjacent seats?

**45.** *Making Paper* A paper rewinding machine takes in paper at a constant rate of 2 ft/sec and winds it onto a roll, as shown in the accompanying figure. Find a formula that expresses the angular velocity of the roll $\omega$ (in rad/sec) in terms of the radius $r$ (in feet). Is angular velocity increasing or decreasing as the radius increases?

**Figure for Exercise 45**

**46.** *Radar Antenna* A radar antenna makes one revolution in 2 sec. What is the angular velocity of the antenna?

**47.** *Belt and Pulleys* A belt connects two pulleys with radii 3 in. and 5 in. as shown in the accompanying diagram. The velocity of point A on the belt is 10 ft/sec. What is the linear velocity (ft/sec) and the angular velocity (rad/sec) for point B? What is the linear velocity (ft/sec) and the angular velocity (rad/sec) for point C?
**HINT** Every point on the belt is moving at the same speed.

**Figure for Exercises 47 and 48**

**48.** *Belt and Pulleys* A belt connects two pulleys with radii 3 in. and 5 in. as shown in the accompanying diagram. Point B is rotating at 1000 rev/min. What is the linear velocity (ft/sec) for points A, B, and C? How many revolutions per minute is point C making?

**49.** *Bicycle Velocity* A bicycle with 27-in.-diameter wheels has a chain ring (by the pedals) with 52 teeth and a cog (by the wheel) with 26 teeth as shown in the accompanying diagram. The cyclist is pedaling at 1 revolution per second. What is the velocity of the bicycle in mi/hr?

**Figure for Exercise 49**

**50.** *Bicycle Velocity* A five-speed bicycle with 26-in.-diameter wheels has a chain ring with 44 teeth and five cogs with 13, 15, 17, 20, and 23 teeth. If a cyclist pedals at 1 revolution per second using any of these cogs, then what is the maximum velocity for this bicycle in mi/hr?

### WRITING/DISCUSSION

**51.** *Writing* Write a paragraph explaining the difference between linear and angular velocity.

**52.** *Cooperative Learning* Work with a group to select a watch that has an hour hand, a minute hand, and a second hand. Determine the linear and angular velocity of the tip of each hand. Use appropriate units. Explain your choices of units. Are these velocities the same for every clock?

## REVIEW

**53.** Perform the indicated operations. Express answers in degrees-minutes-seconds format.

**a.** $8°48'38'' + 4°17'58''$

**b.** $28°18'8'' - 9°19'29''$

**54.** Convert the degree measure $225°$ to radians.

**55.** Convert the radian measure $7\pi/6$ to degrees.

**56.** The length of an arc intercepted by a central angle of $\alpha$ radians in a circle of radius $r$ is _____.

**57.** Express the area $A$ of a circle as a function of its radius $r$.

**58.** Express the radius $r$ of a circle as a function of its area $A$.

## OUTSIDE THE BOX

**59.** *Telling Time* At 12 noon the hour hand, minute hand, and second hand of a clock are all pointing straight upward.
**a.** Find the first time after 12 noon, to the nearest tenth of a second, at which the angle between the hour hand and minute hand is $120°$.

**b.** Is there a time between 12 noon and 1 P.M. at which the three hands divide the face of the clock into thirds (that is, the angles between the hour hand, minute hand, and second hand are equal and each is $120°$)?

**c.** Does the alignment of the three hands described in part (b) ever occur?

**60.** *Finding the Diagonal* The sum of the length, width, and height for the rectangular solid shown in the accompanying figure is 25 in. Its surface area is 225 in.$^2$ Find the distance from $A$ to $B$.

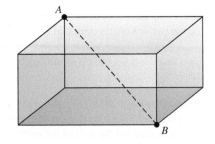

**Figure for Exercise 60**

## 1.3 POP QUIZ

**1.** Convert the angular velocity 300 rad/sec to rad/min.

**2.** Convert the linear velocity 200 mi/hr to ft/sec (to the nearest tenth).

**3.** A fan blade is turning at 240 revolutions per minute. Find the exact angular velocity in radians per second for a point on the tip of the blade.

**4.** What is the linear velocity in miles per hour (to the nearest tenth) for a particle moving in a circular path at 200 revolutions per minute on a circle of radius 60 feet?

**5.** What is the angular velocity in radians per hour for a point on the equator of the earth?

---

## LINKING concepts...    For Individual or Group Explorations

### Constructing the Sine Function

*Imagine that you are riding on a 20-meter-diameter Ferris wheel that is making three revolutions per minute.*

**a)** What is your linear velocity?

**b)** What is your angular velocity?

**c)** If there are eight equally spaced seats on the Ferris wheel, then what is the length of the arc between two adjacent seats?

**d)** Use a compass to draw a circle with diameter 20 cm to represent the Ferris wheel. Let $h(t)$ be your height in meters at time $t$ seconds, where $h(0) = 0$.

Locate your position on the Ferris wheel for each time in the following table and find $h(t)$ by measuring your drawing with a ruler.

| $t$ | 0 | 2.5 | 5 | 7.5 | 10 | 12.5 | 15 | 17.5 | 20 |
|------|---|-----|---|-----|----|------|----|------|----|
| $h(t)$ | | | | | | | | | |

**e)** Sketch the graph of $h(t)$ on graph paper for $t$ ranging from 0 to 60 seconds. The function $h(t)$ is a called a sine function.

**f)** How many solutions are there to $h(t) = 18$ in the interval $[0, 60]$?

**g)** In part (d) you found approximate values for $h(2.5)$ and $h(7.5)$. Find their exact values.

# 1.4 The Trigonometric Functions

The six trigonometric functions are the sine, cosine, tangent, cosecant, secant, and cotangent functions. Abbreviations for the functions are sin, cos, tan, csc, sec, and cot, respectively. There is more than one way to define these foundational functions of trigonometry. In this section we give the trigonometric ratio definition and in Section 2.1 we discuss the unit circle definition.

## Definition of the Trigonometric Functions

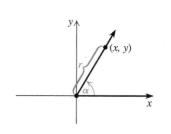

**Figure 1.28**

Suppose that $\alpha$ is an angle in standard position whose terminal side contains the point $(x, y)$ as shown in Fig. 1.28. Let $r$ be the distance between $(x, y)$ and the origin. Using the distance formula, we have $r = \sqrt{x^2 + y^2}$. The **trigonometric ratios** are the six possible ratios that can be formed with the numbers $x, y,$ and $r$. We define the six trigonometric functions as these ratios.

**Definition: The Trigonometric Functions**

If $(x, y)$ is any point other than the origin on the terminal side of an angle $\alpha$ in standard position and $r = \sqrt{x^2 + y^2}$, then

$$\sin \alpha = \frac{y}{r}, \qquad \cos \alpha = \frac{x}{r}, \qquad \tan \alpha = \frac{y}{x},$$

$$\csc \alpha = \frac{r}{y}, \qquad \sec \alpha = \frac{r}{x}, \qquad \cot \alpha = \frac{x}{y},$$

provided that no denominator is zero.

Since $r$ is positive for any angle $\alpha$, $\sin \alpha$ and $\cos \alpha$ are defined for any angle $\alpha$. Since $x = 0$ for any point on the $y$-axis, $\tan \alpha$ and $\sec \alpha$ are undefined for any angle that terminates on the $y$-axis. Since $y = 0$ for any point on the $x$-axis, $\csc \alpha$ and $\cot \alpha$ are undefined for any angle that terminates on the $x$-axis.

The trigonometric functions may be written with or without parentheses, as in $\sin(\alpha)$ or $\sin \alpha$. When an expression is more complicated we always use parentheses, as in $\sin(\alpha + \beta)$.

Note that the cosecant, secant, and cotangent are the reciprocals of the sine, cosine, and tangent, respectively. These relationships are called the **reciprocal identities.** (Recall that an identity is an equation that is true for all values of the variable for which both sides are defined.)

**The Reciprocal Identities**

$$\csc \alpha = \frac{1}{\sin \alpha} \qquad \sec \alpha = \frac{1}{\cos \alpha} \qquad \cot \alpha = \frac{1}{\tan \alpha}$$

**EXAMPLE 1** **Evaluating the trigonometric functions**

Find the values of the six trigonometric functions of the angle $\alpha$ in standard position whose terminal side passes through $(2, -1)$.

**Solution**

Use $x = 2$, $y = -1$, and $r = \sqrt{2^2 + (-1)^2} = \sqrt{5}$ to get

$$\sin \alpha = \frac{y}{r} = \frac{-1}{\sqrt{5}} = \frac{-1 \cdot \sqrt{5}}{\sqrt{5} \cdot \sqrt{5}} = -\frac{\sqrt{5}}{5},$$

$$\cos \alpha = \frac{x}{r} = \frac{2}{\sqrt{5}} = \frac{2 \cdot \sqrt{5}}{\sqrt{5} \cdot \sqrt{5}} = \frac{2\sqrt{5}}{5},$$

$$\tan \alpha = \frac{y}{x} = \frac{-1}{2} = -\frac{1}{2}, \qquad \csc \alpha = \frac{r}{y} = \frac{\sqrt{5}}{-1} = -\sqrt{5},$$

$$\sec \alpha = \frac{r}{x} = \frac{\sqrt{5}}{2}, \qquad \cot \alpha = \frac{x}{y} = \frac{2}{-1} = -2.$$

**TRY THIS.** Find the values of $\sin \alpha$ and $\cos \alpha$ if $\alpha$ is an angle in standard position whose terminal side passes through $(-2, -4)$. ∎

**Figure 1.29**

To evaluate the trigonometric functions for an angle $\alpha$, we must have a point $(x, y)$ on the terminal side of the angle. The trigonometric functions have the same values regardless of which point is used. If $(x, y)$ and $(x', y')$ are two points on the terminal side, as shown in Fig. 1.29, then the two triangles in the figure are similar. Because similar triangles are proportional, we get the same values for the trigonometric functions using $x$, $y$, and $r$ or using $x'$, $y'$, and $r'$. For example,

$$\sin \alpha = \frac{y}{r} = \frac{y'}{r'}, \qquad \cos \alpha = \frac{x}{r} = \frac{x'}{r'}, \qquad \text{and} \qquad \tan \alpha = \frac{y}{x} = \frac{y'}{x'}.$$

In Example 1 we were given a point on the terminal side of the angle. In the next example we are given an angle and we must find a point on the terminal side. If the angle is a quadrantal angle, then it is easy to select a point on its terminal side.

**EXAMPLE 2** **Evaluating trigonometric functions for quadrantal angles**

Find the exact values.

**a.** $\sin(90°)$
**b.** $\tan(180°)$
**c.** $\csc(-5\pi/2)$
**d.** $\cot(0)$

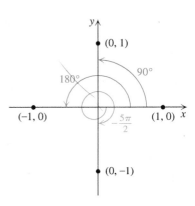

**Figure 1.30**

**Solution**

**a.** The terminal side of 90° lies on the positive $y$-axis as shown in Fig. 1.30. Select any point on the positive $y$-axis. For simplicity we choose $(0, 1)$. Since $r = \sqrt{0^2 + 1^2} = 1$, we have

$$\sin(90°) = \frac{y}{r} = \frac{1}{1} = 1.$$

**b.** The terminal side for 180° lies on the negative $x$-axis as in Fig. 1.30. Choose $(-1, 0)$ on the negative $x$-axis. Then

$$\tan(180°) = \frac{y}{x} = \frac{0}{-1} = 0.$$

**c.** Since $-5\pi/2$ is coterminal with $3\pi/2$ as shown in Fig. 1.30, the terminal side of $-5\pi/2$ lies on the negative $y$-axis and passes through the point $(0, -1)$. Since $r$ is the distance from $(0, -1)$ to the origin, $r = 1$. So

$$\csc(-5\pi/2) = \frac{r}{y} = \frac{1}{-1} = -1.$$

**d.** An angle of 0 radians has terminal side on the positive $x$-axis and passes through the point $(1, 0)$. Since $\cot(\alpha) = \frac{x}{y}$ only when $y \neq 0$, $\cot(0)$ is undefined.

**TRY THIS.** Find the exact values of $\sin(-3\pi/2)$ and $\cos(-3\pi/2)$. ■

Table 1.1 shows the values of the trigonometric functions for the quadrantal angles in the interval $[0°, 360°]$.

**Table 1.1**

| $\alpha$ | 0° | 90° | 180° | 270° | 360° |
|---|---|---|---|---|---|
| $\sin \alpha$ | 0 | 1 | 0 | −1 | 0 |
| $\cos \alpha$ | 1 | 0 | −1 | 0 | 1 |
| $\tan \alpha$ | 0 | undefined | 0 | undefined | 0 |
| $\csc \alpha$ | undefined | 1 | undefined | −1 | undefined |
| $\sec \alpha$ | 1 | undefined | −1 | undefined | 1 |
| $\cot \alpha$ | undefined | 0 | undefined | 0 | undefined |

Quadrant II | Quadrant I

$\sin \alpha > 0$, $\csc \alpha > 0$ | $\sin \alpha > 0$, $\csc \alpha > 0$
$\cos \alpha < 0$, $\sec \alpha < 0$ | $\cos \alpha > 0$, $\sec \alpha > 0$
$\tan \alpha < 0$, $\cot \alpha < 0$ | $\tan \alpha > 0$, $\cot \alpha > 0$

Quadrant III | Quadrant IV

$\sin \alpha < 0$, $\csc \alpha < 0$ | $\sin \alpha < 0$, $\csc \alpha < 0$
$\cos \alpha < 0$, $\sec \alpha < 0$ | $\cos \alpha > 0$, $\sec \alpha > 0$
$\tan \alpha > 0$, $\cot \alpha > 0$ | $\tan \alpha < 0$, $\cot \alpha < 0$

**Figure 1.31**

The *signs* of the trigonometric functions depend on the quadrant in which the angle lies. For any point $(x, y)$ on the terminal side of an angle in quadrant I, the $x$- and $y$-coordinates are positive $(x > 0$ and $y > 0)$. Since $r$ is always positive $(r > 0)$, all six trigonometric functions have positive values in quadrant I. In quadrant II, $x$ is negative while $y$ and $r$ are positive. So in quadrant II $\sin \alpha$ and $\csc \alpha$ are positive and the other four functions are negative. Figure 1.31 gives the signs of all of the trigonometric functions for each of the four quadrants.

A good mnemonic for the basic functions is "**A**ll **S**tudents **T**ake **C**alculus." That is, all are positive in quadrant I, sine is positive in quadrant II, tangent is positive in quadrant III, and cosine is positive in quadrant IV. Since reciprocals have the same sign, the signs of cosecant, secant, and cotangent agree with the signs of sine, cosine, and tangent, respectively.

## Trigonometric Functions at Multiples of 45°

In Example 2, exact values of the trigonometric functions were found for some angles that were multiples of 90° (quadrantal angles). We now find the exact values for any nonquadrantal angle that is a multiple of 45° (or $\pi/4$).

**Figure 1.32**

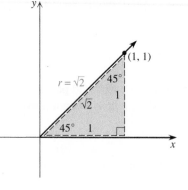

**Figure 1.33**

A triangle that has angles of 45°, 45°, and 90° is called a **45-45-90 triangle.** The 45-45-90 triangle in Fig. 1.32 has legs that are one unit each. Because of the Pythagorean theorem and the fact that $1^2 + 1^2 = (\sqrt{2})^2$, the hypotenuse is $\sqrt{2}$. To evaluate the trigonometric functions for a 45° angle, position a 45-45-90 triangle in the angle as shown in Fig. 1.33. It is then clear that $(1, 1)$ is on the terminal side of 45° and $r = \sqrt{2}$. So

$$\sin 45° = \frac{y}{r} = \frac{1}{\sqrt{2}} = \frac{\sqrt{2}}{2}, \qquad \cos 45° = \frac{x}{r} = \frac{1}{\sqrt{2}} = \frac{\sqrt{2}}{2},$$

$$\tan 45° = \frac{y}{x} = \frac{1}{1} = 1.$$

Since cosecant, secant, and cotangent are the reciprocals of sine, cosine, and tangent, respectively, we have csc 45° = $\sqrt{2}$, sec 45° = $\sqrt{2}$, cot 45° = 1.

We can evaluate the trigonometric functions for any nonquadrantal multiple of 45° by using a 45-45-90 triangle to help us determine $(x, y)$ and $r$.

**EXAMPLE 3** **Evaluating trigonometric functions at a multiple of 45°**

Find the exact value of each function.

**a.** $\sin(135°)$  **b.** $\cos\left(\dfrac{5\pi}{4}\right)$  **c.** $\tan\left(-\dfrac{9\pi}{4}\right)$

**Solution**

**a.** Sketch a 135° angle in quadrant II as in Fig. 1.34. Position a 45-45-90 triangle as shown to determine that the angle passes through $(-1, 1)$ and $r = \sqrt{2}$. So

$$\sin(135°) = \frac{y}{r} = \frac{1}{\sqrt{2}} = \frac{\sqrt{2}}{2}.$$

**b.** Sketch the angle with measure $5\pi/4$ in quadrant III as in Fig. 1.35. Position a 45-45-90 triangle as shown to determine that the angle passes through $(-1, -1)$ and $r = \sqrt{2}$. So

$$\cos(5\pi/4) = \frac{x}{r} = \frac{-1}{\sqrt{2}} = -\frac{\sqrt{2}}{2}.$$

**Figure 1.34**

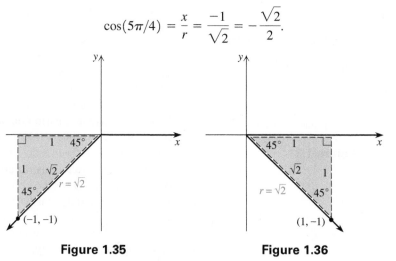

**Figure 1.35**          **Figure 1.36**

**c.** Since $-8\pi/4$ or $-2\pi$ is one clockwise revolution, $-9\pi/4$ is coterminal with $-\pi/4$. Sketch $-\pi/4$ as in Fig. 1.36. Position a 45-45-90 triangle as shown to determine that the angle passes through $(1, -1)$ and $r = \sqrt{2}$. So

$$\tan(-9\pi/4) = \frac{y}{x} = \frac{-1}{1} = -1.$$

**TRY THIS.** Find the exact values of $\sin(-3\pi/4)$ and $\cos(-3\pi/4)$. ∎

## Trigonometric Functions at Multiples of 30°

A triangle with angles of 30°, 60°, and 90° is a **30-60-90 triangle.** Since two congruent 30-60-90 triangles are formed by the altitude of an equilateral triangle, as shown in Fig. 1.37, the side opposite the 30° angle in a 30-60-90 triangle is half the length of the hypotenuse. If the side opposite 30° is 1, then the hypotenuse is 2. Because of the Pythagorean theorem and the fact that $1^2 + \left(\sqrt{3}\right)^2 = 2^2$, the side opposite 60° is $\sqrt{3}$ as shown in Fig. 1.37.

| Figure 1.37 | Figure 1.38 |

To evaluate the trigonometric functions for a 30° angle, position a 30-60-90 triangle as shown in Fig. 1.38. Now use the point $\left(\sqrt{3}, 1\right)$ and $r = 2$ to get

$$\sin 30° = \frac{y}{r} = \frac{1}{2}, \qquad \cos 30° = \frac{x}{r} = \frac{\sqrt{3}}{2},$$

and

$$\tan 30° = \frac{y}{x} = \frac{1}{\sqrt{3}} = \frac{\sqrt{3}}{3}.$$

Since cosecant, secant, and cotangent are the reciprocals of sine, cosine, and tangent, respectively, we have

$$\csc 30° = 2, \qquad \sec 30° = \frac{2}{\sqrt{3}} = \frac{2\sqrt{3}}{3},$$

and

$$\cot 30° = \sqrt{3}.$$

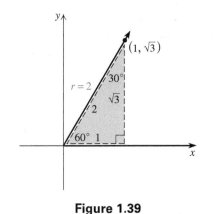

**Figure 1.39**

To evaluate the trigonometric functions for a 60° angle, position a 30-60-90 triangle as shown in Fig. 1.39. Now use the point $\left(1, \sqrt{3}\right)$ and $r = 2$ to get

$$\sin 60° = \frac{\sqrt{3}}{2}, \qquad \cos 60° = \frac{1}{2}, \qquad \text{and} \qquad \tan 60° = \sqrt{3}.$$

To evaluate a trigonometric function for a multiple of 30° (or $\pi/6$) we must find a point on the terminal side. For a nonquadrantal multiple of 30° (or 60°) we can always position a 30-60-90 triangle (as we did for 30° and 60°) to determine a point on the terminal side.

**EXAMPLE 4** **Evaluating trigonometric functions at a multiple of 30°**

Find the exact value of each function.

**a.** $\cos(-60°)$ **b.** $\sin\left(\dfrac{7\pi}{6}\right)$ **c.** $\tan(-240°)$

**Solution**

a. Draw a $-60°$ angle and position a 30-60-90 triangle as shown in Fig. 1.40. The point $\left(1, -\sqrt{3}\right)$ is on the terminal side of $-60°$ and the distance to the origin (or the hypotenuse of the triangle) is $r = 2$. So

$$\cos(-60°) = \frac{x}{r} = \frac{1}{2}.$$

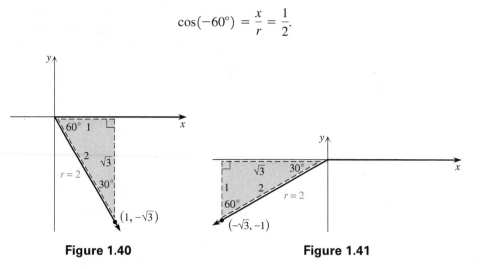

**Figure 1.40**              **Figure 1.41**

b. Since $\pi/6 = 30°$, we have $7\pi/6 = 7(30°) = 210°$. So $7\pi/6$ is $30°$ larger than the straight angle $180°$. Sketch $7\pi/6$ in quadrant III as shown in Fig. 1.41 and position a 30-60-90 triangle as shown in the figure. The point $\left(-\sqrt{3}, -1\right)$ is on the terminal side of $7\pi/6$ and $r = 2$. So

$$\sin(7\pi/6) = \frac{y}{r} = \frac{-1}{2} = -\frac{1}{2}.$$

c. Note that $-240°$ is coterminal with $120°$. Now sketch $-240°$ in quadrant II as shown in Fig. 1.42 and position a 30-60-90 triangle as shown in the figure. The point $\left(-1, \sqrt{3}\right)$ is on the terminal side of $-240°$. So

$$\tan(-240°) = \frac{y}{x} = \frac{\sqrt{3}}{-1} = -\sqrt{3}.$$

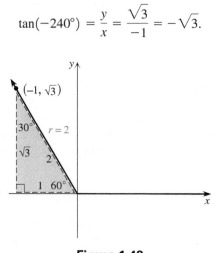

**Figure 1.42**

**TRY THIS.** Find the exact values of $\sin(-\pi/6)$ and $\cos(-\pi/6)$.

## Sines and Cosines for Common Angles

To find the value of any trigonometric function for an angle, we sketch the angle in standard position, find a point $(x, y)$ on the terminal side, then determine the appropriate trigonometric ratio. Sines and cosines of the multiples of $30°$ and $45°$ occur so

frequently that it is important to remember them. There is a nice pattern to the sines and cosines for the common angles. Observe the pattern (before simplifying) in Table 1.2 as you study this table.

**Table 1.2**

| | |
|---|---|
| $\sin 0° = \dfrac{\sqrt{0}}{2} = 0$ | $\cos 0° = \dfrac{\sqrt{4}}{2} = 1$ |
| $\sin 30° = \dfrac{\sqrt{1}}{2} = \dfrac{1}{2}$ | $\cos 30° = \dfrac{\sqrt{3}}{2} = \dfrac{\sqrt{3}}{2}$ |
| $\sin 45° = \dfrac{\sqrt{2}}{2} = \dfrac{\sqrt{2}}{2}$ | $\cos 45° = \dfrac{\sqrt{2}}{2} = \dfrac{\sqrt{2}}{2}$ |
| $\sin 60° = \dfrac{\sqrt{3}}{2} = \dfrac{\sqrt{3}}{2}$ | $\cos 60° = \dfrac{\sqrt{1}}{2} = \dfrac{1}{2}$ |
| $\sin 90° = \dfrac{\sqrt{4}}{2} = 1$ | $\cos 90° = \dfrac{\sqrt{0}}{2} = 0$ |

## Approximate Values of Trigonometric Functions

For quadrantal angles, multiples of 30°, and multiples of 45° we generally use the exact values of the trigonometric functions. For all other angles, we generally use approximate values, which are found with a scientific calculator or a graphing calculator. Calculators can evaluate $\sin \alpha$, $\cos \alpha$, or $\tan \alpha$ where $\alpha$ is a real number (radian) or $\alpha$ is the degree measure of an angle. Generally, there is a MODE key that sets the calculator to degree mode or radian mode. Consult your calculator manual.

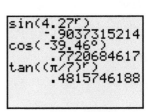

**Figure 1.43**

**EXAMPLE 5**   **Evaluating sine, cosine, and tangent with a calculator**

Find each function value rounded to four decimal places.

**a.** $\sin(4.27)$    **b.** $\cos(-39.46°)$    **c.** $\tan(\pi/7)$

**Solution**

**a.** With the calculator in radian mode we get $\sin(4.27) \approx -0.9037$.
**b.** With the calculator in degree mode we get $\cos(-39.46°) \approx 0.7721$.
**c.** With the calculator in radian mode we get $\tan(\pi/7) \approx 0.4816$.

   If you use the symbol for radians or degrees, as shown in Fig. 1.43, then it is not necessary to change the mode with a graphing calculator.

**TRY THIS.** Find $\sin(55.6)$ and $\cos(34.2°)$ with a calculator.     ∎

To evaluate the cosecant, secant, and cotangent functions with a calculator we must use the reciprocal identities.

**EXAMPLE 6**   **Evaluating cosecant, secant, and cotangent with a calculator**

Find each function value rounded to four decimal places.

**a.** $\csc(3.88)$    **b.** $\sec(-9.24°)$    **c.** $\cot(\pi/5)$

**Solution**

**a.** Set the mode to radian and evaluate as follows:

$$\csc(3.88) = \frac{1}{\sin(3.88)} \approx -1.4856$$

**b.** Set the mode to degrees and evaluate as follows:

$$\sec(-9.24°) = \frac{1}{\cos(-9.24°)} \approx 1.0131$$

**c.** Set the mode to radian and evaluate as follows:

$$\cot(\pi/5) = \frac{1}{\tan(\pi/5)} \approx 1.3764$$

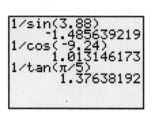

**Figure 1.44**

These computations are shown on a graphing calculator in Fig. 1.44.

**TRY THIS.** Find $\sec(3.5)$ and $\cot(49.2°)$ with a calculator. ∎

## FOR THOUGHT... True or False? Explain.

**1.** $\sin(90°) = 1$

**2.** $\sin(0°) = 0$

**3.** $\cos(45°) = 1/\sqrt{2}$

**4.** $\tan(\pi/4) = 1$

**5.** $\sin(30°) = 1/2$

**6.** $\cos(\pi/2) = 0$

**7.** $\sin(390°) = \sin(30°)$

**8.** $\sin(-\pi/3) = \sin(\pi/3)$

**9.** If $\sin\alpha = 1/5$, then $\csc\alpha = 5$.

**10.** If $\cos\alpha = 2/3$, then $\sec\alpha = 1.5$.

## 1.4 EXERCISES

### CONCEPTS

*Fill in the blank.*

**1.** If $(x, y)$ is any point other than the origin on the terminal side of an angle $\alpha$ in standard position and $r = \sqrt{x^2 + y^2}$, then $y/r = $ _____ and $x/r = $ _____.

**2.** If $(x, y)$ is any point other than the origin on the terminal side of an angle $\alpha$ in standard position and $r = \sqrt{x^2 + y^2}$, then the _____ are the six possible ratios that can be formed with the numbers $x$, $y$, and $r$.

**3.** Cosecant, secant, and cotangent are the _____ of the sine, cosine, and tangent, respectively.

**4.** All six trigonometric functions are positive for angles with terminal sides in _____.

### SKILLS

*Find the exact values of $\sin\alpha$, $\cos\alpha$, $\tan\alpha$, $\csc\alpha$, $\sec\alpha$, and $\cot\alpha$ where $\alpha$ is an angle in standard position whose terminal side contains the given point.*

**5.** $(1, 2)$  **6.** $(-1, -2)$  **7.** $(0, 1)$  **8.** $(1, 0)$

**9.** $(1, 1)$  **10.** $(1, -1)$  **11.** $(-2, 2)$  **12.** $(2, -2)$

**13.** $(-4, -6)$  **14.** $(-5, 10)$

*Evaluate the following without a calculator. Some of these expressions are undefined.*

**15.** $\cos(90°)$  **16.** $\cos(180°)$  **17.** $\sin(-\pi/2)$

**18.** $\cos(-\pi/2)$  **19.** $\tan(180°)$  **20.** $\tan(270°)$

**21.** $\cot(2\pi)$  **22.** $\cot(-\pi)$  **23.** $\csc(3\pi/2)$

**24.** $\csc(5\pi/2)$  **25.** $\sec(\pi)$  **26.** $\sec(-2\pi)$

*Find the exact value of each function without using a calculator.*

**27.** $\sin(\pi/4)$  **28.** $\sin(3\pi/4)$  **29.** $\cos(45°)$

**30.** $\cos(135°)$  **31.** $\tan(-\pi/4)$  **32.** $\tan(3\pi/4)$

**33.** $\sec(7\pi/4)$  **34.** $\csc(-\pi/4)$

*Find the exact value of each function without using a calculator.*

**35.** $\sin(30°)$  **36.** $\sin(60°)$  **37.** $\cos(-60°)$

**38.** $\cos(-120°)$  **39.** $\tan(5\pi/6)$  **40.** $\tan(2\pi/3)$

**41.** $\sec(-4\pi/3)$  **42.** $\sec(-5\pi/6)$  **43.** $\csc(390°)$

**44.** $\csc(-420°)$  **45.** $\cot(13\pi/6)$  **46.** $\cot(-7\pi/3)$

*Find the exact value of each expression without using a calculator. Check your answer with a calculator.*

**47.** $\dfrac{\cos(\pi/3)}{\sin(\pi/3)}$

**48.** $\dfrac{\sin(-5\pi/6)}{\cos(-5\pi/6)}$

**49.** $\dfrac{\sin(7\pi/4)}{\cos(7\pi/4)}$

**50.** $\dfrac{\sin(-3\pi/4)}{\cos(-3\pi/4)}$

**51.** $\sin\left(\dfrac{\pi}{3} + \dfrac{\pi}{6}\right)$

**52.** $\cos\left(\dfrac{\pi}{3} - \dfrac{\pi}{6}\right)$

**53.** $\dfrac{1 - \cos(5\pi/6)}{\sin(5\pi/6)}$

**54.** $\dfrac{\sin(5\pi/6)}{1 + \cos(5\pi/6)}$

**55.** $\sin(\pi/4) + \cos(\pi/4)$

**56.** $\sin(\pi/6) + \cos(\pi/3)$

**57.** $\cos(45°)\cos(60°) - \sin(45°)\sin(60°)$

**58.** $\sin(30°)\cos(135°) + \cos(30°)\sin(135°)$

**59.** $\dfrac{1 - \cos\left(2 \cdot \dfrac{\pi}{6}\right)}{2}$

**60.** $\dfrac{1 + \cos\left(2 \cdot \dfrac{\pi}{8}\right)}{2}$

**61.** $2\cos(210°)$

**62.** $\cos(-270°) - \sin(-270°)$

*Use a calculator to find the value of each function. Round answers to four decimal places.*

**63.** $\sin(43°)$

**64.** $\sin(55°)$

**65.** $\cos(230.46°)$

**66.** $\cos(344.1°)$

**67.** $\tan(-359.4°)$

**68.** $\tan(-269.5°)$

**69.** $\csc(23°48')$

**70.** $\csc(49°13')$

**71.** $\sec(-48°3'12'')$

**72.** $\sec(-9°4'7'')$

**73.** $\cot(\pi/9)$

**74.** $\cot(\pi/10)$

*Use a calculator to evaluate each expression. Round approximate answers to four decimal places.*

**75.** $\dfrac{1 + \cos(44.3°)}{2}$

**76.** $\dfrac{1 - \cos(98.6°)}{\sin(98.6°)}$

**77.** $\dfrac{\sin(\pi/12)}{\cos(\pi/12)}$

**78.** $\dfrac{\sin(9.2)}{\cos(9.2)}$

**79.** $2\sin(4)\cos(4)$

**80.** $2\sin(3)\cos(3)$

**81.** $1 - 2\sin(-88.4°)$

**82.** $2\cos(-9.5°) - 1$

*Find the exact value of each expression for the given value of $\theta$. Do not use a calculator.*

**83.** $\sin(2\theta)$ if $\theta = \pi/4$

**84.** $\sin(2\theta)$ if $\theta = \pi/8$

**85.** $\cos(2\theta)$ if $\theta = \pi/6$

**86.** $\cos(2\theta)$ if $\theta = \pi/3$

**87.** $\sin(\theta/2)$ if $\theta = 3\pi/2$

**88.** $\sin(\theta/2)$ if $\theta = 2\pi/3$

*Solve each problem.*

**89.** *Constructing Manipulatives* Accurately draw a small 45-45-90 triangle on paper or cardboard and cut it out. Label the angles 45°, 45°, and 90°. Label the sides 1, 1, and $\sqrt{2}$. Do the same for a 30-60-90 triangle. Label its angles 30°, 60°, and 90°. Label its sides 1, 2, and $\sqrt{3}$.

**90.** *Using Manipulatives* Use the triangles constructed in the last exercise as templates to assist you in accurately drawing the following angles in standard position and determining $(x, y)$ and $r$ for each angle.

   **a.** 120°         **b.** 135°

   **c.** 210°         **d.** $-\pi/6$

   **e.** $-3\pi/4$     **f.** $5\pi/4$

**91.** If $\sin \alpha = 3/4$, then what is $\csc \alpha$?

**92.** If $\tan \alpha = 1/70$, then what is $\cot \alpha$?

**93.** If $\sec \alpha = 10/3$, then what is $\cos \alpha$?

**94.** If $\sin \alpha = 0$, then what is $\csc \alpha$?

**95.** In each case name the quadrant containing the terminal side of $\alpha$.

   **a.** $\sin \alpha > 0$ and $\cos \alpha < 0$

   **b.** $\sin \alpha < 0$ and $\cos \alpha > 0$

   **c.** $\tan \alpha > 0$ and $\cos \alpha < 0$

   **d.** $\tan \alpha < 0$ and $\sin \alpha > 0$

**96.** In each case name the quadrant containing the terminal side of $2\alpha$.

   **a.** $\alpha = 66°$

   **b.** $\alpha = 2\pi/3$

   **c.** $\alpha = 150°$

   **d.** $\alpha = -5\pi/6$

## MODELING

*Solve each problem.*

**97.** *Steering Geometry* How well a motorcycle or bicycle handles depends on the steering geometry. One measure of stability is the trail, which is usually around 4 to 6 inches. The formula

$$T = \dfrac{r \cos\theta - R}{\sin\theta}$$

gives the trail $T$ as a function of the radius of the wheel $r$, head angle $\theta$, and rake or offset $R$. Find $T$ if $\theta = 60°$, $R = 3$ inches, and $r = 14$ inches. See the accompanying figure on the next page.

**Figure for Exercise 97**

**98.** *Flop factor* Too much trail will cause the front wheel of a motorcycle to turn more than expected (flop over) when the handlebars are rotated away from the straight ahead position. This is why choppers can be difficult to control. The formula

$$F = T \sin \theta \cos \theta$$

gives the flop factor $F$ as a function of the trail $T$ and the head angle $\theta$. Find the flop factor for a motorcycle with a head angle of 39° and a trail of 10 inches.

## REVIEW

**99.** Find the degree measure of the third angle of an isosceles triangle in which the equal angles each measure 9°38′52″.

**100.** Conversion from _____ to radians is based on _____ degrees equals $\pi$ radians.

**101.** Convert $13\pi/12$ radians to degree measure.

**102.** A sector of a circle with radius 8 meters has a central angle of $\pi/8$. Find the area of the sector to the nearest tenth of a square meter.

**103.** A 30-inch lawnmower blade is rotating at 2000 revolutions per minute. Find the linear velocity of the tip of the blade in miles per hour (to the nearest tenth).

**104.** A wheel is rotating at 200 rev/sec. Find the angular velocity in radians per minute (to the nearest tenth).

## OUTSIDE THE BOX

**105.** *Counting Votes* Fifteen experts are voting to determine the best convertible of the year. The choices are a Porsche Carrera, a Chrysler Crossfire, and a Nissan Roadster. The experts will rank the three cars 1st, 2nd, and 3rd. There are three common ways to determine the winner:
  **a.** *Plurality:* The car with the most first-place votes (preferences) is the winner.

  **b.** *Instant runoff:* The car with the least number of preferences is eliminated. Then the ballots for which the eliminated car is first are revised so that the second-place car is moved to first. Finally, the car with the most preferences is the winner.

  **c.** *The point system:* Two points are given for each time a car is ranked first place on a ballot, one point for each time the car appears in second place on a ballot, and no points for third place.

When the ballots were cast, the Porsche won when plurality was used, the Chrysler won when instant runoff was used, and the Nissan won when the point system was used. Determine 15 actual votes for which this result would occur.

**106.** *Range* Find the range of the function
$$y = 1 + \sqrt{12 \cos^3(x) + 37}$$ without using a calculator.

## 1.4 POP QUIZ

**1.** Find $\sin \alpha$, $\cos \alpha$, and $\tan \alpha$ if $\alpha$ is an angle in standard position whose terminal side passes through $(4, 3)$.

*Find the exact value.*

**2.** $\sin(0°)$

**3.** $\sin(-60°)$

**4.** $\cos(3\pi/4)$

**5.** $\cos(90°)$

**6.** $\tan(-2\pi/3)$

**7.** $\tan(11\pi/6)$

**8.** $\csc(\pi/4)$

**9.** $\sec(60°)$

---

# LINKING
# concepts...    For Individual or Group Explorations

```
sin(π/6)
              .5
π/6-(π/6)^3/3!
       .4996741794
```

## Infinite Series for Sine and Cosine

*Like the exponential and logarithmic functions, the sine and cosine functions are transcendental functions. So there are no finite algebraic formulas that will produce the values of these functions. The sine and cosine functions can be evaluated using infinite algebraic formulas called infinite series. It is shown in calculus that*

$$\sin(x) = x - \frac{x^3}{3!} + \frac{x^5}{5!} - \frac{x^7}{7!} + \cdots \quad \text{and}$$

$$\cos(x) = 1 - \frac{x^2}{2!} + \frac{x^4}{4!} - \frac{x^6}{6!} + \cdots$$

*for any real number x. (Recall that n! = $1 \cdot 2 \cdot 3 \cdot \cdots \cdot n$.) You can evaluate only a finite number of terms of this series, but the more terms that you use, the closer the result is to the true value of the function. As you can see from the accompanying figure, even two terms with x = π/6 comes fairly close to the actual value of sin(π/6). The following exercises will give you an idea of what your calculator does when it determines a sine or cosine.*

**a)** Let $x = \pi/6$ and evaluate $x$, $x - x^3/3!$, $x - x^3/3! + x^5/5!$, $x - x^3/3! + x^5/5! - x^7/7!$, and so on. Which is the first expression to give 0.5 on your calculator?

**b)** Repeat part (a) with $x = 13\pi/6$.

**c)** Let $x = \pi/4$ and evaluate $1 - x^2/2!$, $1 - x^2/2! + x^4/4!$, $1 - x^2/2! + x^4/4! - x^6/6!$, and so on. Which expression is the first to agree with your calculator's value for $\cos(\pi/4)$?

**d)** Repeat part (c) with $x = 9\pi/4$.

**e)** What can you conjecture about the size of $x$ and the number of terms needed to give an accurate value for $\sin(x)$ or $\cos(x)$?

**f)** How would you calculate an accurate value for $\sin(601\pi/3)$ using the series for the sine function?

# 1.5 Right Triangle Trigonometry

One reason trigonometry was invented was to determine the measures of the sides and angles of geometric figures without actually measuring them. In this section we will determine the unknown sides and angles of right triangles (triangles with a 90° angle). To accomplish this task, we need to use the trigonometric functions "in reverse." The idea of reversing or inverting a function was discussed in Section P.4. You may wish to review that section now.

## Inverse Trigonometric Functions

A solution to the equation $\sin \alpha = \frac{1}{2}$ is an angle whose sine is $\frac{1}{2}$. Because $\sin 30° = \frac{1}{2}$ and $\sin 150° = \frac{1}{2}$, $\alpha$ could be 30° or 150°. Since any angle with the same terminal side as 30° or 150° is a solution, there are infinitely many solutions. Because right triangles have only acute angles (and one right angle), we are interested only in acute angles here. We will find general solutions to equations in Chapter 4.

**EXAMPLE 1** **Finding acute angles**

Find the angle $\alpha$ that satisfies each equation where $0° \leq \alpha \leq 90°$.

**a.** $\sin \alpha = \dfrac{\sqrt{3}}{2}$ **b.** $\cos \alpha = 1$ **c.** $\tan \alpha = 1$

**Solution**

**a.** Since $\sin 60° = \dfrac{\sqrt{3}}{2}$, $\alpha = 60°$.

**b.** Since $\cos 0° = 1$, $\alpha = 0°$.

**c.** Since $\tan 45° = 1$, $\alpha = 45°$.

**TRY THIS.** Find the angle $\beta$ that satisfies $\cos \beta = 1/2$ where $0 \le \beta \le 90°$. ∎

In Example 1 we found the angle when given its sine. The function whose input is the sine of an angle and whose output is the angle is the **inverse sine** function. Similarly, the **inverse cosine** and **inverse tangent** functions pair numbers with angles. Using inverse function notation from algebra, the results of Example 1 are written as

$$\sin^{-1}\left(\frac{\sqrt{3}}{2}\right) = 60°, \qquad \cos^{-1}(1) = 0°, \qquad \text{and} \qquad \tan^{-1}(1) = 45°.$$

Note that the $-1$ in this notation does not indicate a reciprocal. Because there are infinitely many angles that have a given sine, cosine, or tangent, we must define these functions precisely. The precise definitions follow.

**Definition: Inverse Sine, Cosine, and Tangent Functions**

| | | | | | |
|---|---|---|---|---|---|
| $\sin^{-1}(x) = \alpha$ | provided | $\sin \alpha = x$ | and | $-90° \le \alpha \le 90°$ |
| $\cos^{-1}(x) = \alpha$ | provided | $\cos \alpha = x$ | and | $0° \le \alpha \le 180°$ |
| $\tan^{-1}(x) = \alpha$ | provided | $\tan \alpha = x$ | and | $-90° < \alpha < 90°$ |

The other three trigonometric functions have inverses also, but we will not need them in studying right triangles. They are discussed in Chapter 4.

**EXAMPLE 2** **Evaluating inverse functions**

Evaluate each expression. Give the result in degrees.

**a.** $\cos^{-1}\left(\dfrac{\sqrt{2}}{2}\right)$  **b.** $\sin^{-1}\left(\dfrac{\sqrt{3}}{2}\right)$  **c.** $\tan^{-1}\left(\sqrt{3}\right)$

**Solution**

**a.** Since $\cos 45° = \dfrac{\sqrt{2}}{2}$ and $0° \le 45° \le 180°$, $\cos^{-1}\left(\dfrac{\sqrt{2}}{2}\right) = 45°$.

**b.** Since $\sin 60° = \dfrac{\sqrt{3}}{2}$ and $-90° \le 60° \le 90°$, $\sin^{-1}\left(\dfrac{\sqrt{3}}{2}\right) = 60°$.

**c.** Since $\tan 60° = \sqrt{3}$ and $-90° < 60° < 90°$, $\tan^{-1}\left(\sqrt{3}\right) = 60°$.

**TRY THIS.** Find the exact value in degrees of $\sin^{-1}(1/2)$. ∎

For angles that are not multiples of $30°$ or $45°$, we use a calculator to evaluate an inverse function. Calculators have keys labeled $\sin^{-1}$, $\cos^{-1}$, and $\tan^{-1}$. They are usually the second functions for the sin, cos, and tan keys. When an inverse function key is pressed, your calculator will give the angle in the range specified in the definition of the inverse functions. Of course the angle will be given in degrees or radians depending on the mode setting.

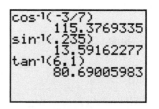

**Figure 1.45**

EXAMPLE **3** **Evaluating inverse trigonometric functions with a calculator**

Evaluate each expression. Give the result in degrees to the nearest tenth.

**a.** $\cos^{-1}(-3/7)$   **b.** $\sin^{-1}(0.235)$   **c.** $\tan^{-1}(6.1)$

**Solution**

**a.** $\cos^{-1}(-3/7) \approx 115.4°$
**b.** $\sin^{-1}(0.235) \approx 13.6°$
**c.** $\tan^{-1}(6.1) \approx 80.7°$

Figure 1.45 shows these results on a graphing calculator.

**TRY THIS.** Find $\sin^{-1}(0.38)$ to the nearest tenth of a degree. ∎

## Right Triangles

Trigonometric functions were defined for angles in standard position in a rectangular coordinate system. We will now define them for the acute angles of a right triangle, without using a coordinate system.

Consider a right triangle with acute angle $\alpha$, legs of length $x$ and $y$, and hypotenuse $r$, as shown in Fig. 1.46. If this triangle is positioned in a coordinate system as in Fig. 1.47 then $(x, y)$ is a point on the terminal side of $\alpha$ and

$$\sin(\alpha) = \frac{y}{r}, \qquad \cos(\alpha) = \frac{x}{r}, \qquad \text{and} \qquad \tan(\alpha) = \frac{y}{x}.$$

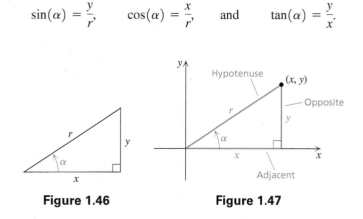

**Figure 1.46**            **Figure 1.47**

The values of the trigonometric functions are simply ratios of the lengths of the sides of the right triangle. It is not necessary to move the triangle to a coordinate system to find them. Notice that $y$ is the length of the side **opposite** the angle $\alpha$, $x$ is the length of the side **adjacent** to $\alpha$, and $r$ is the length of the **hypotenuse.** We use the abbreviations opp, adj, and hyp to represent the lengths of these sides in the following definition.

**Definition: Trigonometric Functions of an Acute Angle of a Right Triangle**

If $\alpha$ is an acute angle of a right triangle, then

$$\sin \alpha = \frac{\text{opp}}{\text{hyp}}, \qquad \cos \alpha = \frac{\text{adj}}{\text{hyp}}, \qquad \text{and} \qquad \tan \alpha = \frac{\text{opp}}{\text{adj}},$$

$$\csc \alpha = \frac{\text{hyp}}{\text{opp}}, \qquad \sec \alpha = \frac{\text{hyp}}{\text{adj}}, \qquad \text{and} \qquad \cot \alpha = \frac{\text{adj}}{\text{opp}}.$$

**Figure 1.48**

**EXAMPLE 4**    **Trigonometric functions in a right triangle**

Find the values of all six trigonometric functions for the angle $\alpha$ of the right triangle with legs of length 1 and 4 as shown in Fig. 1.48.

**Solution**

The length of the hypotenuse, $c$, is $\sqrt{4^2 + 1^2}$ or $\sqrt{17}$. Since the length of the side opposite $\alpha$ is 1 and the length of the adjacent side is 4, we have

$$\sin \alpha = \frac{\text{opp}}{\text{hyp}} = \frac{1}{\sqrt{17}} = \frac{\sqrt{17}}{17}, \qquad \cos \alpha = \frac{\text{adj}}{\text{hyp}} = \frac{4}{\sqrt{17}} = \frac{4\sqrt{17}}{17}$$

$$\tan \alpha = \frac{\text{opp}}{\text{adj}} = \frac{1}{4}, \qquad \csc \alpha = \frac{\text{hyp}}{\text{opp}} = \sqrt{17},$$

$$\sec \alpha = \frac{\text{hyp}}{\text{adj}} = \frac{\sqrt{17}}{4}, \qquad \text{and} \qquad \cot \alpha = \frac{\text{adj}}{\text{opp}} = 4.$$

**TRY THIS.**  A right triangle has legs with lengths 2 and 4. Find $\sin \alpha$, $\cos \alpha$, and $\tan \alpha$ if $\alpha$ is the angle opposite the side of length 4.    ■

## Solving a Right Triangle

A triangle has three sides and three angles. If we know that the angles of a triangle are $30°$, $60°$, and $90°$, we cannot determine the lengths of the sides because there are infinitely many such triangles of different sizes. If we know a combination of sides and/or angles that determines a unique right triangle, then we can use the trigonometric functions to find the unknown parts. Finding all of the unknown parts is called **solving the triangle.**

In solving right triangles, we usually name the acute angles $\alpha$ and $\beta$ (beta) and the lengths of the sides opposite those angles $a$ and $b$. The $90°$ angle is $\gamma$ (gamma), and the length of the side opposite $\gamma$ is $c$.

**Figure 1.49**

**EXAMPLE 5**    **Solving a right triangle**

Solve the right triangle in which $\alpha = 30°$ and $c = 6$.

**Solution**

The triangle is shown in Fig. 1.49. Since $\alpha = 30°$, $\gamma = 90°$, and the sum of the measures of the angles of any triangle is $180°$, we have $\beta = 60°$. Since $\sin \alpha = \text{opp}/\text{hyp}$, we have $\sin 30° = a/6$ and

$$a = 6 \cdot \sin 30° = 6 \cdot \frac{1}{2} = 3.$$

Since $\cos \alpha = \text{adj}/\text{hyp}$, we have $\cos 30° = b/6$ and

$$b = 6 \cdot \cos 30° = 6 \cdot \frac{\sqrt{3}}{2} = 3\sqrt{3}.$$

The angles of the right triangle are $30°$, $60°$, and $90°$, and the sides opposite those angles are $3$, $3\sqrt{3}$, and $6$, respectively.

**TRY THIS.**  Solve the right triangle in which $\alpha = 60°$ and $c = 2$.    ■

In the next example, both acute angles are unknown. In this case we will need an inverse trigonometric function.

**Figure 1.50**

EXAMPLE **6**  **Solving a right triangle**

Solve the right triangle in which $a = 4$ and $b = 6$. Find the acute angles to the nearest tenth of a degree.

**Solution**

The triangle is shown in Fig. 1.50. By the Pythagorean theorem, $c^2 = 4^2 + 6^2$, or $c = \sqrt{52} = 2\sqrt{13}$. To find $\alpha$, first find $\sin \alpha$:

$$\sin \alpha = \frac{\text{opp}}{\text{hyp}} = \frac{4}{2\sqrt{13}} = \frac{2}{\sqrt{13}}$$

Now, $\alpha$ is the angle whose sine is $2/\sqrt{13}$:

$$\alpha = \sin^{-1}\left(\frac{2}{\sqrt{13}}\right) \approx 33.7°$$

Since $\alpha + \beta = 90°$, $\beta \approx 90° - 33.7° = 56.3°$. The angles of the triangle are approximately $33.7°$, $56.3°$, and $90°$, and the sides opposite those angles are 4, 6, and $2\sqrt{13}$, respectively.

**TRY THIS.**   Solve the right triangle in which $a = 2$ and $b = 5$.     ∎

   In Example 6 we could have found $\alpha$ by using $\alpha = \tan^{-1}(4/6) \approx 33.7°$. We could then have found $c$ by using $\cos(33.7°) = 6/c$, or $c = 6/\cos(33.7°) \approx 7.2$. There are many ways to solve a right triangle, but the basic strategy is always the same.

**STRATEGY**

**Solving a Right Triangle**

To solve a right triangle:

   **1.** Use the Pythagorean theorem to find the length of a third side when the lengths of two sides are known.

   **2.** Use the trigonometric ratios to find missing sides or angles.

   **3.** Use the fact that the sum of the measures of the angles of a triangle is 180° to determine a third angle when two are known.

## Applications

Using trigonometry, we can find the size of an object without actually measuring the object. Two common terms used in this regard are **angle of elevation** and **angle of depression.** The angle of elevation $\alpha$ for a point above a horizontal line is the angle formed by the horizontal line and the observer's line of sight through the point as shown in Fig. 1.51. The angle of depression $\beta$ for a point below a horizontal line is the angle formed by the horizontal line and the observer's line of sight through the point as shown in Fig. 1.51. We use these angles and our skills in solving triangles to find the sizes of objects that would be inconvenient to measure.

**Figure 1.51**

**Figure 1.52**

**EXAMPLE 7** **Finding the height of an object**

A guy wire of length 108 meters runs from the top of an antenna to the ground. If the angle of elevation of the top of the antenna, sighting along the guy wire, is 42.3°, then what is the height of the antenna?

**Solution**

Let $y$ represent the height of the antenna as shown in Fig. 1.52. Since $\sin(42.3°) = y/108$,

$$y = 108 \cdot \sin(42.3°) \approx 72.7.$$

The height of the antenna is approximately 72.7 meters.

**TRY THIS.** The angle of elevation of the top of a cell phone tower is 38.2° at a distance of 344 feet from the tower. What is the height of the tower? ▪

In Example 7 we knew the distance to the top of the antenna, and we found the height of the antenna. If we knew the distance on the ground to the base of the antenna and the angle of elevation of the guy wire, we could still have found the height of the antenna. Both cases involve knowing the distance to the antenna either on the ground or through the air. However, one of the biggest triumphs of trigonometry is being able to find the size of an object or the distance to an object (such as the moon) without going to the object. The next example shows one way to find the height of an object without actually going to it. In Section 5.1, Example 6, we will show another (slightly simpler) solution to this same problem using the law of sines.

**EXAMPLE 8** **Finding the height of an object from a distance**

The angle of elevation of the top of a water tower from point $A$ on the ground is 19.9°. From point $B$, 50.0 feet closer to the tower, the angle of elevation is 21.8°. What is the height of the tower?

**Solution**

Let $y$ represent the height of the tower and $x$ represent the distance from point $B$ to the base of the tower as shown in Fig. 1.53.

**Figure 1.53**

At point $B$, $\tan 21.8° = y/x$ or

$$x = \frac{y}{\tan 21.8°}.$$

Since the distance to the base of the tower from point $A$ is $x + 50$,

$$\tan 19.9° = \frac{y}{x + 50}$$

or

$$y = (x + 50) \tan 19.9°.$$

Since $x = y/\tan 21.8°$, we can substitute $y/\tan 21.8°$ for $x$ in the last equation and get an equation involving $y$ only:

$$y = \left(\frac{y}{\tan 21.8°} + 50\right) \tan 19.9°$$

$$y = \frac{y \cdot \tan 19.9°}{\tan 21.8°} + 50 \tan 19.9° \qquad \text{Distributive property}$$

$$y - \frac{y \cdot \tan 19.9°}{\tan 21.8°} = 50 \tan 19.9°$$

$$y\left(1 - \frac{\tan 19.9°}{\tan 21.8°}\right) = 50 \tan 19.9° \qquad \text{Factor out } y.$$

$$y = \frac{50 \tan 19.9°}{1 - \dfrac{\tan 19.9°}{\tan 21.8°}} \approx 191 \text{ ft}$$

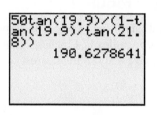

**Figure 1.54**

This computation is shown on a graphing calculator in Fig. 1.54.

**TRY THIS.** At one location, the angle of elevation of the top of an antenna is 44.2°. At a point that is 100 feet closer to the antenna, the angle of elevation is 63.1°. What is the height of the antenna?

In the next example we combine the solution of a right triangle with the arc length of a circle to solve a problem of aerial photography.

**EXAMPLE 9** Photograph from a spy plane

In the late '50s, the Soviets labored to develop a missile that could stop the U-2 spy plane. On May 1, 1960, Nikita S. Khrushchev announced to the world that the Soviets had shot down Francis Gary Powers while Powers was photographing the Soviet Union from a U-2 at an altitude of 14 miles. How wide a path on the surface of the earth could Powers see from that altitude? (Use 3950 miles as the radius of the earth.)

**Figure 1.55**

**Solution**

Figure 1.55 shows the line of sight to the horizon on the left-hand side and right-hand side of the airplane while flying at the altitude of 14 miles. Since a line tangent to a circle (the line of sight) is perpendicular to the radius at the point of tangency, the angle $\alpha$ at the center of the earth in Fig. 1.55 is an acute angle of a right triangle with hypotenuse $3950 + 14$ or 3964. So we have

$$\cos \alpha = \frac{3950}{3964}$$

$$\alpha = \cos^{-1}\left(\frac{3950}{3964}\right) \approx 4.8°.$$

The width of the path seen by Powers is the length of the arc intercepted by the central angle $2\alpha$ or 9.6°. Using the formula $s = \alpha r$ from Section 1.2, where $\alpha$ is in radians, we get

$$s = 9.6 \text{ deg} \cdot \frac{\pi \text{ rad}}{180 \text{ deg}} \cdot 3950 \text{ mi} \approx 661.8 \text{ miles}.$$

From an altitude of 14 miles, Powers could see a path that was about 661.8 miles wide. Actually, he photographed a path that was somewhat narrower, because parts of the photographs near the horizon were not usable.

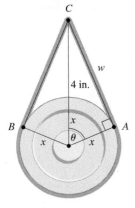

**Figure 1.56**

**TRY THIS.** A sailor spots a small island on the horizon from the top of a 40-ft mast. How far is it to the island? Use 3950 mi for the radius of the earth.

Example 10 involves a circle and tangent lines, and appears to be similar to Example 9. However, in Example 10 we will write an equation that can only be solved using a graphing calculator or computer. Notice that in Example 10 we will use right triangle trigonometry, the arc length formula, the circumference formula, the Pythagorean theorem, and the inverse cosine function.

 **EXAMPLE 10** **Graphing calculator required**

A 20-inch belt connects a pulley with a very small shaft on a motor, as shown in Fig. 1.56. The distance between the shaft and the large pulley is 4 inches. Assuming the shaft is a single point, find the radius of the large pulley to the nearest tenth of an inch.

**Solution**

Let $x$ be the radius of the circle and $w$ be the distance from the shaft to point $A$, as shown in the figure. Since the radius is perpendicular to the belt at $A$, we have $\theta = \cos^{-1}\left(\dfrac{x}{x+4}\right)$. Assuming $\theta$ is in radians, the length of the arc of the circle intercepted by $\theta$ is $x \cdot \cos^{-1}\left(\dfrac{x}{x+4}\right)$. Since the circumference of the circle is $2\pi x$, the length of the long arc from $A$ to $B$ is

$$2\pi x - 2x \cdot \cos^{-1}\left(\frac{x}{x+4}\right).$$

By the Pythagorean theorem, $x^2 + w^2 = (x+4)^2$ or $w = \sqrt{8x+16}$. Since the total length of the belt is 20 inches, we have

$$2\pi x - 2x \cdot \cos^{-1}\left(\frac{x}{x+4}\right) + 2\sqrt{8x+16} = 20.$$

One way to solve this equation with a graphing calculator is to graph

$$y = 2\pi x - 2x \cdot \cos^{-1}\left(\frac{x}{x+4}\right) + 2\sqrt{8x+16} - 20$$

**Figure 1.57**

and find the $x$-intercept, as shown in Fig. 1.57. The radius of the large pulley is approximately 2.2 inches.

**TRY THIS.** It is about 200 miles from Houston to San Antonio on interstate I-10. A pilot at an altitude of 5 miles over Houston spots San Antonio on the horizon. From this information calculate the radius of the earth.

**FOR THOUGHT...** True or False? Explain.

1. If $\sin\alpha = \dfrac{\sqrt{3}}{2}$ and $0° \le \alpha \le 90°$, then $\alpha = 60°$.

2. If $\cos\alpha = \dfrac{\sqrt{3}}{2}$ and $0° \le \alpha \le 90°$, then $\alpha = 45°$.

3. $\sin^{-1}\left(\sqrt{2}/2\right) = 135°$

4. $\cos^{-1}(1/2) = 30°$

5. $\tan^{-1}(1) = 90°$

6. If $a = 2$ and $b = 4$ in a right triangle, then $c = 6$.

7. If $a = 3$ and $b = 4$ in a right triangle, then $c = 5$.

8. If $a = 4$ and $b = 8$ in a right triangle, then $c = \sqrt{12}$.

9. If $\alpha = 55°$ in a right triangle, then $\beta = 35°$.

10. In a right triangle, $\sin(90°) = \text{hyp/adj}$.

## **1.5** EXERCISES

### CONCEPTS

*Fill in the blank.*

1. If a triangle has a 90° angle, then is it is a(n) _____ triangle.

2. The two sides that form an acute angle of a right triangle are the _____ side and the _____.

3. The function whose input is the sine of an angle and whose output is the angle is the _____ function.

4. If $\alpha$ is an acute angle of a right triangle, then $\sin(\alpha)$ is the _____ divided by the _____.

5. If $\alpha$ is an acute angle of a right triangle, then $\cos(\alpha)$ is the _____ divided by the _____.

6. If $\alpha$ is an acute angle of a right triangle, then $\tan(\alpha)$ is the _____ divided by the _____.

### SKILLS

*Find the acute angle $\alpha$ (in degrees) that satisfies each equation. Do not use a calculator.*

7. $\sin \alpha = \dfrac{1}{\sqrt{2}}$

8. $\cos \alpha = \dfrac{\sqrt{3}}{2}$

9. $\tan \alpha = \sqrt{3}$

10. $\cos \alpha = 0$

*Evaluate each expression without using a calculator. Give the result in degrees.*

11. $\cos^{-1}\left(\dfrac{1}{2}\right)$

12. $\sin^{-1}\left(\dfrac{\sqrt{2}}{2}\right)$

13. $\tan^{-1}(0)$

14. $\sin^{-1}(1)$

*Evaluate each expression using a calculator. Give the result in degrees to the nearest tenth.*

15. $\cos^{-1}(1/9)$

16. $\sin^{-1}(2/5)$

17. $\tan^{-1}(2.43)$

18. $\tan^{-1}(5.68)$

*Use a calculator to find the acute angle $\alpha$ (to the nearest tenth of a degree) that satisfies each equation.*

19. $\sin \alpha = 0.44$

20. $\cos \alpha = 0.923$

21. $\tan \alpha = 5/9$

22. $\sin \alpha = 4/5$

*Find the exact values of all six trigonometric functions for the angle $\alpha$ in each given right triangle.*

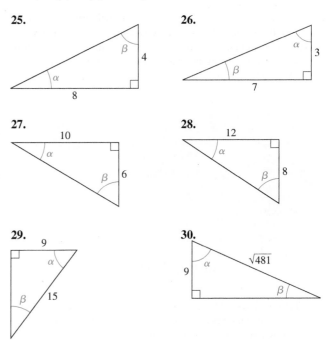

23.

24.

*For each given right triangle, find exact values of $\sin \alpha$, $\cos \alpha$, $\tan \alpha$, $\sin \beta$, $\cos \beta$, and $\tan \beta$.*

25.

26.

27.

28.

29.

30.

*Solve each right triangle with the given sides and angles. In each case, make a sketch. Round approximate answers to the nearest tenth.*

31. $a = 6, b = 8$

32. $a = 10, c = 12$

33. $b = 6, c = 8.3$

34. $\alpha = 32.4°, b = 10$

35. $\alpha = 16°, c = 20$

36. $\beta = 47°, a = 3$

37. $\alpha = 39°9', a = 9$

38. $\beta = 19°12', b = 60$

### MODELING

*Solve each problem.*

39. *Aerial Photography* An aerial photograph from a U-2 spy plane is taken of a building suspected of housing nuclear warheads. The photograph is made when the angle of elevation of the sun is 32°. By comparing the shadow cast by the building

to objects of known size in the photograph, analysts determine that the shadow is 80 ft long. How tall is the building (to the nearest foot)?

**Figure for Exercise 39**

**40.** *Giant Redwood* A hiker stands 80 ft from a giant redwood tree and sights the top with an angle of elevation of 75°. How tall is the tree (to the nearest foot)?

**41.** *Avoiding a Swamp* Muriel was hiking directly toward a long, straight road when she encountered a swamp. She turned 65° to the right and hiked 4 mi in that direction to reach the road. How far was she from the road (to the nearest tenth of a mile) when she encountered the swamp?

**Figure for Exercise 41**

**42.** *Tall Antenna* A 100-ft guy wire is attached to the top of an antenna. The angle between the guy wire and the ground is 62°. How tall is the antenna (to the nearest foot)?

**43.** *Angle of Depression* From a highway overpass, 14.3 m above the road, the angle of depression of an oncoming car is measured at 18.3°. How far is the car from a point on the highway directly below the overpass (to the nearest tenth of a meter)?

**Figure for Exercise 43**

**44.** *Length of a Pipe* A gas pipe under a river is 196.8 ft below the surface at its lowest point as shown in the drawing. If the

angle of depression of the pipe is 4.962°, then what is the distance from point $A$ to point $B$ on the surface of the water? What is the length of the pipe between points $A$ and $B$? Round to the nearest foot.

**Figure for Exercise 44**

**45.** *Height of a Crosswalk* The angle of elevation of a pedestrian crosswalk over a busy highway is 8.34° as shown in the drawing. If the distance between the ends of the crosswalk measured on the ground is 342 ft, then what is the height, $h$, of the crosswalk at the center? Round to the nearest tenth of a foot.

**Figure for Exercise 45**

**46.** *Shortcut to Snyder* To get from Muleshoe to Snyder, Harry drives 50 mph for 178 mi south on route 214 to Seminole, then goes east on route 180 to Snyder. Harriet leaves Muleshoe 1 hr later at 55 mph, but takes US 84, which goes straight from Muleshoe to Snyder through Lubbock. If US 84 intersects route 180 at a 50° angle, then how many more miles (to the nearest mile) does Harry drive? In what state are these towns?

**47.** *Installing a Guy Wire* A 41-m guy wire is attached to the top of a 34.6-m antenna and to a point on the ground. How far is the point on the ground from the base of the antenna (to the nearest meter), and what angle does the guy wire make with the ground (to the nearest tenth of a degree)?

**48.** *Robin and Marian* Robin Hood plans to use a 30-ft ladder to reach the castle window of Maid Marian. Little John, who made the ladder, advised Robin that the angle of elevation of the ladder must be between 55° and 70° for safety. What are the minimum and maximum heights that can safely be reached by the top of the ladder when it is placed against the 50-ft castle wall? Round to the nearest tenth.

**49.** *Motorcycle Trail* The head angle on a motorcycle is the angle between the horizontal and the slanted line through the steering head. The distance along the ground between the line through the steering head and the vertical line through the center of the wheel is the trail. See the accompanying figure. The radius of the wheel on this author's Road King is 13.3 inches and the head angle is 64°. Find the trail.

Head angle
64°

13.3 in.

← Trail

**Figure for Exercise 49**

**50.** *Rake and Trail* To reduce the trail, the front axle is often off-set from the line through the steering head. The amount of off-set is the rake. See the accompanying figure. On this author's Trek 420 bicycle the rake is 5 centimeters, the head angle is 70°, and the diameter of the wheel is 700 millimeters. Find the trail in inches.

Steering head

Rake

Head angle

← Trail

**Figure for Exercise 50**

**51.** *Wooden Rack* According to the World Pool Association, the diameter of a pool ball is 2.25 in. Fifteen balls fit snugly into a wooden rack, as shown in the accompanying figure. The rack is an equilateral triangle whose sides are 0.25 in. thick. Find the out-side perimeter of the rack to the nearest tenth of an inch.

**Figure for Exercise 51**

**52.** *Plastic Rack* The 15 balls from the previous exercise fit snugly into a plastic rack that is shaped to fit the balls in the corners, as shown in the accompanying figure. If the plastic rack is 0.25 in. thick, then what is the outside perimeter of the rack to the nearest tenth of an inch.

**Figure for Exercise 52**

**53.** *Regular Pentagon* The sides of a regular pentagon are each 2 meters in length, as shown in the accompanying figure. Find the distance $h$ to the nearest hundredth of a meter.

2 m

$h$

**Figure for Exercise 53**

**54.** *Metal Star* A copper three-dimensional star that is 36 in. wide is hanging on a wall, as shown in the accompanying figure. The pe-rimeter of the star is flush against the wall, but the center point is 3 in. from the wall. The star is made from 10 triangular pieces of copper. Find the lengths of the three sides of one of them to the nearest hundredth of an inch.

36 in.

**Figure for Exercise 54**

**55.** *Detecting a Speeder* A policewoman has positioned herself 500 ft from the intersection of two roads as shown in the draw-ing. She has carefully measured the angles of the lines of sight to points $A$ and $B$ as shown in the drawing on the next page. If a car passes from $A$ to $B$ in 1.75 sec then what is the speed of the car? Round to the nearest tenth of a mph.

**Figure for Exercise 55**

**56.** *Progress of a Forest Fire*  A forest ranger atop a 3248-ft mesa is watching the progress of a forest fire spreading in her direction. In 5 min the angle of depression of the leading edge of the fire changed from 11.34° to 13.51°. At what speed in miles per hour is the fire spreading in the direction of the ranger? Round to the nearest tenth.

**57.** *Height of a Rock*  Inscription Rock rises almost straight upward from the valley floor. From one point the angle of elevation of the top of the rock is 16.7°. From a point 168 m closer to the rock, the angle of elevation of the top of the rock is 24.1°. How high is Inscription Rock? Round to the nearest tenth of a meter.

**58.** *Height of a Balloon*  A hot air balloon is between two spotters who are 1.2 mi apart. One spotter reports that the angle of elevation of the balloon is 76°, and the other reports that it is 68°. What is the altitude of the balloon to the nearest tenth of a mile?

**59.** *Passing in the Night*  A boat sailing north sights a lighthouse to the east at an angle of 32° from the north as shown in the drawing. After the boat travels one more kilometer, the angle of the lighthouse from the north is 36°. If the boat continues to sail north, then how close will the boat come to the lighthouse to the nearest tenth of a kilometer?

**Figure for Exercise 59**

**60.** *Height of a Skyscraper*  For years the Woolworth skyscraper in New York held the record for the world's tallest office building. If the length of the shadow of the Woolworth building increases by 17.4 m as the angle of elevation of the sun changes from 44° to 42°, then how tall is the building to the nearest tenth of a meter?

**61.** *Parsecs*  In astronomy the light year (abbreviated ly) and the parsec (abbreviated pc) are the two units used to measure distances. The distance in space at which a line from the earth to the sun subtends an angle of 1 sec is 1 parsec. Find the number of miles in 1 parsec by using a right triangle positioned in space with its right angle at the sun as shown in the figure. Use scientific notation and one decimal place. How many years (to the nearest hundredth) does it take light to travel 1 parsec? (One light year is the distance that light travels in one year, one AU is the distance from the earth to the sun, and 1 ly = 63,240 AU.)

**Figure for Exercise 61**

**62.** *View from Landsat*  The satellite Landsat orbits the earth at an altitude of 700 mi, as shown in the figure. What is the width of the path on the surface of the earth that can be seen by the cameras of Landsat? Round to the nearest mile.

**Figure for Exercise 62**

**63.** *Angle of Elevation*  From point *A* the angle of elevation to the top of a building is 30°, as shown in the accompanying figure. From point *B*, 20 meters closer to the building, the angle of elevation is 45°. Find the angle of elevation of the building from point *C*, which is another 20 meters closer to the building.

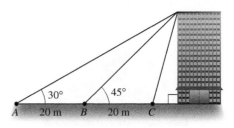

**Figure for Exercise 63**

**64.** *Rate of Ascent* A hot air balloon is rising upward from the earth at a constant rate, as shown in the accompanying figure. An observer 250 meters away spots the balloon at an angle of elevation of 24°. Two minutes later the angle of elevation of the balloon is 58°. At what rate is the balloon ascending? Answer to the nearest tenth of a meter per second.

58°

24°

250 m

**Figure for Exercise 64**

**65.** *Watering the Lawn* Four lawn sprinklers are placed to water a square lawn, as shown in the accompanying figure. The region watered by each sprinkler is circular with a radius of 6 m.
**a.** How many square meters of lawn are not reached by any of the sprinklers? Find the exact answer.

**b.** What percent of the lawn is watered by at least one sprinkler? Round to the nearest tenth of a percent.

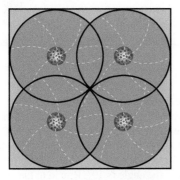

**Figure for Exercise 65**

**66.** *Shipping Pipes* Six identical pipes, each with a radius of 1 foot, are tied tightly together with a metal band, as shown in the accompanying figure. Find the exact length of the metal band.

**Figure for Exercise 66**

**67.** *Communicating via Satellite* A communication satellite is usually put into a synchronous orbit with the earth, which means that it stays above a fixed point on the surface of the earth at all times. The radius of the orbit of such a satellite is 6.5 times the radius of the earth (3950 mi). The satellite is used to relay a signal from one point on the earth to another point on the earth. The sender and receiver of a signal must be in a line of sight with the satellite, as shown in the figure. What is the maximum distance on the surface of the earth between the sender and receiver for this type of satellite? Round to the nearest mile.

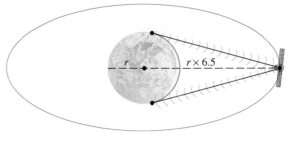

$r$    $r \times 6.5$

**Figure for Exercise 67**

**68.** *Catching a Green Flash* Ed and Diane are on the beach to observe a green flash. As the very last sliver of the sun sinks into the ocean, refraction bends the final ray of sunlight into a rainbow of colors. The red and orange set first, and the green light or *green flash* is the last bit of sunlight seen before sunset.
**a.** If Diane's eyes are 2 ft above the ocean, then what is the degree measure of the angle $\alpha$ shown in the figure? Use four decimal places.

**b.** If Ed's eyes are 6 ft above the ocean, then what is the angle $\alpha$ for Ed? Use degrees and four decimal places.

**c.** Diane sees the green flash first. The difference between the two alphas is how far the earth must rotate before Ed sees it. How many seconds (to the nearest tenth) after Diane sees the green flash does Ed see it?

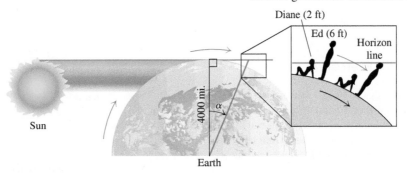

Diane (2 ft)

Ed (6 ft)

Horizon line

4000 mi.

$\alpha$

Sun

Earth

**Figure for Exercise 68**

**69.** *Maximizing Area*  An architect is designing a rectangular parking lot on a triangular lot as shown in the figure. Find the dimensions of the parking lot that will maximize the area of the parking lot. Round to the nearest hundredth of a foot.
HINT Write a quadratic function for the area of the lot.

**Figure for Exercises 69 and 70**

**70.** *Figuring the Setback*  The architect in the previous exercise just remembered that the parking lot must be set back 10 ft from each side of the property and 40 ft from the street. Now find the dimensions of the parking lot that will maximize the area of the parking lot. Round to the nearest tenth of a foot.

☐ *The following problems can be solved by setting up an equation involving inverse trigonometric functions. You will need the equation solver of a graphing calculator or a computer to solve the equation.*

**71.** *Hanging a Pipe*  A contractor wants to pick up a 10-ft-diameter pipe with a chain of length 40 ft. The chain encircles the pipe and is attached to a hook on a crane. What is the distance between the hook and the pipe to the nearest thousandth of a foot? (This is an actual problem that an engineer was asked to solve on the job.)

**Figure for Exercise 71**

**72.** *Circling the Earth*  A rope that is 1 meter longer than the circumference of the earth is placed around the earth at the equator. If the rope is pulled up tightly to point *A*, as shown in the accompanying figure, then what is the distance from point *A* to the surface of the earth. Use 6400 kilometers as the radius of the earth.

**Figure for Exercise 72**

**73.** *Blocking a Pipe*  A large pipe is held in place by using a 1-ft-high block on one side and a 2-ft-high block on the other side. If the length of the arc between the points where the pipe touches the blocks is 6 ft, then what is the radius of the pipe to the nearest thousandth of a foot?
HINT Ignore the thickness of the pipe.

**Figure for Exercise 73**

**74.** *Pipe on an Inclined Plane*  A large pipe on an inclined plane is held in place by sticks of length 2 ft and 6 ft as shown in the figure. The sticks are in line with the center of the pipe. The extended lines of the sticks intersect at an 18° angle at the center of the circle. What is the radius of the pipe? Find the radius of the pipe if the angle is 19°. Round to the nearest thousandth of a foot.

**Figure for Exercise 74**

**75.** *Finding the Radius of the Earth* Assume that it takes 4 sec for the green flash in Exercise 68 to travel from Diane's eyes at 2 ft to Ed's eyes at 6 ft. Find the radius of the earth to the nearest mile.

**76.** *Solving a Lot* The diagonals of the property shown in the figure are 90 ft and 120 ft. The diagonals cross 30 ft from the street. Find *w*, the width of the property to the nearest tenth of a foot.

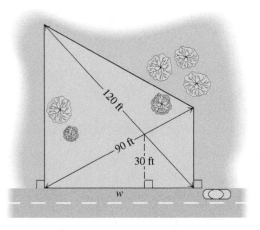

**Figure for Exercise 76**

### WRITING/DISCUSSION

**77.** *Cooperative Learning* Construct a device for measuring angle of elevation or angle of depression by attaching a weighted string to a protractor as shown in the figure. Work with a group to select an object such as a light pole or tree. Measure the angle of elevation to the top of the object and the angle of depression to the bottom of the object while standing at some appropriate distance from the object. Use these angles, the distance to the object, and the distance of your eye from the ground to find the height of the object.

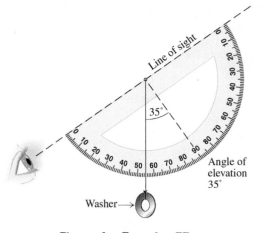

**Figure for Exercise 77**

**78.** *Cooperative Learning* Work in a group to select a local building and find its height without going up to the building. Use the difference in the lengths of its shadow at two different times of day and the technique of Example 8. Use the protractor and string from the last exercise to measure the angle of elevation of the sun, without looking at the sun.

### REVIEW

**79.** Evaluate the trigonometric functions for the angle $\alpha$ in standard position whose terminal side passes through $(4, 3)$.
   **a.** $\sin \alpha$        **b.** $\cos \alpha$        **c.** $\tan \alpha$

**80.** Find the exact value of each function.
   **a.** $\sin(45°)$

   **b.** $\cos(-\pi/4)$

   **c.** $\tan(3\pi/4)$

**81.** Find the exact value of each function.
   **a.** $\sin(60°)$

   **b.** $\cos(-5\pi/6)$

   **c.** $\tan(2\pi/3)$

**82.** Find the negative angle between $0°$ and $-360°$ that is coterminal with $510°$.

**83.** Name the quadrant in which the terminal side of $17\pi/12$ lies.

**84.** A point on the earth's equator makes one revolution in 24 hours. Find the linear velocity in feet per second for such a point, using 3950 miles as the radius of the earth.

### OUTSIDE THE BOX

**85.** *Buckling Bridge* A 100-ft bridge expands 1 in. during the heat of the day. Since the ends of the bridge are embedded in rock, the bridge buckles upward and forms an arc of a circle for which the original bridge is a chord. What is the approximate distance moved by the center of the bridge?

**86.** *Regular Hexagon* A regular hexagon is formed when six equally spaced points on a circle of radius 1 are connected, as shown in the accompanying figure. What is the exact area of the hexagon?

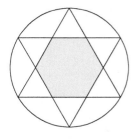

**Figure for Exercise 86**

## 1.5 POP QUIZ

1. If $0° \leq \alpha \leq 90°$ and $\sin \alpha = \sqrt{2}/2$, then what is the degree measure of $\alpha$?

2. Find $\sin^{-1}\left(\sqrt{3}/2\right)$ in degrees.

3. Find $\cos^{-1}(-1/2)$ in degrees.

4. A right triangle has legs with lengths 3 and 6, and $\alpha$ is the acute angle opposite the smallest leg. Find exact values for $\sin \alpha$, $\cos \alpha$, and $\tan \alpha$.

5. At a distance of 1000 feet from a building the angle of elevation to the top of the building is $36°$. Find the height of the building to the nearest foot.

## LINKING concepts...

**For Individual or Group Explorations**

### Discovering $A = \pi r^2$

*Ancient mathematicians discovered the formula for the area of a circle using the formula for the area of a triangle $A = \frac{1}{2}bh$ and trigonometry. Let's see how they did this.*

a)  Suppose a regular pentagon is inscribed in a circle of radius $r$ as shown in the figure. The pentagon is made up of isosceles triangles. Show that the area of the pentagon is $5r^2\sin(36°)\cos(36°)$.

b)  Now suppose that a regular polygon of $n$ sides is inscribed in a circle of radius $r$. Write the area of the $n$-gon in terms of $r$ and $n$.

c)  If $y = kx^2$, we say that $y$ varies directly with the square of $x$ where $k$ is the variation constant. Express the result of part (b) in these terms.

d)  Find the variation constant for a decagon, kilogon, and megagon.

e)  What happens to the shape of the inscribed $n$-gon as $n$ increases? So what can you conclude is the formula for the area of a circle of radius $r$?

f)  Suppose that the $\pi$-key on your calculator is broken. What expression could you use to calculate $\pi$ accurate to nine decimal places?

g)  Use trigonometry to find a formula for the perimeter of an $n$-gon inscribed in a circle of radius $r$.

h)  Use the result of part (g) to find a formula for the circumference of a circle of radius $r$.

# 1.6 The Fundamental Identity and Reference Angles

An identity is an equation that is satisfied for all values of the variable for which both sides are defined. The equations $x + x = 2x$ and $x/x = 1$ are algebraic identities. The fundamental identity of trigonometry comes from the definition of sine and cosine.

## The Fundamental Identity

The fundamental identity involves the squares of the sine and cosine functions. For convenience, we write $(\sin \alpha)^2$ as $\sin^2 \alpha$ and $(\cos \alpha)^2$ as $\cos^2 \alpha$. By definition, $\sin \alpha = \frac{y}{r}$, $\cos \alpha = \frac{x}{r}$, and $r = \sqrt{x^2 + y^2}$. So

$$\sin^2 \alpha + \cos^2 \alpha = \frac{y^2}{r^2} + \frac{x^2}{r^2} = \frac{y^2 + x^2}{r^2} = \frac{r^2}{r^2} = 1.$$

This computation proves the **fundamental identity.**

**The Fundamental Identity of Trigonometry**

> If $\alpha$ is any angle or real number, then
> $$\sin^2 \alpha + \cos^2 \alpha = 1.$$

**Figure 1.58**

The fundamental identity can be illustrated with a graphing calculator as shown in Fig. 1.58. □

If we know the value of the sine or cosine of an angle, then we can use the fundamental identity to find the value of the other function of the angle.

**EXAMPLE 1** Applying the fundamental identity

Find $\cos \alpha$ if $\sin \alpha = 3/5$ and $\alpha$ is an angle in quadrant II.

**Solution**

Use the fundamental identity $\sin^2 \alpha + \cos^2 \alpha = 1$ to find $\cos \alpha$:

$$\left(\frac{3}{5}\right)^2 + \cos^2 \alpha = 1$$

$$\cos^2 \alpha = \frac{16}{25}$$

$$\cos \alpha = \pm\sqrt{\frac{16}{25}} = \pm\frac{4}{5}$$

Since $\cos \alpha < 0$ for any angle in quadrant II, we choose the negative sign and get $\cos \alpha = -4/5$.

**TRY THIS.** Find $\sin \alpha$ given that $\cos \alpha = 1/4$ and $\alpha$ is in quadrant IV. ∎

The fundamental identity can be solved for either sine or cosine to yield additional identities:

$$\sin^2 \alpha + \cos^2 \alpha = 1 \qquad\qquad \sin^2 \alpha + \cos^2 \alpha = 1$$

$$\sin^2 \alpha = 1 - \cos^2 \alpha \qquad\qquad \cos^2 \alpha = 1 - \sin^2 \alpha$$

$$\sin \alpha = \pm\sqrt{1 - \cos^2 \alpha} \qquad\qquad \cos \alpha = \pm\sqrt{1 - \sin^2 \alpha}$$

All of these equations are identities.

# Reference Angles

One way to evaluate the trigonometric functions for an angle is to determine the co-ordinates of a point on the terminal side of the angle. Another way to is to use the corresponding *reference angle.*

| Definition: Reference Angle | If $\theta$ is a nonquadrantal angle in standard position, then the **reference angle** for $\theta$ is the *positive acute angle $\theta'$* (read "theta prime") formed by the terminal side of $\theta$ and the positive or negative *x*-axis. |

Figure 1.59 shows the reference angle $\theta'$ for an angle $\theta$ with terminal side in each of the four quadrants. Notice that in each case $\theta'$ is the acute angle formed by the terminal side of $\theta$ and the *x*-axis. Because the measure of any reference angle is between 0 and $\pi/2$, any reference angle in standard position will be in the first quadrant.

**Figure 1.59**

---

**EXAMPLE 2**  **Finding reference angles**

For each given angle $\theta$, sketch the reference angle $\theta'$ and give the measure of $\theta'$ in both radians and degrees.

**a.** $\theta = 120°$  **b.** $\theta = 7\pi/6$  **c.** $\theta = 690°$  **d.** $\theta = -7\pi/4$

**Solution**

**a.** The terminal side of $120°$ is in quadrant II, as shown in Fig. 1.60. The reference angle $\theta'$ is $180° - 120°$, which is $60°$ or $\pi/3$.

**b.** The terminal side of $7\pi/6$ is in quadrant III, as shown in Fig. 1.61. The reference angle $\theta'$ is $7\pi/6 - \pi$, which is $\pi/6$ or $30°$.

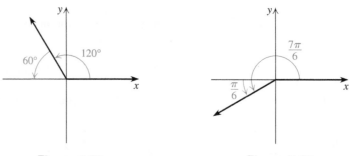

**Figure 1.60**          **Figure 1.61**

**c.** Since $690°$ is coterminal with $330°$, the terminal side is in quadrant IV, as shown in Fig. 1.62. The reference angle $\theta'$ is $360° - 330°$, which is $30°$ or $\pi/6$.

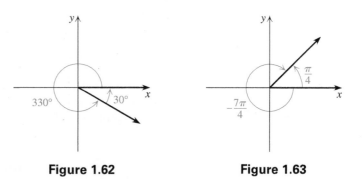

**Figure 1.62**                    **Figure 1.63**

**d.** An angle of $-7\pi/4$ radians has a terminal side in quadrant I as shown in Fig. 1.63 and the reference angle is $\pi/4$ or $45°$.

**TRY THIS.** Find the reference angle for $135°$ in degrees and radians. ∎

Suppose that $\theta$ is an angle with reference angle $\theta'$. Figure 1.64(a) shows the reference angle $\theta'$ in standard position with its terminal side through $(x, y)$. The angle $\theta$ might terminate in any one of the four quadrants. Because the reference angle $\theta'$ goes through $(x, y)$, the angle $\theta$ must pass through $(x, y)$, $(-x, y)$, $(-x, -y)$, or $(x, -y)$ as shown in Fig. 1.64(b). Since these points differ only in sign, $\sin \theta = \pm \sin \theta'$.

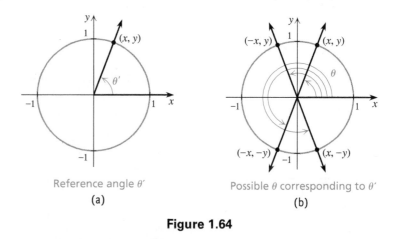

Reference angle $\theta'$                    Possible $\theta$ corresponding to $\theta'$

(a)                                      (b)

**Figure 1.64**

So if we know $\sin \theta'$ we can prefix the appropriate sign to get $\sin \theta$ according to the signs of the trigonometric functions for the four quadrants shown in Fig. 1.65. This same procedure can be used also to evaluate any of the other trigonometric functions.

| Quadrant II | Quadrant I |
|---|---|
| $\sin \alpha > 0$, $\csc \alpha > 0$ | $\sin \alpha > 0$, $\csc \alpha > 0$ |
| $\cos \alpha < 0$, $\sec \alpha < 0$ | $\cos \alpha > 0$, $\sec \alpha > 0$ |
| $\tan \alpha < 0$, $\cot \alpha < 0$ | $\tan \alpha > 0$, $\cot \alpha > 0$ |
| Quadrant III | Quadrant IV |
| $\sin \alpha < 0$, $\csc \alpha < 0$ | $\sin \alpha < 0$, $\csc \alpha < 0$ |
| $\cos \alpha < 0$, $\sec \alpha < 0$ | $\cos \alpha > 0$, $\sec \alpha > 0$ |
| $\tan \alpha > 0$, $\cot \alpha > 0$ | $\tan \alpha < 0$, $\cot \alpha < 0$ |

**Figure 1.65**

**Theorem: Evaluating Trigonometric Functions Using Reference Angles**

For an angle $\theta$ in standard position that is not a quadrantal angle

$$\sin\theta = \pm\sin\theta', \qquad \cos\theta = \pm\cos\theta', \qquad \tan\theta = \pm\tan\theta',$$

$$\csc\theta = \pm\csc\theta', \qquad \sec\theta = \pm\sec\theta', \qquad \cot\theta = \pm\cot\theta'$$

where $\theta'$ is the reference angle for $\theta$ and the sign is determined by the quadrant in which $\theta$ lies.

**Figure 1.66**

**EXAMPLE 3** **Trigonometric functions using reference angles**

Find the sine and cosine for each angle using reference angles.

**a.** $\theta = 120°$ **b.** $\theta = 7\pi/6$ **c.** $\theta = 690°$ **d.** $\theta = -7\pi/4$

**Solution**

**a.** Figure 1.66 shows the terminal side of $120°$ in quadrant II and its reference angle $60°$. Since $\sin\theta > 0$ and $\cos\theta < 0$ for any angle $\theta$ in quadrant II, $\sin(120°) > 0$ and $\cos(120°) < 0$. So to get $\cos(120°)$ from $\cos(60°)$ we must prefix a negative sign. We have

$$\sin(120°) = \sin(60°) = \frac{\sqrt{3}}{2} \qquad \text{and}$$

$$\cos(120°) = -\cos(60°) = -\frac{1}{2}.$$

**Figure 1.67**

**b.** Figure 1.67 shows the terminal side of $7\pi/6$ in quadrant III and its reference angle $\pi/6$. Since $\sin\theta < 0$ and $\cos\theta < 0$ for any angle $\theta$ in quadrant III, we must prefix a negative sign to both $\sin(\pi/6)$ and $\cos(\pi/6)$. So

$$\sin\left(\frac{7\pi}{6}\right) = -\sin\left(\frac{\pi}{6}\right) = -\frac{1}{2} \qquad \text{and}$$

$$\cos\left(\frac{7\pi}{6}\right) = -\cos\left(\frac{\pi}{6}\right) = -\frac{\sqrt{3}}{2}.$$

**Figure 1.68**

**c.** The angle $690°$ terminates in quadrant IV (where $\sin\theta < 0$ and $\cos\theta > 0$) and its reference angle is $30°$, as shown in Fig. 1.68. So

$$\sin(690°) = -\sin(30°) = -\frac{1}{2} \qquad \text{and}$$

$$\cos(690°) = \cos(30°) = \frac{\sqrt{3}}{2}.$$

**Figure 1.69**

**d.** The angle $-7\pi/4$ terminates in quadrant I (where $\sin\theta > 0$ and $\cos\theta > 0$) and its reference angle is $\pi/4$, as shown in Fig. 1.69. So

$$\sin\left(-\frac{7\pi}{4}\right) = \sin\left(\frac{\pi}{4}\right) = \frac{\sqrt{2}}{2} \qquad \text{and}$$

$$\cos\left(-\frac{7\pi}{4}\right) = \cos\left(\frac{\pi}{4}\right) = \frac{\sqrt{2}}{2}.$$

**TRY THIS.** Find $\sin(3\pi/4)$ and $\cos(3\pi/4)$ using reference angles. ■

If you memorize the exact values of sine and cosine for $30°$, $45°$, and $60°$, then you can find the exact value for any nonquadrantal multiple of $30°$ or $45°$ using reference angles.

## Modeling with the Sine Function

The trigonometric functions can be used to model periodic phenomena. The next example shows how the sine function can model the tide in the Bay of Fundy.

**EXAMPLE 4**    Modeling tide

The tide in the Bay of Fundy rises and falls every 13 hours, as shown in Fig. 1.70. The depth of the water at a certain point in the bay is modeled by the function

$$d = 8\sin\left(\frac{2\pi}{13}t\right) + 9$$

where $t$ is time in hours and $d$ is depth in meters. Find the depth at $t = 13/4$ (high tide) and $t = 39/4$ (low tide).

**Figure 1.70**

**Solution**

If $t = 13/4$, then

$$d = 8\sin\left(\frac{2\pi}{13} \cdot \frac{13}{4}\right) + 9 = 8\sin\left(\frac{\pi}{2}\right) + 9 = 8 \cdot 1 + 9 = 17.$$

If $t = 39/4$, then

$$d = 8\sin\left(\frac{2\pi}{13} \cdot \frac{39}{4}\right) + 9 = 8\sin\left(\frac{3\pi}{2}\right) + 9 = 8(-1) + 9 = 1.$$

So the depth at high tide is 17 meters and the depth at low tide is 1 meter.

**TRY THIS.** Demand for a seasonal product can be modeled by the function $d = 200\sin\left(\frac{\pi(t-3)}{6}\right) + 300$, where $d$ is the number of units sold in month $t$ and where $t$ ranges from 1 through 12. Find the demand in March and June.    ■

**FOR THOUGHT...**   True or False? Explain.

1. $\sin^2(\pi/7) + \cos^2(\pi/7) = 1$

2. For any real number $t$, $\sin^2(t) + \cos^2(t) = 1$.

3. If $\sin(\alpha) < 0$ and $\cos(\alpha) > 0$, then $\alpha$ is an angle in quadrant III.

4. If $\sin(\alpha) > 0$ and $\cos(\alpha) < 0$, then $\alpha$ is an angle in quadrant II.

5. If $\sin^2(\alpha) = 1/4$ and $\alpha$ is in quadrant III, then $\sin(\alpha) = 1/2$.

6. If $\cos^2(\alpha) = 1/9$ and $\alpha$ is an angle in quadrant I, then $\cos(\alpha) = 1/3$.

7. The reference angle for $120°$ is $60°$.

8. The reference angle for $-2\pi/3$ is $-\pi/3$.

9. $\cos(120°) = -\cos(60°)$

10. $\sin(7\pi/6) = -\sin(\pi/6)$

## 1.6 EXERCISES

### CONCEPTS

*Fill in the blank.*

1. The identity $\sin^2(\alpha) + \cos^2(\alpha) = 1$ is the _____ identity.

2. If $\theta$ is a nonquadrantal angle in standard position, then the _____ for $\theta$ is the positive acute angle formed by the terminal side of $\theta$ and the positive or negative $x$-axis.

### SKILLS

*Solve each problem.*

3. Find $\sin(\alpha)$, given that $\cos(\alpha) = 1$.

4. Find $\cos(\alpha)$, given that $\sin(\alpha) = 0$ and $\cos(\alpha) < 0$.

5. Find $\cos(\alpha)$, given that $\sin(\alpha) = 5/13$ and $\alpha$ is in quadrant II.

6. Find $\sin(\alpha)$, given that $\cos(\alpha) = -4/5$ and $\alpha$ is in quadrant III.

7. Find $\sin(\alpha)$, given that $\cos(\alpha) = 3/5$ and $\alpha$ is in quadrant IV.

8. Find $\cos(\alpha)$, given that $\sin(\alpha) = -12/13$ and $\alpha$ is in quadrant IV.

9. Find $\cos(\alpha)$, given that $\sin(\alpha) = 1/3$ and $\cos(\alpha) > 0$.

10. Find $\sin(\alpha)$, given that $\cos(\alpha) = 2/5$ and $\sin(\alpha) < 0$.

*Sketch the given angle in standard position and find its reference angle in degrees and radians.*

| | | | |
|---|---|---|---|
| **11.** $210°$ | **12.** $330°$ | **13.** $5\pi/3$ | **14.** $7\pi/6$ |
| **15.** $-300°$ | **16.** $-210°$ | **17.** $5\pi/6$ | **18.** $-13\pi/6$ |
| **19.** $405°$ | **20.** $390°$ | **21.** $-3\pi/4$ | **22.** $-2\pi/3$ |

*Use reference angles to find the exact value of each expression.*

| | | |
|---|---|---|
| **23.** $\sin(135°)$ | **24.** $\sin(420°)$ | **25.** $\cos(5\pi/3)$ |
| **26.** $\cos(11\pi/6)$ | **27.** $\sin(7\pi/4)$ | **28.** $\sin(-13\pi/4)$ |
| **29.** $\cos(-17\pi/6)$ | **30.** $\cos(-5\pi/3)$ | **31.** $\sin(-45°)$ |
| **32.** $\cos(-120°)$ | **33.** $\cos(-240°)$ | **34.** $\sin(-225°)$ |

*Use reference angles to find $\sin\theta$, $\cos\theta$, $\tan\theta$, $\csc\theta$, $\sec\theta$, and $\cot\theta$ for each given angle $\theta$.*

| | | | |
|---|---|---|---|
| **35.** $3\pi/4$ | **36.** $2\pi/3$ | **37.** $4\pi/3$ | **38.** $7\pi/6$ |
| **39.** $300°$ | **40.** $315°$ | **41.** $-135°$ | **42.** $135°$ |

*True or false? Do not use a calculator.*

| | |
|---|---|
| **43.** $\sin(210°) = \sin(30°)$ | **44.** $\sin(150°) = \sin(30°)$ |
| **45.** $\cos(330°) = \cos(30°)$ | **46.** $\cos(150°) = -\cos(30°)$ |
| **47.** $\sin(179°) = -\sin(1°)$ | **48.** $\sin(91°) = -\sin(89°)$ |

49. $\cos(6\pi/7) = -\cos(\pi/7)$

50. $\cos(13\pi/12) = -\cos(\pi/12)$

51. $\sin(23\pi/24) = -\sin(\pi/24)$

52. $\sin(25\pi/24) = \sin(\pi/24)$

53. $\cos(13\pi/7) = \cos(\pi/7)$

54. $\cos(9\pi/5) = \cos(\pi/5)$

### MODELING

*Solve each problem.*

55. *Phoenix Temperature* The temperature in Phoenix for a day in July is modeled by the function

$$T = 18\sin\left(\frac{\pi}{12}(h-12)\right) + 102$$

where $h$ is time in hours and $T$ is degrees Fahrenheit. Find the temperature at $h = 18$ (the daytime high) and at $h = 6$ (the nighttime low).

56. *Bismarck Temperature* The temperature in Bismarck for a day in January is modeled by the function

$$T = 13\cos\left(\frac{\pi}{12}(h+34)\right) - 7$$

where $h$ is time in hours and $T$ is degrees Fahrenheit. Find the temperature at $h = 14$ (the daytime high) and at $h = 2$ (the nighttime low).

57. *Motion of a Spring* A weight is suspended on a vertical spring as shown in the accompanying figure. The weight is set in motion and its position $x$ on the vertical number line in the figure is given by the function

$$x = 4\sin(t) + 3\cos(t)$$

where $t$ is time in seconds.
   **a.** Find the initial position of the weight (its position at times $t = 0$).

   **b.** Find the exact position of the weight at time $t = 5\pi/4$ seconds.

**Figure for Exercises 57 and 58**

**58.** *Motion of a Spring*  A weight is suspended on a vertical spring as in the previous exercise. The weight is set in motion and its position $x$ on the vertical number line in the figure is given by the function

$$x = -\frac{\sqrt{3}}{3}\sin\left(\frac{\pi}{3}t\right) - \cos\left(\frac{\pi}{3}t\right)$$

where $t$ is time in seconds.
  **a.** Find the initial position of the weight.

  **b.** Find the exact position of the weight at time $t = 2$ seconds.

**59.** *Spacing Between Teeth*  The length of an arc intercepted by a central angle of $\theta$ radians in a circle of radius $r$ is $r\theta$. The length of the chord, $c$, joining the endpoints of that arc is given by $c = r\sqrt{2 - 2\cos\theta}$. Find the actual distance between the tips of two adjacent teeth on a 12-in.-diameter carbide-tipped circular saw blade with 22 equally spaced teeth. Compare your answer with the length of a circular arc joining two adjacent teeth on a circle 12 in. in diameter. Round to the nearest thousandth.

**Figure for Exercise 59**

**60.** *Shipping Stop Signs*  A manufacturer of stop signs ships 100 signs in a circular drum. Use the formula from Exercise 59 to find the radius of the inside of the circular drum, given that the length of an edge of the stop sign is 10 in. and the signs just fit as shown in the figure. Round to the nearest hundredth.

**Figure for Exercise 60**

**61.** *Throwing a Javelin*  The formula

$$d = \frac{1}{32}v_0^2\sin(2\theta)$$

gives the distance $d$ in feet that a projectile will travel when its launch angle is $\theta$ and its initial velocity is $v_0$ feet per second. Approximately what initial velocity in miles per hour does it take to throw a javelin 367 ft assuming that $\theta$ is $43°$? Round to the nearest tenth.

**Figure for Exercise 61**

**62.** *Cosine in Terms of Sine*  Use the fundamental identity to write two formulas for $\cos\alpha$ in terms of $\sin\alpha$. Indicate which formula to use for a given value of $\alpha$.

### REVIEW

**63.** Evaluate the trigonometric functions for the angle $\alpha$ in standard position whose terminal side passes through $(3, 4)$.
  **a.** $\sec\alpha$      **b.** $\csc\alpha$      **c.** $\cot\alpha$

**64.** Find the exact value of each function.
  **a.** $\sec(135°)$

  **b.** $\csc(-3\pi/4)$

  **c.** $\cot(\pi/4)$

**65.** Find the exact value of each function.
  **a.** $\sec(60°)$

  **b.** $\csc(-\pi/6)$

  **c.** $\cot(\pi/3)$

**66.** If $\alpha$ is an acute angle of a right triangle, then the sine of $\alpha$ is the length of the _____ side divided by the length of the _____.

**67.** Find $\sin^{-1}(-1/2)$ in degrees.

**68.** At a distance of 2000 feet from a building, the angle of elevation to the top of the building is 30°. Find the height of the building to the nearest foot.

**OUTSIDE THE BOX**

**69.** *Kicking a Field Goal* In professional football the ball must be placed between the left and right hash mark when a field goal is to be kicked. The hash marks are 9.25 ft from the center line of the field. The goal post is 30 ft past the goal line. Suppose that $\theta$ is the angle between the lines of sight from the ball to the left and right uprights, as shown in the figure. The larger the value of $\theta$, the easier it is to kick the ball between the uprights. What is the difference (to the nearest thousandth of a degree) between the values of $\theta$ at the right hash mark and the value of $\theta$ at the center of the field when the ball is 60 ft from the goal line? The two vertical bars on the goal are 18.5 ft apart and the horizontal bar is 10 ft above the ground.

**Figure for Exercise 69**

**70.** *Ruby's Cube* Ruby has 10 white cubes and 17 red cubes that are 1 inch on each side. She arranges them to form a larger cube that is 3 inches on each side. What is the largest possible percent of red surface area on the larger cube?

## 1.6 POP QUIZ

**1.** Find $\sin(\alpha)$, if $\cos(\alpha) = -3/5$ and $\alpha$ is in quadrant III.

**2.** What is the reference angle for 120°?

**3.** What is the reference angle for $11\pi/6$?

*Evaluate each expression using reference angles.*

**4.** $\sin(210°)$

**5.** $\sin(-5\pi/4)$

**6.** $\cos(135°)$

**7.** $\cos(11\pi/6)$

# Highlights

## 1.1 Angles and Degree Measure

| Degree Measure | Divide the circumference of a circle into 360 equal arcs. The degree measure is the number of degrees through which the initial side of an angle is rotated to get to the terminal side (positive for counterclockwise, negative for clockwise). | 90° is 1/4 of a circle. 180° is 1/2 of a circle. |
|---|---|---|
| Degrees-Minutes-Seconds | 1 degree = 60 minutes, 1 minute = 60 seconds $1° = 60'$, $1' = 60''$ | $20.5° = 20°30'$ $5°2'24'' = 5.04°$ |
| Coterminal Angles | Two angles are coterminal if their measures differ by an integer multiple of 360°. | 30°, 390°, and 750° are coterminal. |

## 1.2 Radian Measure, Arc Length, and Area

| Radian Measure | The length of the arc on the unit circle through which the initial side rotates to get to the terminal side. Radian measure is positive if the rotation is counterclockwise, negative if clockwise. | $\pi/2$ is 1/4 of the unit circle. $\pi$ is 1/2 of the unit circle. |
|---|---|---|
| Converting | Use $\pi$ radians = 180° and cancellation of units. | $90° = 90° \cdot \frac{\pi \text{ rad}}{180°} = \frac{\pi}{2}$ rad |
| Arc Length | $s = \alpha r$ where $s$ is the arc intercepted by a central angle of $\alpha$ radians on a circle of radius $r$ | $\alpha = 90°, r = 10$ ft $s = \alpha r = \frac{\pi}{2} \cdot 10$ ft $= 5\pi$ ft |

| **Area of a Sector of a Circle** | Area of a sector of a circle is found by taking a fraction of the area of the circle $(\pi r^2)$, or by the formula $A = \alpha\dfrac{r^2}{2}$, where $\alpha$ is in radians. | Central angle $\pi/2$, radius 4<br>Area of sector $\frac{1}{4} \cdot \pi 4^2$<br>$A = \dfrac{\pi}{2} \cdot \dfrac{4^2}{2} = 4\pi$ |

## 1.3 Angular and Linear Velocity

| **Angular Velocity** | A point in motion on a circle of radius $r$ through an angle of $\alpha$ radians or arc length $s$ in time $t$ has angular velocity $\omega$, where $\omega = \alpha/t$. | $r = 10$ ft, $\alpha = \pi/2$<br>$s = 5\pi$ ft, $t = 2$ sec<br>$\omega = \pi/4$ rad/sec |
| **Linear Velocity** | Linear velocity $v$ is distance divided by time or $v = s/t$. | $v = 2.5\pi$ ft/sec |
| **Relationship** | If $v$ is linear velocity and $\omega$ is angular velocity, then $v = r\omega$ where $r$ is the radius of the circle. | $v = 10\omega$<br>$2.5\pi$ ft/sec $= 10$ ft $\cdot \dfrac{\pi}{4}$ rad/sec |

## 1.4 The Trigonometric Functions

| **Trigonometric Functions** | If $\alpha$ is an angle in standard position and $(x, y)$ is any point other than $(0, 0)$ on the terminal side of $\alpha$, and $r = \sqrt{x^2 + y^2}$, then<br><br>$\sin \alpha = \dfrac{y}{r}, \quad \cos \alpha = \dfrac{x}{r}, \quad \tan \alpha = \dfrac{y}{x}$<br><br>$\csc \alpha = \dfrac{r}{y}, \quad \sec \alpha = \dfrac{r}{x}, \quad \cot \alpha = \dfrac{x}{y}$<br><br>provided no denominator is zero. | $\sin(90°) = 1$<br><br>$\cos(\pi/2) = 0$<br><br>$\sin(0) = 0$ |
| **Reciprocal Identities** | $\csc \alpha = \dfrac{1}{\sin \alpha}, \quad \sec \alpha = \dfrac{1}{\cos \alpha}, \quad \cot \alpha = \dfrac{1}{\tan \alpha}$ | $\csc 30° = \dfrac{1}{\sin 30°}$ |

## 1.5 Right Triangle Trigonometry

| **Inverse Trigonometric Functions** | $\sin^{-1}(x) = \alpha$ if $\sin \alpha = x$ and $-90° \le \alpha \le 90°$<br>$\cos^{-1}(x) = \alpha$ if $\cos \alpha = x$ and $0° \le \alpha \le 180°$<br>$\tan^{-1}(x) = \alpha$ if $\tan \alpha = x$ and $-90° < \alpha < 90°$ | $\sin^{-1}(1/2) = 30°$<br>$\cos^{-1}(1/2) = 60°$<br>$\tan^{-1}(1) = 45°$ |
| **Trigonometric Functions in a Right Triangle** | If $\alpha$ is an acute angle of a right triangle, then<br><br>$\sin \alpha = \dfrac{\text{opp}}{\text{hyp}}, \quad \cos \alpha = \dfrac{\text{adj}}{\text{hyp}}, \quad \tan \alpha = \dfrac{\text{opp}}{\text{adj}},$<br><br>$\csc \alpha = \dfrac{\text{hyp}}{\text{opp}}, \quad \sec \alpha = \dfrac{\text{hyp}}{\text{adj}}, \quad \cot \alpha = \dfrac{\text{adj}}{\text{opp}}.$ | $\sin \alpha = 5/13, \cos \alpha = 12/13$<br><br>$\tan \alpha = 5/12, \csc \alpha = 13/5$<br><br>$\sec \alpha = 13/12, \cot \alpha = 12/5$<br><br> |

## 1.6 The Fundamental Identity and Reference Angles

| **Fundamental Identity** | For any angle $\alpha$ (or real number $\alpha$),<br>$\sin^2(\alpha) + \cos^2(\alpha) = 1$. | $\sin^2(30°) + \cos^2(30°) = 1$<br>$\sin^2\left(\sqrt[3]{2}\right) + \cos^2\left(\sqrt[3]{2}\right) = 1$ |
| **Reference Angle** | The positive acute angle formed by the terminal side of a nonquadrantal angle and the positive or negative $x$-axis | For $120°$ the reference angle is $60°$. |
| **Using Reference Angles** | To find the sine (or cosine) of an angle, first find the sine or cosine of its reference angle and prefix the appropriate sign. | $\sin(2\pi/3) = \sin(\pi/3)$<br>$\cos(2\pi/3) = -\cos(\pi/3)$ |

# Chapter **1** Review Exercises

*Find the measure in degrees of the least positive angle that is coterminal with each given angle.*

**1.** 388°

**2.** −840°

**3.** −153°14′27″

**4.** 455°39′24″

**5.** −π

**6.** −35π/6

**7.** 13π/5

**8.** 29π/12

*Convert each radian measure to degree measure. Do not use a calculator.*

**9.** 5π/3

**10.** −3π/4

**11.** 3π/2

**12.** 5π/6

*Convert each degree measure to radian measure. Do not use a calculator.*

**13.** 330°

**14.** 405°

**15.** −300°

**16.** −210°

*Fill in the tables. Do not use a calculator.*

**17.**

| θ deg | 0 | 30 | 45 | 60 | 90 | 120 | 135 | 150 | 180 |
|---|---|---|---|---|---|---|---|---|---|
| θ rad | | | | | | | | | |
| sin θ | | | | | | | | | |
| cos θ | | | | | | | | | |

**18.**

| θ rad | 0 | $\frac{\pi}{6}$ | $\frac{\pi}{4}$ | $\frac{\pi}{3}$ | $\frac{\pi}{2}$ | $\frac{2\pi}{3}$ | $\frac{3\pi}{4}$ | $\frac{5\pi}{6}$ | π |
|---|---|---|---|---|---|---|---|---|---|
| θ deg | | | | | | | | | |
| sin θ | | | | | | | | | |
| cos θ | | | | | | | | | |

*Give the exact values of each of the following expressions. Do not use a calculator.*

**19.** sin(−π/4)

**20.** cos(−2π/3)

**21.** tan(π/3)

**22.** sec(π/6)

**23.** csc(−120°)

**24.** cot(135°)

**25.** sin(180°)

**26.** tan(0°)

**27.** cos(3π/2)

**28.** csc(5π/6)

**29.** sec(−π)

**30.** cot(−4π/3)

**31.** cot(420°)

**32.** sin(390°)

**33.** cos(−135°)

**34.** tan(225°)

**35.** sec(2π/3)

**36.** csc(−3π/4)

**37.** tan(5π/6)

**38.** sin(7π/6)

*Find an approximate value for each expression. Round to four decimal places.*

**39.** sin(44°)

**40.** sin(−205°)

**41.** cos(4.62)

**42.** cos(3.14)

**43.** tan(π/17)

**44.** tan(2.33)

**45.** sec(105°4′)

**46.** sec(55°3′12″)

**47.** csc(π/9)

**48.** csc(6.88)

**49.** cot(33°44′)

**50.** cot(77°42′9″)

*Find the exact value of each expression in degrees.*

**51.** $\sin^{-1}(1/\sqrt{2})$

**52.** $\tan^{-1}(1)$

**53.** $\cos^{-1}(1)$

**54.** $\cos^{-1}(0)$

**55.** $\cos^{-1}(\sqrt{3}/2)$

**56.** $\sin^{-1}(\sqrt{3}/2)$

**57.** $\sin^{-1}(0.5)$

**58.** $\tan^{-1}(\sqrt{3})$

*For each triangle find the exact values of sin α, cos α, tan α, csc α, sec α, and cot α.*

**59.**

**60.**

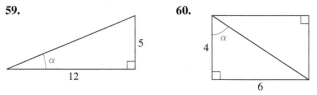

*Solve each right triangle with the given parts.*

**61.** $a = 2, b = 3$

**62.** $a = 3, c = 7$

**63.** $a = 3.2, \alpha = 21.3°$

**64.** $\alpha = 34.6°, c = 9.4$

*Solve each problem.*

**65.** Find sin α if cos α = 1/5 and α is in quadrant IV.

**66.** Find cos α if sin α = 1/3 and α is in quadrant II.

**67.** *Irrigation* A center-pivot irrigation system waters a circular field by rotating about the center of the field. If the system makes one revolution in 8 hr, what is the linear velocity in feet per hour of a nozzle that is 120 ft from the center?

**68.** *Angular Velocity* If a bicycle with a 26-in.-diameter wheel is traveling 16 mph, then what is the angular velocity of the valve stem in radians per hour?

**69.** *Crooked Man* A man is standing 1000 ft from a surveyor. The surveyor measures an angle of 0.4° sighting from the man's feet to his head. Find the height of the man.
**HINT** Assume that the man can be represented by the arc intercepted by a central angle of 0.4° in a circle of radius 1000 ft.

**70.** *Shooting a Target* Judy is standing 200 ft from a circular target with a radius of 3 in. To hit the center of the circle, she must hold the gun perfectly level, as shown in the figure. Will she hit the target if her aim is off by one-tenth of a degree in any direction?

**HINT** Assume that the width of the target is an arc on a circle of radius 200 ft.

**Figure for Exercise 70**

**71.** The two legs of a right triangle are 5 ft and 8 ft. Solve the triangle.

**72.** One leg of a right triangle is 4 in. and the hypotenuse is 9 in. Solve the triangle.

**73.** One of the acute angles of a right triangle is 19.3° and the hypotenuse is 12 ft. Solve the triangle.

**74.** One of the acute angles of a right triangle is 34.6° and the side opposite that angle is 8.4 m. Solve the triangle.

**75.** *Height of Buildings* Two buildings are 300 ft apart. From the top of the shorter building the angle of elevation of the top of the taller building is 23°, and the angle of depression of the base of the taller building is 36°, as shown in the figure. How tall is each building?

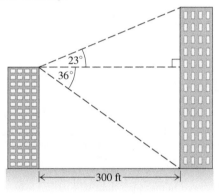

**Figure for Exercise 75**

**76.** *Cloud Height* Visual Flight Rules require that the height of the clouds be more than 1000 ft for a pilot to fly without instrumentation. At night, cloud height can be determined from the ground by using a search light and an observer as shown in the figure. If the beam of light is aimed straight upward and the observer 500 ft away sights the cloud with an angle of elevation of 55°, then what is the height of the cloud cover?

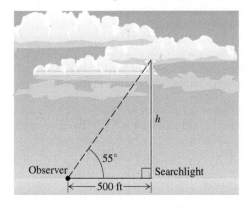

**Figure for Exercise 76**

**77.** *Devil's Tower* A tourist spots a rock climber quite high up at Devil's Tower in Wyoming. The angle of elevation of the climber is 36°. See the accompanying figure. From a point that is 100 ft closer to the climber, the angle of elevation is 44°. What is the height of the climber to the nearest foot?

**Figure for Exercise 77**

**78.** *Eiffel Tower* A tourist wants to determine the height of the Eiffel Tower without looking in her guide book. She observes the angle of elevation of the top of the tower from one point on the street is 31.73°. She moves 200 m closer to the tower and observes an angle of elevation to the top of the tower of 45°. What is the height of the tower to the nearest meter?

**79.** *Two Pulleys and a Belt* The distance between the centers of two circular pulleys is 12 inches. The radii of the pulleys are 3 inches and 6 inches. Find the length (to the nearest tenth of an inch) of the belt that wraps around the two pulleys, as shown in the accompanying figure.

**Figure for Exercise 79**

**80.** *Three Pulleys and a Belt* The centers of three pulleys are located at the vertices of a right triangle whose sides are 5 inches, 12 inches, and 13 inches, as shown in the accompanying diagram. The radii of the pulleys are 1 inch, 2 inches, and 3 inches. Find the length (to the nearest tenth of an inch) of the belt that wraps around all three pulleys, as shown in the diagram.

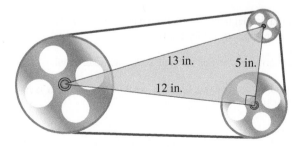

**Figure for Exercise 80**

**81.** *Texas Star* A welder has five steel rods that are each 2 feet in length. She is making a five-pointed star from them, as shown in the accompanying figure. Find the height of the star *h* to the nearest tenth of an inch.

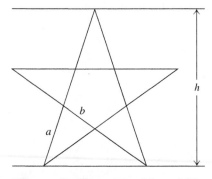

**Figure for Exercises 81 and 82**

**82.** *Texas Triangles* There are five congruent triangles in the star shown in the accompanying figure. Find the dimensions *a* and *b* to the nearest tenth of an inch. Use the length 2 feet from the previous exercise.

**OUTSIDE THE BOX**

**83.** *Clear Sailing* A sailor plans to install a windshield wiper on a porthole that has radius 1 ft. The wiper blade of length *x* ft is to be attached to the edge of the porthole, as shown in the figure. The area cleaned by the blade is a sector of a circle centered at the point of attachment.

**a.** Write the area cleaned by the blade as a function of *x*.

**b.** If the blade cleans half of the window, then what is the exact length of the blade?

**c.** Use a graphing calculator to find the length for the blade that would maximize the area cleaned.

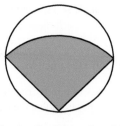

**Figure for Exercise 83**

**84.** *Intersecting Lines* For what values of *m* do the lines $y = x - 5$ and $y = mx + 5$ intersect in the first quadrant?

# Chapter **1** Test

*Find the exact value of each expression without using a calculator.*

**1.** $\cos 420°$

**2.** $\sin(-390°)$

**3.** $\sin(3\pi/4)$

**4.** $\cos(-\pi/3)$

**5.** $\tan(7\pi/6)$

**6.** $\tan(-2\pi/3)$

**7.** $\sec(\pi/2)$

**8.** $\csc(-\pi/2)$

**9.** $\cot(-3\pi)$

**10.** $\cot(225°)$

**11.** $\sin^{-1}\left(\sqrt{2}/2\right)$

**12.** $\cos^{-1}\left(\sqrt{3}/2\right)$

*Solve each problem.*

**13.** Find the arc length intercepted by a central angle of $46°24'6''$ in a circle with a radius of 35.62 m.

**14.** Find the degree measure (to the nearest hundredth of a degree) for an angle of 2.34 radians.

**15.** Determine whether $40°$ and $2200°$ are coterminal.

**16.** Find the exact value of $\cos \alpha$ if $\sin \alpha = 1/4$ and $\alpha$ is in quadrant II.

**17.** If a bicycle wheel with a 26-in. diameter is making 103 revolutions per minute, then what is the angular velocity in radians per minute for a point on the tire?

**18.** At what speed in miles per hour will a bicycle travel if the rider can cause the 26-in.-diameter wheel to rotate 103 revolutions per minute?

**19.** Find the exact values of all six trigonometric functions for an angle $\alpha$ in standard position whose terminal side contains the point $(5, -2)$.

**20.** One of the acute angles of a right triangle is $24°$ and the hypotenuse is 5 ft. Solve the triangle.

**21.** To estimate the blood pressure of *Brachiosaurus,* Professor Ostrom wanted to estimate the height of the head of the famed *Brachiosaurus* skeleton at Humboldt University in Berlin. From a distance of 11 m from a point directly below the head, the angle of elevation of the head was approximately $48°$. Use this information to find the height of the head of *Brachiosaurus.*

**22.** From a point on the street the angle of elevation of the top of the John Hancock Building is $65.7°$. From a point on the street that is 100 ft closer to the building, the angle of elevation is $70.1°$. Find the height of the building.

# 2

# Graphs of the Trigonometric Functions

**2.1** The Unit Circle and Graphing

**2.2** The General Sine Wave

**2.3** Graphs of the Secant and Cosecant Functions

**2.4** Graphs of the Tangent and Cotangent Functions

**2.5** Combining Functions

When will the sun rise and set tomorrow? What percent of the moon will be illuminated? When is the next lunar eclipse? Scientists at the Astronomical Applications Department of the U.S. Naval Observatory answer these questions and more. They compute the position, brightness, and other observable characteristics of celestial bodies, as well as the circumstances of astronomical phenomena.

Many of their data are periodic when measured over time. For example, the sun rises early in the summer and late in the winter, year after year. Because of the circular motion involved in celestial objects, the periodic data can often be modeled with the trigonometric functions.

WHAT YOU WILL **LEARN**

In Exercises 59 and 60 of Section 2.2 you will use trigonometric functions to model some astronomical data obtained from the Web site of the U.S. Naval Observatory (aa.usno.navy.mil).

# 2.1 The Unit Circle and Graphing

We first defined the trigonometric functions as the trigonometric ratios. Later we defined them as the ratios of the sides of a right triangle. When graphing the trigonometric functions it is best to use the unit circle definitions of these functions.

## The Unit Circle Definition

When we defined the trigonometric functions with trigonometric ratios, $(x, y)$ was any point on the terminal side of the angle and $r$ was its distance to the origin. For the unit circle definitions of the trigonometric functions we simply choose $(x, y)$ on the unit circle, where the distance to the origin is 1. So the unit circle definitions are the trigonometric ratios with $r = 1$.

**Definition: The Trigonometric Functions**

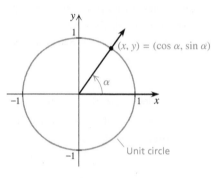

**Figure 2.1**

If $\alpha$ is an angle in standard position whose terminal side intersects the unit circle at point $(x, y)$, then

$$\sin \alpha = y, \qquad \cos \alpha = x, \qquad \tan \alpha = \frac{y}{x},$$

$$\csc \alpha = \frac{1}{y}, \qquad \sec \alpha = \frac{1}{x}, \qquad \cot \alpha = \frac{x}{y},$$

provided no denominator is zero.

Since $\sin \alpha = y$ and $\cos \alpha = x$, the point where $\alpha$ intersects the unit circle is $(\cos \alpha, \sin \alpha)$ as shown in Fig. 2.1. Since we already know the sine and cosine of common angles we can label the points where the common angles intersect the unit circle as in Fig. 2.2. Then we can use Fig. 2.2 to evaluate trigonometric functions with the unit circle definition.

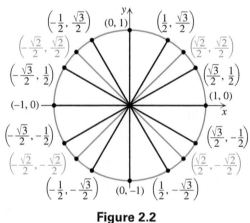

**Figure 2.2**

**EXAMPLE 1** **Evaluating trigonometric functions with the unit circle**

Find the exact value of each trigonometric function.

  **a.** $\sin 45°$   **b.** $\cos(2\pi/3)$   **c.** $\tan(\pi/2)$   **d.** $\sec(\pi/3)$

**Solution**

**a.** A 45° angle intersects the unit circle at $\left(\sqrt{2}/2, \sqrt{2}/2\right)$ as shown in Fig. 2.2. Since sin 45° is the $y$-coordinate, we have sin 45° = $\sqrt{2}/2$.

**b.** The angle $2\pi/3$ intersects the unit circle at $\left(-1/2, \sqrt{3}/2\right)$ as shown in Fig. 2.2. Since $\cos(2\pi/3)$ is the $x$-coordinate, we have $\cos(2\pi/3) = -1/2$.

**c.** Since $\pi/2$ intersects the unit circle at $(0, 1)$, $\tan(\pi/2)$ is undefined.

**d.** Since $\pi/3$ intersects the unit circle at $\left(1/2, \sqrt{3}/2\right)$ in Fig. 2.2 and sec $\alpha = 1/x$, we have $\sec(\pi/3) = 2$.

**TRY THIS.** Find the exact value of $\cos(\pi/4)$ and $\tan(0)$. ∎

We can think of the domain of the trigonometric functions as angles, degree measures, or radian measures. Since the radian measure of an angle $\alpha$ is the same as the length of its intercepted arc $s$ on the unit circle, we can also use the lengths of arcs or the real numbers as the domain. When graphing trigonometric functions we generally think of the real numbers as the domain.

When graphing functions in an $xy$-coordinate system, it is customary to use $x$ as the independent variable and $y$ as the dependent variable, as in $y = \cos(x)$ and $y = \sin(x)$. In this case you should think of $\cos(x)$ as the first coordinate (rather than the $x$-coordinate) and $\sin(x)$ as the second coordinate (rather than the $y$-coordinate) corresponding to an arc of length $x$ on the unit circle. It is confusing to say that $\cos(x)$ is the $x$-coordinate and $\sin(x)$ is the $y$-coordinate.

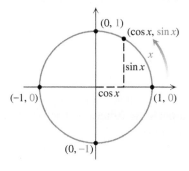

**Figure 2.3**

# Graphing Sine Functions

Any function of the form $y = A \sin[B(x - C)] + D$ with $A \neq 0$ and $B \neq 0$ is a **sine function.** In this section we will graph only sine functions for which $B = 1$. Sine functions with $B \neq 1$ are discussed in the next section. We always assume that $x$ is a real number or radian measure unless it is stated that $x$ is the degree measure of an angle.

In the next example we graph the simplest sine function. Remember that sin $x$ is the second coordinate of the terminal point on the unit circle for an arc of length $x$, as shown in Fig. 2.3.

**EXAMPLE 2**    **Graphing the simplest sine function**

Sketch the graph of $y = \sin x$.

**Solution**

First consider the behavior of $y = \sin(x)$ on the interval $[0, 2\pi]$. As the arc length $x$ increases from length 0 to $\pi/2$ on the unit circle, the second coordinate of the endpoint increases from 0 to 1, its maximum value. From $\pi/2$ to $\pi$, the second coordinate of the endpoint decreases from 1 to 0. From $\pi$ to $2\pi$, the second coordinate decreases to $-1$, its minimum value, and then increases to 0. So there are three $x$-intercepts on the interval $[0, 2\pi]$ and a maximum and minimum point. These five key points on the graph of $y = \sin x$ for $x$ in $[0, 2\pi]$ are given in the following table:

| $x$ | 0 | $\pi/2$ | $\pi$ | $3\pi/2$ | $2\pi$ |
|---|---|---|---|---|---|
| $y = \sin x$ | 0 | 1 | 0 | $-1$ | 0 |

The ordered pairs in the table are graphed in Fig. 2.4 on the next page. To determine the actual shape of the curve through these five points we can either accurately plot many more points by hand or look at the shape of the curve shown in

Fig. 2.5. That figure shows 96 accurately plotted points on the graph of $y = \sin x$. From $2\pi$ to $4\pi$, the values of $\sin x$ again increase from 0 to 1, decrease to $-1$, then increase to 0. Because $\sin(x + 2\pi) = \sin x$, the exact shape that we saw for the interval $[0, 2\pi]$ is repeated for $x$ in intervals such as $[2\pi, 4\pi]$, $[-2\pi, 0]$, $[-4\pi, -2\pi]$, and so on. So the curve in Fig. 2.4 continues indefinitely to the left and right.

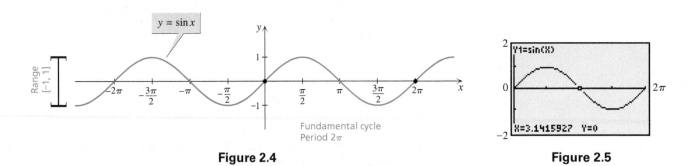

**Figure 2.4**                                    **Figure 2.5**

**TRY THIS.** Graph $y = 4 \sin x$ for $x$ in the interval $[0, 2\pi]$.  ∎

Since the $x$-intercepts and the maximum and minimum values of $y = \sin(x)$ occur at multiples of $\pi/2$, we labeled the $x$-axis in Fig. 2.4 with multiples of $\pi/2$. The domain of $y = \sin x$ is the set of all real numbers $(-\infty, \infty)$ and the range is the interval $[-1, 1]$.

Since the shape of $y = \sin x$ for $x$ in $[0, 2\pi]$ is repeated infinitely often, $y = \sin x$ is called a *periodic function.*

**Definition; Periodic Function**

If $y = f(x)$ is a function and $a$ is a nonzero constant such that $f(x) = f(x + a)$ for every $x$ in the domain of $f$, then $f$ is called a **periodic function.** The smallest such positive constant $a$ is the **period** of the function.

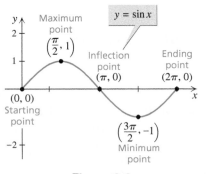

**Figure 2.6**

The smallest value of $a$ such that $\sin(x) = \sin(x + a)$ is $2\pi$. So the period of $y = \sin(x)$ is $2\pi$. The graph of $y = \sin(x)$ over any interval of length $2\pi$ is a **cycle.** The graph of $y = \sin(x)$ over $[0, 2\pi]$ is the **fundamental cycle.** The **five key points** on the fundamental cycle of $y = \sin(x)$ are the starting point, maximum point, inflection point, minimum point, and ending point, as shown in Fig. 2.6. An **inflection point** is a point at which the graph changes from opening downward to opening upward or vice versa. Suppose you are driving on a winding road where the road curves to the left and then to the right. The inflection point is the point at which you momentarily have the steering wheel pointed straight. A more precise definition of inflection is given in calculus.

The graph of any function of the form $y = A \sin[B(x - C)] + D$ is a sine function. The graph of any sine function is a transformation (Section P.3) of the graph of $y = \sin(x)$, and its graph is called a **sine wave, a sinusoidal wave,** or a **sinusoid.** To make an accurate graph of a sine function, we use the idea of transformations to determine what happens to the five key points of the fundamental cycle of $y = \sin(x)$, and then sketch a sine wave between these points.

**EXAMPLE 3**    **Graphing a sine function**

Sketch the graph of $y = 2 \sin x$ for $x$ in the interval $[-2\pi, 2\pi]$ and determine the range of the function.

**Solution**

To graph $y = 2 \sin x$ we stretch the graph of $y = \sin x$ by a factor of 2 (see Section P.3). The points on the $x$-axis of $y = \sin x$ stay put, but the maximum point moves up to $(\pi/2, 2)$ and the minimum point moves down to $(3\pi/2, -2)$. Draw one cycle of $y = 2 \sin x$ through these five key points, as in Fig. 2.7. Draw another cycle of the sine wave for $x$ in $[-2\pi, 0]$ to complete the graph of $y = 2 \sin x$ for $x$ in $[-2\pi, 2\pi]$. Since the minimum value of $y$ is $-2$ and the maximum value of $y$ is 2, the range of the function is $[-2, 2]$.

The calculator graph shown in Fig. 2.8 supports our conclusions.

**Figure 2.7**    **Figure 2.8**

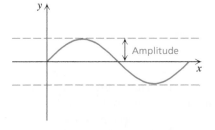

**Figure 2.9**

**TRY THIS.** Graph $y = 3 \sin x$ for $x$ in the interval $[-2\pi, 2\pi]$ and determine the range of the function. ■

The amplitude of a sine wave is a measure of the "height" of the wave as shown in Fig. 2.9. When an oscilloscope is used to get a picture of the sine wave corresponding to a sound, the amplitude of the sine wave corresponds to the intensity or loudness of the sound. To find the amplitude we use the following definition.

**Definition: Amplitude**

> The **amplitude** of a sine wave, or the amplitude of the sine function, is the absolute value of half the difference between the maximum and minimum $y$-coordinates.

**EXAMPLE 4**    **Finding amplitude**

Find the amplitude of the functions $y = \sin x$ and $y = 2 \sin x$.

**Solution**

For $y = \sin x$ the maximum $y$-coordinate is 1, and the minimum is $-1$. Since

$$\left| \frac{1}{2}[1 - (-1)] \right| = 1,$$

the amplitude for $y = \sin x$ is 1. For $y = 2 \sin x$, the maximum $y$-coordinate is 2 and the minimum is $-2$. Since

$$\left| \frac{1}{2}[2 - (-2)] \right| = 2,$$

the amplitude for $y = 2 \sin x$ is 2.

**TRY THIS.** Find the amplitude for $y = 8 \sin(x)$. ■

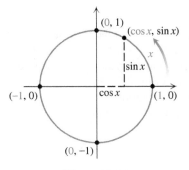

**Figure 2.10**

# Graphing Cosine Functions

Any function of the form $y = A \cos[B(x - C)] + D$ with $A \neq 0$ and $B \neq 0$ is a **cosine function.** In this section we will graph only cosine functions for which $B = 1$ and study cosine functions for which $B \neq 1$ in the next section.

In the next example we graph the simplest cosine function. Recall that $\cos x$ is the first coordinate of the terminal point on the unit circle for an arc of length $x$, as shown in Fig. 2.10.

**EXAMPLE 5** Graphing the simplest cosine function

Sketch the graph of $y = \cos(x)$.

**Solution**

First consider the behavior of $y = \cos(x)$ on the interval $[0, 2\pi]$. As the arc length $x$ increases from length 0 to $\pi/2$, the first coordinate of the endpoint decreases from 1 to 0. From $\pi/2$ to $\pi$, the first coordinate decreases from 0 to $-1$. From $\pi$ to $2\pi$, the first coordinate increases from $-1$ back to 1. So on the interval $[0, 2\pi]$ there are two $x$-intercepts, two maximum points, and one minimum point. These five key points on the graph of $y = \cos x$ for $x$ in $[0, 2\pi]$ are given in the following table:

| $x$ | 0 | $\pi/2$ | $\pi$ | $3\pi/2$ | $2\pi$ |
|---|---|---|---|---|---|
| $y = \cos x$ | 1 | 0 | $-1$ | 0 | 1 |

These five points are graphed in Fig. 2.11. To determine the actual shape of $y = \cos x$ we can either accurately plot many more points by hand or look at the shape of the curve in Fig. 2.12. That figure shows 96 accurately plotted points on the graph of $y = \cos x$. Since $\cos(x) = \cos(x + 2\pi)$, the curve that we see on $[0, 2\pi]$ is repeated on $[2\pi, 4\pi]$, $[-2\pi, 0]$, and so on. So the curve in Fig. 2.11 continues indefinitely to the left and right.

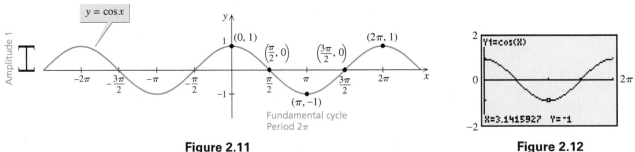

**Figure 2.11**         **Figure 2.12**

**TRY THIS.** Graph $y = 4 \cos x$ for $x$ in $[0, 2\pi]$. ∎

The graph of $y = \cos x$ has exactly the same shape as the graph of $y = \sin x$. It has amplitude 1 and period $2\pi$. If the graph of $y = \sin x$ were shifted a distance of $\pi/2$ to the left, the graphs would coincide. For this reason the graph of $y = \cos x$ is also called a sine wave. The graph of $y = \cos x$ over $[0, 2\pi]$ is called the **fundamental cycle** of $y = \cos x$. Note that the fundamental cycle starts and ends with a maximum point. Midway between them is a minimum point and midway between each maximum point and the minimum point is an inflection point. See Fig. 2.13.

**Figure 2.13**

The graph of any cosine function is similar to the graph of $y = \cos x$. When graphing a cosine function we always identify the five key points for one cycle.

**EXAMPLE 6**  **Graphing a cosine function**

Sketch the graph of $y = -3 \cos x$ for $x$ in the interval $[-2\pi, 2\pi]$. Find the amplitude and range of the function.

**Solution**

We graph $y = -3 \cos x$ by stretching and then reflecting in the $x$-axis the graph of $y = \cos x$. So the starting point $(0, 1)$ is stretched to $(0, 3)$ and then reflected to $(0, -3)$. Similarly, the endpoint is $(2\pi, -3)$. Since the graph is a reflection, the starting and ending points are minimum points on the graph. Midway between these two is the maximum point $(\pi, 3)$. The inflection points are midway between the minimum points and the maximum point at $(\pi/2, 0)$ and $(3\pi/2, 0)$. Note how the five key points divide the interval $[0, 2\pi]$ into four equal parts. Draw one cycle of $y = -3 \cos x$ through these five key points, as shown in Fig. 2.14. Repeat the same shape for $x$ in the interval $[-2\pi, 0]$ to get the graph for $x$ in $[-2\pi, 2\pi]$. Since

$$\left| \frac{1}{2}(3 - (-3)) \right| = 3,$$

the amplitude is 3. Since the minimum $y$-coordinate is $-3$ and the maximum is 3, the range is the interval $[-3, 3]$.

A calculator graph of $y = -3 \cos x$ is shown in Fig. 2.15 and it supports our conclusions.

**TRY THIS.** Graph $y = -8 \cos x$ for $x$ in the interval $[-2\pi, 2\pi]$. Find the amplitude and range of the function. ■

**Figure 2.14**

**Figure 2.15**

## Transformations of Sine and Cosine

In Examples 3 and 6 we saw that slight changes in the function affect the graph of the function. The graph of $y = 2 \sin x$ from Example 3 can be obtained by stretching the graph of $y = \sin x$ by a factor of 2. The graph of $y = -3 \cos x$ in Example 6 can be obtained by stretching (by a factor of 3) and then reflecting the graph of $y = \cos x$ in the $x$-axis. The amplitude of $y = 2 \sin x$ is 2 and the amplitude of $y = -3 \cos x$ is 3. In general, the amplitude is determined by the coefficient of the sine or cosine function.

**Theorem: Amplitude**

The amplitude of $y = A \sin x$ or $y = A \cos x$ is $|A|$.

From algebra (Section P.3) we know that the graph of $y = f(x - C)$ is a horizontal translation of the graph of $y = f(x)$, to the right if $C > 0$ and to the left if $C < 0$. So the graphs of $y = \sin(x - C)$ and $y = \cos(x - C)$ are horizontal translations of $y = \sin x$ and $y = \cos x$, respectively, to the right if $C > 0$ and to the left if $C < 0$. The term *phase shift* is used to describe left or right translations of a sine wave.

**Definition: Phase Shift**

The **phase shift** of the graph of $y = \sin(x - C)$ or $y = \cos(x - C)$ is $C$.

### EXAMPLE 7 Horizontal translation

Graph two cycles of $y = \sin(x + \pi/6)$. Determine the phase shift, amplitude, and range of the function.

**Solution**

Since $x + \pi/6 = x - (-\pi/6)$, $C = -\pi/6$. The phase shift is $-\pi/6$. Because $C < 0$, the graph of $y = \sin(x + \pi/6)$ is obtained by shifting $y = \sin x$ a distance of $\pi/6$ to the left. So the new starting point is $(-\pi/6, 0)$ and the new ending point is $(11\pi/6, 0)$. Label the $x$-axis with multiples of $\pi/6$, as in Fig. 2.16, and divide the interval $[-\pi/6, 11\pi/6]$ into four equal segments. Plot the inflection point in the middle at $(5\pi/6, 0)$ and the maximum $(\pi/3, 1)$ and minimum $(4\pi/3, -1)$. Draw one cycle of the graph through these five key points, as in Fig. 2.16. Continue the pattern for another cycle, as shown in the figure. The second cycle could be drawn to the right or left of the first. Since the minimum $y$-coordinate is $-1$ and the maximum is $1$, the amplitude is $1$ and the range is $[-1, 1]$.

The calculator graph in Fig. 2.17 supports the conclusion that $y_2 = \sin(x + \pi/6)$ is a shift to the left of $y_1 = \sin(x)$.

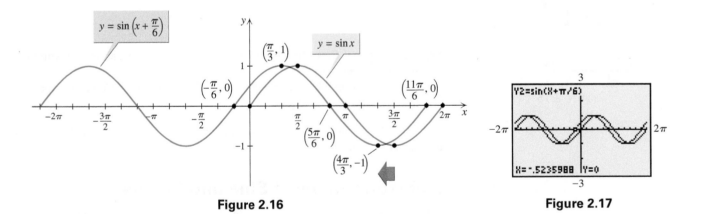

**Figure 2.16**                    **Figure 2.17**

**TRY THIS.** Graph one cycle of $y = \cos(x - \pi/6)$ and find the phase shift, amplitude, and range.

The graphs of $y = \sin(x) + D$ and $y = \cos(x) + D$ are vertical translations of $y = \sin x$ and $y = \cos x$, respectively. The translation is upward for $D > 0$ and downward for $D < 0$. The next example combines a phase shift and a vertical translation. Note how we follow the five key points of the fundamental cycle to see where they go in the transformation.

**EXAMPLE 8** Horizontal and vertical translation

Graph two cycles of $y = \cos(x - \pi/4) + 2$. Determine the phase shift and range.

**Solution**

For $y = \cos(x - \pi/4) + 2$ the phase shift is $\pi/4$ and the vertical translation is 2. Since the phase shift is $\pi/4$, label the $x$-axis with multiples of $\pi/4$, as in Fig. 2.18. Now move the starting and ending points $(0, 1)$ and $(2\pi, 1)$ a distance of $\pi/4$ to the right and up 2. The new starting and ending points are $(\pi/4, 3)$ and $(9\pi/4, 3)$. Next divide the interval $[\pi/4, 9\pi/4]$ into four equal parts and locate the minimum point and the two inflection points as in Fig. 2.18. Draw one cycle of the graph through the five key points as in Fig. 2.18. Continue the pattern for another cycle to the right or left of the first cycle. Since the minimum $y$-coordinate is 1 and the maximum is 3, the range of the function is the interval $[1, 3]$.

A calculator graph of $y = \cos(x - \pi/4) + 2$ is shown in Fig. 2.19 and it supports our conclusions.

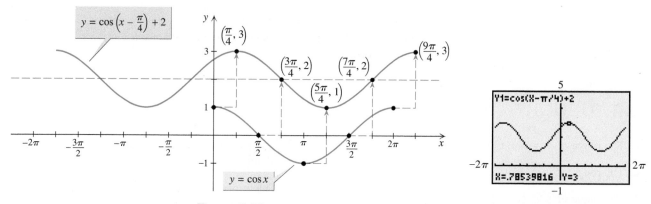

**Figure 2.18**        **Figure 2.19**

**TRY THIS.** Graph one cycle of $y = \sin(x + \pi/4) + 1$ and find the phase shift and range. ∎

In the next example we start with a graph and write a sine function for it.

**EXAMPLE 9** Constructing a sine function from a graph

Write an equation of the form $y = A \sin(x - C) + D$ whose graph is the sine wave shown in Fig. 2.20.

**Figure 2.20**

**Figure 2.21**

### Solution

Because the $y$-coordinates range from $-1$ to $5$, the amplitude of the sine curve is $3$. Now concentrate on the cycle of the function that starts at $(\pi/3, 2)$ and ends at $(7\pi/3, 2)$. We choose this cycle because it duplicates the behavior of $y = \sin(x)$ on its fundamental cycle $[0, 2\pi]$. That is, from $(\pi/3, 2)$ the graph rises to its maximum, then goes down to its minimum, then ends the cycle at $(7\pi/3, 2)$. So if we stretch $y = \sin(x)$ by a factor of 3, then shift it to the right $\pi/3$ and up 2, we will get the given sine wave. So an equation is $y = 3\sin(x - \pi/3) + 2$.

 The calculator graph of $y = 3\sin(x - \pi/3) + 2$ in Fig. 2.21 supports this conclusion.

**TRY THIS.**  The five key points for one cycle of a sine wave are $(\pi/4, 1)$, $(3\pi/4, 6)$, $(5\pi/4, 1)$, $(7\pi/4, -4)$, and $(9\pi/4, 1)$. Find an equation for the curve. ∎

Note that the equation in Example 9 is the simplest equation for the given sine curve, but it is not unique. Any right or left shift by a multiple of $2\pi$ gives an equivalent equation. For example, $y = 3\sin(x - 7\pi/3) + 2$ is an equivalent equation because $\pi/3 + 2\pi = 7\pi/3$. Another equation for the curve could be obtained by using a reflection $(A = -3)$ and then shifting an appropriate amount to the right and up. You should write this equation and then use a graphing calculator to verify that it is correct.

## FOR THOUGHT...  True or False? Explain.

**1.** The range of $y = \sin(x)$ is $[-1, 1]$.

**2.** The range of $y = 4\sin(x) + 3$ is $[-4, 4]$.

**3.** The range of $y = \cos(x) - 5$ is $[-6, -4]$.

**4.** The graph of $y = \sin(x + \pi/6)$ has a phase shift of $\pi/6$.

**5.** The graph of $y = \sin(x + \pi/6)$ lies $\pi/6$ to the right of the graph of $y = \sin(x)$.

**6.** The points $(5\pi/6, 0)$ and $(11\pi/6, 0)$ are on the graph of $y = \cos(x - \pi/3)$.

**7.** The graphs of $y = \sin(x)$ and $y = \cos(x + \pi/2)$ are identical.

**8.** The minimum value of the function $f(x) = 3\sin(x)$ is $-1$.

**9.** The maximum value of the function $y = -2\cos(x) + 4$ is 6.

**10.** If $(-\pi/6, 0)$ is moved $\pi/3$ units to the right its new coordinates are $(\pi/6, 0)$.

## 2.1 EXERCISES

### CONCEPTS

*Fill in the blank.*

**1.** If $\alpha$ is an angle in standard position whose terminal side intersects the unit circle at $(x, y)$ then $y =$ _____ and $x =$ _____.

**2.** Any function of the form $y = A\sin[B(x - C)] + D$ with $A \neq 0$ and $B \neq 0$ is a(n) _____ function.

**3.** The graph of any sine function is a(n) _____.

**4.** If $a$ is a nonzero constant such that $f(x) = f(x + a)$ for every $x$ in the domain of the function $f$, then $f$ is a(n) _____ function.

**5.** The graph of $y = \sin x$ over $[0, 2\pi]$ is the _____ of $y = \sin x$.

**6.** The _____ of a sine wave is the absolute value of half the difference between the maximum and minimum $y$-coordinates on the wave.

**7.** The _____ of the graph of $y = \sin(x - C)$ is $C$.

**8.** Any function of the form $y = A \cos[B(x - C)] + D$ with $A \neq 0$ and $B \neq 0$ is a(n) _____ function.

**9.** The five key points on the graph of the fundamental cycle of $y = \sin x$ are the _____ point, the _____ point, the _____ point, the _____ point, and the _____ point.

**10.** The fundamental cycle of $y = \cos x$ contains two _____ points, two _____ points, and one _____ point.

### SKILLS

*Find the exact value of each trigonometric function using the unit circle.*

**11.** $\sin 0$

**12.** $\cos 0$

**13.** $\tan(\pi/3)$

**14.** $\cot(\pi/6)$

**15.** $\sin(\pi/6)$

**16.** $\cos(\pi/3)$

**17.** $\sec(\pi/3)$

**18.** $\csc(\pi/3)$

**19.** $\cos(\pi/2)$

**20.** $\sin(\pi/2)$

**21.** $\tan(-\pi)$

**22.** $\cot(-3\pi/2)$

**23.** $\csc(-5\pi/6)$

**24.** $\sec(-2\pi/3)$

**25.** $\cos(-3\pi/4)$

**26.** $\sin(-3\pi/2)$

*Find the coordinates of each given point after it is moved $\pi/4$ units to the right.*

**27.** $(0, 0)$

**28.** $(\pi/4, 1)$

**29.** $(\pi/2, 3)$

**30.** $(3\pi/4, -1)$

**31.** $(-\pi/2, -1)$

**32.** $(-\pi/4, -2)$

**33.** $(\pi, 0)$

**34.** $(2\pi, 0)$

*Find the coordinates of each given point after it is moved $\pi/3$ units to the left.*

**35.** $(\pi/3, 0)$

**36.** $(2\pi/3, -1)$

**37.** $(\pi, 1)$

**38.** $(2\pi, 4)$

**39.** $(\pi/2, -1)$

**40.** $(\pi/4, 2)$

**41.** $(-\pi, 1)$

**42.** $(-\pi/2, -1)$

*Find the coordinates of each given point after it is moved $\pi/6$ units to the right and 2 units upward.*

**43.** $(\pi, -1)$

**44.** $(\pi/6, -2)$

**45.** $(\pi/2, 0)$

**46.** $(\pi/3, -1)$

**47.** $(-3\pi/2, 1)$

**48.** $(-\pi/2, 2)$

**49.** $(2\pi, -4)$

**50.** $(-\pi, 5)$

*Determine the point that lies midway between the two given points.*

**51.** $(\pi, 0)$ and $(2\pi, 0)$

**52.** $(0, -2)$ and $(\pi/3, -2)$

**53.** $(0, 2)$ and $(\pi/4, 2)$

**54.** $(\pi/2, 1)$ and $(\pi, 1)$

**55.** $(\pi/6, 1)$ and $(\pi/2, 1)$

**56.** $(\pi/4, 2)$ and $(\pi/2, 2)$

**57.** $(\pi/3, -4)$ and $(\pi/2, -4)$

**58.** $(3\pi/4, 5)$ and $(\pi, 5)$

*Determine the coordinates of points $P$, $Q$, $R$, and $S$ on each given sine wave.*

**59.**

**60.**

**61.**

**62.**

**63.**

**64.**

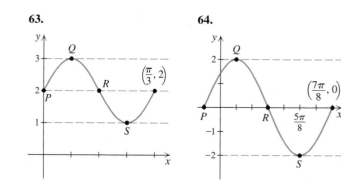

*Determine the amplitude, period, phase shift, and range for each function.*

**65.** $y = -2 \cos x$

**66.** $y = -4 \cos x$

**67.** $f(x) = \cos(x - \pi/2)$

**68.** $f(x) = \sin(x + \pi/2)$

**69.** $y = -2 \sin(x + \pi/3)$

**70.** $y = -3 \sin(x - \pi/6)$

*Determine the amplitude, phase shift, and range for each function. Sketch at least one cycle of the graph and label the five key points on one cycle as done in the examples.*

**71.** $y = -\sin x$

**72.** $y = -\cos x$

**73.** $y = -3 \sin x$

**74.** $y = 4 \sin x$

**75.** $y = \dfrac{1}{2}\cos x$

**76.** $y = \dfrac{1}{3}\cos x$

**77.** $y = \sin(x + \pi)$

**78.** $y = \cos(x - \pi)$

**79.** $y = \cos(x - \pi/3)$

**80.** $y = \cos(x + \pi/4)$

**81.** $f(x) = \cos(x) + 2$

**82.** $f(x) = \cos(x) - 3$

**83.** $y = -\sin(x) - 1$

**84.** $y = -\sin(x) + 2$

**85.** $y = \sin(x + \pi/4) + 2$

**86.** $y = \sin(x - \pi/2) - 2$

**87.** $y = 2\cos(x + \pi/6) + 1$

**88.** $y = 3\cos(x + 2\pi/3) - 2$

**89.** $f(x) = -2\sin(x - \pi/3) + 1$

**90.** $f(x) = -3\cos(x + \pi/3) - 1$

*Write an equation of the form $y = A \sin(x - C) + D$ whose graph is the given sine wave.*

**91.**

**92.**

**93.**

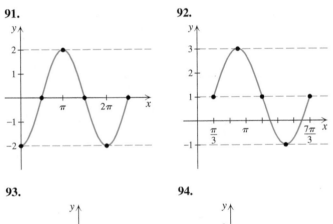

**94.**

*Write an equation of the form $y = A \cos(x - C) + D$ whose graph is the given sine wave.*

**95.**

**96.**

**97.**

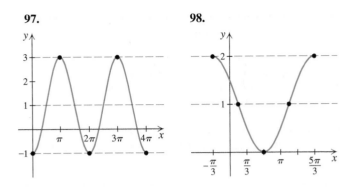

**98.**

*Find the equation of each sine wave in its final position.*

**99.** The graph of $y = \sin(x)$ is shifted $\pi/4$ units to the right.

**100.** The graph of $y = \cos(x)$ is shifted $\pi/6$ units to the right.

**101.** The graph of $y = \sin(x)$ is shifted $\pi/2$ units to the left.

**102.** The graph of $y = \cos(x)$ is shifted $\pi/3$ units to the left.

**103.** The graph of $y = \cos(x)$ is shifted $\pi/5$ units to the right and reflected in the *x*-axis.

**104.** The graph of $y = \sin(x)$ is shifted $\pi/7$ units to the left and reflected in the *x*-axis.

**105.** The graph of $y = \cos(x)$ is shifted $\pi/8$ units to the right, reflected in the *x*-axis, and then translated upward 2 units.

**106.** The graph of $y = \sin(x)$ is reflected in the *x*-axis, shifted $\pi/9$ units to the left, and then translated downward 3 units.

**107.** The graph of $y = \cos(x)$ is stretched by a factor of 3, reflected in the *x*-axis, shifted $\pi/4$ units to the left, and then translated downward 5 units.

**108.** The graph of $y = \sin(x)$ is reflected in the *x*-axis, shrunk by a factor of $\frac{1}{2}$, shifted $\pi/3$ units to the right, and then translated upward 4 units.

## WRITING/DISCUSSION

**109.** *The Role of the Parameters* The numbers $A$, $C$, and $D$ in $y = A \sin(x - C) + D$ are called *parameters*. Explain how each of these parameters affects the location or shape of the graph.

**110.** *Even or Odd* An even function is one for which $f(-x) = f(x)$ and an odd function is one for which $f(-x) = -f(x)$. Determine whether $f(x) = \sin(x)$ and $f(x) = \cos(x)$ are even or odd functions and explain your answers.

## REVIEW

**111.** Find the smallest positive angle in radians that is coterminal with $-23\pi/6$.

**112.** Convert the radian measure $7\pi/4$ to degrees.

**113.** The earth travels around the sun on a path that is roughly circular with a radius of 93 million miles. How fast is the earth moving through space? Round to the nearest thousand miles per hour.

**114.** Evaluate without a calculator. Some of these expressions are undefined.

   **a.** $\cos(\pi)$           **b.** $\sin(3\pi/4)$

   **c.** $\tan(\pi/3)$          **d.** $\tan(\pi/2)$

   **e.** $\sec(2\pi/3)$        **f.** $\csc(\pi)$

   **g.** $\cot(5\pi/6)$        **h.** $\sin(-\pi/4)$

**115.** Use a calculator to find the acute angle $\alpha$ (to the nearest tenth of a degree) that satisfies $\sin(\alpha) = 0.36$.

**116.** Find $\sin(\alpha)$ given that $\cos(\alpha) = 1/3$ and $\alpha$ lies in quadrant IV.

**OUTSIDE THE BOX**

**117.** *Tangent Circles* The three large circles in the accompanying diagram are tangent to each other and each has radius 1. The small circle in the middle is tangent to each of the three large circles. Find its radius.

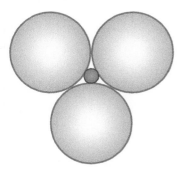

**Figure for Exercise 117**

**118.** *Sines and Cosines* If $\sin x = 3 \cos x$ then what is the value of $\sin x \cos x$?

---

## 2.1 POP QUIZ

**1.** Determine the amplitude, period, phase shift, and range for $y = -5\sin(x + 2\pi/3)$.

**2.** List the coordinates for the five key points for one cycle of $y = 3\sin(x)$.

**3.** If $y = \cos(x)$ is shifted $\pi/2$ to the right, reflected in the $x$-axis, and shifted 3 units upward, then what is the equation of the curve in its final position?

**4.** What is the range of $f(x) = -4\sin(x - 3) + 2$?

**5.** The five key points for one cycle of a sine wave are $(\pi/2, 0)$, $(\pi, -3)$, $(3\pi/2, 0)$, $(2\pi, 3)$, and $(5\pi/2, 0)$. Find an equation for the curve.

---

# LINKING
## concepts...    For Individual or Group Explorations

## Taylor Polynomials

*In calculus it is proved that many functions can be approximated to any degree of accuracy by polynomial functions called Taylor polynomials. The Taylor polynomials for* $\sin(x)$ *are*

$$y = x, \qquad y = x - \frac{x^3}{3!}, \qquad y = x - \frac{x^3}{3!} + \frac{x^5}{5!}, \qquad y = x - \frac{x^3}{3!} + \frac{x^5}{5!} - \frac{x^7}{7!},$$

*and so on. As shown in the accompanying figure, the seventh-degree Taylor polynomial approximates the sine curve very well between* $-\pi$ *and* $\pi$.

   **a)** Graph $y_1 = x$ and $y_2 = \sin(x)$ on a graphing calculator and determine approximately the values for $x$ for which $|y_1 - y_2| < 0.1$.

*(continued)*

**b)** Repeat part (a) using the fifth-degree, ninth-degree, and nineteenth-degree Taylor polynomials for $y_1$.

**c)** The Taylor polynomials for $\cos(x)$ are $y = 1$, $y = 1 - x^2/2!$, $y = 1 - x^2/2! + x^4/4!$, $y = 1 - x^2/2! + x^4/4! - x^6/6!$, and so on. Graph the zero-degree, fourth-degree, eighth-degree, and eighteenth-degree Taylor polynomials and determine approximately the intervals on which each of them differs from $\cos(x)$ by less than 0.1. Compare your answers to the answers obtained for parts (a) and (b).

**d)** Suppose that each operation $(+, -, \times, \div)$ takes one second. Explain how to find $\sin(x)$ to the nearest tenth for any real number $x$ in the shortest amount of time by using Taylor polynomials.

# 2.2 The General Sine Wave

The general sine wave equations are

$$y = A \sin[B(x - C)] + D \quad \text{and} \quad y = A \cos[B(x - C)] + D.$$

In the last section we studied these equations with $B = 1$. In this section we will see what happens to the sine wave when $B \neq 1$.

## Changing the Period

The period of a sine or cosine function with $B = 1$ is $2\pi$. If $B \neq 1$ the period is different.

### EXAMPLE 1 Changing the period

Graph two cycles of $y = \sin(2x)$ and determine the period of the function.

#### Solution

The graph of $y = \sin x$ completes its fundamental cycle for $0 \leq x \leq 2\pi$. To determine the period for $y = \sin(2x)$ we replace $x$ with $2x$ in the inequality $0 \leq x \leq 2\pi$:

$$0 \leq 2x \leq 2\pi$$

$$\frac{0}{2} \leq \frac{2x}{2} \leq \frac{2\pi}{2}$$

$$0 \leq x \leq \pi$$

So one cycle of $y = \sin(2x)$ is completed between 0 and $\pi$. Note that the period $\pi$ can be obtained by dividing the period for $y = \sin(x)$ by 2. Divide the interval $[0, \pi]$ into four equal parts. Midway between the starting point $(0, 0)$ and the ending point $(\pi, 0)$ is an inflection point $(\pi/2, 0)$. Now locate the maximum point at $(\pi/4, 1)$ and the minimum point at $(3\pi/4, -1)$. Draw one cycle $y = \sin(2x)$ through these five key points, as shown in Fig. 2.22. Draw a second cycle to the left or right of the first cycle.

The calculator graph in Fig. 2.23 supports our conclusions. The scale on this graph is $\pi/4$.

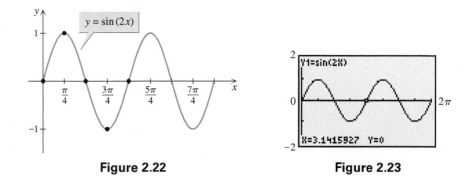

**Figure 2.22**                  **Figure 2.23**

**TRY THIS.** Graph two cycles of $y = \cos(2x)$ and determine the period. ■

Both $y = \sin x$ and $y = \cos x$ have period $2\pi$ and complete one cycle for $0 \le x \le 2\pi$. To determine the period of $y = \sin(Bx)$ or $y = \cos(Bx)$ for $B > 0$ replace $x$ in the inequality $0 \le x \le 2\pi$ with $Bx$:

$$0 \le Bx \le 2\pi$$

$$\frac{0}{B} \le \frac{Bx}{B} \le \frac{2\pi}{B}$$

$$0 \le x \le \frac{2\pi}{B}$$

So one cycle is completed between $0$ and $2\pi/B$.

**Theorem: Period of**
$y = \sin(Bx)$ **and** $y = \cos(Bx)$

The period $P$ of $y = \sin(Bx)$ and $y = \cos(Bx)$ for $B > 0$ is given by

$$P = \frac{2\pi}{B}.$$

If $B$ is a multiple of $\pi$, then $\pi$ cancels out in the expression $2\pi/B$ and we can obtain a period that is a natural number as is shown in the next example.

**EXAMPLE 2** **A period that is not a multiple of $\pi$**

Determine the period of $y = \cos\left(\frac{\pi}{2}x\right)$ and graph two cycles of the function.

**Solution**

We could find the period by solving the inequality $0 \le \frac{\pi}{2}x \le 2\pi$. It is simpler to use $P = 2\pi/B$ with $B = \pi/2$:

$$P = \frac{2\pi}{\pi/2} = 2\pi \cdot \frac{2}{\pi} = 4$$

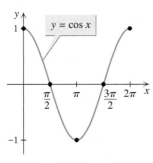

**Figure 2.24**

So one cycle of $y = \cos\left(\frac{\pi}{2}x\right)$ is completed on the interval $[0, 4]$. Divide this interval into four equal parts. Recall the fundamental cycle of $y = \cos x$ in Fig. 2.24. The new starting point is $(0, 1)$ and the new ending point is $(4, 1)$, and these are maximum points. The minimum point is midway between these at $(2, -1)$. Midway between the maximum points and the minimum points are the inflection points $(1, 0)$ and $(3, 0)$. Draw one cycle of $y = \cos\left(\frac{\pi}{2}x\right)$ through these five key points, as shown in Fig. 2.25 on the next page. Continue this pattern from 4 to 8 to get a second cycle.

The calculator graph in Fig. 2.26 supports this conclusion. The scale on this graph is 1.

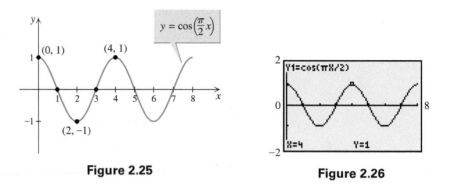

**Figure 2.25**        **Figure 2.26**

**TRY THIS.** Graph two cycles of $y = \cos(\pi x)$ and determine the period. ■

## Graphing a General Sine Wave

We can use any combination of translating, reflecting, phase shifting, stretching, shrinking, or period changing in a single trigonometric function.

**The General Sine Wave**

The graph of

$$y = A \sin[B(x - C)] + D \qquad \text{or} \qquad y = A \cos[B(x - C)] + D$$

is a sine wave with amplitude $|A|$, period $2\pi/B$ ($B > 0$), phase shift $C$, and vertical translation $D$. The phase shift is to the right for $C > 0$ and to the left for $C < 0$. The vertical translation is upward for $D > 0$ and downward for $D < 0$.

We can assume that $B > 0$, because any general sine or cosine function can be rewritten in this form with $B > 0$ using identities, which we will study in Chapter 3. Notice that $A$ and $B$ affect the shape of the curve whereas $C$ and $D$ determine its location. Use the following procedure when graphing a general sine wave.

**PROCEDURE**

### Graphing $y = A \sin[B(x - C)] + D$

1. Determine the period $2\pi/B$, amplitude $|A|$, phase shift $C$, and vertical translation $D$.
2. The starting point is $(C, D)$ and the ending point is $(C + 2\pi/B, D)$.
3. Midway between the starting and ending points is an inflection point.
4. Midway again, locate the maximum and minimum points. If $A < 0$ the graph is a reflection and the minimum point comes first.
5. Sketch one cycle of the curve through the five key points.

Before you start to graph a transformation of $y = \sin x$ you should have a mental picture of the fundamental cycle of $y = \sin x$. If you don't yet remember this image, look back to Fig. 2.4 to get it in mind. See the Function Gallery on page 127 for a visual summary of this procedure.

**EXAMPLE 3** A transformation of $y = \sin(x)$

Determine the amplitude, period, and phase shift, and sketch two cycles of $y = 2 \sin(3x + \pi) + 1$.

**Solution**

First rewrite the function in the form $y = A \sin[B(x - C)] + D$ by factoring 3 out of $3x + \pi$:

$$y = 2 \sin\left[3\left(x + \frac{\pi}{3}\right)\right] + 1$$

From this equation we get $A = 2$, $B = 3$, $C = -\pi/3$, and $D = 1$. The amplitude is 2. The period is $2\pi/3$. The phase shift is $-\pi/3$. The starting point is $(-\pi/3, 1)$ and the ending point is $(-\pi/3 + 2\pi/3, 1)$ or $(\pi/3, 1)$. Midway between these is the inflection point $(0, 1)$. Midway again are the maximum and minimum points $(-\pi/6, 3)$ and $(\pi/6, -1)$. Draw one cycle of the graph through these five points, as shown in Fig. 2.27, and continue this pattern to the right for another cycle. Note that the tick marks on the $x$-axis are at multiples of $\pi/6$ and the dashed line is a new "$x$-axis" corresponding to the vertical translation. Since the amplitude is 2, the graph ranges from 2 units above to 2 units below this new axis.

**Figure 2.27**

**Figure 2.28**

The calculator graph in Fig. 2.28 supports our conclusions about the amplitude, period, and phase shift. Note that it is easier to obtain the amplitude, period, and phase shift from the equation than from the calculator graph.

**TRY THIS.** Determine the amplitude, period, and phase shift, and graph one cycle of $y = 3 \sin(2x - \pi) - 1$. ∎

The procedure for graphing a transformation of $y = \cos x$ is similar to that for $y = \sin x$. Again, you should recall the fundamental cycle of $y = \cos x$ before you begin.

**PROCEDURE**

**Graphing $y = A \cos[B(x - C)] + D$**

1. Determine the period $2\pi/B$, amplitude $|A|$, phase shift $C$, and vertical translation $D$.
2. The starting point is $(C, D + A)$ and the ending point is $(C + 2\pi/B, D + A)$. These are maximum points if $A > 0$ and minimum points if $A < 0$.
3. Midway between the starting and ending points is a minimum point if $A > 0$ or a maximum point if $A < 0$.
4. Midway again, locate the two inflection points.
5. Sketch one cycle of the curve through the five key points.

See the Function Gallery on page 127 for a visual summary of this procedure.

**EXAMPLE 4** ▸ **A transformation of $y = \cos(x)$**

Determine the amplitude, period, and phase shift, and sketch one cycle of $y = -3\cos(2x - \pi) - 1$.

**Solution**

Rewrite the function in the general form $y = A\cos[B(x - C)] + D$ by factoring $2x - \pi$:

$$y = -3\cos\left[2\left(x - \frac{\pi}{2}\right)\right] - 1$$

From this equation we get $A = -3$, $B = 2$, $C = \pi/2$, and $D = -1$. The amplitude is $|-3|$ or 3. The period is $2\pi/2$ or $\pi$. The phase shift is $\pi/2$. The starting point is $(\pi/2, -4)$ and the ending point is $(\pi/2 + \pi, -4)$ or $(3\pi/2, -4)$. Since $A < 0$, these are minimum points. Midway between these is the maximum point $(\pi, 2)$. Midway again are the inflection points $(3\pi/4, -1)$ and $(5\pi/4, -1)$. Draw one cycle of the graph through these five points, as shown in Fig. 2.29. Note that the tick marks on the $x$-axis are at multiples of $\pi/4$ and the dashed line is a new "$x$-axis" corresponding to the vertical translation.

The calculator graph in Fig. 2.30 supports our conclusions.

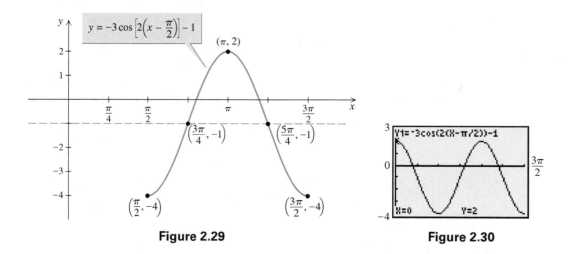

Figure 2.29                                    Figure 2.30

**TRY THIS.** Determine the amplitude, period, and phase shift, and graph one cycle of $y = -2\cos(2x + \pi) + 1$.    ∎

# Frequency

Sine waves are used to model physical phenomena such as radio, sound, or light waves. A high-frequency radio wave is a wave that has a large number of cycles per second. If the $x$-axis is a time axis, then the period of a sine wave is the amount of time required for the wave to complete one cycle, and the reciprocal of the period is the number of cycles per unit of time. For example, the sound wave for middle C on a piano completes 262 cycles per second. The period of the wave is $1/262$ second, which means that one cycle is completed in $1/262$ second.

**Definition: Frequency**    The **frequency** $F$ of a sine wave with period $P$ is defined by $F = 1/P$.

**Figure 2.31**

**Figure 2.32**

EXAMPLE **5**    **Frequency of a sine wave**

Find the frequency of the sine wave given by $y = \sin(524\pi x)$ and sketch one cycle of the graph of the sine wave.

**Solution**

First find the period:

$$P = \frac{2\pi}{524\pi} = \frac{1}{262} \approx 0.004$$

Since $F = 1/P$, the frequency is 262. A sine wave with this frequency completes 262 cycles for $x$ in the interval $[0, 1]$ or one cycle in the interval $[0, 1/262]$. We will not draw a graph showing 262 cycles in an interval of length one, but we can see the cycle that occurs in the interval $[0, 1/262]$ in Fig. 2.31.
The calculator graph in Fig. 2.32 supports our conclusions.

**TRY THIS.** Find the frequency of the sine wave given by $y = \cos(100\pi x)$.    ∎

Note that if $B$ is a large positive number in $y = \sin(Bx)$ or $y = \cos(Bx)$, then the period is short and the frequency is high.

The following Function Gallery summarizes what we have learned about the sine and cosine functions.

**FUNCTION GALLERY**

## THE SINE AND COSINE FUNCTIONS

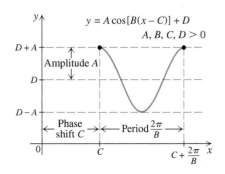

## Sinusoidal Curve Fitting

We now consider the problem of constructing an equation for a sinusoid that passes through some given points. If we know the five key points on one cycle of a sinusoid, then we can write an equation for the curve in the form $y = A \sin[B(x - C)] + D$ or $y = A \cos[B(x - C)] + D$. For simplicity, we will just write a sine equation in the next example.

**EXAMPLE 6** Constructing a sine function

Ten minutes after a furnace is turned on, the temperature in a room reaches 74°F and the furnace turns off. It takes two minutes for the room to cool to 70°F and two minutes for the furnace to bring it back to 74°F as shown in the table.

| Time (minutes) | 10 | 11 | 12 | 13 | 14 | 15 | 16 | 17 | 18 |
|---|---|---|---|---|---|---|---|---|---|
| Temperature (°F) | 74 | 72 | 70 | 72 | 74 | 72 | 70 | 72 | 74 |

Assuming that the temperature (after time 10) is a sine function of the time, construct the function and graph it.

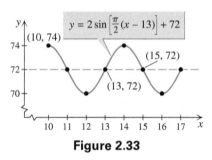

**Figure 2.33**

### Solution

Because the temperature ranges from 70° to 74°, the amplitude of the sine curve is 2. Because the temperature goes from its maximum of 74° back to 74° in 4 minutes, the period is 4. Since the period is $2\pi/B$, we can solve $2\pi/B = 4$ to get $B = \pi/2$. Now concentrate on one cycle of the function. Starting at $(13, 72)$ the temperature increases to its maximum of 74°, decreases to its minimum of 70°, and then ends in the middle at $(17, 72)$. We choose this cycle because it duplicates the behavior of $y = \sin(x)$ on its fundamental cycle $[0, 2\pi]$. So shift $y = \sin(x)$ to the right 13 and up 72 to get $y = 2 \sin\left[\frac{\pi}{2}(x - 13)\right] + 72$. Its graph is shown in Fig. 2.33. Changing this equation with any right or left shift by a multiple of 4 gives an equivalent equation. For example, $y = 2 \sin\left[\frac{\pi}{2}(x - 1)\right] + 72$ is an equivalent equation.

**TRY THIS.** The points $(1, 2)$, $(2, 5)$, $(3, 2)$, $(4, -1)$, and $(5, 2)$ are the five key points on one cycle of a sine wave. Find an equation for the curve. ∎

If we have real data, the points will usually not fit exactly on a sine curve as they did in the last example. In this case we can use the sinusoidal regression feature of a graphing calculator to find a sine curve that approximates the data.

**EXAMPLE 7** Constructing a sine function using a calculator

The times of sunrise in Miami, Florida, on the first of every month for one year are shown in the following table (U.S. Naval Observatory, http://aa.usno.navy.mil). The time is the number of minutes after 5 A.M.

| Month | Time | Month | Time |
|---|---|---|---|
| 1 | 127 | 7 | 33 |
| 2 | 125 | 8 | 47 |
| 3 | 104 | 9 | 61 |
| 4 | 72 | 10 | 73 |
| 5 | 44 | 11 | 90 |
| 6 | 29 | 12 | 111 |

Use the sinusoidal regression feature of a graphing calculator to construct an equation that fits the data. Graph the data and the curve on your graphing calculator. Find the period from the equation.

**Figure 2.34**

**Solution**

Enter the data and use the sinusoidal regression feature (SinReg) to get $y = 49.92 \sin(0.47x + 1.42) + 81.54$ where $x$ is the month and $y$ is the number of minutes after 5 a.m. Fig. 2.34 shows the data and the sine curve. The period is $2\pi/0.47$ or approximately 13.4 months. Since the period should be 12 months, the sine curve does not fit the data very well. See Exercises 59 and 60 for some data that look more like a sine curve.

**TRY THIS.** Use sinusoidal regression to find an equation that fits the points $(1, 2.2)$, $(2, 4.9)$, $(3, 1.9)$, $(4, -1.1)$, and $(5, 2.1)$. ∎

## FOR THOUGHT... True or False? Explain.

**1.** The period of $y = \sin(4x)$ is $\pi/2$.

**2.** The period of $y = \cos(2\pi x)$ is $\pi$.

**3.** The period of $y = -3 \sin(\pi x - \pi) + 1$ is 2.

**4.** The period of $y = \sin(0.1\pi x)$ is 20.

**5.** The graph of $y = \sin(2x + \pi/6)$ has a period of $\pi$ and a phase shift $\pi/6$.

**6.** The graph of $y = \cos(2x + \pi/4)$ has phase shift $\pi/8$.

**7.** The frequency of $y = \sin x$ is $1/(2\pi)$.

**8.** The frequency of $y = \cos(\pi x)$ is $1/2$.

**9.** The graphs of $y = -\cos(x)$ and $y = \sin(x + \pi/2)$ are identical.

**10.** The graphs of $y = -\sin(x)$ and $y = \cos(x - 3\pi/2)$ are identical.

## 2.2 EXERCISES

### CONCEPTS

*Fill in the blank.*

**1.** The _____ of $y = \sin(Bx)$ and $y = \cos(Bx)$ is $2\pi/B$.

**2.** The _____ of a sine wave with period $P$ is $1/P$.

**3.** The graph of $y = A \sin[B(x - C)] + D$ is a sine wave with _____ $|A|$.

**4.** The graph of $y = A \cos[B(x - C)] + D$ is a sine wave with _____ $C$.

### SKILLS

*Determine amplitude, period, and phase shift for each function.*

**5.** $y = 3 \sin(4x)$

**6.** $y = -\cos(x/2) + 3$

**7.** $y = -2 \cos\left(2x + \dfrac{\pi}{2}\right) - 1$

**8.** $y = 4 \cos(3x - 2\pi)$

**9.** $y = -2 \sin(\pi x - \pi)$

**10.** $y = \sin\left(\dfrac{\pi}{2}x + \pi\right)$

*Sketch at least one cycle of the graph of each function. Determine the period, the phase shift, and the range of the function. Label the five key points on the graph of one cycle as done in the examples.*

**11.** $y = \sin(3x)$

**12.** $y = \cos(x/3)$

**13.** $y = -\sin(2x)$

**14.** $y = -\cos(3x)$

**15.** $y = \cos(4x) + 2$

**16.** $y = \sin(3x) - 1$

**17.** $y = 2 - \sin(x/4)$

**18.** $y = 3 - \cos(x/5)$

**19.** $y = \sin\left(\dfrac{\pi}{3}x\right)$

**20.** $y = \sin\left(\dfrac{\pi}{4}x\right)$

**21.** $f(x) = \sin\left[2\left(x - \dfrac{\pi}{2}\right)\right]$

**22.** $f(x) = \sin\left[3\left(x + \dfrac{\pi}{3}\right)\right]$

**23.** $f(x) = \sin\left(\dfrac{\pi}{2}x + \dfrac{3\pi}{2}\right)$

**24.** $f(x) = \cos\left(\dfrac{\pi}{3}x - \dfrac{\pi}{3}\right)$

**25.** $y = 2 \cos\left[2\left(x + \dfrac{\pi}{6}\right)\right] + 1$

**26.** $y = 3 \cos\left[4\left(x - \dfrac{\pi}{2}\right)\right] - 1$

**27.** $y = -\dfrac{1}{2}\sin\left[3\left(x - \dfrac{\pi}{6}\right)\right] - 1$

**28.** $y = -\dfrac{1}{2}\sin\left[4\left(x + \dfrac{\pi}{4}\right)\right] + 1$

*Write an equation of the form $y = A\sin[B(x - C)] + D$ whose graph is the given sine wave.*

**29.**

**30.**

**31.**

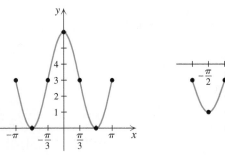

**32.**

*Let $f(x) = \sin(x)$, $g(x) = x - \pi/4$, and $h(x) = 3x$. Find each of the following.*

**33.** $g(\pi/4)$

**34.** $g(\pi/2)$

**35.** $f(g(\pi/4))$

**36.** $f(g(\pi/2))$

**37.** $h(f(g(\pi/4)))$

**38.** $h(f(g(\pi/2)))$

**39.** $f(g(x))$

**40.** $f(h(x))$

**41.** $h(f(g(x)))$

**42.** $f(h(g(x)))$

## MODELING

*Solve each problem.*

**43.** What is the frequency of the sine wave determined by $y = \sin(200\pi x)$, where $x$ is time in seconds?

**44.** What is the frequency of the sine wave determined by $y = \cos(0.001\pi x)$, where $x$ is time in seconds?

**45.** If the period of a sine wave is 0.025 hour, then what is the frequency?

**46.** If the frequency of a sine wave is 40,000 cycles per second, then what is the period?

**47.** *Motion of a Spring* A weight hanging on a vertical spring is set in motion with a downward velocity of 6 cm/sec from its equilibrium position. A formula that gives the location of the weight in centimeters as a function of the time $t$ in seconds is $x = 3\sin(2t)$. Find the amplitude and period of the function and sketch its graph for $t$ in the interval $[0, 2\pi]$.

**48.** *Motion of a Spring* A weight hanging on a vertical spring is set in motion with an upward velocity of 4 cm/sec from its equilibrium position. A formula that gives the location of the weight in centimeters as a function of the time $t$ in seconds is $x = -\dfrac{4}{\pi}\sin(\pi t)$. Find the period of the function and sketch its graph for $t$ in the interval $[0, 4]$.

**49.** *Sunspots* Astronomers have been recording sunspot activity for over 130 years. The number of sunspots per year varies periodically over time as shown in the graph. What is the approximate period of this periodic function?

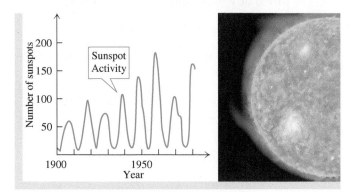

**Figure for Exercise 49**

**50.** *First Pulsar* In 1967, Jocelyn Bell, a graduate student at Cambridge University, England, found the peculiar pattern shown in the graph on a paper chart from a radio telescope. She had made the first discovery of a pulsar, a very small neutron star that emits beams of radiation as it rotates as fast as 1000 times per second. From the graph shown here, estimate the period of the first discovered pulsar, now known as CP 1919.

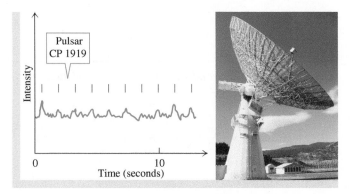

**Figure for Exercise 50**

**51. Lung Capacity** The volume of air $v$ in cubic centimeters in the lungs of a certain distance runner is modeled by the equation $v = 400 \sin(60\pi t) + 900$, where $t$ is time in minutes.
  **a.** What are the maximum and minimum volumes of air in the runner's lungs?

  **b.** How many breaths does the runner take per minute?

**52. Blood Velocity** The velocity $v$ of blood at a valve in the heart of a certain rodent is modeled by the equation

$$v = -4 \cos(6\pi t) + 4,$$

where $v$ is in centimeters per second and $t$ is time in seconds.
  **a.** What are the maximum and minimum velocities of the blood at this valve?

  **b.** What is the rodent's heart rate in beats per minute?

**53. Periodic Revenue** For the past three years, the manager of The Toggery Shop has observed that revenue reaches a high of about $40,000 in December and a low of about $10,000 in June, and that a graph of the revenue looks like a sinusoid. If the months are numbered 1 through 36 with 1 corresponding to January, then what are the period, amplitude, and phase shift for this sinusoid? What is the vertical translation? Write a formula for the curve and find the approximate revenue for April.

**54. Periodic Cost** For the past three years, the manager of The Toggery Shop has observed that the utility bill reaches a high of about $500 in January and a low of about $200 in July, and the graph of the utility bill looks like a sinusoid. If the months are numbered 1 through 36 with 1 corresponding to January, then what are the period, amplitude, and phase shift for this sinusoid? What is the vertical translation? Write a formula for the curve and find the approximate utility bill for November.

**55. Discovering a Planet** In April 26, 1997, astronomers announced the discovery of a new planet orbiting the star Rho Coronae Borealis (Sky & Telescope, July 1997). The astronomers deduced the existence of the planet by measuring the change in line-of-sight velocity of the star over a period of ten months. The measurements appear to fall along a sine wave as shown in the accompanying figure.

  **a.** What are the period, amplitude, and equation for the sine wave?

  **b.** How many Earth days does it take for the planet to orbit Rho Coronae Borealis?

  **c.** Use the equation to estimate the change in line-of-sight velocity on day 36.

  **d.** On day 36 do you think that the planet is between the earth and Rho or that Rho is between the earth and the planet?

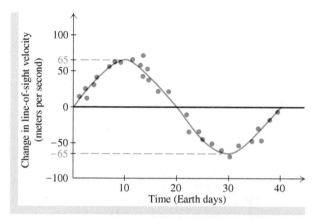

**Figure for Exercise 55**

**56. Jupiter's Satellites** The accompanying graph shows the positions in June of Ganymede, Callisto, Io, and Europa, the four bright satellites of Jupiter (Sky & Telescope, April 1997). Jupiter itself is the center horizontal bar. The paths of the satellites are nearly sine waves because the orbits of the satellites are nearly circular. The distances from Jupiter to Io, Europa, Ganymede, and Callisto are 262,000, 417,000, 666,000, and 1,170,000 miles, respectively.
  **a.** From the graph, estimate the period of revolution to the nearest hour for each satellite. For which satellite can you obtain the period with the most accuracy?

  **b.** What is the amplitude of each sine wave?

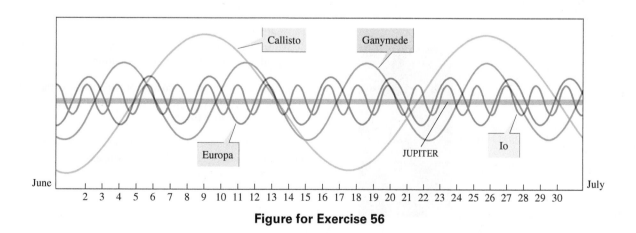

**Figure for Exercise 56**

**57.** *Ocean Waves* Scientists use the same types of terms to describe ocean waves that we use to describe sine waves. The *wave period* is the time between crests and the *wavelength* is the distance between crests. The *wave height* is the vertical distance from the trough to the crest. The accompanying figure shows a *swell* in a coordinate system. Write an equation for the swell, assuming that its shape is that of a sinusoid.

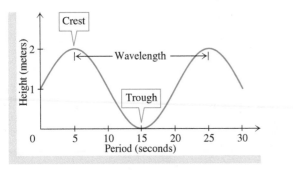

**Figure for Exercise 57**

**58.** *Large Ocean Waves* A *tsunami* is a series of large waves caused by an earthquake. The wavelength for a tsunami can be as long as several hundred kilometers. The accompanying figure shows a tsunami in a coordinate system. Write an equation for the tsunami, assuming that its shape is that of a sinusoid.

**Figure for Exercise 58**

**59.** *Moon Illumination* The accompanying table shows the percentage of the moon that will be illuminated at midnight for the 31 days of January, 2020. (U.S. Naval Observatory, http://aa.usno.navy.mil).

| Day | % | Day | % | Day | % |
|-----|-----|-----|----|-----|----|
| 1 | 32 | 12 | 97 | 23 | 3 |
| 2 | 41 | 13 | 92 | 24 | 1 |
| 3 | 50 | 14 | 85 | 25 | 0 |
| 4 | 60 | 15 | 76 | 26 | 2 |
| 5 | 69 | 16 | 65 | 27 | 5 |
| 6 | 78 | 17 | 54 | 28 | 10 |
| 7 | 85 | 18 | 43 | 29 | 17 |
| 8 | 92 | 19 | 32 | 30 | 24 |
| 9 | 97 | 20 | 22 | 31 | 33 |
| 10 | 100 | 21 | 14 | | |
| 11 | 100 | 22 | 7 | | |

Use the sinusoidal regression feature of a graphing calculator to find the equation for a sine curve that fits the data. Graph the data and the curve on your graphing calculator. Find the period from the equation. Use the equation to predict the percentage of the moon that will be illuminated on February 8, 2020.

**60.** *Amount of Daylight* The accompanying table gives the number of minutes between sunrise and sunset for the first day of each month in the year 2020 in Miami, Florida.

| Month | Time | Month | Time |
|-------|------|-------|------|
| 1 | 634 | 7 | 823 |
| 2 | 659 | 8 | 799 |
| 3 | 699 | 9 | 759 |
| 4 | 746 | 10 | 714 |
| 5 | 788 | 11 | 669 |
| 6 | 819 | 12 | 638 |

Use the sinusoidal regression feature of a graphing calculator to find the equation for a sine curve that fits the data. Graph the data and the curve on your graphing calculator. Find the period from the equation. Use the equation to predict the number of minutes between sunrise and sunset on February 1, 2021.

## WRITING/DISCUSSION

**61.** *Periodic Temperature* Air temperature $T$ generally varies in a periodic manner, with highs during the day and lows during the night. Assume that no drastic changes in the weather are expected and let $t$ be time in hours with $t = 0$ at midnight tonight. Graph $T$ for $t$ in the interval $[0, 48]$ and write a function of the form $T = A \sin[B(t - C)] + D$ for your graph. Explain your choices for $A, B, C,$ and $D$.

**62.** *Equivalent Equations* Find $B, C,$ and $D$ so that the graph of $y = -3 \sin[B(x - C)] + D$ is the same as the graph of $y = 3 \sin[4(x - \pi/2)] + 5$.

## REVIEW

**63.** Determine the amplitude, period, phase shift, and range for the function $f(x) = \frac{1}{2}\sin(x - \pi/2) + 3$.

**64.** The graph of $y = \cos(x)$ is shifted $\pi$ units to the left, reflected in the $x$-axis, and then shifted 2 units upward. What is the equation of the curve in its final position?

**65.** At a distance of 500 feet from a giant redwood tree, the angle of elevation to the top of the tree is 30°. What is the height of the tree to the nearest foot?

**66.** The terminal side of the angle $\beta$ in standard position passes through the point $(-3, 6)$. Find exact values for $\sin(\beta), \cos(\beta),$ and $\tan(\beta)$.

**67.** Two angles of a triangle are $32°37'$ and $48°39'$. Find the third angle in degrees-minutes-seconds format.

**68.** Find the central angle (to the nearest tenth of a degree) that intercepts an arc of length 5 feet on a circle of radius 60 feet.

**OUTSIDE THE BOX**

**69.** *Two Wrongs Make Right* In the following addition problem each letter represents a different digit from 1 through 9, and zero is not allowed. Find digits that make the addition correct.

$$\begin{array}{r} \text{WRONG} \\ +\text{WRONG} \\ \hline \text{RIGHT} \end{array}$$

**70.** *Concentric Circles* A circular fountain has a circular island exactly in the center, as shown in the accompanying figure. A student on vacation finds that the water is 2 feet deep and the distance from $A$ to $B$, measured along the edge of the island, is 80 feet. Find the volume of water in the fountain to the nearest cubic foot.

**Figure for Exercise 70**

**1.** Determine the amplitude, period, and phase shift for $y = 4 \sin(2x - 2\pi/3)$.

**2.** List the coordinates for the five key points for one cycle of $y = -3 \sin(2x)$.

**3.** Determine the period and range for $y = 4 \cos(\pi x) + 2$.

**4.** The five key points for one cycle of a sine wave are $(-\pi/6, 0)$, $(0, 4)$, $(\pi/6, 0)$, $(\pi/3, -4)$, and $(\pi/2, 0)$. Find an equation for the curve.

**5.** What is the frequency of the sine wave determined by $y = \sin(500\pi x)$, where $x$ is time in minutes?

# **2.3** Graphs of the Secant and Cosecant Functions

In this section we graph the *secant* and *cosecant functions*. A **secant function** has the form $y = A \sec[B(x - C)] + D$ and a **cosecant function** has the form $y = A \csc[B(x - C)] + D$. Because $\sec(x) = 1/\cos(x)$ and $\csc(x) = 1/\sin(x)$, the graphs of the secant and cosecant functions are closely related to the graphs of the cosine and sine functions, respectively.

**Figure 2.35**

## Vertical Asymptotes

A **vertical asymptote** is a vertical line that is approached by a graph. You will recall that the graph of the rational function $y = 1/x$ has the $y$-axis as a vertical asymptote as shown in Fig. 2.35. As $x$ approaches 0 from the right, the $y$-coordinates increase without bound $(y \to \infty)$. As $x$ approaches 0 from the left, the $y$-coordinates decrease without bound $(y \to -\infty)$. We read the arrow as "approaches" or "goes to." The graphs of secant and cosecant functions approach vertical asymptotes much as the graph of $y = 1/x$ does.

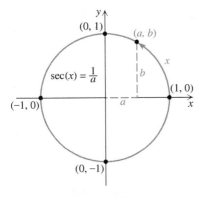

**Figure 2.36**

# Graphing Secant Functions

Suppose $(a, b)$ is the terminal point for an arc of length $x$ on the unit circle as shown in Fig. 2.36. Since $\sec(x) = 1/a$, $\sec(x)$ is undefined if $a = 0$. The only points on the unit circle where $a = 0$ are $(0, 1)$ and $(0, -1)$. Arcs such as $\pi/2, 3\pi/2, 5\pi/2$, and so on, have terminal points $(0, 1)$ and $(0, -1)$. These arcs are of the form $\pi/2 + k\pi$ where $k$ is any integer. So $\sec(\pi/2 + k\pi)$ is undefined for any integer $k$. As $a$ approaches 0 the absolute value of $1/a$ gets larger and larger, which causes the graph of $y = \sec(x)$ to have a vertical asymptote. So $y = \sec(x)$ has a vertical asymptote for every $x$ at which $\sec(x)$ is undefined or $\cos(x) = 0$.

**Domain and Asymptotes of $y = \sec(x)$**

The domain of $y = \sec(x)$ is the set of all real numbers except numbers of the form $\pi/2 + k\pi$ where $k$ is an integer. The equations of the vertical asymptotes are $x = \pi/2 + k\pi$ for any integer $k$.

**EXAMPLE 1** The simplest secant function

Graph $y = \sec(x)$. Determine the period, asymptotes, and range of the function.

**Solution**

Because of the reciprocal identity $\sec x = 1/\cos x$, we first draw the graph of $y = \cos x$ for reference when graphing $y = \sec x$. If $\cos x > 0$ and $\cos x \to 0$, then $\sec x \to \infty$. If $\cos x < 0$ and $\cos x \to 0$, then $\sec x \to -\infty$. So $y = \sec(x)$ has a vertical asymptote at every $x$-intercept of $y = \cos(x)$. If $\cos x = \pm 1$, then $\sec x = \pm 1$. So every maximum or minimum point on the graph of $y = \cos x$ is also on the graph of $y = \sec x$, which is shown in Fig. 2.37. The period of $y = \sec x$ is $2\pi$, the same as the period for $y = \cos x$. The equations of the asymptotes are $x = \pi/2 + k\pi$ for any integer $k$. The range is $(-\infty, -1] \cup [1, \infty)$.

The calculator graph of $y = 1/\cos x$ in Fig. 2.38 shows many points on the graph of $y = \sec x$ and supports our conclusions about the shape and location of the graph.

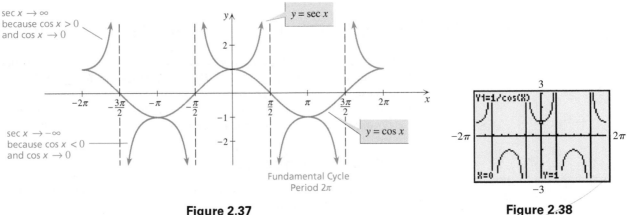

**Figure 2.37**

**Figure 2.38**

**TRY THIS.** Graph one cycle of $y = 3 \sec(x)$ and determine the period, asymptotes, and range.

**EXAMPLE 2** A transformation of $y = \sec x$

Sketch two cycles of $y = 2\sec(x - \pi/2)$. Determine the period, asymptotes, and range of the function.

**Solution**

First graph $y = \cos(x - \pi/2)$ as shown in Fig. 2.39, because

$$y = 2\sec(x - \pi/2) = \frac{2}{\cos(x - \pi/2)}.$$

Now $y = \sec(x - \pi/2)$ goes through the maximum and minimum points of $y = \cos(x - \pi/2)$ and has a vertical asymptote at every $x$-intercept of $y = \cos(x - \pi/2)$. To get $y = 2\sec(x - \pi/2)$, stretch $y = \sec(x - \pi/2)$ by a factor of 2. So the portions of the curve that open up do not go lower than 2, and the portions that open down do not go higher than $-2$, as in the figure. The period is $2\pi$. The asymptotes are $x = k\pi$ for any integer $k$. The range is $(-\infty, -2] \cup [2, \infty)$. The calculator graph of $y = 2/\cos(x - \pi/2)$ in Fig. 2.40 supports our conclusions.

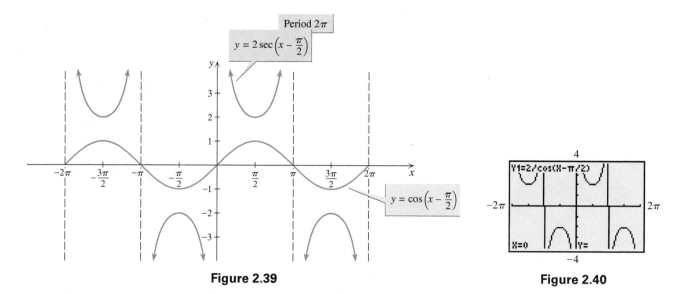

Period $2\pi$

$y = 2\sec\left(x - \dfrac{\pi}{2}\right)$

$y = \cos\left(x - \dfrac{\pi}{2}\right)$

**Figure 2.39**

Y1=2/cos(X-π/2)

**Figure 2.40**

**TRY THIS.** Graph one cycle of $y = 3\sec(2x)$ and determine the period, asymptotes, and range.

## Graphing Cosecant Functions

Suppose $(a, b)$ is the terminal point for an arc of length $x$ on the unit circle as shown in Fig. 2.41. Since $\csc(x) = 1/b$, $\csc(x)$ is undefined if $b = 0$. The only points on the unit circle where $b = 0$ are $(1, 0)$ and $(-1, 0)$. Arcs such as $0$, $\pi$, $2\pi$, and so on, have terminal points $(1, 0)$ and $(-1, 0)$. These arcs are of the form $k\pi$ where $k$ is any integer. So $\csc(k\pi)$ is undefined for any integer $k$. Like the secant function, the graph of $y = \csc(x)$ has a vertical asymptote wherever $\csc(x)$ is undefined.

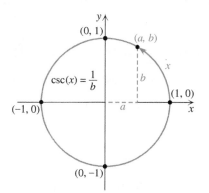

$(0, 1)$

$(a, b)$

$\csc(x) = \dfrac{1}{b}$

$(1, 0)$

$(-1, 0)$

$(0, -1)$

**Figure 2.41**

**Domain and Asymptotes of $y = \csc(x)$**

The domain of $y = \csc(x)$ is the set of all real numbers except numbers of the form $k\pi$ where $k$ is an integer. The equations of the vertical asymptotes are $x = k\pi$ for any integer $k$.

The graph of a cosecant function is very similar to the graph of a secant function and we graph them in the same manner.

### EXAMPLE 3 The simplest cosecant function

Graph $y = \csc(x)$. Determine the period, asymptotes, and range of the function.

**Solution**

Because of the reciprocal identity $\csc x = 1/\sin x$, we first draw the graph of $y = \sin x$ for reference when graphing $y = \csc x$. The graph of $y = \csc x$ has a vertical asymptote at each $x$-intercept of $y = \sin(x)$. If $\sin x = \pm 1$, then $\csc x = \pm 1$. So every maximum or minimum point on the graph of $y = \sin x$ is also on the graph of $y = \csc x$, which is shown in Fig. 2.42. The period for $y = \csc x$ is $2\pi$, the same as the period for $y = \sin x$. The equations of the vertical asymptotes are $x = k\pi$ for any integer $k$. The range is $(-\infty, -1] \cup [1, \infty)$.

The calculator graph of $y = 1/\sin x$ in Fig. 2.43 shows many points on the graph of $y = \csc x$ and it supports our conclusions.

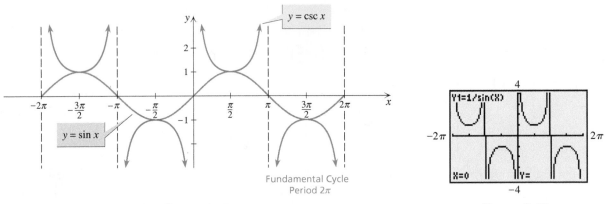

**Figure 2.42**          **Figure 2.43**

**TRY THIS.** Graph one cycle of $y = 2\csc(x)$ and determine the period, asymptotes, and range.                                                                                   ∎

The cosecant function in the next example has a shorter period and horizontal translation.

### EXAMPLE 4 A cosecant function with a transformation

Sketch two cycles of the graph of $y = \csc(2x - 2\pi/3)$. Determine the period, asymptotes, and the range of the function.

**Solution**

First graph $y = \sin[2(x - \pi/3)]$ because

$$y = \csc(2x - 2\pi/3) = \frac{1}{\sin[2(x - \pi/3)]}.$$

The period for $y = \sin[2(x - \pi/3)]$ is $\pi$ with phase shift of $\pi/3$. So the fundamental cycle of $y = \sin x$ is transformed to occur on the interval $[\pi/3, 4\pi/3]$. Draw two cycles of $y = \sin[2(x - \pi/3)]$. The graph of $y = \csc[2(x - \pi/3)]$ has a vertical asymptote at every $x$-intercept of the sine curve, as shown in Fig. 2.44. Each portion of $y = \csc[2(x - \pi/3)]$ that opens up has a minimum value of 1, and each portion that opens down has a maximum value of $-1$. The period of the

function is $\pi$. The vertical asymptotes are $x = \pi/3 + k\pi/2$ for any integer $k$ and the range is $(-\infty, -1] \cup [1, \infty)$.

The calculator graph of $y = 1/\sin[2(x - \pi/3)]$ in Fig. 2.45 supports these conclusions.

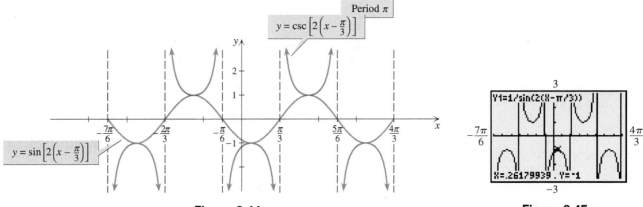

**Figure 2.44**                                           **Figure 2.45**

**TRY THIS.** Graph one cycle of $y = 3\csc(2x - \pi)$ and determine the period, asymptotes, and range.

**FOR THOUGHT...** True or False? Explain.

**1.** $\sec(\pi/4) = 1/\sin(\pi/4)$

**2.** $\csc(1.24) = 1/\sin(1.24)$

**3.** $\csc(\pi/2) = 1$

**4.** $\sec(\pi/2) = 0$

**5.** The period of $y = \csc(2x)$ is $\pi$.

**6.** The period of $y = \sec(\pi x)$ is 2.

**7.** The graphs of $y = 2\csc x$ and $y = 1/(2\sin x)$ are identical.

**8.** The range of $y = 0.5\csc(13x - 5\pi)$ is $(-\infty, -0.5] \cup [0.5, \infty)$.

**9.** The lines $x = \pm\pi/4$ are vertical asymptotes for $y = \sec(2x)$.

**10.** The line $x = 0$ is a vertical asymptote for $y = \csc(4x)$.

Scan this code to link to section lecture videos.

# 2.3 EXERCISES

## CONCEPTS

*Fill in the blank.*

**1.** The _____ of $y = \sec(x)$ is the set of all real numbers except numbers of the form $\pi/2 + k\pi$ where $k$ is an integer.

**2.** The _____ of $y = \csc(x)$ is the set of all real numbers except numbers of the form $k\pi$, where $k$ is an integer.

**3.** A vertical _____ is a vertical line that is approached by a graph.

**4.** The vertical asymptotes of $y = \sec(x)$ pass through the _____ of $y = \cos(x)$.

## SKILLS

*Find the exact value of each expression.*

**5.** $\sec(\pi/3)$      **6.** $\sec(\pi/4)$      **7.** $\csc(-\pi/4)$

**8.** $\csc(\pi/6)$      **9.** $\sec(\pi/2)$      **10.** $\sec(3\pi/2)$

**11.** $\csc(\pi)$      **12.** $\csc(0)$

*Find the approximate value of each expression to the nearest tenth.*

**13.** $\sec(1.56)$      **14.** $\sec(1.58)$      **15.** $\csc(0.01)$

**16.** $\csc(-0.002)$      **17.** $\csc(3.14)$      **18.** $\csc(6.28)$

**19.** $\sec(4.71)$      **20.** $\sec(4.72)$

*Determine the period of each function.*

**21.** $y = 5\sec(2x)$      **22.** $y = 3\sec(4x)$

**23.** $y = \csc(3x/2)$      **24.** $y = 2\csc(x/2)$

**25.** $y = \sec(\pi x)$      **26.** $y = \sec(2\pi x)$

*Determine the range of each function.*

**27.** $y = 2 \sec(x)$

**28.** $y = 4 \sec(x)$

**29.** $y = \dfrac{1}{2} \csc(x)$

**30.** $y = \dfrac{1}{3} \csc(x)$

**31.** $y = \sec(\pi x - 3\pi) - 1$

**32.** $y = \sec(3x + \pi/3) + 1$

*Sketch at least one cycle of the graph of each secant function. Determine the period, asymptotes, and range of each function.*

**33.** $y = 2 \sec x$

**34.** $y = \dfrac{1}{2} \sec x$

**35.** $y = \sec(3x)$

**36.** $y = \sec(4x)$

**37.** $y = \sec(x + \pi/4)$

**38.** $y = \sec(x - \pi/6)$

**39.** $y = \sec(x/2)$

**40.** $y = \sec(x/3)$

**41.** $y = 2 \sec(\pi x)$

**42.** $y = 3 \sec(\pi x/2)$

**43.** $y = 2 + 2 \sec(2x)$

**44.** $y = 2 - 2 \sec\left(\dfrac{x}{2}\right)$

*Sketch at least one cycle of the graph of each cosecant function. Determine the period, asymptotes, and range of each function.*

**45.** $y = -2 \csc(x)$

**46.** $y = -3 \csc(x)$

**47.** $y = -\csc(x + \pi/2)$

**48.** $y = -\csc(x - \pi)$

**49.** $y = \csc(2x - \pi/2)$

**50.** $y = \csc(3x + \pi)$

**51.** $y = \csc\left(\dfrac{\pi}{2}x - \dfrac{\pi}{2}\right)$

**52.** $y = \csc\left(\dfrac{\pi}{4}x + \dfrac{3\pi}{4}\right)$

**53.** $y = -\csc\left(\dfrac{\pi}{2}x + \dfrac{\pi}{2}\right)$

**54.** $y = -2 \csc(\pi x - \pi)$

*Write an equation for each curve in its final position.*

**55.** The graph of $y = \sec(x)$ is shifted $\pi/2$ units to the right and 1 unit upward.

**56.** The graph of $y = \sec(x)$ is shifted $\pi$ units to the left, reflected in the $x$-axis, then shifted 2 units upward.

**57.** The graph of $y = \csc(x)$ is reflected in the $x$-axis, shifted 1 unit to the left, then shifted 4 units upward.

**58.** The graph of $y = \csc(x)$ is reflected in the $x$-axis, shifted 2 units to the right, and then shifted 3 units downward.

*Find the equations for all vertical asymptotes for each function.*

**59.** $y = -\sec(x)$

**60.** $y = -\csc(x)$

**61.** $y = \csc(2x)$

**62.** $y = \csc(4x)$

**63.** $y = \sec(x - \pi/2)$

**64.** $y = \sec(x + \pi)$

**65.** $y = \csc(2x - \pi)$

**66.** $y = \csc(4x + \pi)$

**67.** $y = \dfrac{1}{2} \csc(2x) + 4$

**68.** $y = \dfrac{1}{3} \csc(3x) - 6$

**69.** $y = \sec(\pi x + \pi)$

**70.** $y = \sec(\pi x/2 - \pi/2)$

## WRITING/DISCUSSION

**71.** What is the range of the function $y = A \sec[B(x - C)] + D$?

**72.** What is the range of the function $y = A \csc[B(x - C)] + D$?

## REVIEW

**73.** If $\alpha$ is an angle in standard position whose terminal side intersects the unit circle at the point $(x, y)$, then $\sin(\alpha) = $ _____ and $\cos(\alpha) = $ _____.

**74.** The graph of $y = 3 \cos(5(x - \pi)) + 7$ is a _____ wave.

**75.** Find the amplitude, period, phase shift, and range for the function $f(x) = 5 \cos(2x - \pi) + 3$.

**76.** If the period of a sine wave is 0.125 second, then what is the frequency?

**77.** If $\beta$ is an angle in standard position such that $\sin(\beta) = 1$, then what is $\cos(\beta)$?

**78.** Evaluate each function. Give the result in degrees.
 **a.** $\sin^{-1}(-1/2)$   **b.** $\cos^{-1}(-1/2)$   **c.** $\tan^{-1}(-1)$

## OUTSIDE THE BOX

**79.** *Cutting Cardboard* A cardboard tube has height 1 and diameter 1. Suppose that it is cut at a 45° angle as shown in the first figure. The tube is then laid flat in the first quadrant as shown in the second figure. What is the equation of the curve? (Assume that the curve is a sine wave.)

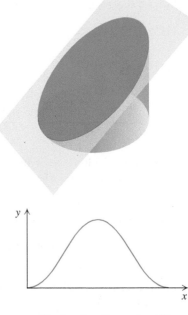

**Figure for Exercise 79**

**80.** *Lightest Kid* Mitt has 5 kids. He weighs them in groups of 4 and gets 354, 314, 277, 265, and 254 pounds. What is the weight of the lightest kid?

**1.** Find the exact value of $\sec(\pi/4)$.

**2.** Find the exact value of $\csc(\pi)$.

**3.** Find the equations of all asymptotes for $y = \sec(x) + 3$.

**4.** Find the equations of all asymptotes for $y = \csc(x) - 1$.

**5.** Find the equations of all asymptotes for $y = \sec(2x)$.

**6.** What is the range of $y = 3\csc(2x)$?

## LINKING
## concepts...    For Individual or Group Explorations

### Modeling a Guitar Note

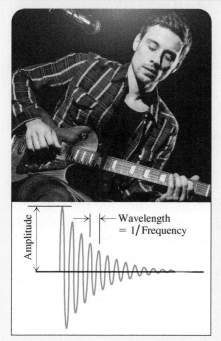

*The waveform of a guitar note is characterized by an initial sharp peak that falls off rapidly as shown in the figure. Guitar effects pedals electronically modify that waveform. We can mathematically modify the oscillation of the sine wave to get the waveform of a guitar note as well as many others. The basic sine wave oscillates between the two horizontal lines $y = 1$ and $y = -1$. If we multiply $\sin(x)$ by any other function $g(x)$, we will get a curve that oscillates between the graphs of $y = g(x)$ and $y = -g(x)$.*

**a)** Graph $y_1 = x\sin(x), y_2 = x$, and $y_3 = -x$. For what exact values of $x$ is $x\sin(x) = x$? For what exact values of $x$ is $x\sin(x) = -x$?

**b)** For what exact values of $x$ is $x^2\sin(x) = x^2$? For what exact values of $x$ is $x^2\sin(x) = -x^2$? Support your conclusions with a graph.

**c)** Graph $y_1 = \frac{1}{x}\sin(x)$ for $0 \le x \le 10$ and $-2 \le y \le 2$. Is it true that for $x > 0, -\frac{1}{x} < \frac{1}{x}\sin(x) < \frac{1}{x}$? Prove your answer.

**d)** Graph $f(x) = \frac{1}{x}\sin(x)$ for $-2 \le x \le 2$ and $-2 \le y \le 2$. What is $f(0)$? Is it true that for all $x$ in the interval $[-0.1, 0.1]$ for which $f(x)$ is defined, $f(x)$ satisfies $0.99 < f(x) < 1$? Explain.

**e)** Experiment with functions until you get one that looks like the one in the figure.

# 2.4  Graphs of the Tangent and Cotangent Functions

**Figure 2.46**

So far we have graphed four of the six trigonometric functions. In this section we will graph functions of the form $y = A \tan[B(x - C)] + D$ and $y = A \cot[B(x - C)] + D$, which are the **tangent functions** and the **cotangent functions.** Like the secant and cosecant functions, the tangent and cotangent functions have vertical asymptotes.

## Graphing Tangent Functions

Suppose $(a, b)$ is the terminal point for an arc of length $x$ on the unit circle as shown in Fig. 2.46. Since $\tan(x) = b/a$, $\tan(x)$ is undefined if $a = 0$. The only points on the unit circle where $a = 0$ are $(0, 1)$ and $(0, -1)$. Arcs such as $\pi/2, 3\pi/2, 5\pi/2$, and so on, terminate at either $(0, 1)$ or $(0, -1)$. These arcs are of the form $\pi/2 + k\pi$ where $k$ is any integer. So $\tan(x)$ is undefined for these arcs. As $x$ approaches $\pi/2 + k\pi$ for any $k$, $a$ approaches 0 and $\tan(x)$ approaches $\infty$ or $-\infty$. So $y = \tan x$ has a vertical asymptote wherever it is undefined.

**Domain and Asymptotes of $y = \tan(x)$**

The domain of $y = \tan(x)$ is the set of all real numbers except numbers of the form $\pi/2 + k\pi$ where $k$ is an integer. The equations of the vertical asymptotes are $x = \pi/2 + k\pi$ for any integer $k$.

> **EXAMPLE 1**  **The simplest tangent function**
>
> Graph $y = \tan(x)$.

**Solution**

As $x \to -\pi/2$ from the right, $y \to -\infty$. For example, $\tan(-1.56) \approx -92.6$. So $x = -\pi/2$ is a vertical asymptote. As $x \to \pi/2$ from the left, $y \to \infty$. For example, $\tan(1.56) \approx 92.6$. So $x = \pi/2$ is a vertical asymptote. The numbers $-\pi/4, 0,$ and $\pi/4$ divide the interval between the asymptotes $(-\pi/2, \pi/2)$ into four equal parts. So we evaluate the function at these points as shown in the following table:

| $x$ | $-\pi/2$ | $-\pi/4$ | $0$ | $\pi/4$ | $\pi/2$ |
|---|---|---|---|---|---|
| $y = \tan x$ | undefined | $-1$ | $0$ | $1$ | undefined |

Graph the vertical asymptotes $x = \pm\pi/2$ and the three points as shown in Fig. 2.47. The tangent curve is increasing on the interval $(-\pi/2, \pi/2)$ and approaches its

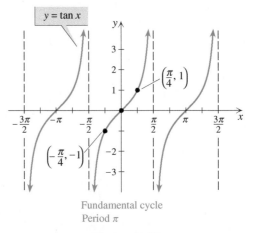

Fundamental cycle
Period $\pi$

**Figure 2.47**

**Figure 2.48**

asymptotes as shown in the figure. The graph between $-\pi/2$ and $\pi/2$ is repeated between each pair of consecutive asymptotes, which are the lines $x = \pi/2 + k\pi$ for any integer $k$. Note that we draw the asymptotes and then draw the curve so that it approaches its asymptotes.

The calculator graph of $y = \tan(x)$ in Fig. 2.48 shows many more points than we could plot by hand and it confirms the shape of the curve in Fig. 2.47. In connected mode the calculator connects two points on opposite sides of each asymptote and appears to draw the vertical asymptotes.

**TRY THIS.** Graph two cycles of $y = 2\tan(x)$. Determine the period and equations of the asymptotes. ∎

The period of $y = \tan(x)$ is $\pi$ and the fundamental cycle is the portion of the graph between $-\pi/2$ and $\pi/2$. The points $(-\pi/4, -1)$, $(0, 0)$, and $(\pi/4, 1)$ are the **key points** for the graph of $y = \tan(x)$. The point $(0, 0)$ is an inflection point. The inflection point, the other two key points, and the asymptotes determine the shape of the graph. Since the range of $y = \tan(x)$ is $(-\infty, \infty)$, the concept of amplitude is not applicable.

To graph $y = A\tan[B(x - C)] + D$, a transformation of $y = \tan(x)$, we use the following procedure. Remember that the asymptotes are essential, but they are not part of the graph of the function. Think of them as boundary lines for the graph. Note how we divide one cycle into four equal parts, just as we did for the sine and cosine functions.

---

**PROCEDURE**

### Graphing $y = A\tan[B(x - C)] + D$

1. Determine the period $\pi/B$, phase shift $C$, and vertical translation $D$.
2. The asymptotes are $x = C \pm \pi/(2B)$.
3. Midway between the asymptotes is the inflection point $(C, D)$.
4. Locate the other two key points midway between the inflection point and the asymptotes.
5. Sketch one cycle of the curve through the three key points and approaching the asymptotes. The function is increasing for $A > 0$ and decreasing for $A < 0$.

---

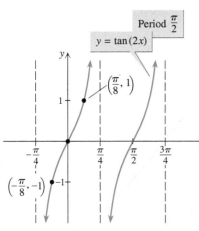

Period $\frac{\pi}{2}$

$y = \tan(2x)$

$\left(\frac{\pi}{8}, 1\right)$

$\left(-\frac{\pi}{8}, -1\right)$

**Figure 2.49**

**Figure 2.50**

**EXAMPLE 2**   **Changing the period**

Sketch two cycles of the function $y = \tan(2x)$. Determine the period and equations of the asymptotes.

**Solution**

For $y = \tan(2x)$ the period is $\pi/2$. Since $C = 0$ and $D = 0$, the asymptotes are $x = \pm\pi/4$ and the inflection point is $(0, 0)$. Midway between the asymptotes and the inflection point we find the points $(-\pi/8, -1)$ and $(\pi/8, 1)$. Sketch one cycle of $y = \tan(2x)$ through the three key points and approaching its asymptotes as shown in Fig. 2.49. Continue the same pattern to draw another cycle to the right of the first cycle. Because the period is $\pi/2$ the equations of all of the asymptotes have the form $y = \pi/4 \pm k\pi/2$ for any integer $k$.

The calculator graph shown in Fig. 2.50 is drawn in dot mode and thus does not connect the separate portions of the tangent curve.

**TRY THIS.** Graph two cycles of $y = 6\tan(2x)$. Determine the period and equations of the asymptotes. ∎

The period of $y = \tan x$ is $\pi$ and the fundamental cycle occurs when $-\frac{\pi}{2} < x < \frac{\pi}{2}$. To determine the period for $y = \tan Bx$ we replace $x$ with $Bx$:

$$-\frac{\pi}{2} < Bx < \frac{\pi}{2}$$

$$-\frac{\pi}{2B} < x < \frac{\pi}{2B} \qquad \text{Divide each part of the inequality by } B.$$

Since the length of this interval is $\frac{\pi}{2B} - \left(-\frac{\pi}{2B}\right)$ or $\frac{\pi}{B}$, the period of $\tan Bx$ is $\frac{\pi}{B}$. So we divide the old period $\pi$ by $B$ to get the new period. This is the same rule that we used to find the periods for $y = \sin Bx$ and $y = \cos Bx$. The following Function Gallery illustrates the period change for all three of these trigonometric functions for $B > 1$.

**FUNCTION GALLERY**

### Periods of Sine, Cosine, and Tangent ($B > 1$)

**Fundamental cycles**

In Example 2 the transformation involved changing only the period. In Example 3 we will perform a more complicated transformation.

**EXAMPLE 3** **A transformation of $y = \tan x$**

Sketch two cycles of the function $f(x) = \frac{1}{2}\tan(3x + \pi/2) + 1$. Determine the period and equations of the asymptotes.

**Solution**

First rewrite the function as $f(x) = \frac{1}{2}\tan\left[3\left(x + \frac{\pi}{6}\right)\right] + 1$ to determine that $A = \frac{1}{2}$, $B = 3$, $C = -\frac{\pi}{6}$, and $D = 1$. So the period is $\pi/3$, the phase shift is $-\pi/6$, and the vertical translation is 1. The asymptotes for one cycle are $x = -\pi/6 \pm \pi/6$, which is $x = -\pi/3$ and $x = 0$. The inflection point is $(-\pi/6, 1)$. The other two key points are midway between the inflection point and the asymptotes at $(-\pi/4, 1/2)$ and $(-\pi/12, 3/2)$. Sketch one cycle of the curve through the three key points and approaching the asymptotes, as shown in Fig. 2.51. Repeat the same pattern for another cycle to the right. Since the period is $\pi/3$ the equations of all asymptotes are of the form $x = k\pi/3$ for any integer $k$.

**Figure 2.51**                    **Figure 2.52**

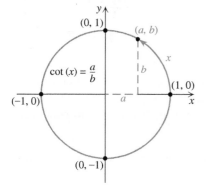

**Figure 2.53**

The calculator graph shown in Fig. 2.52 supports these conclusions.

**TRY THIS.** Graph two cycles of $y = 8\tan(2x + \pi) - 5$. Determine the period and equations of the asymptotes. ▪

## Graphing Cotangent Functions

Suppose $(a, b)$ is the terminal point for an arc of length $x$ on the unit circle as shown in Fig. 2.53. Since $\cot(x) = a/b$, $\cot(x)$ is undefined if $b = 0$. The only points on the unit circle where $b = 0$ are $(1, 0)$ and $(-1, 0)$. Arcs that terminate at those points are multiples of $\pi$, such as $0, \pi, 2\pi$, and so on. These arcs are of the form $k\pi$ where $k$ is an integer. Just like $y = \tan(x)$, $y = \cot(x)$ has a vertical asymptote wherever it is undefined.

**Domain and Asymptotes of $y = \cot(x)$**

> The domain of $y = \cot(x)$ is the set of all real numbers except numbers of the form $k\pi$ where $k$ is an integer. The equations of the vertical asymptotes are $x = k\pi$ for any integer $k$.

**EXAMPLE 4**  **The simplest cotangent function**

Graph $y = \cot x$.

**Solution**

Consider some ordered pairs that satisfy $y = \cot x$.

| $x$ | 0 | $\pi/4$ | $\pi/2$ | $3\pi/4$ | $\pi$ |
|---|---|---|---|---|---|
| $y = \cot x$ | undefined | 1 | 0 | $-1$ | undefined |

As $x \to 0$ from the right, $y \to \infty$. For example $\cot(0.01) \approx 99.997$. So $x = 0$ is a vertical asymptote. As $x \to \pi$ from the left, $y \to -\infty$. For example, $\cot(3.13) \approx -86.26$. So $x = \pi$ is a vertical asymptote. The graph of $y = \cot x$ is decreasing on the interval $(0, \pi)$, as shown in Fig. 2.54 on the next page. The cycle of the curve in the interval $(0, \pi)$ is repeated between each pair of consecutive asymptotes, which are the lines $x = k\pi$ for any integer $k$.

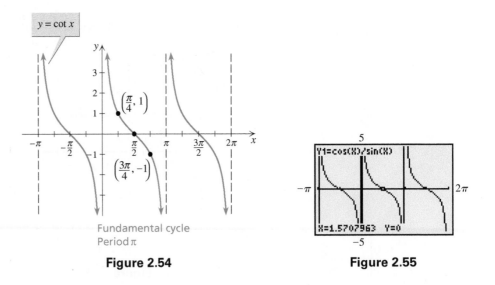

**Figure 2.54**          **Figure 2.55**

Graph $y = \cot(x)$ on a graphing calculator by graphing $y = \cos(x)/\sin(x)$ as in Fig. 2.55. The graph of $y = 1/\tan(x)$ looks the same as $y = \cot(x)$, but it is technically incorrect. When $\cot(x) = 0$, $\tan(x)$ is undefined. So there are no $x$-intercepts on the graph of $y = 1/\tan(x)$. Graph it and see.

**TRY THIS.** Graph two cycles of $y = 5\cot(x)$. Determine the period and the equations of the asymptotes. ◼

The point $(\pi/2, 0)$ is an inflection point for $y = \cot(x)$. The asymptotes, inflection point, and the other two key points determine the shape of the curve. Since the range of $y = \cot(x)$ is $(-\infty, \infty)$, the concept of amplitude is not applicable.

The period of $y = \cot(x)$ is $\pi$, and the fundamental cycle occurs for $0 < x < \pi$. To determine the period of $y = \cot(Bx)$ for $B > 0$, replace $x$ with $Bx$:

$$0 < Bx < \pi$$

$$0 < x < \frac{\pi}{B} \qquad \text{\footnotesize Divide each part of the inequality by } B.$$

The period for $y = \cot(Bx)$ with $B > 0$ is $\pi/B$ and $y = \cot(Bx)$ completes one cycle for $0 < x < \pi/B$ with asymptotes $x = 0$ and $x = \pi/B$.

To graph $y = A\cot[B(x - C)] + D$, a transformation of $y = \cot(x)$, we use the following procedure. Remember to start with the asymptotes, then locate the inflection point midway between them. Then midway again, locate the other two key points.

---

**PROCEDURE**

**Graphing $y = A\cot[B(x - C)] + D$**

1. Determine the period $\pi/B$, phase shift $C$, and vertical translation $D$.
2. The asymptotes for one cycle are $x = C$ and $x = C + \pi/B$.
3. Midway between the asymptotes is the inflection point $(C + \pi/(2B), D)$.
4. Locate the other two key points midway between the inflection point and the asymptotes.
5. Sketch one cycle of the curve through the three key points and approaching the asymptotes. The function is decreasing for $A > 0$ and increasing for $A < 0$.

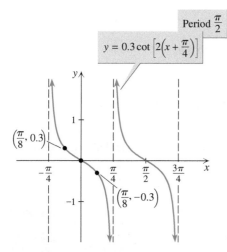

Period $\frac{\pi}{2}$

$y = 0.3 \cot \left[2\left(x + \frac{\pi}{4}\right)\right]$

$\left(\frac{\pi}{8}, 0.3\right)$

$\left(\frac{\pi}{8}, -0.3\right)$

**Figure 2.56**

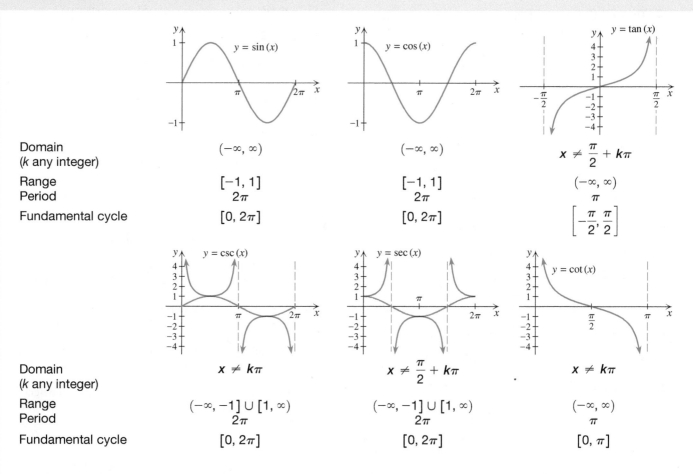

Y1=.3/tan(2X+π/2)

X=0     Y=

**Figure 2.57**

**EXAMPLE 5** **A transformation of $y = \cot x$**

Sketch two cycles of the function $f(x) = 0.3 \cot(2x + \pi/2)$. Determine the period and equations of the asymptotes.

**Solution**

First rewrite the function as $f(x) = 0.3 \cot\left[2\left(x + \frac{\pi}{4}\right)\right]$ to determine that $A = 0.3$, $B = 2$, $C = -\frac{\pi}{4}$, and $D = 0$. So the period is $\pi/2$, the phase shift is $-\pi/4$, and the vertical translation is 0. The asymptotes for one cycle are $x = -\pi/4$ and $x = \pi/4$. In the middle is the inflection point $(0, 0)$. The other two key points are midway between the inflection point and the asymptotes at $(-\pi/8, 0.3)$ and $(\pi/8, -0.3)$. Sketch one cycle of the curve through the three key points and approaching the asymptotes as shown in Fig. 2.56. Repeat the same pattern for another cycle to the right. Since the period is $\pi/2$, the equations of all asymptotes are of the form $x = -\pi/4 + k\pi/2$ for any integer $k$.

The calculator graph shown in Fig. 2.57 supports these conclusions.

**TRY THIS.** Graph two cycles of $y = 3 \cot(2x + \pi)$. Determine the period and the equations of the asymptotes. ∎

Because $\cot x$ is the reciprocal of $\tan x$, $\cot x$ is large when $\tan x$ is small, and vice versa. The graph of $y = \cot x$ has an $x$-intercept wherever $y = \tan x$ has a vertical asymptote, and a vertical asymptote wherever $y = \tan x$ has an $x$-intercept.

The following Function Gallery summarizes some of the facts that we have learned about the six trigonometric functions. Also, one cycle of the graph of each trigonometric function is shown.

**FUNCTION GALLERY** **TRIGONOMETRIC FUNCTIONS**

| | $y = \sin(x)$ | $y = \cos(x)$ | $y = \tan(x)$ |
|---|---|---|---|
| Domain (k any integer) | $(-\infty, \infty)$ | $(-\infty, \infty)$ | $x \neq \frac{\pi}{2} + k\pi$ |
| Range | $[-1, 1]$ | $[-1, 1]$ | $(-\infty, \infty)$ |
| Period | $2\pi$ | $2\pi$ | $\pi$ |
| Fundamental cycle | $[0, 2\pi]$ | $[0, 2\pi]$ | $\left[-\frac{\pi}{2}, \frac{\pi}{2}\right]$ |

| | $y = \csc(x)$ | $y = \sec(x)$ | $y = \cot(x)$ |
|---|---|---|---|
| Domain (k any integer) | $x \neq k\pi$ | $x \neq \frac{\pi}{2} + k\pi$ | $x \neq k\pi$ |
| Range | $(-\infty, -1] \cup [1, \infty)$ | $(-\infty, -1] \cup [1, \infty)$ | $(-\infty, \infty)$ |
| Period | $2\pi$ | $2\pi$ | $\pi$ |
| Fundamental cycle | $[0, 2\pi]$ | $[0, 2\pi]$ | $[0, \pi]$ |

## FOR THOUGHT... True or False? Explain.

1. $\tan(\pi/4) = \sin(\pi/4)/\cos(\pi/4)$

2. $\cot(1.24) = 1/\tan(1.24)$

3. $\cot(\pi/2) = 1/\tan(\pi/2)$

4. $\tan(0) = 0$

5. $\tan(\pi/2) = 1$

6. $\tan(5\pi/2) = 0$

7. The range of $y = \tan x$ is $(-\infty, \infty)$.

8. The range of $y = \cot x$ is $(0, \pi)$.

9. The graph of $y = \tan(3x)$ has vertical asymptotes at $x = \pm\pi/6$.

10. The graph of $y = \cot(4x)$ has vertical asymptotes at $x = \pm\pi/4$.

Scan this code
to link to section
lecture videos.

## 2.4 EXERCISES

### CONCEPTS

*Fill in the blank.*

1. Any function of the form $y = A\tan[B(x - C)] + D$ with $A \neq 0$ and $B \neq 0$ is a _____ function.

2. A vertical line approached by a graph is a(n) _____.

3. The _____ of $y = \tan(x)$ is the set of all real numbers except numbers of the form $\pi/2 + k\pi$, where $k$ is an integer.

4. The _____ of $y = \cot(x)$ is the set of all real numbers except numbers of the form $k\pi$, where $k$ is an integer.

5. The point $(0, 0)$ on the graph of $y = \tan x$ is a(n) _____ point.

6. The _____ of $y = \tan(Bx)$ or $y = \cot(Bx)$ is $\pi/B$.

### SKILLS

*Find the exact value of each expression.*

7. $\tan(\pi/3)$

8. $\tan(\pi/4)$

9. $\tan(\pi/2)$

10. $\tan(3\pi/2)$

11. $\tan(\pi)$

12. $\tan(2\pi)$

13. $\cot(\pi/4)$

14. $\cot(\pi/3)$

15. $\cot(0)$

16. $\cot(\pi)$

17. $\cot(\pi/2)$

18. $\cot(3\pi/2)$

*Find the approximate value of each expression to the nearest tenth.*

19. $\tan(1.56)$

20. $\tan(1.57)$

21. $\tan(1.58)$

22. $\tan(1.575)$

23. $\cot(0.002)$

24. $\cot(0.003)$

25. $\cot(-0.002)$

26. $\cot(-0.003)$

*Determine the period of each function.*

27. $y = \tan(8x)$

28. $y = \tan(2x)$

29. $y = \cot(\pi x)$

30. $y = \cot(\pi x/2)$

31. $y = 2\tan(\pi x/3)$

32. $y = 3\tan(\pi x + 1)$

33. $y = \cot(3x + \pi)$

34. $y = \cot(2x - \pi)$

*Sketch at least one cycle of the graph of each function. Determine the period and the equations of the vertical asymptotes.*

35. $y = \tan(3x)$

36. $y = \tan(4x)$

37. $y = \tan(\pi x)$

38. $y = \tan(\pi x/2)$

39. $y = -2\tan(x) + 1$

40. $y = 3\tan(x) - 2$

41. $y = -\tan(x - \pi/2)$

42. $y = \tan(x + \pi/2)$

43. $y = \tan\left(\dfrac{\pi}{2}x - \dfrac{\pi}{2}\right)$

44. $y = \tan\left(\dfrac{\pi}{4}x + \dfrac{3\pi}{4}\right)$

45. $y = \cot(x + \pi/4)$

46. $y = \cot(x - \pi/6)$

47. $y = \cot(x/2)$

48. $y = \cot(x/3)$

49. $y = -\cot(x + \pi/2)$

50. $y = 2 + \cot x$

51. $y = \cot(2x - \pi/2) - 1$

52. $y = \cot(3x + \pi) + 2$

*Write the equation of each curve in its final position.*

53. The graph of $y = \tan(x)$ is shifted $\pi/4$ units to the right, stretched by a factor of 3, then translated 2 units upward.

54. The graph of $y = \tan(x)$ is shifted $\pi/2$ units to the left, shrunk by a factor of $\frac{1}{2}$, then translated 5 units downward.

55. The graph of $y = \cot(x)$ is shifted $\pi/2$ units to the left, reflected in the $x$-axis, then translated 1 unit upward.

56. The graph of $y = \cot(x)$ is shifted $\pi/3$ units to the right, stretched by a factor of 2, then translated 2 units downward.

*Write an equation for the given tangent curve in the form*
$y = A \tan[B(x - C)] + D.$

**57.**

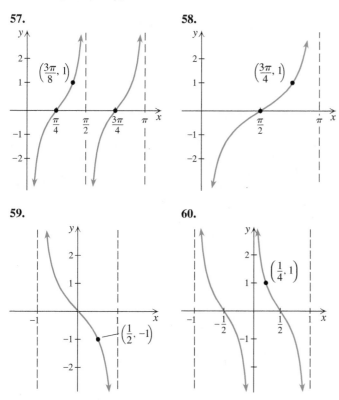

**58.**

**59.**

**60.**

*Let* $f(x) = \tan(x)$, $g(x) = x + 3$, *and* $h(x) = 2x$. *Find the following.*

**61.** $f(g(-3))$

**62.** $g(f(0))$

**63.** $g(h(f(\pi/2)))$

**64.** $g(f(h(\pi/6)))$

**65.** $f(g(h(x)))$

**66.** $g(f(h(x)))$

**67.** $g(h(f(x)))$

**68.** $h(f(g(x)))$

*Let* $\theta$ *be the "angle of inclination" for the line through* $(x_1, y_1)$ *as shown in the accompanying figure. For* $-\pi/2 < \theta < \pi/2$, $\tan \theta$ *is the slope of the line and the equation of the line is*

$$y - y_1 = (\tan \theta)(x - x_1).$$

*Find the equation in the form* $y = mx + b$ *for each line described below.*

**69.** The line through $(2, 3)$ with angle of inclination $\pi/4$

**70.** The line through $(-1, 2)$ with angle of inclination $-\pi/4$

**71.** The line through $(3, -1)$ with angle of inclination $\pi/3$

**72.** The line through $(-2, -1)$ with angle of inclination $\pi/6$

**Figure for Exercises 69 through 72**

## MODELING

*Solve each problem.*

**73.** *Discovering Planets* On October 23, 1996, astronomers announced that they had discovered a planet orbiting the star 16 Cygni B (*Sky & Telescope,* January 1997). They did not see the planet itself. Rather, for eight years they detected the periodic wobbling motion toward and away from earth that the planet induced in the star as it orbited the star. See the accompanying figure.
   **a.** Estimate the period of the function shown in the figure.

   **b.** Which trigonometric function has a graph similar to the graph in the figure?

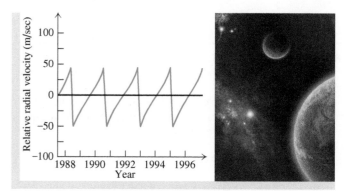

**Figure for Exercise 73**

**74.** *Average Rate of Change* The average rate of change of a function on a short interval $[x, x + h]$ for a fixed value of $h$ is a function itself. Sometimes it is a function that we can recognize by its graph.
   **a.** Graph $y_1 = \sin(x)$ on a graphing calculator and its average rate of change

   $$y_2 = (y_1(x + 0.1) - y_1(x))/0.1$$

   for $-2\pi \le x \le 2\pi$. What familiar function does $y_2$ look like?

   **b.** Repeat part (a) for $y_1 = \cos(x)$, $y_1 = e^x$, $y_1 = \ln(x)$, and $y_1 = x^2$.

## REVIEW

**75.** If $\alpha$ is an angle in standard position whose terminal side intersects the unit circle at $(x, y)$, then $\csc \alpha = $ _____, $\sec \alpha = $ _____, and $\cot \alpha = $ _____, provided no denominator is zero.

**76.** The five key points on one cycle of a sine wave are $(-\pi/4, 0)$, $(0, -2)$, $(\pi/4, 0)$, $(\pi/2, 2)$, and $(3\pi/4, 0)$. Write the equation of the sine wave in the form $y = A \sin(B(x - C)) + D$.

**77.** Determine the period, asymptotes, and range for the graph of $y = 3 \sec(2x - \pi)$.

**78.** If $\beta$ is an angle in standard position such that $\sin(\beta) = 1/4$ and $\beta$ terminates in quadrant II, then what is the exact value of $\cos(\beta)$?

**79.** A worker on top of a 432-foot building spots the boss on the ground. The angle of depression for the line of sight to the boss is 28°. How far from the building (to the nearest foot) is the boss?

**80.** Evaluate each trigonometric function if possible.
   a. $\sin(7\pi/6)$        b. $\cos(3\pi)$

   c. $\tan(-3\pi/4)$        d. $\sec(\pi/2)$

   e. $\csc(\pi/3)$        f. $\cot(-\pi/4)$

### OUTSIDE THE BOX

**81.** *Tiling a Room* Tile Mart is selling new T-shaped ceramic tiles as shown in the figure. Each tile is 3 ft long and 2 ft wide, and covers four square ft.
   a. Is it possible to tile completely an 8-ft × 8-ft room with these T-shaped tiles? No cutting, breaking, or overlapping of the tiles is allowed. Explain.

   b. Is it possible to tile a 6-ft × 6-ft room with these tiles?

**Figure for Exercise 81**

**82.** *Arranging Digits* It is possible to rearrange the digits 1 through 9 in the cells shown in the accompanying figure so that the sum of the 5 digits that run vertically is 27 and the sum of the 5 digits that run horizontally is 24. What digit must be place in the shared corner cell?

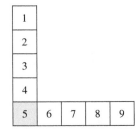

**Figure for Exercise 82**

## 2.4 POP QUIZ

**1.** What is the exact value of $\tan(-3\pi/4)$?

**2.** What is the exact value of $\cot(2\pi/3)$?

**3.** What is the period for $y = \tan(3x)$?

**4.** Find the equations of all asymptotes for $y = \cot(2x)$.

**5.** The graph of $y = \tan(x)$ is shifted $\pi/4$ to the left, stretched by a factor of 3, then translated 1 unit downward. What is the equation of the curve in its final position?

# 2.5 Combining Functions

In this section we will investigate functions of the form $y = y_1 + y_2$ where $y_1$ and $y_2$ are familiar functions. We will learn to sketch the graph of the combined function from the graphs of $y_1$ and $y_2$. Of course, in a textbook we say "sketch" and then show a nice computer-drawn graph. Hand-drawn graphs of the combined functions will most likely be quite rough. Concentrate on understanding why these graphs appear the way they do. If you have a graphing calculator, make a rough sketch and check it with your calculator. See if you can anticipate the graph that appears on your calculator.

## Combining Trigonometric and Algebraic Functions

We graphed functions such as $y = 1 + \sin x$ in Section 2.1 by using transformations. We can also view $y = 1 + \sin x$ as the sum of two functions, $y_1 = 1$ and $y_2 = \sin x$. For any $x$, the $y$-coordinate on $y = 1 + \sin x$ is the sum of the $y$-coordinates on the

graphs of $y_1$ and $y_2$. So if we graph $y_1$ and $y_2$, then we can visually add the $y$-coordinates to sketch the graph of $y$ as shown in Fig. 2.58. When done by hand, this method will not produce a very accurate graph. However, we can use this idea to get a quick sketch of the graph of a combined function and to better understand the graph of a combined function.

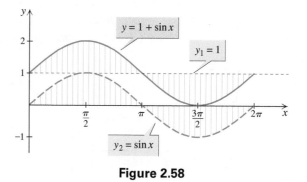

**Figure 2.58**

We use this idea of visually adding $y$-coordinates to sketch the graph of $y = x + \sin x$ in the next example.

**EXAMPLE 1** A sum of a linear and a trigonometric function

Graph $y = x + \sin x$ by adding $y$-coordinates.

**Solution**

The graphs of $y_1 = x$ and $y_2 = \sin x$ are shown in Fig. 2.59. For each $x$-coordinate, the $y$-coordinate on $y = x + \sin x$ is the sum of the $y$-coordinates on $y_1$ and $y_2$. Since $\sin x$ oscillates between $-1$ and $1$, the graph of $y = x + \sin x$ will oscillate about the line $y_1 = x$ as shown in the figure, crossing $y_1 = x$ whenever $\sin x$ is zero. The combined function is not periodic.

The calculator graph of $y = x + \sin x$ in Fig. 2.60 shows many points on the graph and supports our conclusions about the shape and location of the graph.

**Figure 2.59**                    **Figure 2.60**

**TRY THIS.** Graph $y = x - \sin(x)$.

**EXAMPLE 2** The sum of a rational and a trigonometric function

Sketch the graph of $y = \frac{1}{x} + \cos x$ by adding $y$-coordinates.

**Solution**

First graph the functions $y_1 = \frac{1}{x}$ and $y_2 = \cos x$ as shown in Fig. 2.61. For each $x$-coordinate, we obtain the $y$-coordinate on $y = \frac{1}{x} + \cos x$ by adding the corresponding values of $y_1$ and $y_2$. Note that $y = \frac{1}{x} + \cos x$ has a vertical asymptote at $x = 0$ because $y_1$ has a vertical asymptote at $x = 0$. Near zero the graph $y = \frac{1}{x} + \cos x$ looks like the graph of $y_1$ because $\frac{1}{x}$ is large and $\cos x$ is relatively small. When $x$ is large, $\frac{1}{x}$ is small and the graph looks more like the graph of $y_2$.

The calculator graph of $y = \frac{1}{x} + \cos x$ in Fig. 2.62 supports our conclusions.

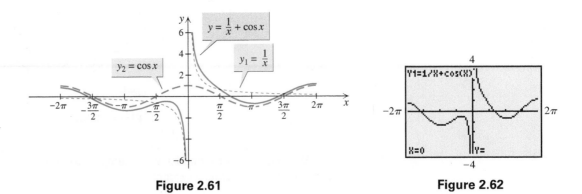

Figure 2.61                    Figure 2.62

**TRY THIS.**  Graph $y = 1/x - \cos(x)$. ∎

## Combining Trigonometric Functions

If the sum or difference of two periodic functions is a periodic function, then we can usually identify the period from the graph. However, it is not easy to determine the exact amplitude. The topic of amplitude for functions of the form $y = A \sin x + B \cos x$ is addressed in Section 3.6 with the reduction formula.

**EXAMPLE 3**   **A sum of two trigonometric functions**

Graph $y = \sin x + \cos x$ by adding $y$-coordinates.

**Solution**

The graphs of $y_1 = \sin x$ and $y_2 = \cos x$ are shown in Fig. 2.63. For each $x$-coordinate, the $y$-coordinate on $y = \sin x + \cos x$ is the sum of the $y$-coordinates on $y_1$ and $y_2$. When the $y$-coordinate on one graph is negative, it will reduce the $y$-coordinate on the other. Since $\sin x$ oscillates between $-1$ and $1$, the graph of $y = \sin x + \cos x$ oscillates about the curve $y_2 = \cos x$ as shown in the figure. The combined function appears to have period $2\pi$. The amplitude and period of this function can be determined exactly using the reduction formula in Section 3.6.

Figure 2.63

**Figure 2.64**

The calculator graph of $y = \sin x + \cos x$ in Fig. 2.64 shows many points on the graph and supports our conclusions about the shape and location of the graph.

**TRY THIS.** Graph $y = \sin(x) + 2\cos(x)$. ∎

To graph a function of the form $y = y_1 - y_2$ we could graph $y_1$ and $-y_2$ and then add the $y$-coordinates, or we can graph $y_1$ and $y_2$ and then subtract the $y$-coordinates. In the next example we graph by subtracting $y$-coordinates. The function in the next example is a difference of two simpler functions.

**EXAMPLE 4** A difference of two trigonometric functions

Sketch the graph of $y = \cos x - \sin(2x)$ by subtracting $y$-coordinates.

**Solution**

First graph the functions $y_1 = \cos x$ and $y_2 = \sin 2x$ as shown in Fig. 2.65. For each $x$-coordinate, we obtain the $y$-coordinate on $y = \cos x - \sin 2x$ by visually subtracting $y_2$ from $y_1$. Since $\sin 2x$ has a smaller period than $\cos x$, it is best to think of $y_2 = \sin 2x$ as being subtracted from $\cos x$ and causing $y = \cos x - \sin 2x$ to oscillate about $y_1$. The graph of $y = \cos x - \sin 2x$ crosses $y_1 = \cos x$ whenever $y_2$ is zero.

The calculator graph of $y = \cos x - \sin 2x$ in Fig. 2.66 supports our conclusions.

**Figure 2.65**

**Figure 2.66**

**TRY THIS.** Graph $y = \cos(2x) - \sin(x)$. ∎

# Modeling the Motion of a Spring

The sine and cosine functions are used in modeling the motion of a spring. If a weight is at rest while hanging from a spring, as shown in Fig. 2.67, then it is at the **equilibrium** position, or 0 on a vertical number line. If the weight is set in motion with an initial velocity $v_0$ from location $x_0$, then the location at time $t$ is given by

$$x = \frac{v_0}{\omega}\sin(\omega t) + x_0\cos(\omega t).$$

The letter $\omega$ (omega) is a constant that depends on the stiffness of the spring and the amount of weight on the spring. For positive values of $x$ the weight is below equilibrium, and for negative values it is above equilibrium. The initial velocity is considered to be positive if it is in the downward direction and negative if it is upward.

**Figure 2.67**

**EXAMPLE 5** Motion of a spring

A weight on a certain spring is set in motion with an upward velocity of 3 centimeters per second from a position 2 centimeters below equilibrium. Assume that for this spring and weight combination the constant $\omega$ has a value of 1. Write a formula

that gives the location of the weight in centimeters as a function of the time $t$ in seconds, and find the location of the weight 2 seconds after the weight is set in motion. Graph the function and estimate the period and amplitude.

**Solution**

Since upward velocity is negative and locations below equilibrium are positive, use $v_0 = -3$, $x_0 = 2$, and $\omega = 1$ in the formula for the motion of a spring:

$$x = -3\sin(t) + 2\cos(t)$$

If $t = 2$ seconds, then $x = -3\sin(2) + 2\cos(2) \approx -3.6$ centimeters, which is 3.6 centimeters above the equilibrium position. To graph the function, make a table of values for $x$ between 0 and $2\pi$:

| $t$ | 0 | $\pi/2$ | $\pi$ | $3\pi/2$ | $2\pi$ |
|---|---|---|---|---|---|
| $x = -3\sin(t) + 2\cos(t)$ | 2 | $-3$ | $-2$ | 3 | 2 |

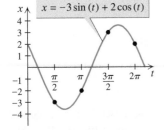

**Figure 2.68**

Sketch the graph through these points as shown in Fig. 2.68. From the graph, it appears that the period of the function is $2\pi$ and the amplitude is about 3.6. In Section 3.6 we will see how to determine the exact period and exact amplitude of this function.

The calculator graph in Fig. 2.69 can be used to estimate the period and amplitude of this function.

**Figure 2.69**

**TRY THIS.** If $x = -2\sin(2t) + 3\cos(2t)$ gives the location in centimeters of a weight on a spring $t$ seconds after it is set in motion, then what is the location at time $t = 3$? Graph the function and estimate the period and amplitude. ∎

**FOR THOUGHT...** True or False? Explain.

1. The function $y = x + \sin x$ is a periodic function.

2. The graph of $y = x + \sin x$ oscillates about the line $y = x$.

3. The graph of $y = \frac{1}{x}$ has a vertical asymptote at $x = 0$.

4. The $x$-axis is the horizontal asymptote for the graph of $y = \frac{1}{x}$.

5. The graph of $y = \frac{1}{x} + \sin x$ oscillates about the graph of $y = \frac{1}{x}$.

6. The graph of $y = \frac{1}{x} + \sin x$ has infinitely many $x$-intercepts.

7. The maximum $y$-coordinate on $y = \cos x + \cos 2x$ is 1.

8. The maximum $y$-coordinate on $y = \sin x + \cos x$ is 2.

9. On the interval $[0, 2\pi]$, $y = \sin x + \cos x$ crosses the graph of $y = \cos x$ at $x = 0$, $\pi$, and $2\pi$.

10. The period of the function $y = \sin \pi x$ is 2.

## 2.5 EXERCISES

**SKILLS**

*Use addition (or subtraction) of y-coordinates to sketch a graph of each function.*

1. $y = x + \cos x$

2. $y = -x + \sin x$

3. $y = \frac{1}{x} - \sin x$

4. $y = \frac{1}{x} + \sin x$

5. $y = \frac{1}{2}x + \sin x$

6. $y = \frac{1}{2}x - \cos x$

7. $y = x^2 + \sin x$

8. $y = x^2 + \cos x$

9. $y = \sqrt{x} + \cos x$

10. $y = \sqrt{x} - \sin x$

11. $y = |x| + 2\sin x$

12. $y = |x| - 2\cos x$

*Use addition (or subtraction) of y-coordinates to sketch a graph of each function between $-2\pi$ and $2\pi$.*

**13.** $y = \cos x + 2 \sin x$

**14.** $y = 2 \cos x + \sin x$

**15.** $y = \sin x - \cos x$

**16.** $y = \cos x - \sin x$

**17.** $y = \sin x + \cos 2x$

**18.** $y = \cos x + \sin 2x$

**19.** $y = 2 \sin x - \cos 2x$

**20.** $y = 3 \cos x - \sin 2x$

**21.** $y = \sin x + \sin 2x$

**22.** $y = \cos x + \cos 2x$

**23.** $y = \sin x + \cos \frac{1}{2}x$

**24.** $y = \cos x + \sin \frac{1}{2}x$

*Sketch the graph of each function between $-2$ and $2$.*

**25.** $y = x + \cos \pi x$

**26.** $y = x - \sin \pi x$

**27.** $y = \frac{1}{x} + \cos \pi x$

**28.** $y = \frac{1}{x} - \sin 2\pi x$

**29.** $y = \sin \pi x - \cos \pi x$

**30.** $y = \cos \pi x - \sin 2\pi x$

**31.** $y = \sin \pi x + \sin 2\pi x$

**32.** $y = \cos \pi x + \cos 2\pi x$

**33.** $y = \cos \frac{\pi}{2}x + \sin \pi x$

**34.** $y = \sin \frac{\pi}{2}x + \cos \pi x$

**35.** $y = \sin \pi x + \sin \frac{\pi}{2}x$

**36.** $y = \cos \pi x + \cos \frac{\pi}{2}x$

*Match each equation with its graph (a)–(f). Do not use a graphing calculator.*

**37.** $y = x + \sin(6x)$

**38.** $y = \frac{1}{x} + \sin(6x)$

**39.** $y = -x + \cos(0.5x)$

**40.** $y = \cos(x) + \cos(10x)$

**41.** $y = \sin(x) + \cos(10x)$

**42.** $y = x^2 + \cos(6x)$

**a.**

**b.**

**c.**

**d.**

**e.**

**f.**

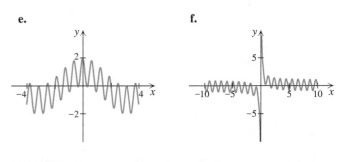

## MODELING

*Solve each problem.*

**43.** *Motion of a Spring* A weight on a vertical spring is given an initial downward velocity of 4 cm/sec from a point 3 cm above equilibrium. Assuming that the constant $\omega$ has a value of 1, write the formula for the location of the weight at time $t$, and find its location 3 sec after it is set in motion. Use a graph of the function to estimate the period and amplitude.

**44.** *Motion of a Spring* A weight on a vertical spring is given an initial upward velocity of 3 in./sec from a point 1 in. below equilibrium. Assuming that the constant $\omega$ has a value of $\sqrt{3}$, write the formula for the location of the weight at time $t$, and find its location 2 sec after it is set in motion. Use a graph of the function to estimate the period and amplitude.

**45.** *Total Profit* A person owns two businesses. For one the profit is growing exponentially at a rate of 1% per month. The other business is cyclical, with a higher profit in the summer than in the winter. The function $P(x) = 1000(1.01)^x + 500 \sin\left(\frac{\pi}{6}(x - 4)\right) + 2000$ gives the total profit as a function of the month with $x = 1$ corresponding to January of 2006.
   **a.** Graph the function for 60 months.

   **b.** What does the graph look like if the domain is 600 months (50 years)?

**46.** *Maximum Profit* What is the maximum profit for 2008 in the previous exercise?

## FOR WRITING/DISCUSSION

**47.** Graph $y = x + \sin x$ on a graphing calculator for $-100 \le x \le 100$ and $-100 \le y \le 100$. Explain your results.

**48.** Graph $y = x + \sin(50x)$ on a graphing calculator for $-10 \le x \le 10$ and $-10 \le y \le 10$. Explain your results.

**49.** Graph $y = x + \tan x$ on a graphing calculator for $-6 \le x \le 6$ and $-10 \le y \le 10$. Explain your results.

**50.** *Cooperative Learning* Work in a small group to combine some basic trigonometric functions using addition, subtraction, multiplication, or division and graph the combined function. (For instance, graph $y = \sin x + \cos x - \tan x$.) A graphing calculator would be useful here. Are these combined functions periodic? Can you determine the period of a combined function from the periods of the basic functions that are used to form it?

## REVIEW

**51.** Find the amplitude, period, phase shift, and range for the function $y = -3 \sin(\pi x/2 - \pi/2) + 7$.

**52.** The five key points on one cycle of a sine wave are $(\pi/4, 2)$, $(\pi/2, 5)$, $(3\pi/4, 2)$, $(\pi, -1)$, and $(5\pi/4, 2)$. Write the equation of the sine wave in the form $y = A \cos(B(x - C)) + D$.

**53.** Find the period and range for the function $y = 5 \sec(\pi x)$.

**54.** Find the domain of the function $y = \csc(2x)$.

**55.** Find the period and range of the function $y = \cot(\pi x)$.

**56.** Find the equations of all asymptotes to the graph of $y = -6 \tan(2x) + 1$.

## OUTSIDE THE BOX

**57.** *Renumbering Elm Street* The city council in Perfect City has changed the numbering scheme for the 200 houses on Elm Street. The houses will be renumbered with the natural numbers from 1 through 200. A city worker is given a box containing 1000 metal numbers, 100 of each digit, and told to distribute new house numbers in order of the addresses starting with 1 Elm Street. What address is the first one for which she will not have the correct digits?

**58.** *Real Solutions* Find all real solutions to
$$\left(\frac{x - 5}{3}\right)^{x^2 + x} = 1.$$

## 2.5 POP QUIZ

*Match each function with its graph (a)–(f).*

**1.** $y = x - \sin x$

**2.** $y = \sin(x) - 2\cos(x)$

**3.** $y = \dfrac{1}{x} + \sin(2x)$

**4.** $y = x^3 + \cos(9x)$

**5.** $y = \sin(x) - \cos(8x)$

**6.** $y = \sqrt{x} + \sin(\pi x)$

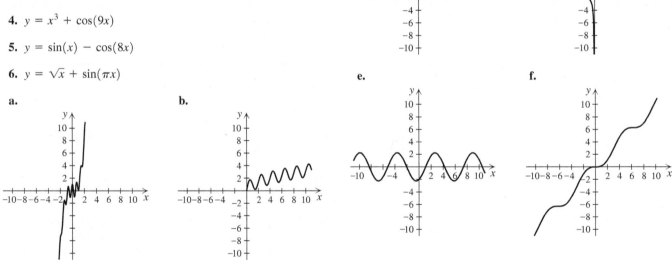

# Highlights

## 2.1 The Unit Circle and Graphing

| | | |
|---|---|---|
| **Unit Circle Definition** | If $\alpha$ is an angle in standard position and $(x, y)$ is the point of intersection of the terminal side and the unit circle, then $$\sin \alpha = y, \ \cos \alpha = x, \ \tan \alpha = \frac{y}{x},$$ $$\csc \alpha = \frac{1}{y}, \ \sec \alpha = \frac{1}{x}, \ \cot \alpha = \frac{x}{y}.$$ | $\sin(90°) = 1$ $\cos(\pi/2) = 0$ $\sin(30°) = 1/2$ $\sin(\pi/4) = \sqrt{2}/2$ $\sin(72.6) \approx -0.3367$ |
| **Graphs** | The graphs of $y = \sin x$ and $y = \cos x$ are sine waves with domain $(-\infty, \infty)$, range $[-1, 1]$, and period $2\pi$. | |

## 2.2 The General Sine Wave

| | | |
|---|---|---|
| **General Sine and Cosine Functions** | The graph of $y = A \sin[B(x - C)] + D$ or $y = A \cos[B(x - C)] + D$ is a sine wave with amplitude $|A|$, period $2\pi/|B|$, phase shift $C$, and vertical translation $D$. | $y = -3 \sin(2(x - \pi)) + 1$ amplitude 3, period $\pi$, phase shift $\pi$, vertical translation 1 |
| **Graphing a General Sine Function** | Start with the five key points on $y = \sin x$: $$(0, 0), \left(\frac{\pi}{2}, 1\right), (\pi, 0), \left(\frac{3\pi}{2}, -1\right), (2\pi, 0)$$ Divide each $x$-coordinate by $B$ and add $C$. Multiply each $y$-coordinate by $A$ and add $D$. Sketch one cycle of the general function through the five new points. Graph a general cosine function in the same manner. | $y = 3 \sin\left(2\left(x - \frac{\pi}{2}\right)\right) + 1$ Five new points: $\left(\frac{\pi}{2}, 1\right), \left(\frac{3\pi}{4}, 4\right),$ $(\pi, 1), \left(\frac{5\pi}{4}, -2\right), \left(\frac{3\pi}{2}, 1\right)$ Draw one cycle through these points. |
| **Frequency** | $F = 1/P$, $P$ is the period and $F$ is the frequency | $y = \sin(24\pi x)$, $F = 12$ |

## 2.3 Graphs of the Secant and Cosecant Functions

| | | |
|---|---|---|
| $y = \sec(x)$ | Domain $x \neq \pi/2 + k\pi$, range $(-\infty, -1] \cup [1, \infty)$, vertical asymptotes $x = \pi/2 + k\pi$ | See the Function Gallery in Section 2.4 on page 145. |
| $y = \csc(x)$ | Domain $x \neq k\pi$, range $(-\infty, -1] \cup [1, \infty)$, vertical asymptotes $x = k\pi$ | |

## 2.4 Graphs of the Tangent and Cotangent Functions

| | |
|---|---|
| $y = \tan(x)$ | Domain $x \neq \pi/2 + k\pi$, range $(-\infty, \infty)$, vertical asymptotes $x = \pi/2 + k\pi$, period $\pi$ |
| $y = \cot(x)$ | Domain $x \neq k\pi$, range $(-\infty, \infty)$, vertical asymptotes $x = k\pi$, period $\pi$ |

## 2.5 Combining Functions

| | |
|---|---|
| **Adding Functions** | The graph of $y = y_1 + y_2$ is sketched by adding the $y$-coordinates of the graphs of $y_1$ and $y_2$. If $y_1$ is a sine wave, then the graph of $y$ oscillates about the graph of $y_2$. |

# Chapter 2 Review Exercises

*Determine the exact value of each expression.*

**1.** $\sin(\pi/2)$  **2.** $\cos(\pi/3)$  **3.** $\tan(\pi/4)$

**4.** $\cot(\pi/4)$  **5.** $\sec(\pi)$  **6.** $\csc(\pi/2)$

**7.** $\cos(-3\pi/2)$  **8.** $\tan(-3\pi)$

*Determine the period of each function.*

**9.** $y = 2\sin(x)$  **10.** $y = 3\cos(x)$

**11.** $y = 6\tan(2x)$  **12.** $y = \cot(3x - 3)$

**13.** $y = \sec(\pi x) + 2$  **14.** $y = 4\csc(\pi x/2)$

**15.** $y = \cos(x/2 - 1)$  **16.** $y = 4\tan(3\pi x)$

*Determine the domain and range of each function.*

**17.** $y = 2\sin(3x)$  **18.** $y = 5\cos(4x)$

**19.** $y = \tan(2x)$  **20.** $y = \cot(3x)$

**21.** $y = \sec(x) - 2$  **22.** $y = \csc(x/2)$

**23.** $y = \cot(\pi x)$  **24.** $y = \tan(\pi x)$

*Determine the asymptotes for the graph of each function.*

**25.** $y = \tan(2x)$  **26.** $y = \tan(4x)$

**27.** $y = \cot(\pi x) + 1$  **28.** $y = 3\cot(\pi x/2)$

**29.** $y = \sec(x - \pi/2)$  **30.** $y = \sec(x + \pi/2)$

**31.** $y = \csc(\pi x + \pi)$  **32.** $y = \csc(2x - \pi)$

*Sketch at least one cycle of the graph of each function. Label the five key points on one cycle and determine the amplitude, period, phase shift, and range.*

**33.** $y = 3\sin(2x)$  **34.** $y = 2\cos(3x)$

**35.** $y = \cos(x - \pi/6)$  **36.** $y = \sin(x + \pi/4)$

**37.** $y = 1 + \cos(x + \pi/4)$  **38.** $y = \sin(x - \pi/2) - 3$

**39.** $y = \cos\left(\dfrac{\pi}{2}x\right)$  **40.** $y = \sin\left(\dfrac{\pi}{3}x\right)$

**41.** $y = \sin(2x + \pi)$  **42.** $y = \sin(3x - \pi)$

**43.** $y = -\dfrac{1}{2}\cos(2x)$  **44.** $y = 1 - 2\sin(x)$

**45.** $y = -2\sin(2x + \pi/3) + 1$

**46.** $y = -3\sin(2x + \pi/2) + 3$

**47.** $y = \dfrac{1}{3}\cos(\pi x + \pi) + 1$  **48.** $y = 1 + 2\cos(\pi x - \pi)$

*For each of the following curves find an equation of the form*
$y = A\sin[B(x - C)] + D.$

**49.**  **50.**

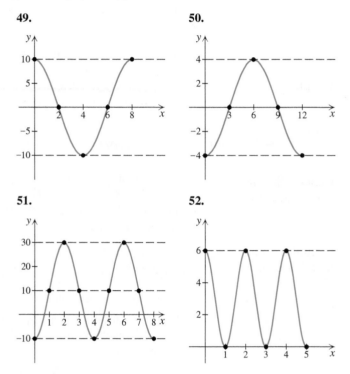

**51.**  **52.**

*Sketch at least one cycle of the graph of each function. Determine the period, asymptotes, and range of the function.*

**53.** $y = \tan(3x)$  **54.** $y = 1 + \tan(x + \pi/4)$

**55.** $y = \tan(2x + \pi)$  **56.** $y = \cot(x - \pi/4)$

**57.** $y = \sec\left(\dfrac{1}{2}x\right)$  **58.** $y = \csc\left(\dfrac{\pi}{2}x\right)$

**59.** $y = \dfrac{1}{2}\cot(2x)$  **60.** $y = 1 - \tan(x - \pi/3)$

**61.** $y = \cot(2x + \pi/3)$  **62.** $y = 2\tan(x + \pi/4)$

**63.** $y = \dfrac{1}{3}\csc(2x + \pi)$

**64.** $y = 1 + 2\sec(x - \pi/4)$

**65.** $y = 2\sec(x + \pi/4) - 1$

**66.** $y = \dfrac{1}{2}\csc(x - \pi) + 1$

*Sketch the graph of each function between $-2\pi$ and $2\pi$ by using the technique of adding y-coordinates.*

**67.** $y = \dfrac{1}{2}x + \sin x$

**68.** $y = 2x + \cos x$

**69.** $y = x^2 - \sin x$

**70.** $y = -x^2 + \cos x$

**71.** $y = \sin x - \sin 2x$

**72.** $y = \cos x - \cos 2x$

**73.** $y = 3 \sin x + \sin 2x$

**74.** $y = 3 \cos x + \cos 2x$

*Solve each problem.*

**75.** *Broadcasting the Oldies* If radio station Q92 is broadcasting its oldies at 92.3 FM, then it is broadcasting at a frequency of 92.3 megahertz, or $92.3 \times 10^6$ cycles per second. What is the period of a wave with this frequency? (FM stands for frequency modulation.)

**76.** *AM Radio* WWL in New Orleans is broadcasting at 870 AM (amplitude modulation), and its signal has a frequency of 870 kilohertz, or $870 \times 10^3$ cycles per second. What is the period of a wave with this frequency?

**77.** *Oscillating Depth* The depth of water in a tank oscillates between 12 ft and 16 ft. It takes 10 min for the depth to go from 12 to 16 ft and 10 min for the depth to go from 16 to 12 ft. Express the depth as a function of time in the form

$y = A \sin[B(x - C)] + D$ where the depth is 16 ft at time 0. Graph one cycle of the function.

**78.** *Oscillating Temperature* The temperature of the water in a tank oscillates between 100°F and 120°F. It takes 30 min for the temperature to go from 100° to 120° and 30 min for the temperature to go from 120° to 100°. Express the temperature as a function of time in the form $y = A \sin[B(x - C)] + D$ where the temperature is 100° at time 20 min. Graph one cycle of the function.

## OUTSIDE THE BOX

**79.** *Nines* Nine people applying for credit at the Highway 99 Loan Company listed nine different incomes, each containing a different number of digits. Each of the nine incomes was a whole number of dollars and the maximum income was a nine-digit number. The loan officer found that the arithmetic mean of the nine incomes was $123,456,789. What are the nine incomes?

**80.** *Tortoise and Hare* A scientist has trained a tortoise and a hare to travel around a circular track. The tortoise takes 10 min for one revolution and the hare takes 6 min. They start traveling simultaneously around the track in the same direction from the same starting point. In how many minutes will the hare pass the tortoise for the first time?

# Chapter **2** Test

*Sketch at least one cycle of the graph of each function. Determine the period, range, and amplitude for each function. Label the five key points on one cycle of the curve.*

**1.** $y = \sin(3x)$

**2.** $y = \cos(x + \pi/4)$

**3.** $y = \sin(\pi x/2)$

**4.** $f(x) = -2 \cos(x - \pi)$

**5.** $f(x) = -\dfrac{1}{2} \sin(x - \pi/2)$

**6.** $y = -\cos(2x - \pi/3) + 3$

**7.** Determine the amplitude and period for the sine curve in the accompanying graph. Write its equation in the form $y = A \sin[B(x - C)] + D$.

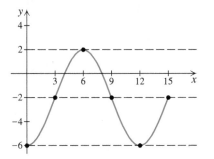

*Sketch at least one cycle of the graph of each function. Determine the period, asymptotes, and range for each function.*

**8.** $y = \tan(3x) - 2$

**9.** $y = \cot(x + \pi/2)$

**10.** $y = 2 \sec(x - \pi)$

**11.** $y = \csc(x - \pi/2) + 1$

*Solve each problem.*

**12.** Graph the function $y = 3 \sin x + \cos(2x)$ for $x$ between $-2\pi$ and $2\pi$ using the technique of adding the $y$-coordinates.

**13.** The pH of a water supply oscillates between 7.2 and 7.8. It takes 2 days for the pH to go from 7.2 to 7.8 and 2 days for the pH to go from 7.8 to 7.2. Express the pH as a function of time in the form $y = A \sin[B(x - C)] + D$ where the pH is 7.2 on day 13. Graph one cycle of the function.

# TYING IT ALL TOGETHER

## Chapters P–2

*Fill in the tables. Do not use a calculator.*

**1.**

| $\theta$ deg | 0 | 30 | 45 | 60 | 90 | 120 | 135 | 150 | 180 |
|---|---|---|---|---|---|---|---|---|---|
| $\theta$ rad | | | | | | | | | |
| $\sin\theta$ | | | | | | | | | |
| $\cos\theta$ | | | | | | | | | |
| $\tan\theta$ | | | | | | | | | |
| $\csc\theta$ | | | | | | | | | |
| $\sec\theta$ | | | | | | | | | |
| $\cot\theta$ | | | | | | | | | |

**2.**

| $\theta$ rad | $\pi$ | $7\pi/6$ | $5\pi/4$ | $4\pi/3$ | $3\pi/2$ | $5\pi/3$ | $7\pi/4$ | $11\pi/6$ | $2\pi$ |
|---|---|---|---|---|---|---|---|---|---|
| $\theta$ deg | | | | | | | | | |
| $\sin\theta$ | | | | | | | | | |
| $\cos\theta$ | | | | | | | | | |
| $\tan\theta$ | | | | | | | | | |
| $\csc\theta$ | | | | | | | | | |
| $\sec\theta$ | | | | | | | | | |
| $\cot\theta$ | | | | | | | | | |

*Sketch at least one cycle of the graph of each function. State the domain, range, and period of each function.*

**3.** $y = \sin(2x)$  **4.** $y = \cos(2x)$  **5.** $y = \tan(2x)$  **6.** $y = \sec(2x)$

**7.** $y = \csc(2x)$  **8.** $y = \sin(\pi x)$  **9.** $y = \cos(\pi x)$  **10.** $y = \tan(\pi x)$

*If $f(-x) = f(x)$ for all x then f is an even function, and if $f(-x) = -f(x)$ then f is an odd function.*
*Determine whether each function is even, odd, or neither by examining its graph.*

**11.** $f(x) = \sin(x)$  **12.** $f(x) = \cos(x)$  **13.** $f(x) = \tan(x)$  **14.** $f(x) = \cot(x)$

**15.** $f(x) = \sec(x)$  **16.** $f(x) = \csc(x)$  **17.** $f(x) = \sin^2(x)$  **18.** $f(x) = \cos^2(x)$

*Determine whether each function is increasing or decreasing on the given interval.*

**19.** $f(x) = \sin x,\ (0, \pi/2)$  **20.** $f(x) = \cos(x),\ (2\pi, 5\pi/2)$  **21.** $f(x) = \tan x,\ (3\pi/2, 5\pi/2)$

**22.** $f(x) = \cos(3x),\ (2\pi/3, \pi)$  **23.** $f(x) = \cot(2x),\ (3\pi/2, 2\pi)$  **24.** $f(x) = \sec(x),\ (\pi/2, \pi)$

FUNCTION GALLERY

**SOME BASIC FUNCTIONS OF TRIGONOMETRY**

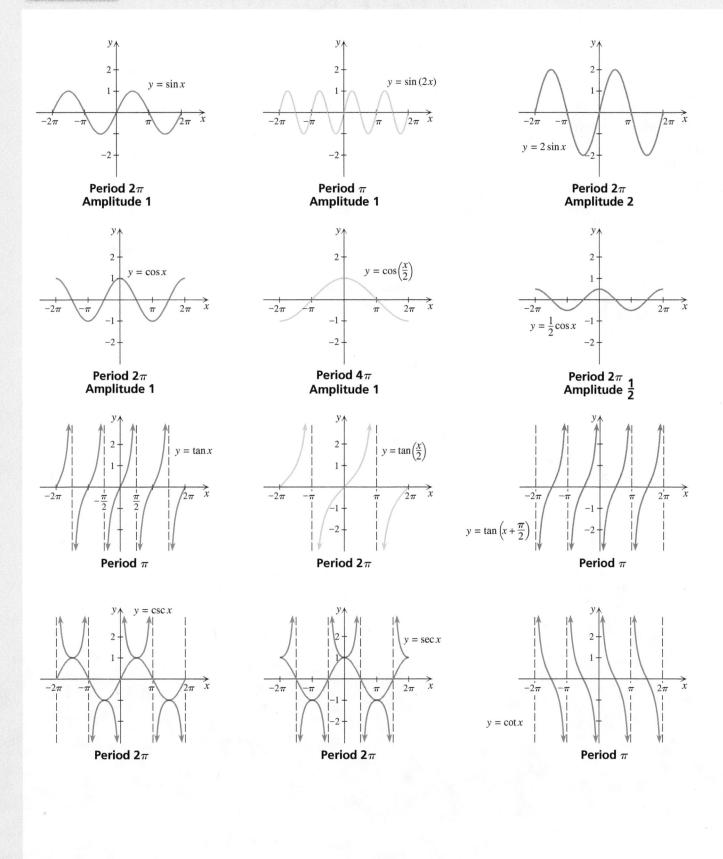

**Period $2\pi$**
**Amplitude 1**

**Period $\pi$**
**Amplitude 1**

**Period $2\pi$**
**Amplitude 2**

**Period $2\pi$**
**Amplitude 1**

**Period $4\pi$**
**Amplitude 1**

**Period $2\pi$**
**Amplitude $\frac{1}{2}$**

**Period $\pi$**

**Period $2\pi$**

**Period $\pi$**

**Period $2\pi$**

**Period $2\pi$**

**Period $\pi$**

# 3

# Trigonometric Identities

**3.1** Basic Identities

**3.2** Verifying Identities

**3.3** Sum and Difference Identities for Cosine

**3.4** Sum and Difference Identities for Sine and Tangent

**3.5** Double-Angle and Half-Angle Identities

**3.6** Product and Sum Identities

People have been hurling missiles of one kind or another since the Ice Age. Whether a ball, a rock, or a bullet is thrown, we've learned through trial and error that a projectile's path is determined by its initial velocity and the angle at which it's launched.

Athletes achieve a balance between accuracy and distance through years of practice. Operators of modern artillery achieve accuracy by calculating the velocity and angle of trajectory with mathematical precision.

**In this chapter, we will use trigonometric equations to analyze the flight of a projectile. Using trigonometry we can find the velocity and angle of trajectory to achieve the maximum distance for a projectile.**

# 3.1 Basic Identities

Recall that an identity is an equation that is satisfied by *every* number for which both sides are defined. Identities are used to simplify expressions and determine whether expressions are equivalent. There are infinitely many trigonometric identities, but only the most common identities should be memorized. We start by reviewing the most basic ones.

## Reciprocal Identities

We first mentioned the reciprocal relationships between the trigonometric functions in Section 1.4. Because of the definitions of the trigonometric functions, every trigonometric function is the reciprocal of another.

**Reciprocal Identities**

The following equations are identities:

$$\sin x = \frac{1}{\csc x} \qquad \cos x = \frac{1}{\sec x} \qquad \tan x = \frac{1}{\cot x}$$

$$\csc x = \frac{1}{\sin x} \qquad \sec x = \frac{1}{\cos x} \qquad \cot x = \frac{1}{\tan x}$$

If $x$ is an arc on the unit circle that terminates at $(a, b)$, then $\sin x = b$, $\cos x = a$, $\tan x = b/a$, and $\cot x = a/b$. Replacing $a$ and $b$ with $\sin x$ and $\cos x$ yields the following identities.

**Tangent and Cotangent in Terms of Sine and Cosine**

The following equations are identities:

$$\tan x = \frac{\sin x}{\cos x} \qquad \cot x = \frac{\cos x}{\sin x}$$

In Section 1.6 we used $r^2 = x^2 + y^2$ (which comes from the Pythagorean theorem) to prove that $\sin^2 \alpha + \cos^2 \alpha = 1$ for any angle $\alpha$. Dividing each side of $\sin^2 \alpha + \cos^2 \alpha = 1$ by $\sin^2 \alpha$ (provided $\sin \alpha \neq 0$) yields a new identity:

$$\frac{\sin^2 \alpha}{\sin^2 \alpha} + \frac{\cos^2 \alpha}{\sin^2 \alpha} = \frac{1}{\sin^2 \alpha}$$

$$1 + \left(\frac{\cos \alpha}{\sin \alpha}\right)^2 = \left(\frac{1}{\sin \alpha}\right)^2$$

$$1 + \cot^2 \alpha = \csc^2 \alpha$$

If we divide each side of the fundamental identity by $\cos^2 \alpha$ (provided $\cos \alpha \neq 0$), we get another new identity:

$$\frac{\sin^2 \alpha}{\cos^2 \alpha} + \frac{\cos^2 \alpha}{\cos^2 \alpha} = \frac{1}{\cos^2 \alpha}$$

$$\tan^2 \alpha + 1 = \sec^2 \alpha$$

Since these three identities are related to the Pythagorean theorem, they are called the **Pythagorean identities.** Note that we discussed these identities in terms of an angle $\alpha$, but the identities are true for any angle or real number $x$.

**Pythagorean Identities**

The following equations are identities:

$$\sin^2 x + \cos^2 x = 1 \qquad \cot^2 x + 1 = \csc^2 x \qquad \tan^2 x + 1 = \sec^2 x$$

We can support our conclusions that an equation is an identity by examining a graph. For example, the calculator graph of $y = \sin^2 x + \cos^2 x$ in Fig. 3.1 agrees with the graph of $y = 1$ for the 95 points that are shown on the calculator screen. This graph supports our conclusion that $\sin^2 x + \cos^2 x = 1$ is correct for all real numbers. However, you cannot use a calculator graph to prove that an equation is an identity, because a calculator plots only a finite number of points on a curve. For example, Fig. 3.2 shows 95 points on the graph of $y = \sin(x)$. The viewing window was carefully chosen so that the calculator would plot only the maximum points for 95 cycles of the function, and those points all have $y$-coordinates of 1. Between any two consecutive points plotted by the calculator in Fig. 3.2 there is really one whole cycle of the sine curve, but the calculator does not have enough resolution to show it. So the graph of $y = \sin x$ in Fig. 3.2 looks like the graph of $y = 1$, but $\sin x = 1$ is certainly not an identity.

**Figure 3.1**                    **Figure 3.2**

## Using Identities

Identities are used in many ways. One way is to simplify expressions involving trigonometric functions. Since all of the trigonometric functions are related to the sine and cosine functions, we can simplify an expression by first writing it in terms of sines and cosines only. In the next example we follow that strategy.

> **EXAMPLE 1** Using identities to simplify
>
> Write each expression in terms of sines and/or cosines, and then simplify.
>
> **a.** $\dfrac{\tan x}{\sec x}$    **b.** $\sin x + \cot x \cos x$

**Solution**

**a.** Replace $\tan x$ with $\frac{\sin x}{\cos x}$ and $\sec x$ with $\frac{1}{\cos x}$:

$$\frac{\tan x}{\sec x} = \frac{\dfrac{\sin x}{\cos x}}{\dfrac{1}{\cos x}}$$

$$= \frac{\sin x}{\cos x} \cdot \frac{\cos x}{1} \qquad \text{Invert and multiply.}$$

$$= \sin x$$

**b.** $\sin x + \cot x \cos x = \sin x + \dfrac{\cos x}{\sin x} \cdot \cos x$     Rewrite using sines and cosines.

$$= \sin x + \frac{\cos^2 x}{\sin x}$$

$$= \frac{\sin^2 x}{\sin x} + \frac{\cos^2 x}{\sin x}$$     Multiply $\sin x$ by $\frac{\sin x}{\sin x}$.

$$= \frac{\sin^2 x + \cos^2 x}{\sin x}$$     Add the fractions.

$$= \frac{1}{\sin x}$$     Since $\sin^2 x + \cos^2 x = 1$

$$= \csc x$$     Definition of cosecant

**TRY THIS.** Simplify $\frac{\tan x \csc x}{\sec x}$. ∎

Using identities we can write any one of the six trigonometric functions in terms of any other. For example, the identity

$$\sin x = \frac{1}{\csc x}$$

expresses the sine function in terms of the cosecant. To write sine in terms of cosine, we rewrite the fundamental identity:

$$\sin^2 x + \cos^2 x = 1$$

$$\sin^2 x = 1 - \cos^2 x$$

$$\sin x = \pm\sqrt{1 - \cos^2 x}$$

The $\pm$ symbol in this identity means that if $\sin x > 0$ then the identity is $\sin x = \sqrt{1 - \cos^2 x}$, and if $\sin x < 0$ then $\sin x = -\sqrt{1 - \cos^2 x}$.

 Note that the graph of $y = \sqrt{1 - \cos^2 x}$ in Fig. 3.3 is the same as the graph of $y = \sin x$ only when $\sin x > 0$. □

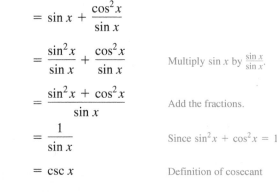

**Figure 3.3**

**EXAMPLE 2**  **Writing one function in terms of another**

Write an identity that expresses the tangent function in terms of the sine function.

**Solution**

The Pythagorean identity $\sin^2 x + \cos^2 x = 1$ yields

$$\cos x = \pm\sqrt{1 - \sin^2 x}.$$

Since $\tan x = \sin x / \cos x$, we can write

$$\tan x = \pm \frac{\sin x}{\sqrt{1 - \sin^2 x}}.$$

We use the positive or negative sign depending on the value of $x$.

**TRY THIS.** Write an identity that expresses the cotangent function in terms of the sine function. ∎

Because all of the trigonometric functions are related by identities, we can find their values for an angle if we know the value of any one of them and the quadrant of the angle.

EXAMPLE **3** **Using identities to find function values**

Given that $\tan \alpha = -2/3$ and $\alpha$ is in quadrant IV, find the values of the remaining five trigonometric functions at $\alpha$ by using identities.

**Solution**

Use $\tan \alpha = -2/3$ in the identity $\sec^2 \alpha = 1 + \tan^2 \alpha$:

$$\sec^2 \alpha = 1 + \left(-\frac{2}{3}\right)^2 = \frac{13}{9}$$

$$\sec \alpha = \pm\frac{\sqrt{13}}{3}$$

Since $\alpha$ is in quadrant IV, $\cos \alpha > 0$ and $\sec \alpha > 0$. So $\sec \alpha = \sqrt{13}/3$. Since cosine is the reciprocal of secant and cotangent is the reciprocal of tangent,

$$\cos \alpha = \frac{3}{\sqrt{13}} = \frac{3\sqrt{13}}{13} \quad \text{and} \quad \cot \alpha = -\frac{3}{2}.$$

Since $\sin \alpha$ is negative in quadrant IV, $\sin \alpha = -\sqrt{1 - \cos^2 \alpha}$:

$$\sin \alpha = -\sqrt{1 - \frac{9}{13}} = -\sqrt{\frac{4}{13}} = -\frac{2\sqrt{13}}{13}$$

Since cosecant is the reciprocal of sine, $\csc \alpha = -\sqrt{13}/2$.

**TRY THIS.** Suppose that $\cot \alpha = -1/3$ and $\alpha$ is in quadrant II. Find $\sin \alpha$ and $\cos \alpha$. ∎

Example 3 can be checked with right triangle trigonometry. If $\alpha$ is an angle of a right triangle, $\tan(\alpha) = \text{opp/adj}$. Since $\tan(\alpha) = -2/3$, draw a right triangle with angle $\alpha$ and label the opposite side 2 and the adjacent side 3 as in Fig. 3.4, ignoring the negative sign. By the Pythagorean theorem, the hypotenuse is $\sqrt{13}$. In quadrant IV only cosine and secant are positive. Now use the right triangle to find the other five values, affixing the appropriate sign for quadrant IV:

$$\sin(\alpha) = -\frac{2}{\sqrt{13}} = -\frac{2\sqrt{13}}{13}, \quad \csc(\alpha) = -\frac{\sqrt{13}}{2},$$

$$\cos(\alpha) = -\frac{3}{\sqrt{13}} = \frac{3\sqrt{13}}{13}, \quad \sec(\alpha) = \frac{\sqrt{13}}{3}, \quad \cot(\alpha) = -\frac{3}{2}.$$

**Figure 3.4**

**Figure 3.5**

## Odd and Even Identities

An *odd function* is one for which $f(-x) = -f(x)$, and an *even function* is one for which $f(-x) = f(x)$. We can classify each of the six trigonometric functions as either odd or even, based on the definitions of the functions in terms of the unit circle.

Figure 3.5 shows the real numbers $s$ and $-s$ as arcs on the unit circle. If the terminal point of $s$ is $(x, y)$, then the terminal point of $-s$ is $(x, -y)$. The $y$-coordinate for the terminal point of $-s$ is the opposite of the $y$-coordinate for the terminal point of $s$. So $\sin(-s) = -\sin(s)$, and sine is an odd function. Since the $x$-coordinates for $s$ and $-s$ are equal, $\cos(-s) = \cos(s)$, and cosine is an even function. Because sine is an odd function, cosecant is also an odd function:

$$\csc(-s) = \frac{1}{\sin(-s)} = \frac{1}{-\sin(s)} = -\csc(s)$$

Likewise, we can establish that secant is an even function and tangent and cotangent are both odd functions. These arguments establish the following identities.

**Odd and Even Identities**

| | | |
|---|---|---|
| **Odd:** | $\sin(-x) = -\sin(x)$ | $\csc(-x) = -\csc(x)$ |
| | $\tan(-x) = -\tan(x)$ | $\cot(-x) = -\cot(x)$ |
| **Even:** | $\cos(-x) = \cos(x)$ | $\sec(-x) = \sec(x)$ |

Note that only cosine and secant are even and the rest are odd functions. The graph of an odd function is symmetric about the origin and the graph of an even function is symmetric about the $y$-axis. The graphs of $y = \sin x$, $y = \csc x$, $y = \tan x$, and $y = \cot x$ are symmetric about the origin, while $y = \cos x$ and $y = \sec x$ are symmetric about the $y$-axis.

**EXAMPLE 4** Using odd and even identities

Simplify each expression.

**a.** $\sin(-x)\cot(-x)$ **b.** $\dfrac{1}{1 + \cos(-x)} + \dfrac{1}{1 - \cos x}$

**Solution**

**a.**
$$\sin(-x)\cot(-x) = (-\sin x)(-\cot x)$$
$$= \sin x \cdot \cot x$$
$$= \sin x \cdot \frac{\cos x}{\sin x}$$
$$= \cos x$$

**b.** First note that $\cos(-x) = \cos(x)$. Then find a common denominator and add the expressions.

$$\frac{1}{1 + \cos(-x)} + \frac{1}{1 - \cos x} = \frac{1}{1 + \cos x} + \frac{1}{1 - \cos x}$$

$$= \frac{1(1 - \cos x)}{(1 + \cos x)(1 - \cos x)} + \frac{1(1 + \cos x)}{(1 - \cos x)(1 + \cos x)}$$

$$= \frac{1 - \cos x}{1 - \cos^2 x} + \frac{1 + \cos x}{1 - \cos^2 x} \qquad \text{Product of a sum and difference}$$

$$= \frac{1 - \cos x}{\sin^2 x} + \frac{1 + \cos x}{\sin^2 x} \qquad \text{Pythagorean identity}$$

$$= \frac{1 - \cos x + 1 + \cos x}{\sin^2 x} \qquad \text{Add the fractions.}$$

$$= \frac{2}{\sin^2 x} = 2 \cdot \frac{1}{\sin^2 x}$$

$$= 2\csc^2 x \qquad \csc x = 1/\sin x$$

**TRY THIS.** Simplify $\csc(-x)\tan(-x)$.

We can check Example 4(b) by making a table for $y_1 = 1/(1 + \cos(-x)) + 1/(1 - \cos(x))$ and $y_2 = 2/(\sin(x))^2$ as shown in Fig. 3.6. The identical values for $y_1$ and $y_2$ support the result. What are the exact values listed in the $x$-column in Fig. 3.6?

| X | Y₁ | Y₂ |
|---|---|---|
| 0 | ERROR | ERROR |
| .7854 | 4 | 4 |
| 1.5708 | 2 | 2 |
| 2.3562 | 4 | 4 |
| 3.1416 | ERROR | ERROR |
| 3.927 | 4 | 4 |
| 4.7124 | 2 | 2 |

X=0

**Figure 3.6**

**EXAMPLE 5** Odd or even functions

Determine whether each function is odd, even, or neither.

**a.** $f(x) = \dfrac{\cos(2x)}{x}$ **b.** $g(t) = \sin t + \cos t$

**Solution**

**a.** Replace $x$ by $-x$ and see whether $f(-x)$ is equal to $f(x)$ or $-f(x)$:

$$f(-x) = \frac{\cos(2(-x))}{-x} \quad \text{Replace } x \text{ by } -x.$$

$$= \frac{\cos(-2x)}{-x}$$

$$= \frac{\cos(2x)}{-x} \quad \text{Since cosine is an even function}$$

$$= -\frac{\cos(2x)}{x} \quad \text{This is } -f(x).$$

**Figure 3.7**

Since $f(-x) = -f(x)$, the function $f$ is an odd function.

 The graph of $y = \cos(2x)/x$ in Fig. 3.7 appears to be symmetric about the origin, which supports the conclusion that the function is odd. □

**b.** Replace $t$ by $-t$ and see whether $g(-t)$ is equal to $g(t)$ or $-g(t)$:

$$g(-t) = \sin(-t) + \cos(-t) \quad \text{Replace } t \text{ by } -t.$$

$$= -\sin t + \cos t \quad \text{Sine is odd and cosine is even.}$$

Since $g(-t) \neq g(t)$ and $g(-t) \neq -g(t)$, the function is neither odd nor even.

 Since the graph of $y = \sin(x) + \cos(x)$ in Fig. 3.8 is neither symmetric about the origin nor symmetric about the $y$-axis, the graph supports the conclusion that the function is neither odd nor even.

**Figure 3.8**

**TRY THIS.** Determine whether $f(x) = \csc(x) + \tan(x)$ is an odd or even function. ■

# Identity or Not?

In this section we have seen the reciprocal identities, the Pythagorean identities, and the odd-even identities and we are just getting started. With so many identities in trigonometry it might seem difficult to determine whether a given equation is an identity. Remember that an identity is satisfied by every value of the variable for which both sides are defined. If you can find one number for which the left-hand side of the equation has a value different from the right-hand side, then the equation is not an identity. An equation that is not an identity might have many solutions. So you might have to try more than one value for the variable to find one that fails to satisfy the equation.

**EXAMPLE 6**    **Proving that an equation is not an identity**

Show that $\sin(2t) = 2\sin(t)$ is not an identity.

**Solution**

We select a value for $t$ for which we can easily evaluate each side of the equation. If $t = \pi/4$, then

$$\sin(2t) = \sin\left(2 \cdot \frac{\pi}{4}\right) = \sin\left(\frac{\pi}{2}\right) = 1$$

and

$$2\sin(t) = 2\sin\left(\frac{\pi}{4}\right) = 2 \cdot \frac{\sqrt{2}}{2} = \sqrt{2}.$$

**Figure 3.9**

Since the values of $\sin(2t)$ and $2\sin(t)$ are unequal for $t = \pi/4$, the equation is not an identity. Note, if $t = 0$, the equation is satisfied. So $t = 0$ is a bad choice.

 The graphs of $y_1 = \sin(2x)$ and $y_2 = 2\sin(x)$ shown in Fig. 3.9 appear to be different. This supports the conclusion that $\sin(2t) = 2\sin(t)$ is not an identity.

**TRY THIS.** Show that $\cos(3t) = 3\cos(t)$ is not an identity. ■

## FOR THOUGHT... True or False? Explain.

1. The equation $\sin x = \cos x$ is an identity.

2. If we simplify the expression $(\tan x)(\cot x)$, we get 1.

3. If $\sin(2x) = 2\sin(x)\cos(x)$ is an identity, then so is $\dfrac{\sin(2x)}{\sin(x)} = 2\cos(x)$.

4. The equation $(\sin x + \cos x)^2 = \sin^2 x + \cos^2 x$ is an identity.

5. $\tan 1 = \sqrt{1 - \sec^2 1}$

6. $\sin^2(-6) = -\sin^2(6)$

7. $\cos^3(-9) = -\cos^3(9)$

8. $\sin(-3)\csc(-3) = 1$

9. $\sin^2(-4) + \cos^2(-4) = -1$

10. $(1 - \sin 2)(1 + \sin 2) = 1 - \sin^2 2$

Scan this code
to link to section
lecture videos.

## **3.1** EXERCISES

### CONCEPTS

*Fill in the blank.*

1. The identity $1 + \cot^2 x = \csc^2 x$ is one of the _____ identities.

2. If $f(-x) = f(x)$, then $f$ is a(n) _____ function.

3. If $f(-x) = -f(x)$, then $f$ is a(n) _____ function.

4. An equation that is true for all values of the variable for which both sides are defined is a(n) _____ .

5. Sine, cosecant, tangent, and cotangent are _____ functions.

6. Cosine and secant are _____ functions.

7. The sine function is the _____ of the cosecant function.

8. The tangent function is the _____ function divided by the _____ function.

### SKILLS

*Write each expression in terms of sines and/or cosines, and then simplify.*

9. $\dfrac{\sec x}{\tan x}$

10. $\dfrac{\cot x}{\csc x}$

11. $\dfrac{\sin x}{\csc x} + \dfrac{\cos x}{\sec x}$

12. $\dfrac{1}{\sin^2 x} - \dfrac{1}{\tan^2 x}$

13. $(1 - \sin \alpha)(1 + \sin \alpha)$

14. $(\sec \alpha - 1)(\sec \alpha + 1)$

15. $(\cos \beta \tan \beta + 1)(\sin \beta - 1)$

16. $(1 + \cos \beta)(1 - \cot \beta \sin \beta)$

17. $\dfrac{1 + \cos \alpha \tan \alpha \csc \alpha}{\csc \alpha}$

18. $\dfrac{(\cos \alpha \tan \alpha + 1)(\sin \alpha - 1)}{\cos^2 \alpha}$

*Write an identity that expresses the first function in terms of the second.*

19. $\cot(x)$, in terms of $\csc(x)$

20. $\sec(x)$, in terms of $\tan(x)$

21. $\sin(x)$, in terms of $\cot(x)$

22. $\cos(x)$, in terms of $\tan(x)$

23. $\tan(x)$, in terms of $\csc(x)$

24. $\cot(x)$, in terms of $\sec(x)$

*In each exercise, use identities to find the exact values at $\alpha$ for the remaining five trigonometric functions.*

25. $\tan \alpha = 1/2$ and $0 < \alpha < \pi/2$

26. $\sin \alpha = 3/4$ and $\pi/2 < \alpha < \pi$

27. $\cos \alpha = -\sqrt{3}/5$ and $\alpha$ is in quadrant III

28. $\sec \alpha = -4\sqrt{5}/5$ and $\alpha$ is in quadrant II

29. $\cot \alpha = -1/3$ and $-\pi/2 < \alpha < 0$

30. $\csc \alpha = \sqrt{3}$ and $0 < \alpha < \pi/2$

*Simplify each expression.*

31. $\sin(-x)\cot(-x)$

32. $\sec(-x) - \sec(x)$

33. $\sin(y) + \sin(-y)$

34. $\cos(y) + \cos(-y)$

35. $\dfrac{\sin(x)}{\cos(-x)} + \dfrac{\sin(-x)}{\cos(x)}$

36. $\dfrac{\cos(-x)}{\sin(-x)} - \dfrac{\cos(-x)}{\sin(x)}$

37. $(1 + \sin(\alpha))(1 + \sin(-\alpha))$

38. $(1 - \cos(-\alpha))(1 + \cos(\alpha))$

39. $\sin(-\beta)\cos(-\beta)\csc(\beta)$

40. $\tan(-\beta)\csc(-\beta)\cos(\beta)$

*Determine whether each function is odd, even, or neither.*

41. $f(x) = \sin(2x)$

42. $f(x) = \cos(2x)$

43. $f(x) = \cos x + \sin x$

44. $f(x) = 2\sin x \cos x$

**45.** $f(t) = \sec^2(t) - 1$      **46.** $f(t) = 2 + \tan(t)$

**47.** $f(\alpha) = 1 + \sec \alpha$      **48.** $f(\beta) = 1 + \csc \beta$

**49.** $f(x) = \dfrac{\sin x}{x}$      **50.** $f(x) = x \cos x$

**51.** $f(x) = x + \sin x$      **52.** $f(x) = \csc(x^2)$

*Show that each equation is not an identity. Write your explanation in paragraph form.*

**53.** $(\sin \gamma + \cos \gamma)^2 = \sin^2 \gamma + \cos^2 \gamma$

**54.** $\tan^2 x - 1 = \sec^2 x$

**55.** $(1 + \sin \beta)^2 = 1 + \sin^2 \beta$      **56.** $\sin(2\alpha) = \sin \alpha \cos \alpha$

**57.** $\sin \alpha = \sqrt{1 - \cos^2 \alpha}$      **58.** $\tan \alpha = \sqrt{\sec^2 \alpha - 1}$

**59.** $\sin(y) = \sin(-y)$      **60.** $\cos(-y) = -\cos y$

**61.** $\cos^2(y) - \sin^2(y) = \sin(2y)$   **62.** $\cos(2x) = 2 \cos x \sin x$

*Use identities to simplify each expression.*

**63.** $1 - \dfrac{1}{\cos^2 x}$      **64.** $\dfrac{\sin^4 x - \sin^2 x}{\sec x}$

**65.** $\dfrac{-\tan^2 t - 1}{\sec^2 t}$      **66.** $\dfrac{\cos w \sin^2 w + \cos^3 w}{\sec w}$

**67.** $\dfrac{\sin^2 \alpha - \cos^2 \alpha}{1 - 2 \cos^2 \alpha}$      **68.** $\dfrac{\sin^3(-\theta)}{\sin^3 \theta - \sin \theta}$

**69.** $\dfrac{\tan^3 x - \sec^2 x \tan x}{\cot(-x)}$      **70.** $\sin x + \dfrac{\cos^2 x}{\sin x}$

**71.** $\dfrac{1}{\sin^3 x} - \dfrac{\cot^2 x}{\sin x}$      **72.** $1 - \dfrac{\sec^2 x}{\tan^2 x}$

**73.** $\sin^4 x - \cos^4 x$      **74.** $\csc^4 x - \cot^4 x$

## MODELING

*Solve each problem.*

**75.** *Rake and Trail* How well a motorcycle or bicycle handles depends on the steering geometry. The formula

$$T = \frac{r \cos \theta - R}{\sin \theta}$$

gives the trail $T$ as a function of the radius of the wheel $r$, head angle $\theta$, and rake or offset $R$. Show that this formula is correct. See the accompanying figure.

**Figure for Exercise 75**

**76.** *Finding the Head Angle* Find the head angle $\theta$, in degrees, that would produce a trail of 4.9 inches when $R = 3$ in. and $r = 13.5$ in. Use the formula from the previous exercise along with a graphing calculator.

## WRITING/DISCUSSION

**77.** *Algebraic Identities* List as many algebraic identities as you can and explain why each one is an identity.

**78.** *Trigonometric Identities* List as many trigonometric identities as you can and explain why each one is an identity.

## REVIEW

**79.** Find the smallest positive angle that is coterminal with $-35°$.

**80.** Convert each degree measure to radian measure.
   **a.** $225°$           **b.** $-210°$

   **c.** $270°$

**81.** Find the exact area of the sector of the circle with radius 6 feet and central angle $15°$.

**82.** A race car is averaging 180 mph on a circular track with radius 1/4 mile. Find its angular velocity in radians per minute.

**83.** Evaluate without a calculator. Some of these expressions are undefined.
   **a.** $\cos(-3\pi/2)$       **b.** $\sin(9\pi/4)$

   **c.** $\tan(-3\pi/2)$       **d.** $\tan(3\pi)$

   **e.** $\sec(13\pi/6)$       **f.** $\csc(5\pi/2)$

   **g.** $\cot(-5\pi/3)$       **h.** $\sin(3\pi/2)$

**84.** The lengths of the legs of a right triangle are 5 meters and 8 meters. Find the length of the hypotenuse and the degree measures of the acute angles. Round to the nearest tenth.

**OUTSIDE THE BOX**

85. *Army of Ants* An army of ants is marching across the kitchen floor. If they form columns with 10 ants in each column, then there are 6 ants left over. If they form columns with 7, 11, or 13 ants in each column, then there are 2 ants left over. What is the smallest number of ants that could be in this army?

86. *Four Pipes* Four pipes with circular cross sections are placed in a V-shaped trench so that they all just fit as shown in the accompanying figure. If the radius of the smallest pipe is 16 inches and the radius of the largest is 54 inches, then what are the radii of the two pipes in between?

**Figure for Exercise 86**

## 3.1 POP QUIZ

1. Simplify the expression $\cot x \sec x$.

2. Find exact values for $\cos \alpha$ and $\cot \alpha$ if $\sin \alpha = 1/3$ and $0 < \alpha < \pi/2$.

3. Is $f(x) = \cos(3x)$ even or odd?

4. Simplify $\dfrac{2}{\sec^2 \alpha} + \dfrac{2}{\csc^2 \alpha}$.

5. Is $\sin(2x) = 2 \sin(x)$ an identity?

## LINKING concepts...

**For Individual or Group Explorations**

### Modeling the Motion of a Projectile

*A projectile is fired with initial velocity of $v_0$ feet per second. The projectile can be pictured as being fired from the origin into the first quadrant, making an angle $\theta$ with the positive x-axis, as shown in the figure. If there is no air resistance, then at t seconds the coordinates of the projectile (in feet) are $x = v_0 t \cos \theta$ and $y = -16t^2 + v_0 t \sin \theta$. Suppose a projectile leaves the gun at 100 ft/sec and $\theta = 60°$.*

a) What are the coordinates of the projectile at time $t = 4$ sec?

b) For how many seconds is the projectile in the air?

c) How far from the gun does the projectile land?

*(continued)*

    **d)** What is the maximum height attained by the projectile?

    **e)** Find an expression in terms of $v_0$ and $\theta$ for the time in the air.

    **f)** Find an expression in terms of $v_0$ and $\theta$ for the distance from the gun.

    **g)** Find an expression in terms of $v_0$ and $\theta$ for the maximum height.

    **h)** Show that $y = -\dfrac{16 \sec^2\theta}{v_0^2} x^2 + x \tan \theta$.

# 3.2 Verifying Identities

In Section 3.1 we saw that any trigonometric function can be expressed in terms of any other trigonometric function. So an expression involving one or more trigonometric functions could be written in many different equivalent forms. We must often decide whether two expressions are equivalent. The fact that two expressions are equivalent is expressed as an identity. In this section we concentrate on techniques for verifying that a given equation is an identity.

## Developing a Strategy

To prove or verify that an equation is an identity, we start with the expression on one side of the equation and use known identities and properties of algebra to convert it into the expression on the other side of the equation. It is the same process as simplifying expressions, except that in this case we know exactly what we want as the final expression. If one side appears to be more complicated than the other, we usually start with the more complicated side and simplify it. An expression that involves fewer symbols is generally considered simpler.

**EXAMPLE 1**   **Verifying an identity**

Verify that $1 + \sec x \sin x \tan x = \sec^2 x$ is an identity.

**Solution**

We start with the left-hand side, the more complicated side, and write it in terms of sine and cosine. Our goal is to get $\sec^2 x$ as the final simplified expression.

$$1 + \sec x \sin x \tan x = 1 + \frac{1}{\cos x} \sin x \frac{\sin x}{\cos x} \qquad \text{Convert to sines and cosines.}$$

$$= 1 + \frac{\sin^2 x}{\cos^2 x} \qquad \text{Simplify.}$$

$$= 1 + \tan^2 x \qquad \text{Definition of tangent}$$

$$= \sec^2 x \qquad \text{Pythagorean identity}$$

So $1 + \sec x \sin x \tan x = \sec^2 x$ is an identity.

**TRY THIS.** Verify that $1 - \sec(x)\csc(x)\tan(x) = -\tan^2(x)$ is an identity. ∎

    In verifying identities it is often necessary to multiply binomials or to factor trinomials involving trigonometric functions.

EXAMPLE **2** **Multiplying binomials**

Find each product.

**a.** $(1 + \tan x)(1 - \tan x)$  **b.** $(2 \sin \alpha + 1)^2$

**Solution**

**a.** The product of a sum and a difference is equal to the difference of two squares:

$$(1 + \tan x)(1 - \tan x) = 1 - \tan^2 x$$

Note that $(1 + \tan x)(1 - \tan x) = 1 - \tan^2 x$ is an identity.

**b.** Use $(a + b)^2 = a^2 + 2ab + b^2$ with $a = 2 \sin \alpha$ and $b = 1$:

$$(2 \sin \alpha + 1)^2 = (2 \sin \alpha)^2 + 2 \cdot 2 \sin \alpha \cdot 1 + 1^2$$
$$= 4 \sin^2 \alpha + 4 \sin \alpha + 1$$

Note that $(2 \sin \alpha + 1)^2 = 4 \sin^2 \alpha + 4 \sin \alpha + 1$ is an identity.

**TRY THIS.** Find the product $(1 - 2 \sin x)(1 + 2 \sin x)$.  ■

EXAMPLE **3** **Factoring with trigonometric functions**

Factor each expression.

**a.** $\sec^2 x - \tan^2 x$  **b.** $\sin^2 \beta + \sin \beta - 2$

**Solution**

**a.** The expression $\sec^2 x - \tan^2 x$ is a difference of two squares. A difference of two squares is equal to a product of a sum and a difference, $a^2 - b^2 = (a + b)(a - b)$. So

$$\sec^2 x - \tan^2 x = (\sec x + \tan x)(\sec x - \tan x).$$

**b.** If $x = \sin \beta$, then $\sin^2 \beta + \sin \beta - 2$ has the form $x^2 + x - 2$, which factors as $(x + 2)(x - 1)$. So

$$\sin^2 \beta + \sin \beta - 2 = (\sin \beta + 2)(\sin \beta - 1).$$

**TRY THIS.** Factor $\sin^2 x + 4 \sin x + 4$.  ■

The equations that appear in Examples 2 and 3 are identities. They are not the kind of identities that are memorized because they can be obtained at any time by the principles of multiplying or factoring. The next example uses the identity $(1 - \sin \alpha)(1 + \sin \alpha) = 1 - \sin^2 \alpha$.

EXAMPLE **4** **An identity with equal fractions**

Prove that $\dfrac{\cos \alpha}{1 - \sin \alpha} = \dfrac{1 + \sin \alpha}{\cos \alpha}$ is an identity.

**Solution**

To prove that the equation is an identity, multiply the numerator and denominator of the left-hand side by $1 + \sin \alpha$ because $1 + \sin \alpha$ appears in the numerator of the right-hand side.

$$\frac{\cos \alpha}{1 - \sin \alpha} = \frac{\cos \alpha(1 + \sin \alpha)}{(1 - \sin \alpha)(1 + \sin \alpha)}$$

$$= \frac{\cos \alpha(1 + \sin \alpha)}{1 - \sin^2 \alpha} \qquad \text{Leave numerator in factored form.}$$

$$= \frac{\cos \alpha(1 + \sin \alpha)}{\cos^2 \alpha} \qquad \text{Pythagorean identity}$$

$$= \frac{1 + \sin \alpha}{\cos \alpha} \qquad \text{Reduce.}$$

**Figure 3.10**

So $\frac{\cos \alpha}{1 - \sin \alpha} = \frac{1 + \sin \alpha}{\cos \alpha}$ is an identity.

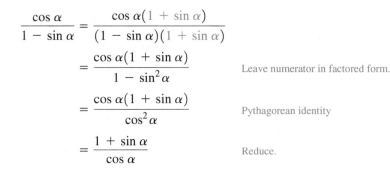 In Fig. 3.10, the graphs of $y_1 = \cos(x)/(1 - \sin(x))$ and $y_2 = (1 + \sin(x))/\cos(x)$ appear to coincide. This supports our conclusion that the equation is an identity.

**TRY THIS.** Prove that $\frac{\csc x - 1}{\cot x} = \frac{\cot x}{\csc x + 1}$ is an identity. ■

In Example 4 the numerator and denominator were multiplied by $1 + \sin \alpha$ because the expression on the right-hand side has $1 + \sin \alpha$ in its numerator. We could have multiplied the numerator and denominator of the left-hand side by $\cos \alpha$ because the expression on the right-hand side has $\cos \alpha$ in its denominator. You should prove that the equation of Example 4 is an identity by multiplying the numerator and denominator by $\cos \alpha$.

Another strategy to use when a fraction has a sum or difference in its numerator is to write the fraction as a sum or difference of two fractions, as in the next example.

**EXAMPLE 5**  **Separating one fraction into two**

Prove that $\frac{\csc x - \sin x}{\sin x} = \cot^2 x$ is an identity.

**Solution**

First rewrite the left-hand side as a difference of two rational expressions.

$$\frac{\csc x - \sin x}{\sin x} = \frac{\csc x}{\sin x} - \frac{\sin x}{\sin x} \qquad \frac{a}{b} - \frac{c}{b} = \frac{a - c}{b}$$

$$= \csc x \cdot \frac{1}{\sin x} - 1 \qquad \frac{a}{b} = a \cdot \frac{1}{b}$$

$$= \csc^2 x - 1 \qquad \text{Since } 1/\sin x = \csc x$$

$$= \cot^2 x \qquad \text{Pythagorean identity}$$

So $\frac{\csc x - \sin x}{\sin x} = \cot^2 x$ is an identity.

**TRY THIS.** Prove that $\frac{\sec x - \cos x}{\cos x} = \tan^2 x$ is an identity. ■

In verifying an identity, you can start with the expression on either side of the equation and use known identities and properties of algebra to get the expression on the other side. In the next example the expression on the right-hand side appears to be more complicated. So we start with it and simplify it.

**EXAMPLE 6**  **Subtracting fractions in an identity**

Prove that $2 \tan^2 x = \frac{1}{\csc x - 1} - \frac{1}{\csc x + 1}$ is an identity.

**Solution**

In this equation, the right-hand side is the more complicated one. We will simplify the right-hand side by getting a common denominator and combining the fractions, keeping $2\tan^2 x$ in mind as our goal.

$$\frac{1}{\csc x - 1} - \frac{1}{\csc x + 1} = \frac{1(\csc x + 1)}{(\csc x - 1)(\csc x + 1)} - \frac{1(\csc x - 1)}{(\csc x + 1)(\csc x - 1)}$$

$$= \frac{\csc x + 1}{\csc^2 x - 1} - \frac{\csc x - 1}{\csc^2 x - 1}$$

$$= \frac{2}{\csc^2 x - 1} \qquad \text{Subtract the numerators.}$$

$$= \frac{2}{\cot^2 x} \qquad \text{Pythagorean identity}$$

$$= 2 \cdot \frac{1}{\cot^2 x} \qquad \frac{a}{b} = a \cdot \frac{1}{b}$$

$$= 2\tan^2 x \qquad \text{Reciprocal identity.}$$

**TRY THIS.** Prove that $-2\cot^2 x = \frac{1}{1 - \sec x} + \frac{1}{1 + \sec x}$ is an identity. ∎

In every example so far, we started with one side and converted it into the other side. Although all identities can be verified by that method, sometimes it is simpler to convert both sides into the same expression. If both sides are shown to be equivalent to a common expression, then the identity is certainly verified. This technique works best when both sides are rather complicated and we simply set out to simplify them.

**EXAMPLE 7** Verifying an identity

Prove that $\frac{1 - \sin^2 t}{1 - \csc(-t)} = \frac{1 + \sin(-t)}{\csc t}$ is an identity.

**Solution**

Use $\csc(-t) = -\csc t$ to simplify the left-hand side:

$$\frac{1 - \sin^2 t}{1 - \csc(-t)} = \frac{1 - \sin^2 t}{1 + \csc t} = \frac{\cos^2 t}{1 + \csc t}$$

Now look for a way to get the right-hand side of the original equation equivalent to $(\cos^2 t)/(1 + \csc t)$. Since $\sin(-t) = -\sin t$, we can write $1 + \sin(-t)$ as $1 - \sin t$ and then get $\cos^2 t$ in the numerator by multiplying by $1 + \sin t$:

$$\frac{1 + \sin(-t)}{\csc t} = \frac{1 - \sin t}{\csc t} \qquad \text{Since } \sin(-t) = -\sin t$$

$$= \frac{(1 - \sin t)(1 + \sin t)}{\csc t(1 + \sin t)} \qquad \begin{array}{l}\text{Multiply numerator and denominator by}\\ 1 + \sin t.\end{array}$$

$$= \frac{1 - \sin^2 t}{\csc t + \csc t \sin t} \qquad (a - b)(a + b) = a^2 - b^2$$

$$= \frac{\cos^2 t}{\csc t + \dfrac{1}{\sin t}\sin t} \qquad \text{Pythagorean identity}$$

$$= \frac{\cos^2 t}{1 + \csc t}$$

Since both sides of the equation are equivalent to the same expression, the equation is an identity.

**TRY THIS.** Prove that $\dfrac{1 - \cos^2(-t)}{\sin(-t)} = \tan(-t)\cos(-t)$ is an identity. ∎

## The Strategy

Verifying identities takes practice. There are many different ways to verify a particular identity, so just go ahead and get started. If you do not seem to be getting anywhere, try another approach. The strategy that we used in the examples is stated below. Other techniques are sometimes needed to prove identities, but you can prove most of the identities in this text using the following strategy.

**STRATEGY**

### Verifying Identities

1. Work on one side of the equation (usually the more complicated side), keeping in mind the expression on the other side as your goal.

2. Some expressions can be simplified quickly if they are rewritten in terms of sines and cosines only. (See Example 1.)

3. To convert one rational expression into another, multiply the numerator and denominator of the first by either the numerator or the denominator of the desired expression. (See Example 4.)

4. If the numerator of a rational expression is a sum or difference, convert the rational expression into a sum or difference of two rational expressions. (See Example 5.)

5. If a sum or difference of two rational expressions occurs on one side of the equation, then find a common denominator and combine them into one rational expression. (See Example 6.)

6. Sometimes it is easiest to simplify both sides of the equation. If each side is equivalent to a third expression, then the equation is an identity. (See Example 7.)

**FOR THOUGHT...** True or False? Answer true if the equation is an identity and false if it is not. If false, find a value for $x$ for which the two sides have different values.

1. $\dfrac{\sin x}{\csc x} = \sin^2 x$

2. $\dfrac{\cot x}{\tan x} = \tan^2 x$

7. $\dfrac{1}{1 - \sin x} = \dfrac{1 + \sin x}{\cos^2 x}$

3. $\dfrac{\sec x}{\csc x} = \tan x$

4. $\sin x \sec x = \tan x$

8. $\tan x \cot x = 1$

5. $\dfrac{\cos x + \sin x}{\cos x} = 1 + \tan x$

9. $(1 - \cos x)(1 - \cos x) = \sin^2 x$

6. $\sec x + \dfrac{\sin x}{\cos x} = \dfrac{1 + \sin x \cos x}{\cos x}$

10. $(1 - \csc x)(1 + \csc x) = \cot^2 x$

Scan this code to link to section lecture videos.

## 3.2 EXERCISES

### CONCEPTS

*Fill in the blank.*

1. To verify an identity, we often simplify the more _____ side.

2. Sometimes we write all expressions in terms of _____ and _____ .

3. To convert one rational expression into another, multiply the _____ and _____ of the first by the numerator or denominator of the second.

4. If each side of an equation can be converted into the same expression, then the equation is a(n) _____ .

## SKILLS

*Match each expression on the left with one on the right that completes each equation as an identity.*

**5.** $\cos x \tan x = ?$

    **A.** 1

**6.** $\sec x \cot x = ?$

    **B.** $-\tan^2 x$

**7.** $(\csc x - \cot x)(\csc x + \cot x) = ?$

    **C.** $\cot^2 x$

**8.** $\dfrac{\sin x + \cos x}{\sin x} = ?$

    **D.** $\sin x$

**9.** $(1 - \sec x)(1 + \sec x) = ?$

    **E.** $-\sin^2 x$

**10.** $(\csc x - 1)(\csc x + 1) = ?$

    **F.** $\dfrac{1}{\sin x \cos x}$

**11.** $\dfrac{\csc x - \sin x}{\csc x} = ?$

    **G.** $1 + \cot^2 x$

**12.** $\dfrac{\cos x - \sec x}{\sec x} = ?$

    **H.** $\cos^2 x$

**13.** $\dfrac{\csc x}{\sin x} = ?$

    **I.** $\csc x$

**14.** $\dfrac{\sin x}{\cos x} + \dfrac{\cos x}{\sin x} = ?$

    **J.** $1 + \cot x$

*Find the products.*

**15.** $(\sin \alpha + 1)(\sin \alpha - 1)$    **16.** $(\tan \alpha + 2)(\tan \alpha - 2)$

**17.** $(2\cos \beta + 1)(\cos \beta - 1)$    **18.** $(2\csc \beta - 1)(\csc \beta - 3)$

**19.** $(\csc x + \sin x)^2$    **20.** $(2\cos x - \sec x)^2$

**21.** $(2\sin \theta - 1)(2\sin \theta + 1)$    **22.** $(3\sec \theta - 2)(3\sec \theta + 2)$

**23.** $(3\sin \theta + 2)^2$    **24.** $(3\cos \theta - 2)^2$

**25.** $(2\sin^2 y - \csc^2 y)^2$    **26.** $(\tan^2 y + \cot^2 y)^2$

*Factor each trigonometric expression.*

**27.** $2\sin^2 \gamma - 5\sin \gamma - 3$    **28.** $\cos^2 \gamma - \cos \gamma - 6$

**29.** $\tan^2 \alpha - 6\tan \alpha + 8$    **30.** $2\cot^2 \alpha + \cot \alpha - 3$

**31.** $4\sec^2 \beta + 4\sec \beta + 1$    **32.** $9\csc^2 \theta - 12\csc \theta + 4$

**33.** $\tan^2 \alpha - \sec^2 \beta$    **34.** $\sin^4 y - \cos^4 x$

**35.** $\sin^2 \beta \cos \beta + \sin \beta \cos \beta - 2\cos \beta$

**36.** $\cos^2 \theta \tan \theta - 2\cos \theta \tan \theta - 3\tan \theta$

**37.** $4\sec^4 x - 4\sec^2 x + 1$    **38.** $\cos^4 x - 2\cos^2 x + 1$

**39.** $\sin \alpha \cos \alpha + \cos \alpha + \sin \alpha + 1$

**40.** $2\sin^2 \theta + \sin \theta - 2\sin \theta \cos \theta - \cos \theta$

*Prove that each of the following equations is an identity.*

**41.** $\sin(x)\cot(x) = \cos(x)$

**42.** $\cos^2(x)\tan^2(x) = \sin^2(x)$

**43.** $1 - \sec(x)\cos^3(x) = \sin^2(x)$

**44.** $1 - \csc(x)\sin^3(x) = \cos^2(x)$

**45.** $1 + \sec^2(x)\sin^2(x) = \sec^2(x)$

**46.** $1 + \csc^2(x)\cos^2(x) = \csc^2(x)$

**47.** $\dfrac{\sin^3(x) + \sin(x)\cos^2(x)}{\cos(x)} = \tan(x)$

**48.** $\dfrac{\cos(x)\sin^2(x) + \cos^3(x)}{\sin(x)} = \cot(x)$

**49.** $\dfrac{\sin(x)}{\csc(x)} + \dfrac{\cos(x)}{\sec(x)} = 1$

**50.** $\sin^3(x)\csc(x) + \cos^3(x)\sec(x) = 1$

**51.** $\tan x \cos x + \csc x \sin^2 x = 2\sin x$

**52.** $\cot x \sin x - \cos^2 x \sec x = 0$

**53.** $(1 + \sin \alpha)^2 + \cos^2 \alpha = 2 + 2\sin \alpha$

**54.** $(1 + \cot \alpha)^2 - 2\cot \alpha = \dfrac{1}{(1 - \cos \alpha)(1 + \cos \alpha)}$

**55.** $2 - \csc \beta \sin \beta = \sin^2 \beta + \cos^2 \beta$

**56.** $(1 - \sin^2 \beta)(1 + \sin^2 \beta) = 2\cos^2 \beta - \cos^4 \beta$

**57.** $\tan x + \cot x = \sec x \csc x$

**58.** $\dfrac{\csc x}{\cot x} - \dfrac{\cot x}{\csc x} = \dfrac{\tan x}{\csc x}$    **59.** $\dfrac{\sec x}{\tan x} - \dfrac{\tan x}{\sec x} = \cos x \cot x$

**60.** $\dfrac{1 - \sin^2 x}{1 - \sin x} = \dfrac{\csc x + 1}{\csc x}$    **61.** $\sec^2 x = \dfrac{\csc x}{\csc x - \sin x}$

**62.** $\dfrac{\sin x}{\sin x + 1} = \dfrac{\csc x - 1}{\cot^2 x}$

**63.** $1 + \csc x \sec x = \dfrac{\cos(-x) - \csc(-x)}{\cos(x)}$

**64.** $\tan^2(-x) - \dfrac{\sin(-x)}{\sin x} = \sec^2 x$

**65.** $\dfrac{1}{\csc \theta - \cot \theta} = \dfrac{1 + \cos \theta}{\sin \theta}$    **66.** $\dfrac{-1}{\tan \theta - \sec \theta} = \dfrac{1 + \sin \theta}{\cos \theta}$

**67.** $\dfrac{\csc y + 1}{\csc y - 1} = \dfrac{1 + \sin y}{1 - \sin y}$

**68.** $\dfrac{1 - 2\cos^2 y}{1 - 2\cos y \sin y} = \dfrac{\sin y + \cos y}{\sin y - \cos y}$

**69.** $\dfrac{\cot x + \tan x}{\csc x} = \dfrac{1}{\cos x}$     **70.** $\dfrac{\sin(-x)}{-x} = \dfrac{\sin x}{x}$

**71.** $\dfrac{1 - \sin^2(-x)}{1 - \sin(-x)} = 1 - \sin x$

**72.** $\dfrac{1 - \sin^2 x \csc^2 x + \sin^2 x}{\cos^2 x} = \tan^2 x$

**73.** $\dfrac{1 - \cot^2 w + \cos^2 w \cot^2 w}{\csc^2 w} = \sin^4 w$

**74.** $\dfrac{\sec^2 z - \csc^2 z + \csc^2 z \cos^2 z}{\cot^2 z} = \tan^4 z$

*Prove that each of the following equations is an identity.*
**HINT** $\ln(a/b) = \ln(a) - \ln(b)$ *and* $\ln(ab) = \ln(a) + \ln(b)$
*for* $a > 0$ *and* $b > 0$.

**75.** $\ln(\sec \theta) = -\ln(\cos \theta)$

**76.** $\ln(\tan \theta) = \ln(\sin \theta) + \ln(\sec \theta)$

**77.** $\ln|\sec \alpha + \tan \alpha| = -\ln|\sec \alpha - \tan \alpha|$

**78.** $\ln|\csc \alpha + \cot \alpha| = -\ln|\csc \alpha - \cot \alpha|$

*The equation $f_1(x) = f_2(x)$ is an identity if and only if the graphs of $y = f_1(x)$ and $y = f_2(x)$ coincide at all values of $x$ for which both sides are defined. Graph $y = f_1(x)$ and $y = f_2(x)$ on the same screen of your calculator for each of the following equations. From the graphs, make a conjecture as to whether each equation is an identity, then prove your conjecture.*

**79.** $\dfrac{\sin \theta + \cos \theta}{\sin \theta} = 1 + \cot \theta$     **80.** $\dfrac{\sin \theta + \cos \theta}{\cos \theta} = \cot \theta + 1$

**81.** $(\sin x + \csc x)^2 = \sin^2 x + \csc^2 x$

**82.** $\tan x + \sec x = \dfrac{\sin^2 x + 1}{\cos x}$

**83.** $\cot x + \sin x = \dfrac{1 + \cos x - \cos^2 x}{\sin x}$

**84.** $1 - 2\cos^2 x + \cos^4 x = \sin^4 x$

**85.** $\dfrac{\sin x}{\cos x} - \dfrac{\cos x}{\sin x} = \dfrac{2\cos^2 x - 1}{\sin x \cos x}$

**86.** $\dfrac{1}{1 - \sin x} + \dfrac{1}{1 + \sin x} = \dfrac{2}{\cos^2 x}$

**87.** $\dfrac{\cos(-x)}{1 - \sin x} = \dfrac{1 - \sin(-x)}{\cos x}$     **88.** $\dfrac{\sin^2 x}{1 - \cos x} = 0.99 + \cos x$

### WRITING/DISCUSSION

**89.** Find functions $f_1(x)$ and $f_2(x)$ such that $f_1(x) = f_2(x)$ for infinitely many values of $x$, but $f_1(x) = f_2(x)$ is not an identity. Explain your example.

**90.** Find functions $f_1(x)$ and $f_2(x)$ such that $f_1(x) \neq f_2(x)$ for infinitely many values of $x$, but $f_1(x) = f_2(x)$ is an identity. Explain your example.

**91.** Explain how the graphs on a calculator of $y = f_1(x)$ and $y = f_2(x)$ could appear identical when $f_1(x) = f_2(x)$ is not an identity.

**92.** Explain how the graphs on a calculator of $y = f_1(x)$ and $y = f_2(x)$ could appear different when $f_1(x) = f_2(x)$ is an identity.

### REVIEW

**93.** The formula $d = \frac{1}{32} v_0^2 \sin(2\theta)$ gives the distance $d$ in feet that a projectile will travel when its launch angle is $\theta$ and its initial velocity is $v_0$ feet per second. What initial velocity in miles per hour does it take to throw a baseball 200 feet with $\theta = 33°$? Round to the nearest tenth.

**94.** Find the point that lies midway between $(\pi/3, 1)$ and $(\pi/2, 1)$.

**95.** Determine the amplitude, period, and phase shift for the function $y = -4\sin(2\pi x/3 - \pi/3)$.

**96.** Determine the period, asymptotes, and range for the function $y = 2\sec(x/4)$.

**97.** State the three Pythagorean identities.

**98.** Use identities to simplify the expression $\dfrac{1}{\cos^2 x} - \tan^2 x$.

### OUTSIDE THE BOX

**99.** *Counting Coworkers* Chris and Pat work at Tokyo Telemarketing. One day Chris said to Pat, "19/40 of my co-workers are female." Pat replied, "That's strange, 12/25 of my coworkers are female." If both are correct, then how many workers are there at Tokyo Telemarketing and what are the genders of Chris and Pat?

**100.** *Logs and Tangents* Find the value of
$\ln(\tan 1°) + \ln(\tan 2°) + \ln(\tan 3°) + \cdots$
$+ \ln(\tan 88°) + \ln(\tan 89°)$.

### **3.2** POP QUIZ

**1.** Find the product $(2\sin x + 1)(\sin x - 1)$.

**2.** Factor $2\cos^2 x + \cos x - 1$.

**3.** Simplify $\dfrac{1}{\cos(-x)} - \dfrac{\sin^2(-x)}{\cos(x)}$.

**4.** Prove that $\dfrac{\cos(-x) - \sec(-x)}{\sec(x)} = -\sin^2(x)$ is an identity.

### Constructing an Eyebrow Window

*A company that manufactures custom aluminum windows makes an eyebrow window that is placed on top of a rectangular window as shown in the diagram. When a customer orders an eyebrow window, the customer gives the width w and the height h of the eyebrow. To make the window, the shop needs to know the radius of the circular arc and the length of the circular arc. (This is an actual problem given to the author by a former student who worked for the company.)*

**a)** If the width and height of the eyebrow are 36 in. and 10 in., respectively, then what is the radius of the circular arc?

**b)** Find the length of the circular arc for a width of 36 in. and a height of 10 in.

**c)** Find a formula that expresses the radius of the circular arc $r$ in terms of the width of the eyebrow $w$ and the height of the eyebrow $h$.

**d)** Find a formula that expresses the length of the circular arc $L$ in terms of $w$ and $h$.

# 3.3 Sum and Difference Identities for Cosine

In every identity discussed in Sections 3.1 and 3.2, the trigonometric functions were functions of only a single variable, such as $\theta$ or $x$. In this section we establish identities for the cosine of a sum or difference of two variables. They do not follow from the known identities but rather from the geometry of the unit circle.

## The Cosine of a Sum

With so many identities showing relationships between the trigonometric functions, we might be fooled into thinking that almost any equation is an identity. For example, consider the equation

$$\cos(\alpha + \beta) = \cos\alpha + \cos\beta.$$

This equation looks nice, but is it an identity? It is easy to check with a calculator that if $\alpha = 30°$ and $\beta = 45°$, we get

$$\cos(30° + 45°) = \cos(75°) \approx 0.2588$$

and

$$\cos(30°) + \cos(45°) \approx 1.573.$$

So the equation $\cos(\alpha + \beta) = \cos\alpha + \cos\beta$ is *not* an identity.

We will now derive an identity for $\cos(\alpha + \beta)$. Consider angles $\alpha$, $\alpha + \beta$, and $-\beta$ in standard position, as shown in Fig. 3.11 on the next page. Assume that $\alpha$ and $\beta$ are acute angles, although any angles $\alpha$ and $\beta$ may be used. The terminal side of $\alpha$ intersects the unit circle at the point $A(\cos\alpha, \sin\alpha)$. The terminal side of

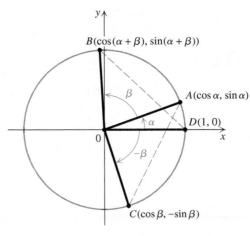

**Figure 3.11**

$\alpha + \beta$ intersects the unit circle at $B(\cos(\alpha + \beta), \sin(\alpha + \beta))$. The terminal side of $-\beta$ intersects the unit circle at $C(\cos(-\beta), \sin(-\beta))$ or $C(\cos\beta, -\sin\beta)$. Let $D$ be the point $(1, 0)$. Since $\angle BOD = \alpha + \beta$ and $\angle AOC = \alpha + \beta$, the chords $\overline{BD}$ and $\overline{AC}$ are equal in length. Using the distance formula to find the lengths, we get the following equation:

$$\sqrt{(\cos(\alpha + \beta) - 1)^2 + (\sin(\alpha + \beta) - 0)^2}$$
$$= \sqrt{(\cos\alpha - \cos\beta)^2 + (\sin\alpha - (-\sin\beta))^2}$$

Square each side and simplify:

$$(\cos(\alpha + \beta) - 1)^2 + \sin^2(\alpha + \beta) = (\cos\alpha - \cos\beta)^2 + (\sin\alpha + \sin\beta)^2$$

Square the binomials on both sides of the equation:

$$\cos^2(\alpha + \beta) - 2\cos(\alpha + \beta) + 1 + \sin^2(\alpha + \beta)$$
$$= \cos^2\alpha - 2\cos\alpha\cos\beta + \cos^2\beta + \sin^2\alpha + 2\sin\alpha\sin\beta + \sin^2\beta$$

Using the identity $\cos^2 x + \sin^2 x = 1$ once on the left-hand side and twice on the right-hand side, we get the following equation:

$$2 - 2\cos(\alpha + \beta) = 2 - 2\cos\alpha\cos\beta + 2\sin\alpha\sin\beta$$

Subtract 2 from each side and divide each side by $-2$ to get the identity for the cosine of a sum:

**Identity: Cosine of a Sum**

$$\cos(\alpha + \beta) = \cos\alpha\cos\beta - \sin\alpha\sin\beta$$

If an angle is the sum of two angles for which we know the exact values of the trigonometric functions, we can use this new identity to find the exact value of the cosine of the sum. For example, we can find the exact values of expressions such as $\cos(75°)$, $\cos(7\pi/12)$, and $\cos(195°)$ because

$$75° = 45° + 30°, \qquad \frac{7\pi}{12} = \frac{\pi}{3} + \frac{\pi}{4}, \qquad \text{and} \qquad 195° = 150° + 45°.$$

**EXAMPLE 1** **The cosine of a sum**

Find the exact value of $\cos(75°)$.

**Solution**

Use $75° = 30° + 45°$ and the identity for the cosine of a sum.

$$\cos 75° = \cos(30° + 45°)$$
$$= \cos(30°)\cos(45°) - \sin(30°)\sin(45°)$$
$$= \frac{\sqrt{3}}{2} \cdot \frac{\sqrt{2}}{2} - \frac{1}{2} \cdot \frac{\sqrt{2}}{2}$$
$$= \frac{\sqrt{6} - \sqrt{2}}{4}$$

To check, evaluate $\cos 75°$ and $\left(\sqrt{6} - \sqrt{2}\right)/4$ using a calculator.

**TRY THIS.** Find the exact value of $\cos(105°)$ using the cosine of a sum identity. ▪

Note that we found the exact value of $\cos(75°)$ only to illustrate an identity. If we need $\cos(75°)$ in an application, we can use the calculator's value for it.

## The Cosine of a Difference

To derive an identity for the cosine of a difference we write $\alpha - \beta = \alpha + (-\beta)$ and use the identity for the cosine of a sum:

$$\cos(\alpha - \beta) = \cos(\alpha + (-\beta))$$
$$= \cos\alpha\cos(-\beta) - \sin\alpha\sin(-\beta)$$

Now use the identities $\cos(-\beta) = \cos\beta$ and $\sin(-\beta) = -\sin\beta$ to get the identity for the cosine of a difference:

**Identity: Cosine of a Difference**

$$\cos(\alpha - \beta) = \cos\alpha\cos\beta + \sin\alpha\sin\beta$$

If an angle is the difference of two angles for which we know the exact values of sine and cosine, then we can find the exact value for the cosine of the angle.

**EXAMPLE 2** The cosine of a difference

Find the exact value of $\cos(\pi/12)$.

**Solution**

Since $\frac{\pi}{3} = \frac{4\pi}{12}$ and $\frac{\pi}{4} = \frac{3\pi}{12}$, we can use $\frac{\pi}{12} = \frac{\pi}{3} - \frac{\pi}{4}$ and the identity for the cosine of a difference:

$$\cos\left(\frac{\pi}{12}\right) = \cos\left(\frac{\pi}{3} - \frac{\pi}{4}\right)$$
$$= \cos\frac{\pi}{3}\cos\frac{\pi}{4} + \sin\frac{\pi}{3}\sin\frac{\pi}{4}$$
$$= \frac{1}{2} \cdot \frac{\sqrt{2}}{2} + \frac{\sqrt{3}}{2} \cdot \frac{\sqrt{2}}{2}$$
$$= \frac{\sqrt{2} + \sqrt{6}}{4}$$

To check, evaluate $\cos(\pi/12)$ and $\left(\sqrt{2} + \sqrt{6}\right)/4$ using a calculator.

**TRY THIS.** Find the exact value of $\cos(75°)$ using the cosine of a difference identity. ▪

In the next example we simplify expressions by using the identities in reverse.

**EXAMPLE 3** Simplifying with sum and difference identities

Use an appropriate identity to simplify each expression.

**a.** $\cos(49°)\cos(4°) + \sin(49°)\sin(4°)$   **b.** $\cos(2)\cos(-3) - \sin(-2)\sin(3)$

**Solution**

**a.** This expression is the right-hand side of the identity for the cosine of a difference, $\cos(\alpha - \beta) = \cos \alpha \cos \beta + \sin \alpha \sin \beta$. Let $\alpha = 49°$ and $\beta = 4°$ to get

$$\cos(49°)\cos(4°) + \sin(49°)\sin(4°) = \cos(49° - 4°)$$

$$= \cos(45°)$$

$$= \frac{\sqrt{2}}{2}.$$

**b.** First note that $\cos(-3) = \cos(3)$ because cosine is an even function. Also, $\sin(-2) = -\sin(2)$ because sine is an odd function:

$$\cos(2)\cos(-3) - \sin(-2)\sin(3) = \cos(2)\cos(3) - (-\sin(2))\sin(3)$$

$$= \cos(2)\cos(3) + \sin(2)\sin(3)$$

$$= \cos(2 - 3) \quad \text{Cosine of a difference identity}$$

$$= \cos(-1)$$

$$= \cos(1) \quad \text{Cosine is even.}$$

**TRY THIS.** Simplify $\cos x \cos 5x + \sin x \sin 5x$. ■

## Cofunction Identities

Sine and cosine are cofunctions, tangent and cotangent are cofunctions, and secant and cosecant are cofunctions. The identities for the cosine of a sum or a difference can be used to get identities that relate each trigonometric function and its **cofunction.**

If $\alpha = \pi/2$ is substituted into the identity for $\cos(\alpha - \beta)$, we get another identity:

$$\cos\left(\frac{\pi}{2} - \beta\right) = \cos\frac{\pi}{2}\cos\beta + \sin\frac{\pi}{2}\sin\beta$$

$$= 0 \cdot \cos\beta + 1 \cdot \sin\beta$$

$$= \sin\beta$$

The identity $\cos(\pi/2 - \beta) = \sin\beta$ is a **cofunction identity.**

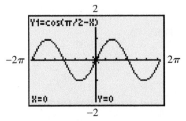

**Figure 3.12**

The graphing calculator graph in Fig. 3.12 supports this conclusion because $y_1 = \cos(\pi/2 - x)$ appears to coincide with $y_2 = \sin(x)$. □

If $u = \pi/2 - \beta$, then $\beta = \pi/2 - u$. If we replace $u$ with $\pi/2 - \beta$ and $\beta$ with $\pi/2 - u$ in $\sin\beta = \cos(\pi/2 - \beta)$, we get the identity

$$\sin\left(\frac{\pi}{2} - u\right) = \cos u.$$

Now

$$\tan\left(\frac{\pi}{2} - u\right) = \frac{\sin\left(\frac{\pi}{2} - u\right)}{\cos\left(\frac{\pi}{2} - u\right)} = \frac{\cos(u)}{\sin(u)} = \cot(u).$$

So $\tan(\pi/2 - u) = \cot(u)$. Using these identities, we can establish the cofunction identities for secant, cosecant, and cotangent.

**Cofunction Identities**

$$\sin\left(\frac{\pi}{2} - u\right) = \cos u \qquad \cos\left(\frac{\pi}{2} - u\right) = \sin u$$

$$\tan\left(\frac{\pi}{2} - u\right) = \cot u \qquad \cot\left(\frac{\pi}{2} - u\right) = \tan u$$

$$\sec\left(\frac{\pi}{2} - u\right) = \csc u \qquad \csc\left(\frac{\pi}{2} - u\right) = \sec u$$

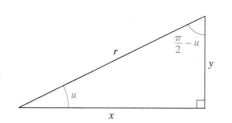

**Figure 3.13**

The cofunction identities simply indicate that *the value of any trigonometric function at u is equal to the value of its cofunction at* $(\pi/2 - u)$.

If $0 < u < \pi/2$, then $u$ and $(\pi/2 - u)$ are the measures of the acute angles of a right triangle as shown in Fig. 3.13. The term *cofunction* comes from the fact that these angles are complementary. The cofunction identities indicate that the value of a trigonometric function of one acute angle of a right triangle is equal to the value of the cofunction of the other. Since the cofunction identities hold also for complementary angles measured in degrees, we can write equations such as

$$\sin(20°) = \cos(70°), \quad \cot(89°) = \tan(1°), \quad \text{and} \quad \sec(50°) = \csc(40°).$$

The cofunction identities are consistent with the values obtained for the trigonometric functions for an acute angle using the ratios of the sides of a right triangle. For example, the cofunction identity $\sin(u) = \cos(\pi/2 - u)$ is correct for angles $u$ and $(\pi/2 - u)$ shown in Fig. 3.13, because

$$\sin(u) = \frac{\text{opp}}{\text{hyp}} = \frac{y}{r} \quad \text{and} \quad \cos\left(\frac{\pi}{2} - u\right) = \frac{\text{adj}}{\text{hyp}} = \frac{y}{r}.$$

**EXAMPLE 4** **Using the cofunction identities**

Use a cofunction identity to find the exact value of $\sin(5\pi/12)$.

**Solution**

Since $\sin(u) = \cos(\pi/2 - u)$,

$$\sin\left(\frac{5\pi}{12}\right) = \cos\left(\frac{\pi}{2} - \frac{5\pi}{12}\right) = \cos\left(\frac{\pi}{12}\right).$$

From Example 2, $\cos(\pi/12) = \left(\sqrt{2} + \sqrt{6}\right)/4$. So

$$\sin\frac{5\pi}{12} = \frac{\sqrt{2} + \sqrt{6}}{4}.$$

**TRY THIS.** Find the exact value of $\sin(15°)$ using a cofunction identity.    ■

In the next example we use the odd/even identities in conjunction with the cofunction identities. Remember that only cosine and secant are even; the other four trigonometric functions are odd. Also recall that $a - b = -(b - a)$.

**EXAMPLE 5** **Using the odd/even and cofunction identities**

Determine whether the equation $\cos(x - \pi/2) = \sin(x)$ is an identity.

**Solution**

Note that $x - \pi/2 = -(\pi/2 - x)$. Because cosine is an even function, $\cos(x - \pi/2) = \cos(\pi/2 - x)$. So

$$\cos\left(x - \frac{\pi}{2}\right) = \cos\left(\frac{\pi}{2} - x\right) \quad \text{Because } \cos(\alpha) = \cos(-\alpha)$$

$$= \sin(x) \qquad \text{Cofunction identity}$$

So $\cos(x - \pi/2) = \sin(x)$ is an identity.

**TRY THIS.** Determine whether $\tan(x - \pi/2) = \cot(x)$ is an identity.    ■

## FOR THOUGHT...   True or False? Explain.

**1.** $\frac{\pi}{4} + \frac{\pi}{2} = \frac{\pi}{6}$

**2.** $\frac{\pi}{4} - \frac{\pi}{3} = \frac{\pi}{12}$

**3.** $\cos(3°)\cos(2°) - \sin(3°)\sin(2°) = \cos(1°)$

**4.** $\cos(4)\cos(5) + \sin(4)\sin(5) = \cos(-1)$

**5.** $\cos(\pi/2 - 5) = \sin(5)$

**6.** $\cos(\pi/7 - 3) = \cos(3 - \pi/7)$.

**7.** $\sin(\pi/7 - 3) = \sin(3 - \pi/7)$.

**8.** For any real number $x$, $\sin(x - \pi/2) = \cos x$.

**9.** $\sec(\pi/3) = \csc(\pi/6)$

**10.** $\tan(21°30'5'') = \cot(68°29'55'')$

Scan this code
to link to section
lecture videos.

## 3.3 EXERCISES

### CONCEPTS

*Fill in the blank.*

**1.** Sine and _____ are cofunctions.

**2.** Cosecant and _____ are cofunctions.

**3.** Tangent and _____ are cofunctions.

**4.** The value of any trigonometric function at $x$ is equal to the value of its _____ at $(\pi/2 - x)$.

### SKILLS

*Find the exact values of the following sums or differences.*

**5.** $\frac{\pi}{4} + \frac{\pi}{3}$

**6.** $\frac{\pi}{3} + \frac{\pi}{6}$

**7.** $\frac{3\pi}{4} + \frac{\pi}{3}$

**8.** $\frac{\pi}{6} + \frac{\pi}{4}$

**9.** $\frac{\pi}{4} - \frac{\pi}{3}$

**10.** $\frac{\pi}{2} - \frac{\pi}{6}$

**11.** $\frac{\pi}{4} - \frac{\pi}{6}$

**12.** $\frac{\pi}{3} - \frac{\pi}{2}$

*Express each given angle as $\alpha + \beta$ or $\alpha - \beta$, where $\cos \alpha$ and $\cos \beta$ are known exactly.*

**13.** $75°$

**14.** $15°$

**15.** $165°$

**16.** $195°$

**17.** $\frac{\pi}{12}$

**18.** $\frac{11\pi}{12}$

**19.** $\frac{7\pi}{12}$

**20.** $\frac{5\pi}{12}$

*Use appropriate identities to find the exact value of each expression.*

**21.** $\cos(15°)$

**22.** $\cos(75°)$

**23.** $\cos(105°)$

**24.** $\cos(165°)$

**25.** $\cos(5\pi/12)$

**26.** $\cos(7\pi/12)$

**27.** $\cos(13\pi/12)$

**28.** $\cos(17\pi/12)$

**29.** $\cos(-\pi/12)$

**30.** $\cos(-5\pi/12)$

**31.** $\cos(-13\pi/12)$

**32.** $\cos(-7\pi/12)$

*Simplify each expression by using appropriate identities. Do not use a calculator.*

**33.** $\cos(23°)\cos(67°) + \sin(23°)\sin(67°)$

**34.** $\cos(34°)\cos(13°) + \sin(34°)\sin(13°)$

**35.** $\cos(5)\cos(6) - \sin(5)\sin(6)$

**36.** $\cos(7.1)\cos(1.4) - \sin(7.1)\sin(1.4)$

**37.** $\cos 2k \cos k + \sin 2k \sin k$

**38.** $\cos 3y \cos y - \sin 3y \sin y$

**39.** $\cos(-\pi/2)\cos(\pi/5) + \sin(\pi/2)\sin(\pi/5)$

**40.** $\cos(12°)\cos(-3°) - \sin(12°)\sin(-3°)$

**41.** $\cos(-\pi/5)\cos(\pi/3) + \sin(-\pi/5)\sin(-\pi/3)$

**42.** $\cos(-6)\cos(4) - \sin(6)\sin(-4)$

*Match each expression with an equivalent expression from (a)–(h). Do not use a calculator.*

**43.** $\sin(20°)$

**44.** $\tan(85°)$

**45.** $\cos(90°)$

**46.** $\cot(40°)$

**47.** $\sec(\pi/6)$

**48.** $\csc(\pi/8)$

**49.** $\sin(5\pi/12)$

**50.** $\cot(\pi/3)$

**a.** $\csc(\pi/3)$

**b.** $\sec(3\pi/8)$

**c.** $\sin(0°)$

**d.** $\tan(50°)$

**e.** $\cos(70°)$

**f.** $\tan(\pi/6)$

**g.** $\cos(\pi/12)$

**h.** $\cot(5°)$

*Match each expression with an equivalent expression from (a)–(h). Do not use a calculator.*

**51.** $\cos(44°)$

**52.** $\sin(-46°)$

**53.** $\cot(134°)$

**54.** $\sin(136°)$

**55.** $\sec(1)$

**56.** $\tan\left(\dfrac{\pi}{7}\right)$

**57.** $\csc\left(\dfrac{\pi}{2}\right)$

**58.** $\sin(-44°)$

**a.** $\cos(0)$

**b.** $-\cos(44°)$

**c.** $-\tan(44°)$

**d.** $\cot\left(\dfrac{5\pi}{14}\right)$

**e.** $-\cos(46°)$

**f.** $\csc\left(\dfrac{\pi - 2}{2}\right)$

**g.** $\sin(46°)$

**h.** $\sin(44°)$

*Simplify each expression by applying the odd/even identities, cofunction identities, and cosine of a sum or difference identities. Do not use a calculator.*

**59.** $\cos(14°)\cos(29°) + \sin(14°)\cos(61°)$

**60.** $\cos(4°)\cos(9°) + \cos(86°)\cos(81°)$

**61.** $\cos(10°)\cos(20°) + \sin(-10°)\cos(70°)$

**62.** $\sin(85°)\sin(40°) + \sin(-5°)\sin(-50°)$

**63.** $\cos(\pi/2 - \alpha)\cos(-\alpha) - \sin(-\alpha)\sin(\alpha - \pi/2)$

**64.** $\sin(\pi/2 - z)\cos(-z) - \cos(\pi/2 - z)\sin(-z)$

**65.** $\cos(-3k)\cos(-k) - \cos(\pi/2 - 3k)\sin(-k)$

**66.** $\cos(y - \pi/2)\cos(y) + \sin(\pi/2 - y)\sin(-y)$

*Solve each problem.*

**67.** Find the exact value of $\cos(\alpha + \beta)$ if $\sin \alpha = 3/5$ and $\sin \beta = 5/13$, with $\alpha$ in quadrant II and $\beta$ in quadrant I.

**68.** Find the exact value of $\cos(\alpha - \beta)$ if $\sin \alpha = -4/5$ and $\cos \beta = 12/13$, with $\alpha$ in quadrant III and $\beta$ in quadrant IV.

**69.** Find the exact value of $\cos(\alpha + \beta)$ if $\sin \alpha = -7/25$ and $\sin \beta = 8/17$, with $\alpha$ in quadrant IV and $\beta$ in quadrant II.

**70.** Find the exact value of $\cos(\alpha - \beta)$ if $\sin \alpha = 24/25$ and $\cos \beta = 8/17$, with $\alpha$ in quadrant II and $\beta$ in quadrant IV.

**71.** Find the exact value of $\cos(\alpha - \beta)$ if $\sin \alpha = \sqrt{3}/2$ and $\cos \beta = -\sqrt{2}/2$, with $\alpha$ in quadrant I and $\beta$ in quadrant II.

**72.** Find the exact value of $\cos(\alpha + \beta)$ if $\sin \alpha = \sqrt{2}/2$ and $\sin \beta = 1/2$, with $\alpha$ in quadrant II and $\beta$ in quadrant I.

**73.** Find the exact value of $\cos(\alpha + \beta)$ if $\sin \alpha = 2/3$ and $\sin \beta = -1/2$, with $\alpha$ in quadrant I and $\beta$ in quadrant III.

**74.** Find the exact value of $\cos(\alpha - \beta)$ if $\cos \alpha = \sqrt{3}/4$ and $\cos \beta = -\sqrt{2}/3$, with $\alpha$ in quadrant I and $\beta$ in quadrant II.

*Write each expression as a function of $\alpha$ alone.*

**75.** $\cos(\pi/2 + \alpha)$

**76.** $\cos(\alpha - \pi/2)$

**77.** $\cos(180° - \alpha)$

**78.** $\cos(180° + \alpha)$

**79.** $\cos(3\pi/2 - \alpha)$

**80.** $\cos(2\pi - \alpha)$

**81.** $\cos(90° + \alpha)$

**82.** $\cos(\alpha - 360°)$

*Verify that each equation is an identity.*

**83.** $\cos(x - \pi/2) = \cos x \tan x$

**84.** $\dfrac{\cos(x + y)}{\cos x \cos y} = 1 - \tan x \tan y$

**85.** $\cos(\alpha + \beta)\cos(\alpha - \beta) = \cos^2\beta - \sin^2\alpha$

**86.** $\cos(2x) = \cos^2 x - \sin^2 x$
> HINT $2x = x + x$

**87.** $\cos(x - y) + \cos(y - x) = 2\cos x \cos y + 2\sin x \sin y$

**88.** $\dfrac{\cos(x + y)}{\cos(x - y)} = \dfrac{\cot y - \tan x}{\cot y + \tan x}$

**89.** $\dfrac{\cos(\alpha + \beta)}{\cos \alpha + \sin \beta} = \dfrac{\cos \alpha - \sin \beta}{\cos(\beta - \alpha)}$

**90.** $\sec(v + t) = \dfrac{\cos v \cos t + \sin v \sin t}{\cos^2 v - \sin^2 t}$

## WRITING/DISCUSSION

**91.** Verify the three cofunction identities that were not proven in the text.

**92.** Show that $\sin(\alpha + \beta) = \sin \alpha + \sin \beta$ is not an identity.

**93.** Show that $\cos(\alpha - \beta) = \cos \alpha - \cos \beta$ is not an identity.

**94.** Explain why $\cos(180° - \alpha) = -\cos \alpha$ using the unit circle.

## REVIEW

**95.** Use identities to simplify the expression $\frac{\csc x}{\sec x}$.

**96.** Suppose that $\sin \alpha = 1/4$ and $\alpha$ is in quadrant II. Use identities to find the exact values of the other five trigonometric functions.

**97.** Simplify $(1 - \sin \alpha)(1 + \sin \alpha)$.

**98.** Simplify $\dfrac{\sin x}{\csc x - \sin x}$.

**99.** Find the period and equations of the asymptotes for the function $y = \cot(2x) + 1$.

**100.** Let $f(x) = \sin(x)$, $g(x) = x + 2$, and $h(x) = 3x$. Find $g(f(h(x)))$ and $h(g(f(x)))$.

## OUTSIDE THE BOX

**101.** *Overlapping Circles* Two lawn sprinklers that each water a circular region with radius $a$ are placed $a$ ft apart. What is the total area watered by the sprinklers?

**102.** *Solving a Triangle* In triangle $ABC$, $AB = 24$, $BC = 7$, and $AC = 25$. $M$ is the midpoint of $\overline{AB}$. Find the exact length of $\overline{CM}$.

---

### **3.3** POP QUIZ

**1.** Use an identity to find the exact value of $\cos(135° - 120°)$.

**2.** Use a cofunction identity to find an angle $\alpha$ for which $\sin(10°) = \cos(\alpha)$.

**3.** Simplify $\cos(x)\cos(3x) + \sin(-x)\sin(3x)$.

**4.** Verify that $\cos(6y) = \cos^2(3y) - \sin^2(3y)$ is an identity.

---

# **3.4** Sum and Difference Identities for Sine and Tangent

In the last section we learned the identities for the cosine of a sum or difference. In this section we learn identities for the sine or tangent of a sum or difference.

## Sine of a Sum or a Difference

We can use a cofunction identity from the last section to find an identity for $\sin(\alpha + \beta)$:

$$\sin(\alpha + \beta) = \cos\left(\frac{\pi}{2} - (\alpha + \beta)\right) \qquad \text{Cofunction identity}$$

$$= \cos\left(\left(\frac{\pi}{2} - \alpha\right) - \beta\right) \qquad \text{Associative property}$$

$$= \cos\left(\frac{\pi}{2} - \alpha\right)\cos \beta + \sin\left(\frac{\pi}{2} - \alpha\right)\sin \beta \qquad \text{Cosine of a difference}$$

$$= \sin \alpha \cos \beta + \cos \alpha \sin \beta \qquad \text{Cofunction identities}$$

To find an identity for $\sin(\alpha - \beta)$, use the identity for $\sin(\alpha + \beta)$:

$$\sin(\alpha - \beta) = \sin(\alpha + (-\beta))$$
$$= \sin\alpha\cos(-\beta) + \cos\alpha\sin(-\beta) \qquad \text{Sine of a sum}$$
$$= \sin\alpha\cos\beta - \cos\alpha\sin\beta \qquad \text{Odd/even identities}$$

The identities for the sine of a sum or difference are stated as follows.

**Identities: Sine of a Sum or Difference**

$$\sin(\alpha + \beta) = \sin\alpha\cos\beta + \cos\alpha\sin\beta$$
$$\sin(\alpha - \beta) = \sin\alpha\cos\beta - \cos\alpha\sin\beta$$

 **EXAMPLE 1** The sine of a sum

Find the exact value of $\sin(195°)$.

**Solution**

Use $195° = 150° + 45°$ and the identity for the sine of a sum:

$$\sin(195°) = \sin(150° + 45°)$$
$$= \sin(150°)\cos(45°) + \cos(150°)\sin(45°)$$
$$= \frac{1}{2} \cdot \frac{\sqrt{2}}{2} + \left(-\frac{\sqrt{3}}{2}\right) \cdot \frac{\sqrt{2}}{2}$$
$$= \frac{\sqrt{2} - \sqrt{6}}{4}$$

To check, evaluate $\sin(195°)$ and $\left(\sqrt{2} - \sqrt{6}\right)/4$ using a calculator.

**TRY THIS.** Find the exact value of $\sin(75°)$ using the sine of a sum identity. ∎

## Tangent of a Sum or Difference

We can use the identities for $\sin(\alpha + \beta)$ and $\cos(\alpha + \beta)$ to find an identity for $\tan(\alpha + \beta)$:

$$\tan(\alpha + \beta) = \frac{\sin(\alpha + \beta)}{\cos(\alpha + \beta)}$$
$$= \frac{\sin\alpha\cos\beta + \cos\alpha\sin\beta}{\cos\alpha\cos\beta - \sin\alpha\sin\beta}$$

To express the right-hand side in terms of tangent, multiply the numerator and denominator by $1/(\cos\alpha\cos\beta)$ and use the identity $\tan x = \sin x/\cos x$:

$$\tan(\alpha + \beta) = \frac{\dfrac{\sin\alpha\cos\beta}{\cos\alpha\cos\beta} + \dfrac{\cos\alpha\sin\beta}{\cos\alpha\cos\beta}}{\dfrac{\cos\alpha\cos\beta}{\cos\alpha\cos\beta} - \dfrac{\sin\alpha\sin\beta}{\cos\alpha\cos\beta}}$$
$$= \frac{\tan\alpha + \tan\beta}{1 - \tan\alpha\tan\beta}$$

Use the identity $\tan(-\beta) = -\tan\beta$ to get a similar identity for $\tan(\alpha - \beta)$.

**Identities: Tangent of a Sum or Difference**

$$\tan(\alpha + \beta) = \frac{\tan\alpha + \tan\beta}{1 - \tan\alpha\tan\beta} \qquad \tan(\alpha - \beta) = \frac{\tan\alpha - \tan\beta}{1 + \tan\alpha\tan\beta}$$

EXAMPLE **2** The tangent of a difference

Find the exact value of $\tan(\pi/12)$.

**Solution**

Use $\pi/12 = \pi/3 - \pi/4$ and the identity for the tangent of a difference:

$$\tan\left(\frac{\pi}{12}\right) = \tan\left(\frac{\pi}{3} - \frac{\pi}{4}\right)$$

$$= \frac{\tan\dfrac{\pi}{3} - \tan\dfrac{\pi}{4}}{1 + \tan\dfrac{\pi}{3}\tan\dfrac{\pi}{4}} = \frac{\sqrt{3} - 1}{1 + \sqrt{3}\cdot 1}$$

$$= \frac{\left(\sqrt{3} - 1\right)\left(\sqrt{3} - 1\right)}{\left(\sqrt{3} + 1\right)\left(\sqrt{3} - 1\right)}$$

$$= \frac{3 - 2\sqrt{3} + 1}{2} = \frac{4 - 2\sqrt{3}}{2} = 2 - \sqrt{3}$$

**TRY THIS.** Find the exact value of $\tan(75°)$ using the tangent of a sum identity. ∎

In the next example we use sum and difference identities in reverse to simplify an expression.

EXAMPLE **3** Simplifying with sum and difference identities

Use an appropriate identity to simplify each expression.

**a.** $\sin(7°)\cos(2°) + \cos(7°)\sin(2°)$   **b.** $\sin(-t)\cos 2t - \cos(-t)\sin(-2t)$

**Solution**

**a.** This expression is the right-hand side of the identity for the sine of a sum, $\sin(\alpha + \beta) = \sin\alpha\cos\beta + \cos\alpha\sin\beta$. Let $\alpha = 7°$ and $\beta = 2°$ to get

$$\sin(7°)\cos(2°) + \cos(7°)\sin(2°) = \sin(7° + 2°)$$
$$= \sin(9°)$$

**b.** First use the fact that sine is odd and cosine is even:

$$\sin(-t)\cos 2t - \cos(-t)\sin(-2t)$$
$$= -\sin t\cos 2t - \cos t(-\sin 2t)$$
$$= -(\sin t\cos 2t - \cos t\sin 2t)$$
$$= -\sin(t - 2t) \qquad \text{Sine of a difference identity}$$
$$= -\sin(-t)$$
$$= \sin t \qquad \text{Sine is an odd function.}$$

**TRY THIS.** Simplify $\sin\alpha\cos 3\alpha + \cos\alpha\sin 3\alpha$. ∎

In the next example we use identities to find $\sin(\alpha + \beta)$ without knowing $\alpha$ or $\beta$.

EXAMPLE **4** Using identities

Find the exact value of $\sin(\alpha + \beta)$ if $\sin\alpha = -3/5$ and $\cos\beta = -1/3$, with $\alpha$ in quadrant IV and $\beta$ in quadrant III.

**Solution**

To use the identity for $\sin(\alpha + \beta)$, we need $\cos \alpha$ and $\sin \beta$ in addition to the given values. Use $\sin \alpha = -3/5$ in the identity $\sin^2 x + \cos^2 x = 1$ to find $\cos \alpha$:

$$\left(-\frac{3}{5}\right)^2 + \cos^2\alpha = 1$$

$$\cos^2\alpha = \frac{16}{25}$$

$$\cos \alpha = \pm\frac{4}{5}$$

Since cosine is positive in quadrant IV, $\cos \alpha = 4/5$. Use $\cos \beta = -1/3$ in $\sin^2 x + \cos^2 x = 1$ to find $\sin \beta$:

$$\sin^2\beta + \left(-\frac{1}{3}\right)^2 = 1$$

$$\sin^2\beta = \frac{8}{9}$$

$$\sin \beta = \pm\frac{2\sqrt{2}}{3}$$

Since sine is negative in quadrant III, $\sin \beta = -2\sqrt{2}/3$. Now use the appropriate values in the identity $\sin(\alpha + \beta) = \sin \alpha \cos \beta + \cos \alpha \sin \beta$:

$$\sin(\alpha + \beta) = -\frac{3}{5}\left(-\frac{1}{3}\right) + \frac{4}{5}\left(-\frac{2\sqrt{2}}{3}\right)$$

$$= \frac{3}{15} - \frac{8\sqrt{2}}{15}$$

$$= \frac{3 - 8\sqrt{2}}{15}$$

To check, use a calculator to find approximate values for $\alpha$, $\beta$, and $\sin(\alpha + \beta)$.

**TRY THIS.** Find the exact value of $\cos(\alpha - \beta)$ given that $\sin \alpha = 1/3$, $\cos \beta = -1/3$, and both angles terminate in quadrant II. ∎

## FOR THOUGHT... True or False? Explain.

**1.** $\sin(3°)\cos(2°) - \cos(3°)\sin(2°) = \sin(1°)$

**2.** $\sin(4)\cos(5) + \cos(4)\sin(5) = \sin(-1)$

**3.** $\sin(7\pi/12) = \sin(\pi/3)\cos(\pi/4) + \cos(\pi/3)\sin(\pi/4)$

**4.** For any angles $\alpha$ and $\beta$, $\sin(\alpha + \beta) = \sin \alpha + \sin \beta$.

**5.** $\sin(2\pi) = \sin(\pi) + \sin(\pi)$

**6.** $\sin(\pi/2) = \sin(\pi/4) + \sin(\pi/4)$

**7.** $\tan(10°) = \tan(2°) + \tan(8°)$

**8.** $\tan(7) = \dfrac{\tan 3 + \tan 4}{1 - \tan 4 \tan 3}$

**9.** $\tan(-1) = \dfrac{\tan 3 - \tan 4}{1 + \tan 3 \tan 4}$

**10.** $\dfrac{\tan(13°) + \tan(-20°)}{1 - \tan(13°)\tan(-20°)} = \dfrac{\tan(45°) + \tan(-52°)}{1 - \tan(45°)\tan(-52°)}$

# 3.4 EXERCISES

## SKILLS

*Find the exact values of the following sums or differences.*

**1.** $\dfrac{\pi}{4} + \dfrac{\pi}{12}$  **2.** $\dfrac{\pi}{3} + \dfrac{\pi}{12}$

**3.** $\dfrac{3\pi}{4} - \dfrac{\pi}{3}$  **4.** $\dfrac{\pi}{6} - \dfrac{\pi}{4}$

*Express each given angle as $\alpha + \beta$ or $\alpha - \beta$, where $\sin \alpha$ and $\sin \beta$ are known exactly.*

**5.** $105°$  **6.** $15°$  **7.** $-15°$  **8.** $-75°$

*Use appropriate identities to find the exact value of each expression.*

**9.** $\sin(7\pi/12)$  **10.** $\sin(5\pi/12)$

**11.** $\tan(75°)$  **12.** $\tan(-15°)$

**13.** $\sin(-15°)$  **14.** $\sin(165°)$

**15.** $\tan(-13\pi/12)$  **16.** $\tan(7\pi/12)$

*Simplify each expression by using appropriate identities. Do not use a calculator.*

**17.** $\sin(23°)\cos(67°) + \cos(23°)\sin(67°)$

**18.** $\sin(2°)\cos(7°) - \cos(2°)\sin(7°)$

**19.** $\sin(34°)\cos(13°) + \cos(-34°)\sin(-13°)$

**20.** $\sin(12°)\cos(-3°) - \cos(12°)\sin(-3°)$

**21.** $\sin(-\pi/2)\cos(\pi/5) + \cos(\pi/2)\sin(-\pi/5)$

**22.** $\sin(-\pi/6)\cos(-\pi/3) + \cos(-\pi/6)\sin(-\pi/3)$

**23.** $\sin(14°)\cos(35°) + \cos(-14°)\cos(55°)$

**24.** $\cos(10°)\cos(20°) + \cos(-80°)\sin(-20°)$

**25.** $\dfrac{\tan(\pi/9) + \tan(\pi/6)}{1 - \tan(\pi/9)\tan(\pi/6)}$  **26.** $\dfrac{\tan(\pi/3) - \tan(\pi/5)}{1 + \tan(\pi/3)\tan(\pi/5)}$

**27.** $\dfrac{\tan(\pi/7) + \cot(\pi/2 - \pi/6)}{1 + \tan(-\pi/7)\tan(\pi/6)}$

**28.** $\dfrac{\cot(\pi/2 - \pi/3) + \tan(-\pi/6)}{1 + \tan(\pi/3)\cot(\pi/2 - \pi/6)}$

*Solve each problem.*

**29.** Find the exact value of $\sin(\alpha + \beta)$ if $\sin \alpha = 3/5$ and $\sin \beta = 5/13$, with $\alpha$ in quadrant II and $\beta$ in quadrant I.

**30.** Find the exact value of $\sin(\alpha - \beta)$ if $\sin \alpha = -4/5$ and $\cos \beta = 12/13$, with $\alpha$ in quadrant III and $\beta$ in quadrant IV.

**31.** Find the exact value of $\sin(\alpha + \beta)$ if $\sin \alpha = 7/25$ and $\sin \beta = -8/17$, with $\alpha$ in quadrant II and $\beta$ in quadrant III.

**32.** Find the exact value of $\sin(\alpha - \beta)$ if $\sin \alpha = -24/25$ and $\cos \beta = -8/17$, with $\alpha$ in quadrant III and $\beta$ in quadrant II.

**33.** Find the exact value of $\sin(\alpha - \beta)$ if $\sin \alpha = \sqrt{3}/2$ and $\cos \beta = \sqrt{2}/2$, with $\alpha$ in quadrant II and $\beta$ in quadrant I.

**34.** Find the exact value of $\sin(\alpha + \beta)$ if $\sin \alpha = -\sqrt{2}/2$ and $\sin \beta = 1/2$, with $\alpha$ in quadrant IV and $\beta$ in quadrant II.

**35.** Find the exact value of $\sin(\alpha + \beta)$ if $\sin \alpha = 2/3$ and $\sin \beta = -1/2$, with $\alpha$ in quadrant I and $\beta$ in quadrant III.

**36.** Find the exact value of $\sin(\alpha - \beta)$ if $\cos \alpha = \sqrt{3}/4$ and $\cos \beta = -\sqrt{2}/3$, with $\alpha$ in quadrant I and $\beta$ in quadrant II.

*Write each expression as a function of $\alpha$ alone.*

**37.** $\sin(\alpha - \pi)$  **38.** $\sin(180° - \alpha)$

**39.** $\sin(360° - \alpha)$  **40.** $\sin(90° + \alpha)$

**41.** $\tan(\pi/4 + \alpha)$  **42.** $\tan(\pi/4 - \alpha)$

**43.** $\tan(180° + \alpha)$  **44.** $\tan(360° - \alpha)$

*Verify that each equation is an identity.*

**45.** $\sin(180° - \alpha) = \dfrac{1 - \cos^2\alpha}{\sin \alpha}$

**46.** $\dfrac{\sin(x + y)}{\sin x \cos y} = 1 + \cot x \tan y$

**47.** $\sin(\alpha + \beta)\sin(\alpha - \beta) = \sin^2\alpha - \sin^2\beta$

**48.** $\sin(2x) = 2 \sin x \cos x$
HINT $2x = x + x$

**49.** $\sin(x - y) - \sin(y - x) = 2 \sin x \cos y - 2 \cos x \sin y$

**50.** $\tan(s + t)\tan(s - t) = \dfrac{\tan^2 s - \tan^2 t}{1 - \tan^2 s \tan^2 t}$

**51.** $\tan(\pi/4 + x) = \cot(\pi/4 - x)$

**52.** $\dfrac{\cos(\alpha + \beta)}{\sin(\alpha - \beta)} = \dfrac{1 - \tan \alpha \tan \beta}{\tan \alpha - \tan \beta}$

**53.** $\dfrac{\cos(\alpha - \beta)}{\sin(\alpha + \beta)} = \dfrac{1 + \tan \alpha \tan \beta}{\tan \alpha + \tan \beta}$

**54.** $\csc(v - t) = \dfrac{\sin v \cos t + \cos v \sin t}{\sin^2 v - \sin^2 t}$

**55.** $\dfrac{\sin(x + y)}{\sin(x - y)} = \dfrac{\cot y + \cot x}{\cot y - \cot x}$

**56.** $\dfrac{\sin(\alpha + \beta)}{\sin \alpha + \sin \beta} = \dfrac{\sin \alpha - \sin \beta}{\sin(\alpha - \beta)}$

**57.** Verify the identity for the tangent of a difference.

**58.** Explain why $\sin(180° - \alpha) = \sin\alpha$ using the unit circle.

REVIEW

**59.** Complete the following odd and even identities.

    **a.** $\sin(-x) = $ _____

    **b.** $\cos(-x) = $ _____

    **c.** $\tan(-x) = $ _____

    **d.** $\csc(-x) = $ _____

    **e.** $\sec(-x) = $ _____

    **f.** $\cot(-x) = $ _____

**60.** Simplify $\dfrac{1}{1 + \sin(-x)} + \dfrac{1}{1 + \sin(x)}$.

**61.** Simplify $\dfrac{\cos^3(x) + \cos(x)\sin^2(x)}{\sin(x)}$.

**62.** Complete the sum and difference identities.

    **a.** $\cos(x + y) = $ _____

    **b.** $\cos(x - y) = $ _____

**63.** One of the acute angles of a right triangle is 26° and its hypotenuse is 38.6 inches. Find the lengths of its legs to the nearest tenth of an inch.

**64.** Evaluate each trigonometric function if possible.

    **a.** $\sin(5\pi/6)$         **b.** $\cos(\pi)$

    **c.** $\tan(3\pi/2)$       **d.** $\sec(-\pi/3)$

    **e.** $\csc(-\pi/2)$      **f.** $\cot(5\pi/4)$

OUTSIDE THE BOX

**65.** *Two Common Triangles* An equilateral triangle with sides of length 1 and an isosceles right triangle with legs of length 1 are positioned as shown in the accompanying diagram. Find the exact area of the shaded triangle.

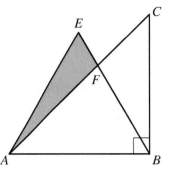

**Figure for Exercise 65**

**66.** *Interesting Angle* Find the degree measure of the acute angle between the lines $y = \frac{2}{3}x$ and $y = 5x - 13$.

**3.4** POP QUIZ

**1.** Find the exact value of $\sin(\alpha - \beta)$ if $\sin\alpha = 4/5$ and $\cos\beta = 1/2$, with $\alpha$ and $\beta$ in quadrant I.

**2.** Simplify $\sin(x)\cos(3x) + \cos(-x)\sin(3x)$.

**3.** Simplify $\dfrac{\tan(3\pi/20) + \tan(\pi/10)}{1 - \tan(3\pi/20)\tan(\pi/10)}$.

**4.** Verify that $\sin(10y) = 2\sin(5y)\cos(5y)$ is an identity.

# **3.5** Double-Angle and Half-Angle Identities

In Sections 3.3 and 3.4 we studied identities for functions of sums and differences of angles. The double-angle and half-angle identities, which we develop next, are special cases of those identities. These special cases occur so often that they are remembered as separate identities.

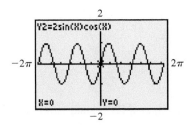

**Figure 3.14**

## **Double-Angle Identities**

To get an identity for $\sin 2x$, replace both $\alpha$ and $\beta$ by $x$ in the identity for $\sin(\alpha + \beta)$:

$$\sin 2x = \sin(x + x) = \sin x \cos x + \cos x \sin x = 2\sin x \cos x$$

To get visual support for this identity, graph $y_1 = \sin(2x)$ and $y_2 = 2\sin(x)\cos(x)$. The graphs in Fig. 3.14 appear to coincide. □

To find an identity for $\cos 2x$, replace both $\alpha$ and $\beta$ by $x$ in the identity for $\cos(\alpha + \beta)$:

$$\cos 2x = \cos(x + x) = \cos x \cos x - \sin x \sin x = \cos^2 x - \sin^2 x$$

To get a second form of the identity for $\cos 2x$, replace $\sin^2 x$ with $1 - \cos^2 x$:

$$\cos(2x) = \cos^2 x - \sin^2 x = \cos^2 x - (1 - \cos^2 x) = 2\cos^2 x - 1$$

Replacing $\cos^2 x$ with $1 - \sin^2 x$ produces a third form of the identity for $\cos 2x$:

$$\cos(2x) = 1 - 2\sin^2 x$$

To get an identity for $\tan 2x$, we can replace both $\alpha$ and $\beta$ by $x$ in the identity for $\tan(\alpha + \beta)$:

$$\tan 2x = \tan(x + x) = \frac{\tan x + \tan x}{1 - \tan x \tan x} = \frac{2\tan x}{1 - \tan^2 x}$$

These identities, which are summarized below, are known as the **double-angle identities.**

**Double-Angle Identities**

$$\sin 2x = 2\sin x \cos x \qquad \tan 2x = \frac{2\tan x}{1 - \tan^2 x}$$

$$\cos 2x = \cos^2 x - \sin^2 x$$

$$\cos 2x = 2\cos^2 x - 1$$

$$\cos 2x = 1 - 2\sin^2 x$$

Be careful to learn the double-angle identities exactly as they are written. A "nice looking" equation such as $\cos 2x = 2\cos x$ could be mistaken for an identity if you are not careful. [Since $\cos(\pi/2) \neq 2\cos(\pi/4)$, the "nice looking" equation is not an identity.] Remember that an equation is not an identity if at least one permissible value of the variable fails to satisfy the equation.

**EXAMPLE 1** **Using the double-angle identities**

Find exact values of $\sin(120°)$, $\cos(120°)$, and $\tan(120°)$ by using double-angle identities.

**Solution**

Note that $120° = 2 \cdot 60°$ and use the values $\sin(60°) = \sqrt{3}/2$, $\cos(60°) = 1/2$, and $\tan(60°) = \sqrt{3}$ in the appropriate identities:

$$\sin(120°) = 2\sin(60°)\cos(60°) = 2 \cdot \frac{\sqrt{3}}{2} \cdot \frac{1}{2} = \frac{\sqrt{3}}{2}$$

$$\cos(120°) = \cos^2(60°) - \sin^2(60°) = \left(\frac{1}{2}\right)^2 - \left(\frac{\sqrt{3}}{2}\right)^2 = -\frac{1}{2}$$

$$\tan(120°) = \frac{2\tan(60°)}{1 - \tan^2(60°)} = \frac{2 \cdot \sqrt{3}}{1 - (\sqrt{3})^2} = \frac{2\sqrt{3}}{-2} = -\sqrt{3}$$

These results are the well-known values of $\sin(120°)$, $\cos(120°)$, and $\tan(120°)$.

**TRY THIS.** Find $\sin(60°)$ using a double-angle identity. ∎

EXAMPLE **2**  **A triple-angle identity**

Prove that the following equation is an identity:

$$\sin(3x) = \sin x(3\cos^2 x - \sin^2 x)$$

**Solution**

Write $3x$ as $2x + x$ and use the identity for the sine of a sum:

$$
\begin{aligned}
\sin(3x) &= \sin(2x + x) \\
&= \sin 2x \cos x + \cos 2x \sin x && \text{Sine of a sum identity} \\
&= 2\sin x \cos x \cos x + (\cos^2 x - \sin^2 x)\sin x && \text{Double-angle identities} \\
&= 2\sin x \cos^2 x + \sin x \cos^2 x - \sin^3 x && \text{Distributive property} \\
&= 3\sin x \cos^2 x - \sin^3 x \\
&= \sin x(3\cos^2 x - \sin^2 x) && \text{Factor.}
\end{aligned}
$$

**TRY THIS.** Prove $\cos(3x) = \cos^3 x - 3\cos x \sin^2 x$ is an identity. ∎

Note that in Example 2 we used identities to expand the simpler side of the equation. In this case, simplifying the more complicated side is more difficult.

The double-angle identities can be used to get identities for $\sin(x/2)$, $\cos(x/2)$, and $\tan(x/2)$. These identities are known as **half-angle identities.**

## Half-Angle Identities

To get an identity for $\cos(x/2)$, start by solving the double-angle identity $\cos 2x = 2\cos^2 x - 1$ for $\cos x$:

$$
\begin{aligned}
2\cos^2 x - 1 &= \cos 2x \\
\cos^2 x &= \frac{1 + \cos 2x}{2} \\
\cos x &= \pm\sqrt{\frac{1 + \cos 2x}{2}}
\end{aligned}
$$

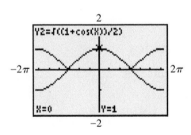

**Figure 3.15**

Because the last equation is correct for any value of $x$, it is also correct if $x$ is replaced by $x/2$. Replacing $x$ by $x/2$ yields the half-angle identity

$$\cos\frac{x}{2} = \pm\sqrt{\frac{1 + \cos x}{2}}.$$

So for any value of $x$, $\cos(x/2)$ is equal to either the positive or the negative square root of $(1 + \cos x)/2$.

The graphs of $y_1 = \cos(x/2)$ and $y_2 = \sqrt{(1 + \cos x)/2}$ in Fig. 3.15 illustrate this identity. Graph these curves on your calculator to see that the graph of $y_2$ does not go below the $x$-axis. Then use TRACE to see that $y_1 = y_2$ for $y_1 > 0$ and $y_1 = -y_2$ for $y_1 < 0$. □

EXAMPLE **3**  **Using half-angle identities**

Use the half-angle identity to find the exact value of $\cos(\pi/8)$.

**Solution**

Use $x = \pi/4$ in the half-angle identity for cosine:

$$\cos\frac{\pi}{8} = \cos\frac{\pi/4}{2}$$

$$= \sqrt{\frac{1 + \cos\pi/4}{2}}$$

$$= \sqrt{\frac{1 + \dfrac{\sqrt{2}}{2}}{2}}$$

$$= \sqrt{\frac{2 + \sqrt{2}}{4}}$$

$$= \frac{\sqrt{2 + \sqrt{2}}}{2}$$

The positive square root is used because the angle $\pi/8$ is in the first quadrant, where cosine is positive.

**TRY THIS.** Find the exact value of cos 75° using a half-angle identity. ■

To get an identity for $\sin(x/2)$, solve $1 - 2\sin^2 x = \cos 2x$ for $\sin x$:

$$1 - 2\sin^2 x = \cos 2x$$

$$-2\sin^2 x = \cos 2x - 1$$

$$\sin^2 x = \frac{1 - \cos 2x}{2}$$

$$\sin x = \pm\sqrt{\frac{1 - \cos 2x}{2}}$$

Replacing $x$ by $x/2$ yields the half-angle identity for sine:

$$\sin\frac{x}{2} = \pm\sqrt{\frac{1 - \cos x}{2}}$$

The half-angle identity for sine can be used to find the exact value for any angle that is one-half of an angle for which we know the exact value of cosine.

### EXAMPLE 4 Using half-angle identities

Use the half-angle identity to find the exact value of $\sin(75°)$.

**Solution**

Use $x = 150°$ in the half-angle identity for sine:

$$\sin(75°) = \sin\left(\frac{150°}{2}\right) = \pm\sqrt{\frac{1 - \cos(150°)}{2}}$$

Since 75° is in quadrant I, $\sin(75°)$ is positive. The reference angle for 150° is 30°, and $\cos(150°)$ is negative because 150° is in quadrant II. So

$$\cos(150°) = -\cos(30°) = -\frac{\sqrt{3}}{2}$$

and

$$\sin(75°) = \sqrt{\dfrac{1 + \dfrac{\sqrt{3}}{2}}{2}} = \sqrt{\dfrac{2 + \sqrt{3}}{4}} = \dfrac{\sqrt{2 + \sqrt{3}}}{2}.$$

**TRY THIS.** Find the exact value of $\sin(-22.5°)$ using a half-angle identity. ■

To get a half-angle identity for tangent, use the half-angle identities for sine and cosine:

$$\tan\dfrac{x}{2} = \dfrac{\sin\dfrac{x}{2}}{\cos\dfrac{x}{2}} = \dfrac{\pm\sqrt{\dfrac{1 - \cos x}{2}}}{\pm\sqrt{\dfrac{1 + \cos x}{2}}} = \pm\sqrt{\dfrac{1 - \cos x}{1 + \cos x}}$$

To get another form of this identity, multiply the numerator and denominator inside the radical by $1 + \cos x$. Use the fact that $\sqrt{\sin^2 x} = |\sin x|$ but $\sqrt{(1 + \cos x)^2} = 1 + \cos x$ because $1 + \cos x$ is nonnegative:

$$\tan\dfrac{x}{2} = \pm\sqrt{\dfrac{(1 - \cos x)(1 + \cos x)}{(1 + \cos x)^2}}$$

$$= \pm\sqrt{\dfrac{\sin^2 x}{(1 + \cos x)^2}}$$

$$= \pm\dfrac{|\sin x|}{1 + \cos x}$$

We use the positive or negative sign depending on the sign of $\tan(x/2)$. It can be shown that $\sin x$ has the same sign as $\tan(x/2)$. (See Exercise 75.) With this fact we can omit the absolute value and the $\pm$ symbol and write the identity as

$$\tan\dfrac{x}{2} = \dfrac{\sin x}{1 + \cos x}.$$

To get a third form of the identity for $\tan(x/2)$, multiply the numerator and denominator of the right-hand side by $1 - \cos x$:

$$\tan\dfrac{x}{2} = \dfrac{\sin x(1 - \cos x)}{(1 + \cos x)(1 - \cos x)}$$

$$= \dfrac{\sin x(1 - \cos x)}{\sin^2 x}$$

$$= \dfrac{1 - \cos x}{\sin x}$$

The half-angle identities are summarized as follows.

**Half-Angle Identities**

$$\sin\dfrac{x}{2} = \pm\sqrt{\dfrac{1 - \cos x}{2}} \qquad \cos\dfrac{x}{2} = \pm\sqrt{\dfrac{1 + \cos x}{2}}$$

$$\tan\dfrac{x}{2} = \pm\sqrt{\dfrac{1 - \cos x}{1 + \cos x}} \qquad \tan\dfrac{x}{2} = \dfrac{\sin x}{1 + \cos x} \qquad \tan\dfrac{x}{2} = \dfrac{1 - \cos x}{\sin x}$$

**EXAMPLE 5** **Using half-angle identities**

Use the half-angle identity to find the exact value of $\tan(-15°)$.

**Solution**

Use $x = -30°$ in the half-angle identity $\tan \dfrac{x}{2} = \dfrac{\sin x}{1 + \cos x}$.

$$\tan(-15°) = \tan\left(\frac{-30°}{2}\right)$$

$$= \frac{\sin(-30°)}{1 + \cos(-30°)}$$

$$= \frac{-\dfrac{1}{2}}{1 + \dfrac{\sqrt{3}}{2}}$$

$$= \frac{-1}{2 + \sqrt{3}}$$

$$= \frac{-1\left(2 - \sqrt{3}\right)}{\left(2 + \sqrt{3}\right)\left(2 - \sqrt{3}\right)}$$

$$= -2 + \sqrt{3}$$

**TRY THIS.** Find the exact value of $\tan(-22.5°)$ using a half-angle identity. ■

### EXAMPLE 6  Verifying identities involving half-angles

Prove that the following equation is an identity:

$$\tan \frac{x}{2} + \cot \frac{x}{2} = 2 \csc x$$

**Solution**

Write the left-hand side in terms of $\tan(x/2)$ and then use two different half-angle identities for $\tan(x/2)$:

$$\tan \frac{x}{2} + \cot \frac{x}{2} = \tan \frac{x}{2} + \frac{1}{\tan \dfrac{x}{2}}$$

$$= \frac{\sin x}{1 + \cos x} + \frac{\sin x}{1 - \cos x}$$

$$= \frac{\sin x(1 - \cos x)}{(1 + \cos x)(1 - \cos x)} + \frac{\sin x(1 + \cos x)}{(1 - \cos x)(1 + \cos x)}$$

$$= \frac{2 \sin x}{1 - \cos^2 x}$$

$$= \frac{2 \sin x}{\sin^2 x} = 2\,\frac{1}{\sin x} = 2 \csc x$$

**TRY THIS.** Prove $\sin^2\left(\frac{x}{2}\right) \cos^2\left(\frac{x}{2}\right) = \frac{\sin^2 x}{4}$ is an identity. ■

### EXAMPLE 7  Using the identities

In each case find $\sin \alpha$, $\cos \alpha$, and $\tan \alpha$.

**a.** $\cos(2\alpha) = -1/3$ and $\pi < 2\alpha < 3\pi/2$
**b.** $\sin(\alpha/2) = 4/5$ and $\pi/4 < \alpha/2 < \pi/2$

**Solution**

**a.** Since $\alpha$ is one-half of $2\alpha$, use the half-angle identity $\sin\frac{x}{2} = \pm\sqrt{\frac{1-\cos x}{2}}$ or $\sin\alpha = \pm\sqrt{\frac{1-\cos 2\alpha}{2}}$ to find $\sin\alpha$. Divide each part of $\pi < 2\alpha < 3\pi/2$ by 2 to get $\pi/2 < \alpha < 3\pi/4$. Since $\sin\alpha$ is positive for these values of $\alpha$ and $\cos(2\alpha) = -1/3$,

$$\sin\alpha = \sqrt{\frac{1 - \left(-\frac{1}{3}\right)}{2}} = \sqrt{\frac{2}{3}} = \frac{\sqrt{6}}{3}.$$

From $\cos^2\alpha + \sin^2\alpha = 1$ we have $\cos\alpha = \pm\sqrt{1 - \sin^2\alpha}$. Since $\sin\alpha = \sqrt{2/3}$, and since $\cos\alpha$ is negative for $\pi/2 < \alpha < 3\pi/4$,

$$\cos\alpha = -\sqrt{1 - \sin^2\alpha} = -\sqrt{1 - \frac{2}{3}} = -\frac{\sqrt{3}}{3}.$$

Since $\tan\alpha = \frac{\sin\alpha}{\cos\alpha}$,

$$\tan\alpha = \frac{\sqrt{6}/3}{-\sqrt{3}/3} = -\sqrt{2}.$$

**b.** Since $\alpha$ is twice $\alpha/2$, we use the double-angle identity $\cos 2x = 1 - 2\sin^2 x$ or $\cos\alpha = 1 - 2\sin^2(\alpha/2)$ to find $\cos\alpha$.

$$\cos\alpha = 1 - 2\left(\frac{4}{5}\right)^2 = -\frac{7}{25}$$

From $\sin^2\alpha + \cos^2\alpha = 1$ we have $\sin\alpha = \pm\sqrt{1 - \cos^2\alpha}$. Multiply each part of $\pi/4 < \alpha/2 < \pi/2$ by 2 to get $\pi/2 < \alpha < \pi$. Since $\cos\alpha = -7/25$, and since $\sin\alpha$ is positive for these values of $\alpha$,

$$\sin\alpha = \sqrt{1 - \left(-\frac{7}{25}\right)^2} = \frac{24}{25}.$$

Since $\tan\alpha = \frac{\sin\alpha}{\cos\alpha}$,

$$\tan\alpha = \frac{24/25}{-7/25} = -\frac{24}{7}.$$

**TRY THIS.** Find the exact value of $\cos(\alpha/2)$ given that $\sin\alpha = 1/3$ and $\pi/2 < \alpha < \pi$.  ∎

# FOR THOUGHT...   True or False? Explain.

**1.** $\frac{\sin 42°}{2} = \sin 21° \cos 21°$

**2.** $\cos\left(\sqrt{8}\right) = 2\cos^2\left(\sqrt{2}\right) - 1$

**3.** $\sin 150° = \sqrt{\frac{1 - \cos 75°}{2}}$

**4.** $\sin 200° = -\sqrt{\frac{1 - \cos 40°}{2}}$

**5.** $\tan\frac{7\pi}{8} = \sqrt{\frac{1 - \cos(7\pi/4)}{1 + \cos(7\pi/4)}}$

**6.** $\tan(-\pi/8) = \frac{1 - \cos(\pi/4)}{\sin(-\pi/4)}$

**7.** For any real number $x$, $\frac{\sin 2x}{2} = \sin x$.

**8.** The equation $\cos x = \sqrt{\frac{1 + \cos 2x}{2}}$ is an identity.

**9.** The equation $\sqrt{(1 - \cos x)^2} = 1 - \cos x$ is an identity.

**10.** If $180° < \alpha < 360°$, then $\sin\alpha < 0$ and $\tan(\alpha/2) < 0$.

Scan this code
to link to section
lecture videos.

## 3.5 EXERCISES

### SKILLS

*Find the exact value of each expression using double-angle identities.*

1. $\sin(90°)$

2. $\cos(60°)$

3. $\tan(60°)$

4. $\cos(180°)$

5. $\sin(3\pi/2)$

6. $\cos(4\pi/3)$

7. $\tan(4\pi/3)$

8. $\sin(2\pi/3)$

*Find the exact value of each expression using the half-angle identities.*

9. $\sin(15°)$

10. $\cos(15°)$

11. $\tan(15°)$

12. $\sin(-\pi/6)$

13. $\cos(\pi/8)$

14. $\tan(3\pi/8)$

15. $\sin(22.5°)$

16. $\tan(75°)$

17. $\cos(7\pi/8)$

18. $\sin(5\pi/6)$

*For each equation determine whether the positive or negative sign makes the equation correct. Do not use a calculator.*

19. $\sin 118.5° = \pm\sqrt{\dfrac{1 - \cos 237°}{2}}$

20. $\sin 222.5° = \pm\sqrt{\dfrac{1 - \cos 445°}{2}}$

21. $\cos 100° = \pm\sqrt{\dfrac{1 + \cos 200°}{2}}$

22. $\cos\dfrac{9\pi}{7} = \pm\sqrt{\dfrac{1 + \cos(18\pi/7)}{2}}$

23. $\tan\dfrac{-5\pi}{12} = \pm\sqrt{\dfrac{1 - \cos(-5\pi/6)}{1 + \cos(-5\pi/6)}}$

24. $\tan\dfrac{17\pi}{12} = \pm\sqrt{\dfrac{1 - \cos(17\pi/6)}{1 + \cos(17\pi/6)}}$

*Use identities to simplify each expression. Do not use a calculator.*

25. $2 \sin 13° \cos 13°$

26. $\sin^2\left(\dfrac{\pi}{5}\right) - \cos^2\left(\dfrac{\pi}{5}\right)$

27. $2 \cos^2(22.5°) - 1$

28. $1 - 2 \sin^2\left(-\dfrac{\pi}{8}\right)$

29. $\dfrac{\tan 15°}{1 - \tan^2(15°)}$

30. $\dfrac{2}{\cot 5(1 - \tan^2 5)}$

31. $\dfrac{\tan 30°}{1 - \tan^2 30°}$

32. $\cos^2\left(\dfrac{\pi}{9}\right) - \sin^2\left(\dfrac{\pi}{9}\right)$

33. $2 \sin\left(\dfrac{\pi}{9} - \dfrac{\pi}{2}\right)\cos\left(\dfrac{\pi}{2} - \dfrac{\pi}{9}\right)$

34. $2 \cos^2\left(\dfrac{\pi}{5} - \dfrac{\pi}{2}\right) - 1$

35. $\dfrac{\sin 12°}{1 + \cos 12°}$

36. $\csc 8°(1 - \cos 8°)$

*Prove that each equation is an identity.*

37. $\cos^4 s - \sin^4 s = \cos 2s$

38. $\sin 2s = -2 \sin s \sin(s - \pi/2)$

39. $\dfrac{\sin 4t}{4} = \cos^3 t \sin t - \sin^3 t \cos t$

40. $\cos 3t = \cos^3 t - 3 \sin^2 t \cos t$

41. $\dfrac{\cos 2x + \cos 2y}{\sin x + \cos y} = 2 \cos y - 2 \sin x$

42. $(\sin \alpha - \cos \alpha)^2 = 1 - \sin 2\alpha$

43. $\dfrac{\cos 2x}{\sin^2 x} = \csc^2 x - 2$

44. $\dfrac{\cos 2s}{\cos^2 s} = \sec^2 s - 2 \tan^2 s$

45. $2 \sin^2\left(\dfrac{u}{2}\right) = \dfrac{\sin^2 u}{1 + \cos u}$

46. $\cos 2y = \dfrac{1 - \tan^2 y}{1 + \tan^2 y}$

47. $\tan^2\left(\dfrac{x}{2}\right) = \dfrac{\sec x + \cos x - 2}{\sec x - \cos x}$

48. $\sec^2\left(\dfrac{x}{2}\right) = \dfrac{2 \sec x + 2}{\sec x + 2 + \cos x}$

49. $\dfrac{1 - \sin^2\left(\dfrac{x}{2}\right)}{1 + \sin^2\left(\dfrac{x}{2}\right)} = \dfrac{1 + \cos x}{3 - \cos x}$

50. $\dfrac{1 - \cos^2\left(\dfrac{x}{2}\right)}{1 - \sin^2\left(\dfrac{x}{2}\right)} = \dfrac{1 - \cos x}{1 + \cos x}$

*In each case, find $\sin \alpha$, $\cos \alpha$, $\tan \alpha$, $\csc \alpha$, $\sec \alpha$, and $\cot \alpha$.*

51. $\cos 2\alpha = 3/5$ and $0° < 2\alpha < 90°$

52. $\cos 2\alpha = 1/3$ and $360° < 2\alpha < 450°$

53. $\sin(2\alpha) = 5/13$ and $0° < \alpha < 45°$

54. $\sin(2\alpha) = -8/17$ and $180° < 2\alpha < 270°$

55. $\cos(\alpha/2) = -1/4$ and $\pi/2 < \alpha/2 < 3\pi/4$

56. $\sin(\alpha/2) = -1/3$ and $7\pi/4 < \alpha/2 < 2\pi$

57. $\sin(\alpha/2) = 4/5$ and $\alpha/2$ is in quadrant II

58. $\sin(\alpha/2) = 1/5$ and $\alpha/2$ is in quadrant II

*Solve each problem.*

**59.** Find the exact value of $\sin(2\alpha)$ given that $\sin(\alpha) = 3/5$ and $\alpha$ is in quadrant II.

**60.** Find the exact value of $\sin(2\alpha)$ given that $\tan(\alpha) = -8/15$ and $\alpha$ is in quadrant IV.

**61.** Find the exact value of $\cos(2\alpha)$ given that $\sin(\alpha) = 8/17$ and $\alpha$ is in quadrant II.

**62.** Find the exact value of $\tan(2\alpha)$ given that $\sin(\alpha) = -4/5$ and $\alpha$ is in quadrant III.

**63.** In the figure, $\angle B$ is a right angle and the line segment $\overline{AD}$ bisects $\angle A$. If $AB = 5$ and $BC = 3$, find the exact value of $BD$.
    **HINT** Use an appropriate identity.

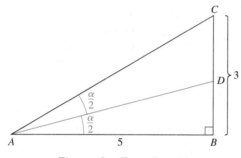

**Figure for Exercise 63**

**64.** In the figure, $\angle B$ is a right angle and the line segment $\overline{AD}$ bisects $\angle A$. If $AB = 10$ and $BD = 2$, find the exact value of $CD$.

**Figure for Exercise 64**

**65.** *Viewing Area* Find a formula for the viewing area of a television screen in terms of its diagonal and the angle $\alpha$ shown in the figure. Rewrite the formula using a single trigonometric function.

**Figure for Exercises 65 and 66**

**66.** Use the formula from the previous exercise to find the viewing area for a 32-in.-diagonal television for which $\alpha = 37.2°$.

*For each equation, either prove that it is an identity or prove that it is not an identity.*

**67.** $\sin(2x) = 2 \sin x$

**68.** $\dfrac{\cos 2x}{2} = \cos x$

**69.** $\tan\left(\dfrac{x}{2}\right) = \dfrac{1}{2} \tan x$

**70.** $\tan\left(\dfrac{x}{2}\right) = \sqrt{\dfrac{1 - \cos x}{1 + \cos x}}$

**71.** $\sin(2x) \cdot \sin\left(\dfrac{x}{2}\right) = \sin^2 x$

**72.** $\tan x + \tan x = \tan 2x$

**73.** $\cot\dfrac{x}{2} - \tan\dfrac{x}{2} = \dfrac{\sin 2x}{\sin^2 x}$

**74.** $\csc^2\left(\dfrac{x}{2}\right) + \sec^2\left(\dfrac{x}{2}\right) = 4 \csc^2 x$

## WRITING/DISCUSSION

**75.** Show that $\tan(x/2)$ has the same sign as $\sin x$ for any real number $x$.

**76.** Explain why $1 + \cos x \geq 0$ for any real number $x$.

**77.** Explain why $\sin 2x = 2 \sin x$ is not an identity by using graphs and by using the definition of the sine function.

**78.** Explain why $\tan(2\alpha) = 2 \tan(\alpha)$ is not an identity by using graphs and by using the definition of the tangent function.

## REVIEW

**79.** Complete the sum and difference identities for sine.

   **a.** $\sin(x + y) = $ _____

   **b.** $\sin(x - y) = $ _____

**80.** Complete the sum and difference identities for tangent.

   **a.** $\tan(x + y) = $ _____

   **b.** $\tan(x - y) = $ _____

**81.** Find the exact value of $\cos \beta$ if $\sin \beta = 2/3$ with $\beta$ in quadrant II.

**82.** Find the exact value of $\sin(\alpha + \beta)$ if $\sin \alpha = 1/3$ and $\sin \beta = 1/2$ with $\alpha$ in quadrant II and $\beta$ in quadrant I.

**83.** Find $\sin(\pi/2 - x)$, if $\cos x = 3/4$.

**84.** Find $\sin(x - \pi/2)$, if $\cos x = 2/3$.

## OUTSIDE THE BOX

**85.** *Completely Saturated* Four lawn sprinklers are positioned at the vertices of a square that is 2 m on each side. If each sprinkler waters a circular area with radius 2 m, then what is the exact area of the region that gets watered by all four sprinklers?

**86.** *Sines and Cosines* Without using a calculator find the exact value of $\sin\left(\dfrac{\pi}{24}\right)\cos\left(\dfrac{\pi}{24}\right)\cos\left(\dfrac{\pi}{12}\right)\cos\left(\dfrac{\pi}{6}\right)\cos\left(\dfrac{\pi}{3}\right)$.

**3.5** POP QUIZ

1. Simplify $2 \sin(\pi/8)\cos(\pi/8)$.

2. Simplify $\sqrt{\dfrac{1 - \cos(6x)}{2}}$ for $0 \le x \le \pi/3$.

3. Find $\sin(2\alpha)$ exactly if $\sin\alpha = 1/4$ and $\pi/2 < \alpha < \pi$.

4. Find $\sin(\alpha/2)$ exactly if $\sin\alpha = -4/5$ and $\pi < \alpha < 3\pi/2$.

5. Prove that $\sin^4 x - \cos^4 x = -\cos(2x)$ is an identity.

---

## LINKING concepts...

**For Individual or Group Explorations**

### Modeling the Motion of a Football

*A good field-goal kicker must learn through experience how to get the maximum distance in a kick. However, without touching a football, we can find the angle at which the football should be kicked to achieve the maximum distance. Assuming no air resistance, the distance in feet that a projectile (such as a football) travels when launched from the ground with an initial velocity of $v_0$ ft/sec is given by*

$$x = (v_0^2 \sin\theta \cos\theta)/16,$$

*where $\theta$ is the angle between the trajectory and the ground as shown in the figure. (See Section 3.1, Linking Concepts.)*

a) Use an identity to write the distance $x$ as a function of $2\theta$.

b) Graph the function that you found in part (a) using an initial velocity of 50 ft/sec.

c) From your graph, determine the value of $\theta$ that maximizes $x$. Does this value of $\theta$ maximize $x$ for any velocity? Explain.

d) When a player kicks a football 55 yd using the angle determined in part (c), what is the initial velocity of the football in miles per hour?

e) Do you think that the actual initial velocity for a 55-yd field goal is larger or smaller than that found in part (d)?

---

## 3.6 Product and Sum Identities

In this section we will develop identities that involve sums of trigonometric functions. We will also show how the combination of the sine and cosine functions in the spring equation (from Section 2.5) can work together to produce a periodic motion.

## Product-to-Sum Identities

In Sections 3.3 and 3.4 we learned identities for sine and cosine of a sum or difference:

$$\sin(A + B) = \sin A \cos B + \cos A \sin B$$

$$\sin(A - B) = \sin A \cos B - \cos A \sin B$$

$$\cos(A + B) = \cos A \cos B - \sin A \sin B$$

$$\cos(A - B) = \cos A \cos B + \sin A \sin B$$

We can add the first two of these identities to get a new identity:

$$\sin(A + B) = \sin A \cos B + \cos A \sin B$$

$$\underline{\sin(A - B) = \sin A \cos B - \cos A \sin B}$$

$$\sin(A + B) + \sin(A - B) = 2 \sin A \cos B$$

When we divide each side by 2, we get an identity expressing a product in terms of a sum:

$$\sin A \cos B = \frac{1}{2}[\sin(A + B) + \sin(A - B)]$$

We can produce three other similar identities from the sum and difference identities. The four identities, known as the **product-to-sum identities,** are listed below. These identities are not used as often as the other identities that we study in this chapter. It is not necessary to memorize these identities. Just remember them by name and look them up as necessary.

**Product-to-Sum Identities**

$$\sin A \cos B = \frac{1}{2}[\sin(A + B) + \sin(A - B)]$$

$$\sin A \sin B = \frac{1}{2}[\cos(A - B) - \cos(A + B)]$$

$$\cos A \sin B = \frac{1}{2}[\sin(A + B) - \sin(A - B)]$$

$$\cos A \cos B = \frac{1}{2}[\cos(A - B) + \cos(A + B)]$$

**EXAMPLE 1** **Expressing a product as a sum**

Use the product-to-sum identities to rewrite each expression:

**a.** $\sin 12° \cos 9°$   **b.** $\sin(\pi/12)\sin(\pi/8)$

**Solution**

**a.** Use the product-to-sum identity for $\sin A \cos B$:

$$\sin 12° \cos 9° = \frac{1}{2}[\sin(12° + 9°) + \sin(12° - 9°)]$$

$$= \frac{1}{2}[\sin 21° + \sin 3°]$$

**b.** Use the product-to-sum identity for $\sin A \sin B$:

$$\sin\left(\frac{\pi}{12}\right)\sin\left(\frac{\pi}{8}\right) = \frac{1}{2}\left[\cos\left(\frac{\pi}{12} - \frac{\pi}{8}\right) - \cos\left(\frac{\pi}{12} + \frac{\pi}{8}\right)\right]$$

$$= \frac{1}{2}\left[\cos\left(-\frac{\pi}{24}\right) - \cos\left(\frac{5\pi}{24}\right)\right]$$

$$= \frac{1}{2}\left[\cos\frac{\pi}{24} - \cos\frac{5\pi}{24}\right]$$

 Use a calculator to check as in Fig. 3.16.

**Figure 3.16**

**TRY THIS.** Use a product-to-sum identity to rewrite $\sin(4x)\cos(3x)$. ▪

**EXAMPLE 2** **Evaluating a product**

Use a product-to-sum identity to find the exact value of

$$\cos(67.5°)\sin(112.5°).$$

**Solution**

Use the identity $\cos A \sin B = \frac{1}{2}[\sin(A + B) - \sin(A - B)]$:

$$\cos(67.5°)\sin(112.5°) = \frac{1}{2}[\sin(67.5° + 112.5°) - \sin(67.5° - 112.5°)]$$

$$= \frac{1}{2}[\sin(180°) - \sin(-45°)]$$

$$= \frac{1}{2}[0 + \sin(45°)]$$

$$= \frac{1}{2} \cdot \frac{\sqrt{2}}{2} = \frac{\sqrt{2}}{4}$$

**TRY THIS.** Find the exact value of $\sin(52.5°)\cos(7.5°)$. ▪

## Sum-to-Product Identities

It is sometimes useful to write a sum of two trigonometric functions as a product. Consider the product-to-sum formula

$$\sin A \cos B = \frac{1}{2}[\sin(A + B) + \sin(A - B)].$$

To make the right-hand side look simpler we let $A + B = x$ and $A - B = y$. From these equations we get $x + y = 2A$ and $x - y = 2B$, or

$$A = \frac{x + y}{2} \quad \text{and} \quad B = \frac{x - y}{2}.$$

Substitute these values into the identity for $\sin A \cos B$:

$$\sin\left(\frac{x + y}{2}\right)\cos\left(\frac{x - y}{2}\right) = \frac{1}{2}[\sin x + \sin y]$$

Multiply each side by 2 to get

$$\sin x + \sin y = 2\sin\left(\frac{x + y}{2}\right)\cos\left(\frac{x - y}{2}\right).$$

The other three product-to-sum formulas can also be rewritten as sum-to-product identities by using similar procedures.

**Sum-to-Product Identities**

$$\sin x + \sin y = 2 \sin\left(\frac{x+y}{2}\right)\cos\left(\frac{x-y}{2}\right)$$

$$\sin x - \sin y = 2 \cos\left(\frac{x+y}{2}\right)\sin\left(\frac{x-y}{2}\right)$$

$$\cos x + \cos y = 2 \cos\left(\frac{x+y}{2}\right)\cos\left(\frac{x-y}{2}\right)$$

$$\cos x - \cos y = -2 \sin\left(\frac{x+y}{2}\right)\sin\left(\frac{x-y}{2}\right)$$

**EXAMPLE 3** **Expressing a sum or difference as a product**

Use the sum-to-product identities to rewrite each expression.

**a.** $\sin 8° + \sin 6°$    **b.** $\cos(\pi/5) - \cos(\pi/8)$    **c.** $\sin(6t) - \sin(4t)$

**Solution**

**a.** Use the sum-to-product identity for $\sin x + \sin y$:

$$\sin 8° + \sin 6° = 2 \sin\left(\frac{8° + 6°}{2}\right)\cos\left(\frac{8° - 6°}{2}\right)$$

$$= 2 \sin 7° \cos 1°$$

**b.** Use the sum-to-product identity for $\cos x - \cos y$:

$$\cos\left(\frac{\pi}{5}\right) - \cos\left(\frac{\pi}{8}\right) = -2 \sin\left(\frac{\pi/5 + \pi/8}{2}\right)\sin\left(\frac{\pi/5 - \pi/8}{2}\right)$$

$$= -2 \sin\left(\frac{13\pi}{80}\right)\sin\left(\frac{3\pi}{80}\right)$$

**c.** Use the sum-to-product identity for $\sin x - \sin y$:

$$\sin 6t - \sin 4t = 2 \cos\left(\frac{6t + 4t}{2}\right)\sin\left(\frac{6t - 4t}{2}\right)$$

$$= 2 \cos 5t \sin t$$

**TRY THIS.** Use a sum-to-product identity to rewrite $\cos(4x) + \cos(2x)$.  ■

**EXAMPLE 4** **Evaluating a sum**

Use a sum-to-product identity to find the exact value of

$$\cos(112.5°) + \cos(67.5°).$$

**Solution**

Use the sum-to-product identity $\cos x + \cos y = 2 \cos\left(\frac{x+y}{2}\right)\cos\left(\frac{x-y}{2}\right)$:

$$\cos(112.5°) + \cos(67.5°) = 2 \cos\left(\frac{112.5° + 67.5°}{2}\right)\cos\left(\frac{112.5° - 67.5°}{2}\right)$$

$$= 2 \cos(90°)\cos(22.5°)$$

$$= 0 \qquad\qquad \text{Because } \cos(90°) = 0$$

**TRY THIS.** Use a sum-to-product identity to evaluate $\sin(105°) + \sin(15°)$.  ■

## The Function $y = a \sin x + b \cos x$

The function $y = a \sin x + b \cos x$ involves an expression similar to those in the sum-to-product identities, but it is not covered by the sum-to-product identities. Functions of this type occur in applications such as the position of a weight in motion due to the force of a spring, the position of a swinging pendulum, and the current in an electrical circuit. In these applications it is important to express this function in terms of a single trigonometric function.

To get an identity for $a \sin x + b \cos x$ for nonzero values of $a$ and $b$, first multiply $a$ and $b$ in $a \sin x + b \cos x$ by $\dfrac{\sqrt{a^2 + b^2}}{\sqrt{a^2 + b^2}}$:

$$a \sin x + b \cos x = a \cdot \frac{\sqrt{a^2 + b^2}}{\sqrt{a^2 + b^2}} \sin x + b \cdot \frac{\sqrt{a^2 + b^2}}{\sqrt{a^2 + b^2}} \cos x$$

Next, factor out $\sqrt{a^2 + b^2}$:

$$a \sin x + b \cos x = \sqrt{a^2 + b^2} \left( \frac{a}{\sqrt{a^2 + b^2}} \sin x + \frac{b}{\sqrt{a^2 + b^2}} \cos x \right)$$

If $\alpha$ is an angle in standard position whose terminal side contains the point $(a, b)$, then we have the trigonometric ratios

$$\cos \alpha = \frac{a}{r} = \frac{a}{\sqrt{a^2 + b^2}} \quad \text{and} \quad \sin \alpha = \frac{b}{r} = \frac{b}{\sqrt{a^2 + b^2}}.$$

By substitution we get

$$\begin{aligned} a \sin x + b \cos x &= \sqrt{a^2 + b^2} \, (\cos \alpha \sin x + \sin \alpha \cos x) \\ &= \sqrt{a^2 + b^2} \, (\sin x \cos \alpha + \cos x \sin \alpha) \\ &= \sqrt{a^2 + b^2} \, \sin(x + \alpha). \qquad \text{\small Sine of a sum identity} \end{aligned}$$

This identity is called the **reduction formula** because it reduces two trigonometric functions to one.

**Theorem: Reduction Formula**

If $\alpha$ is an angle in standard position whose terminal side contains $(a, b)$, then

$$a \sin x + b \cos x = \sqrt{a^2 + b^2} \, \sin(x + \alpha)$$

for any real number $x$.

To rewrite an expression of the form $a \sin x + b \cos x$ by using the reduction formula, we need to find $\alpha$ so that the terminal side of $\alpha$ goes through $(a, b)$. By using trigonometric ratios, we have

$$\sin \alpha = \frac{b}{\sqrt{a^2 + b^2}}, \qquad \cos \alpha = \frac{a}{\sqrt{a^2 + b^2}}, \qquad \text{and} \qquad \tan \alpha = \frac{b}{a}.$$

Since we know $a$ and $b$, we can find $\sin \alpha$, $\cos \alpha$, or $\tan \alpha$, and then use an inverse trigonometric function to find $\alpha$. However, because of the ranges of the inverse functions, the angle obtained from an inverse function might not have its terminal side through $(a, b)$ as required. We will address this problem in the next example.

**EXAMPLE 5**  **Using the reduction formula**

Use the reduction formula to rewrite $-3 \sin x - 3 \cos x$ in the form $A \sin(x + C)$.

**Solution**

Because $a = -3$ and $b = -3$, we have

$$\sqrt{a^2 + b^2} = \sqrt{18} = 3\sqrt{2}.$$

**Figure 3.17**

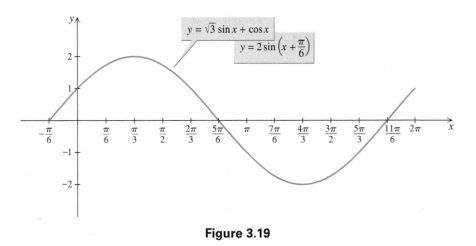

**Figure 3.18**

Since the terminal side of $\alpha$ must go through $(-3, -3)$, we have

$$\cos \alpha = \frac{-3}{3\sqrt{2}} = -\frac{\sqrt{2}}{2}.$$

Now $\cos^{-1}\left(-\sqrt{2}/2\right) = 3\pi/4$, but the terminal side of $3\pi/4$ is in quadrant II, as shown in Fig. 3.17. However, we also have $\cos(5\pi/4) = -\sqrt{2}/2$ and the terminal side for $5\pi/4$ does pass through $(-3, -3)$ in quadrant III. So $\alpha = 5\pi/4$. By the reduction formula

$$-3 \sin x - 3 \cos x = 3\sqrt{2} \sin\left(x + \frac{5\pi}{4}\right).$$

The reduction formula explains why the calculator graph of $y = -3 \sin x - 3 \cos x$ in Fig. 3.18 appears to be a sine wave with amplitude $3\sqrt{2}$.

**TRY THIS.** Rewrite $-2 \sin x + 2 \cos x$ in the form $A \sin(x + C)$. ■

By the reduction formula, the graph of $y = a \sin x + b \cos x$ is a sine wave. After the function is rewritten as $y = \sqrt{a^2 + b^2} \sin(x + \alpha)$, the amplitude, period, and phase shift can be determined.

**EXAMPLE 6** Using the reduction formula in graphing

Graph one cycle of the function

$$y = \sqrt{3} \sin x + \cos x$$

and state the amplitude, period, and phase shift.

**Solution**

Because $a = \sqrt{3}$ and $b = 1$, we have $\sqrt{a^2 + b^2} = \sqrt{4} = 2$. The terminal side of $\alpha$ must go through $\left(\sqrt{3}, 1\right)$. So $\sin \alpha = 1/2$, and $\sin^{-1}(1/2) = \pi/6$. Since the terminal side of $\pi/6$ is in quadrant I and goes through $\left(\sqrt{3}, 1\right)$, we can rewrite the function as

$$y = \sqrt{3} \sin x + \cos x = 2 \sin\left(x + \frac{\pi}{6}\right).$$

From the new form, we see that the graph is a sine wave with amplitude 2, period $2\pi$, and phase shift $-\pi/6$, as shown in Fig. 3.19.

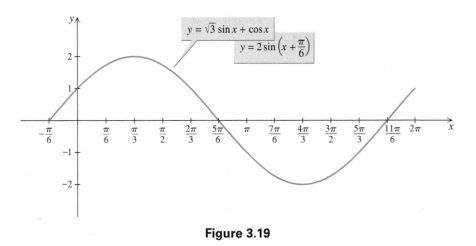

**Figure 3.19**

**TRY THIS.** Determine the amplitude, period, and phase shift for $y = \sin x - \cos x$. ■

Since $y = a \sin x + b \cos x$ and $y = \sqrt{a^2 + b^2} \sin(x + \alpha)$ are the same function, the graph of $y = a \sin x + b \cos x$ is a sine wave with amplitude $\sqrt{a^2 + b^2}$, period $2\pi$, and phase shift $-\alpha$.

## Modeling the Motion of a Spring

The function

$$x = \frac{v_0}{\omega} \sin(\omega t) + x_0 \cos(\omega t)$$

gives the position $x$ at time $t$ for a weight in motion on a vertical spring. In this formula, $v_0$ is the initial velocity, $x_0$ is the initial position, and $\omega$ is a constant. The same equation is used for a weight attached to a horizontal spring and set in motion on a frictionless surface, as shown in Fig. 3.20. In the next example we use the reduction formula on the spring equation to determine how far a spring actually stretches and compresses when the block is set in motion with a particular velocity.

**Figure 3.20**

**EXAMPLE 7** Using the reduction formula with springs

At time $t = 0$ the block shown in Fig. 3.20 is moved (and the spring compressed) to a position 1 meter to the left of the resting position. From this position the block is given a velocity of 2 meters per second to the right. Use the reduction formula to find the maximum distance from rest that is reached by the block. Assume that $\omega = 1$.

**Solution**

Since $v_0 = 2$, $x_0 = -1$, and $\omega = 1$, the position of the block at time $t$ is given by

$$x = 2 \sin t - \cos t.$$

Use the reduction formula to rewrite this equation. If $a = 2$ and $b = -1$ then

$$\sqrt{a^2 + b^2} = \sqrt{2^2 + (-1)^2} = \sqrt{5}.$$

If the terminal side of $\alpha$ goes through $(2, -1)$, then $\tan \alpha = -1/2$ and

$$\tan^{-1}(-1/2) \approx -0.46.$$

An angle of $-0.46$ radians lies in quadrant IV and goes through $(2, -1)$. Use $\alpha = -0.46$ in the reduction formula to get

$$x = \sqrt{5} \sin(t - 0.46).$$

Since the amplitude of this function is $\sqrt{5}$, the block oscillates between $x = \sqrt{5}$ meters and $x = -\sqrt{5}$ meters. The maximum distance from $x = 0$ is $\sqrt{5}$ meters. The maximum $y$ value on the calculator graph in Fig. 3.21 supports the conclusion that the amplitude is $\sqrt{5}$.

**TRY THIS.** The location in centimeters of a block on a spring is given by $x = \sin t - 3 \cos t$. Find the maximum distance reached by the block from rest. ∎

**Figure 3.21**

## FOR THOUGHT...  True or False? Explain.

**1.** $\sin 45° \cos 15° = 0.5(\sin 60° + \sin 30°)$

**2.** $\cos(\pi/8)\sin(\pi/4) = 0.5[\sin(3\pi/8) - \sin(\pi/8)]$

**3.** $2 \cos 6° \cos 8° = \cos 2° + \cos 14°$

**4.** $\sin 5° - \sin 9° = 2 \cos 7° \sin 2°$

**5.** $\cos 4 + \cos 12 = 2 \cos 8 \cos 4$

**6.** $\cos(\pi/3) - \cos(\pi/2) = -2 \sin(5\pi/12)\sin(\pi/12)$

**7.** $\sin(\pi/6) + \cos(\pi/6) = \sqrt{2} \sin(\pi/6 + \pi/4)$

**8.** $\frac{1}{2} \sin(\pi/6) + \frac{\sqrt{3}}{2} \cos(\pi/6) = \sin(\pi/2)$

**9.** The graph of $y = \frac{1}{2} \sin x + \frac{\sqrt{3}}{2} \cos x$ is a sine wave with amplitude 1.

**10.** The equation $\frac{1}{\sqrt{2}} \sin x + \frac{1}{\sqrt{2}} \cos x = \sin\left(x + \frac{\pi}{4}\right)$ is an identity.

Scan this code to link to section lecture videos.

## 3.6 EXERCISES

### SKILLS

*Use the product-to-sum identities to rewrite each expression.*

**1.** $\sin 13° \sin 9°$

**2.** $\cos 34° \cos 39°$

**3.** $\sin 16° \cos 20°$

**4.** $\cos 9° \sin 8°$

**5.** $\cos\left(\frac{\pi}{6}\right)\cos\left(\frac{\pi}{5}\right)$

**6.** $\sin\left(\frac{2\pi}{9}\right)\sin\left(\frac{3\pi}{4}\right)$

**7.** $\cos(5y^2)\cos(7y^2)$

**8.** $\cos 3t \sin 5t$

**9.** $\sin(2s - 1)\cos(s + 1)$

**10.** $\sin(3t - 1)\sin(2t + 3)$

*Find the exact value of each product.*

**11.** $\sin(52.5°)\sin(7.5°)$

**12.** $\cos(105°)\cos(75°)$

**13.** $\sin\left(\frac{13\pi}{24}\right)\cos\left(\frac{5\pi}{24}\right)$

**14.** $\cos\left(\frac{5\pi}{24}\right)\sin\left(-\frac{\pi}{24}\right)$

*Use the sum-to-product identities to rewrite each expression.*

**15.** $\sin 12° - \sin 8°$

**16.** $\sin 7° + \sin 11°$

**17.** $\cos 80° - \cos 87°$

**18.** $\cos 44° + \cos 31°$

**19.** $\sin 3.6 - \sin 4.8$

**20.** $\sin 5.1 + \sin 6.3$

**21.** $\cos(5y - 3) - \cos(3y + 9)$

**22.** $\cos(6t^2 - 1) + \cos(4t^2 - 1)$

**23.** $\sin 5\alpha - \sin 8\alpha$

**24.** $\sin 3s + \sin 5s$

**25.** $\cos\left(\frac{\pi}{3}\right) - \cos\left(\frac{\pi}{5}\right)$

**26.** $\cos\left(\frac{1}{2}\right) + \cos\left(\frac{2}{3}\right)$

*Find the exact value of each sum.*

**27.** $\sin(75°) + \sin(15°)$

**28.** $\sin(285°) - \sin(15°)$

**29.** $\cos\left(-\frac{\pi}{24}\right) - \cos\left(\frac{7\pi}{24}\right)$

**30.** $\cos\left(\frac{5\pi}{24}\right) + \cos\left(\frac{\pi}{24}\right)$

*Rewrite each expression in the form $A \sin(x + C)$.*

**31.** $\sin x - \cos x$

**32.** $2 \sin x + 2 \cos x$

**33.** $-\frac{1}{2} \sin x + \frac{\sqrt{3}}{2} \cos x$

**34.** $\frac{\sqrt{2}}{2} \sin x - \frac{\sqrt{2}}{2} \cos x$

**35.** $\frac{\sqrt{3}}{2} \sin x - \frac{1}{2} \cos x$

**36.** $-\frac{\sqrt{3}}{2} \sin x - \frac{1}{2} \cos x$

*Write each function in the form $y = A \sin(x + C)$. Then graph at least one cycle and state the amplitude, period, and phase shift.*

**37.** $y = -\sin x + \cos x$

**38.** $y = \sin x + \sqrt{3} \cos x$

**39.** $y = \sqrt{2} \sin x - \sqrt{2} \cos x$

**40.** $y = 2 \sin x - 2 \cos x$

**41.** $y = -\sqrt{3} \sin x - \cos x$

**42.** $y = -\frac{1}{2} \sin x - \frac{\sqrt{3}}{2} \cos x$

*For each function, determine the exact amplitude and find the phase shift in radians (to the nearest tenth).*

**43.** $y = 3 \sin x + 4 \cos x$

**44.** $y = \sin x + 5 \cos x$

**45.** $y = -6 \sin x + \cos x$

**46.** $y = -\sqrt{5} \sin x + 2 \cos x$

**47.** $y = -3 \sin x - 5 \cos x$

**48.** $y = -\sqrt{2} \sin x - \sqrt{7} \cos x$

*Prove that each equation is an identity.*

**49.** $\frac{\sin 3t - \sin t}{\cos 3t + \cos t} = \tan t$

**50.** $\frac{\sin 3x + \sin 5x}{\sin 3x - \sin 5x} = -\frac{\tan 4x}{\tan x}$

**51.** $\frac{\cos x - \cos 3x}{\cos x + \cos 3x} = \tan 2x \tan x$

**52.** $\frac{\cos 5y + \cos 3y}{\cos 5y - \cos 3y} = -\cot 4y \cot y$

**53.** $\cos^2 x - \cos^2 y = -\sin(x + y)\sin(x - y)$

**54.** $\sin^2 x - \sin^2 y = \sin(x + y)\sin(x - y)$

**55.** $\left(\sin \dfrac{x + y}{2} + \cos \dfrac{x + y}{2}\right)\left(\sin \dfrac{x - y}{2} + \cos \dfrac{x - y}{2}\right) = \sin x + \cos y$

**56.** $\sin 2A \sin 2B = \sin^2(A + B) - \sin^2(A - B)$

**57.** $\cos^2(A - B) - \cos^2(A + B) = \sin^2(A + B) - \sin^2(A - B)$

**58.** $(\sin A + \cos A)(\sin B + \cos B) = \sin(A + B) + \cos(A - B)$

## MODELING

*Solve each problem.*

**59.** *Motion of a Spring* A block is attached to a spring and set in motion, as in Example 7. For this block and spring, the location on the surface at any time $t$ in seconds is given in meters by $x = \sqrt{3} \sin t + \cos t$. Find the maximum distance reached by the block from the resting position.
HINT Use the reduction formula.

**60.** *Motion of a Spring* A block hanging from a spring, as shown in the figure, oscillates in the same manner as the block of Example 7. If a block attached to a certain spring is given an upward velocity of 0.3 m/sec from a point 0.5 m below its resting position, then its position at any time $t$ in seconds is given in meters by $x = -0.3 \sin t + 0.5 \cos t$. Find the maximum distance that the block travels from the resting position.
HINT Use the reduction formula.

**Figure for Exercise 60**

## WRITING/DISCUSSION

**61.** Derive the identity $\cos(2x) = \cos^2 x - \sin^2 x$ from a product-to-sum identity.

**62.** Derive the identity $\sin(2x) = 2 \sin x \cos x$ from a product-to-sum identity.

**63.** Prove the three product-to-sum identities that were not proved in the text.

**64.** Prove the three sum-to-product identities that were not proved in the text.

## REVIEW

**65.** Find the exact value of $\tan(x/2)$ given that $\sin(x) = -\sqrt{8/9}$ and $\pi < x < 3\pi/2$.

**66.** Find the exact value of $\sin(x/2)$ given that $\cos(x) = -1/4$ and $\pi/2 < x < \pi$.

**67.** Use an identity to simplify each expression.
**a.** $\sin 3.5 \cos 2.1 + \cos 3.5 \sin 2.1$

**b.** $\sin(2x)\cos(x) - \cos(2x)\sin(x)$

**c.** $2 \sin(4.8)\cos(4.8)$

**68.** Find the exact value of $\sin(2y)$ given that $\sin(y) = -4/5$ and $270° < y < 360°$.

**69.** Find the acute angles (to the nearest tenth of a degree) for a right triangle whose sides are 5 miles, 12 miles, and 13 miles.

**70.** Find the exact value of each trigonometric function.
**a.** $\sin(0)$       **b.** $\sin(\pi/6)$

**c.** $\sin(\pi/4)$       **d.** $\sin(\pi/3)$

**e.** $\sin(\pi/2)$

## OUTSIDE THE BOX

**71.** *As the Crow Flies* In Perfect City the avenues run east and west, the streets run north and south, and all of the blocks are square. A crow flies from the corner of 1st Avenue and 1st Street to the corner of $m$th Avenue and $n$th Street, "as the crow flies." Assume that $m$ and $n$ are positive integers greater than 1 and that the streets and avenues are simply lines on a map. If the crow flies over an intersection, then he flies over only two of the blocks that meet at the intersection.
**a.** If $m - 1$ and $n - 1$ are relatively prime (no common factors), then how many city blocks does the crow fly over?

**b.** If $d$ is the greatest common factor for $m - 1$ and $n - 1$, then how many city blocks does the crow fly over?

**72.** *Circle and Square* A circle is drawn so that it is tangent to one side of a square and passes through two vertices, as shown in the accompanying figure. The sides of the square are length 2 feet. Find the total area of the two regions that are inside the square but outside the circle. Round to the nearest tenth of a square foot.

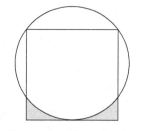

**Figure for Exercise 72**

## 3.6 POP QUIZ

1. Simplify $\frac{1}{2}[\sin(2\alpha + \beta) - \sin(2\alpha - \beta)]$.

2. Rewrite $\cos(2\alpha) + \cos(4\alpha)$ as a product using a sum-to-product identity.

3. Write $y = \sqrt{3}\sin x - \cos x$ in the form $y = A\sin(x + C)$.

4. Find the amplitude for $y = 3\sin x - 5\cos x$.

# LINKING
# concepts...
### For Individual or Group Explorations

## Maximizing the Total Profit

*Profits at The Christmas Store vary periodically with a high of $50,000 in December and a low of $10,000 in June, as shown in the accompanying graph. The owner of the Christmas Store also owns The Pool Store, where profits reach a high of $80,000 in August and a low of $20,000 in February. Assume that the profit function for each store is a sine wave.*

a) Write the profit function for The Christmas Store as a function of the month and sketch its graph.

b) Write the profit function for The Pool Store as a function of the month and sketch its graph.

c) Write the total profit as a function of the month and sketch its graph. What is the period?

d) Use the maximum feature of a graphing calculator to find the owner's maximum total profit and the month in which it occurs.

e) Find the owner's minimum total profit and the month in which it occurs.

f) We know that $y = a\sin x + b\cos x$ is a sine function. However, the sum of two arbitrary sine or cosine functions is not necessarily a sine function. Find an example in which the graph of the sum of two sine functions does not look like a sine curve.

g) Do you think that the sum of any two sine functions is a periodic function? Explain.

# Highlights

## 3.1 Basic Identities

| | | |
|---|---|---|
| **Identity** | An equation that is satisfied by every number for which both sides are defined | $\sin^2 x + \cos^2 x = 1$ <br> $\tan x = \sin(x)/\cos(x)$ |
| **Reciprocal Identities** | $\sin x = \dfrac{1}{\csc x} \qquad \cos x = \dfrac{1}{\sec x} \qquad \tan x = \dfrac{1}{\cot x}$ <br> $\csc x = \dfrac{1}{\sin x} \qquad \sec x = \dfrac{1}{\cos x} \qquad \cot x = \dfrac{1}{\tan x}$ | |
| **Tangent and Cotangent** | $\tan x = \dfrac{\sin x}{\cos x} \qquad \cot x = \dfrac{\cos x}{\sin x}$ | |
| **Pythagorean Identities** | $\sin^2 x + \cos^2 x = 1$ <br> $\cot^2 x + 1 = \csc^2 x$ <br> $\tan^2 x + 1 = \sec^2 x$ | |
| **Odd Identities** | $\sin(-x) = -\sin(x) \qquad \csc(-x) = -\csc(x)$ <br> $\tan(-x) = -\tan(x) \qquad \cot(-x) = -\cot(x)$ | |
| **Even Identities** | $\cos(-x) = \cos(x) \qquad \sec(-x) = \sec(x)$ | |

## 3.2 Verifying Identities

| | |
|---|---|
| **Strategy** | 1. Work on the more complicated side. <br> 2. Rewrite in terms of sines and cosines only. <br> 3. Multiply the numerator and denominator of one rational expression by either the numerator or denominator of the other. <br> 4. Write a single rational expression as a sum of two rational expressions. <br> 5. Combine a sum of two rational expressions into a single rational expression. <br> 6. If both sides simplify to a third expression, then the equation is an identity. |

## 3.3 Sum and Difference Identities for Cosine

| | |
|---|---|
| **Cosine of a Sum or Difference** | $\cos(\alpha + \beta) = \cos \alpha \cos \beta - \sin \alpha \sin \beta$ <br> $\cos(\alpha - \beta) = \cos \alpha \cos \beta + \sin \alpha \sin \beta$ |
| **Cofunction Identities** | The value of any trigonometric function at $u$ equals its cofunction value at $\pi/2 - u$. |

## 3.4 Sum and Difference Identities for Sine and Tangent

| | |
|---|---|
| **Sine of a Sum or Difference** | $\sin(\alpha + \beta) = \sin \alpha \cos \beta + \cos \alpha \sin \beta$ <br> $\sin(\alpha - \beta) = \sin \alpha \cos \beta - \cos \alpha \sin \beta$ |
| **Tangent of a Sum or Difference** | $\tan(\alpha + \beta) = \dfrac{\tan \alpha + \tan \beta}{1 - \tan \alpha \tan \beta}$ <br><br> $\tan(\alpha - \beta) = \dfrac{\tan \alpha - \tan \beta}{1 + \tan \alpha \tan \beta}$ |

## 3.5 Double-Angle and Half-Angle Identities

**Double-Angle Identities**

$\sin 2x = 2 \sin x \cos x$

$\cos 2x = \cos^2 x - \sin^2 x$

$\qquad = 2 \cos^2 x - 1$

$\qquad = 1 - 2 \sin^2 x$

$\tan 2x = \dfrac{2 \tan x}{1 - \tan^2 x}$

**Half-Angle Identities**

$\sin \dfrac{x}{2} = \pm \sqrt{\dfrac{1 - \cos x}{2}}$

$\cos \dfrac{x}{2} = \pm \sqrt{\dfrac{1 + \cos x}{2}}$

$\tan \dfrac{x}{2} = \pm \sqrt{\dfrac{1 - \cos x}{1 + \cos x}}$

$\qquad = \dfrac{\sin x}{1 + \cos x}$

$\qquad = \dfrac{1 - \cos x}{\sin x}$

## 3.6 Product and Sum Identities

**Product-to-Sum Identities**

$\sin A \cos B = \frac{1}{2}[\sin(A + B) + \sin(A - B)]$

$\sin A \sin B = \frac{1}{2}[\cos(A - B) - \cos(A + B)]$

$\cos A \sin B = \frac{1}{2}[\sin(A + B) - \sin(A - B)]$

$\cos A \cos B = \frac{1}{2}[\cos(A - B) + \cos(A + B)]$

**Sum-to-Product Identities**

$\sin x + \sin y = 2 \sin\left(\dfrac{x + y}{2}\right)\cos\left(\dfrac{x - y}{2}\right)$

$\sin x - \sin y = 2 \cos\left(\dfrac{x + y}{2}\right)\sin\left(\dfrac{x - y}{2}\right)$

$\cos x + \cos y = 2 \cos\left(\dfrac{x + y}{2}\right)\cos\left(\dfrac{x - y}{2}\right)$

$\cos x - \cos y = -2 \sin\left(\dfrac{x + y}{2}\right)\sin\left(\dfrac{x - y}{2}\right)$

**Reduction Formula** | If $\alpha$ is an angle in standard position whose terminal side contains $(a, b)$ and $x$ is a real number, then

$$a \sin x + b \cos x = \sqrt{a^2 + b^2} \sin(x + \alpha).$$

$\alpha = \tan^{-1}(4/3)$

$3 \sin x + 4 \cos x$

$= 5 \sin(x + \tan^{-1}(4/3))$

# Chapter 3 Review Exercises

*Simplify each expression.*

**1.** $(1 - \sin \alpha)(1 + \sin \alpha)$

**2.** $\csc x \tan x + \sec(-x)$

**3.** $(1 - \csc x)(1 - \csc(-x))$

**4.** $\dfrac{\cos^2 x - \sin^2 x}{\sin 2x}$

**5.** $\dfrac{1}{1 + \sin \alpha} - \dfrac{\sin(-\alpha)}{\cos^2 \alpha}$

**6.** $2 \sin\left(\dfrac{\pi}{2} - \alpha\right)\cos\left(\dfrac{\pi}{2} - \alpha\right)$

**7.** $\dfrac{2 \tan 2s}{1 - \tan^2 2s}$

**8.** $\dfrac{\tan 2w - \tan 4w}{1 + \tan 2w \tan 4w}$

**9.** $\sin 3\theta \cos 6\theta - \cos 3\theta \sin 6\theta$

**10.** $\dfrac{\sin 2y}{1 + \cos 2y}$

**11.** $\dfrac{1 - \cos 2z}{\sin 2z}$

**12.** $\cos^2\left(\dfrac{x}{2}\right) - \sin^2\left(\dfrac{x}{2}\right)$

*Use identities to find the exact values of the remaining five trigonometric functions at α.*

**13.** $\cos \alpha = -5/13$ and $\pi/2 < \alpha < \pi$

**14.** $\tan \alpha = 5/12$ and $\pi < \alpha < 3\pi/2$

**15.** $\sin\left(\dfrac{\pi}{2} - \alpha\right) = -3/5$ and $\pi < \alpha < 3\pi/2$

**16.** $\csc\left(\dfrac{\pi}{2} - \alpha\right) = 3$ and $0 < \alpha < \pi/2$

**17.** $\sin(\alpha/2) = 3/5$ and $3\pi/4 < \alpha/2 < \pi$

**18.** $\cos(\alpha/2) = -1/3$ and $\pi/2 < \alpha/2 < 3\pi/4$

*Determine whether each equation is an identity. Prove your answer.*

**19.** $(\sin x + \cos x)^2 = 1 + \sin 2x$

**20.** $\cos(A - B) = \cos A \cos B - \sin A \sin B$

**21.** $\csc^2 x - \cot^2 x = \tan^2 x - \sec^2 x$

**22.** $\sin^2\left(\dfrac{x}{2}\right) = \dfrac{\sin^2 x}{2 + \sin 2x \csc x}$

*Determine whether each function is odd, even, or neither.*

**23.** $f(x) = \dfrac{\sin x - \tan x}{\cos x}$

**24.** $f(x) = 1 + \sin^2 x$

**25.** $f(x) = \dfrac{\cos x - \sin x}{\sec x}$

**26.** $f(x) = \csc^3 x - \tan^3 x$

**27.** $f(x) = \dfrac{\sin x \tan x}{\cos x + \sec x}$

**28.** $f(x) = \sin x + \cos x$

*Prove that each of the following equations is an identity.*

**29.** $\sec 2\theta = \dfrac{1 + \tan^2 \theta}{1 - \tan^2 \theta}$

**30.** $\tan^2 \theta = \dfrac{1 - \cos 2\theta}{1 + \cos 2\theta}$

**31.** $\sin^2\left(\dfrac{x}{2}\right) = \dfrac{\csc^2 x - \cot^2 x}{2 \csc^2 x + 2 \csc x \cot x}$

**32.** $\cot(-x) = \dfrac{1 - \sin^2 x}{\cos(-x)\sin(-x)}$

**33.** $\cot(\alpha - 45°) = \dfrac{1 + \tan \alpha}{\tan \alpha - 1}$

**34.** $\cos(\alpha + 45°) = \dfrac{\cos \alpha - \sin \alpha}{\sqrt{2}}$

**35.** $\dfrac{\sin 2\beta}{2 \csc \beta} = \sin^2 \beta \cos \beta$

**36.** $\sin(45° - \beta) = \dfrac{\cos 2\beta}{\sqrt{2}(\cos \beta + \sin \beta)}$

**37.** $\dfrac{\cot^3 y - \tan^3 y}{\sec^2 y + \cot^2 y} = 2 \cot 2y$

**38.** $\dfrac{\sin^3 y - \cos^3 y}{\sin y - \cos y} = \dfrac{2 + \sin 2y}{2}$

**39.** $\dfrac{1}{\sin(-x)} + \dfrac{1}{\cos(-x)} = \dfrac{2\sin(x) - 2\cos(x)}{\sin(2x)}$

**40.** $\dfrac{1}{\cot^2(-x)} + \dfrac{1}{\sec^2(-x)} = \dfrac{\sec^2(x) + \cot^2(x)}{\csc^2(x)}$

**41.** $\dfrac{\cos(-x) + \sin(-x)}{\cos(-x) - \sin(-x)} = \dfrac{\cos(2x)}{1 + \sin(2x)}$

**42.** $\dfrac{\cot(-x) - \tan(-x)}{\tan(x) + \cot(x)} = \dfrac{\sec^2(x) - \csc^2(x)}{\sec^2(x) + \csc^2(x)}$

**43.** $\cos 4x = 8 \sin^4 x - 8 \sin^2 x + 1$

**44.** $\cos 3x = \cos x(1 - 4 \sin^2 x)$

**45.** $\sin^4 2x = 16 \sin^4 x - 32 \sin^6 x + 16 \sin^8 x$

**46.** $1 - \cos^6 x = 3 \sin^2 x - 3 \sin^4 x + \sin^6 x$

*Use an appropriate identity to find the exact value of each expression.*

**47.** $\tan(-\pi/12)$

**48.** $\sin(-\pi/8)$

**49.** $\sin(-75°)$

**50.** $\cos(105°)$

*Write each function in the form $y = A \sin(x + C)$, and graph at least one cycle of the function. Determine the amplitude, period, and phase shift.*

**51.** $y = 4 \sin x + 4 \cos x$

**52.** $y = \sqrt{3} \sin x + 3 \cos x$

**53.** $y = -2 \sin x + \cos x$

**54.** $y = -2 \sin x - \cos x$

*Use the sum-to-product identities to rewrite each expression as a product.*

**55.** $\cos 15° + \cos 19°$

**56.** $\cos 4° - \cos 6°$

**57.** $\sin(\pi/4) - \sin(-\pi/8)$

**58.** $\sin(-\pi/6) + \sin(\pi/12)$

*Use the product-to-sum identities to write each expression as a sum or difference.*

**59.** $2 \sin 11° \cos 13°$

**60.** $2 \sin 8° \sin 12°$

**61.** $2 \cos(x/4)\cos(x/3)$

**62.** $2 \cos s \sin 3s$

*Solve each problem.*

**63.** *Motion of a Car*  A car with worn shock absorbers hits a bump. The height in inches above or below its normal position for a point on the front bumper is given by $y = 2 \sin t + \cos t$, where $t$ is time in seconds. Find the maximum height above its normal position that is reached by the point on the bumper.
HINT  Use the reduction formula.

**64.** *Oscillating Light*  The intensity of a light source is given by $I = 6 \sin t + 4 \cos t + 10$, where $t$ is time in hours and $I$ is measured in lumens. Find the maximum and minimum intensity of the light source to one decimal place.
HINT  Use the reduction formula.

**OUTSIDE THE BOX**

**65.** *One in a Million*  If you write the integers from 1 through 1,000,000 inclusive, then how many ones will you write?

**66.** *Large Integer*  The number $10^{40}$ is a very large integer. How many factors does it have?

# Chapter **3** Test

*Use identities to simplify each expression.*

**1.** $\sec x \cot x \sin 2x$

**2.** $\sin 2t \cos 5t + \cos 2t \sin 5t$

**3.** $\dfrac{1}{1 - \cos y} + \dfrac{1}{1 + \cos y}$

**4.** $\dfrac{\tan(\pi/5) + \tan(\pi/10)}{1 - \tan(\pi/5)\tan(\pi/10)}$

*Prove that each of the following equations is an identity.*

**5.** $\dfrac{\sin \beta \cos \beta}{\tan \beta} = 1 - \sin^2 \beta$

**6.** $\dfrac{1}{\sec \theta - 1} - \dfrac{1}{\sec \theta + 1} = 2 \cot^2 \theta$

**7.** $\cos\left(\dfrac{\pi}{2} - x\right)\cos(-x) = \dfrac{\sin(2x)}{2}$

**8.** $\tan \dfrac{t}{2} \cos^2 t - \tan \dfrac{t}{2} = \dfrac{\sin t}{\sec t} - \sin t$

*Solve each problem.*

**9.** Write $y = \sin x - \sqrt{3} \cos x$ in the form $y = A \sin(x + C)$ and graph the function. Determine the period, amplitude, and phase shift.

**10.** If $\csc \alpha = 2$ and $\alpha$ is in quadrant II, then find the exact values at $\alpha$ for the remaining five trigonometric functions.

**11.** Determine whether the function $f(x) = x \sin x$ is odd, even, or neither.

**12.** Use an appropriate identity to find the exact value of $\sin(-\pi/12)$.

**13.** Prove that the equation $\tan x + \tan y = \tan(x + y)$ is not an identity.

## TYING IT ALL TOGETHER

### Chapters P–3

*Evaluate each expression without using a calculator.*

**1.** $\sin(\pi/4)$

**2.** $\sin(3\pi/4)$

**3.** $\cos(\pi/3)$

**4.** $\cos(5\pi/3)$

**5.** $\sin(\pi/6)$

**6.** $\sin(5\pi/6)$

**7.** $\sin(55\pi/6)$

**8.** $\sin(-23\pi/2)$

**9.** $\tan(3\pi/4)$

**10.** $\tan(-77\pi)$

*Determine whether each function is even or odd.*

**11.** $f(x) = x^3 + \sin x$

**12.** $f(x) = x^3 \sin x$

**13.** $f(x) = \dfrac{\sin x}{x}$

**14.** $f(x) = |\sin x|$

**15.** $f(x) = \cos^5 x + \cos^3 x - 2\cos x$

**16.** $f(x) = x^3 \sin^4 x + x \sin^2 x$

*Determine whether each equation is an identity. Prove your answer.*

**17.** $\sin(\alpha + \beta) = \sin \alpha + \sin \beta$

**18.** $\cos(\alpha - \beta) = \cos \alpha - \cos \beta$

**19.** $\sin^{-1}(x) = \dfrac{1}{\sin x}$

**20.** $\sin^2 x = \sin(x^2)$

*Solve each right triangle that has the given parts.*

**21.** $\alpha = 30°$ and $a = 4$

**22.** $a = \sqrt{3}$ and $b = 1$

**23.** $\cos \beta = 0.3$ and $b = 5$

**24.** $\sin \alpha = 0.6$ and $a = 2$

*Fill in the blanks.*

**25.** A(n) _____ is a union of two rays with a common endpoint.

**26.** An angle whose vertex is the center of a circle is called a(n) _____ angle.

**27.** A(n) _____ angle has a degree measure between 0° and 90°.

**28.** An angle in standard position whose terminal side lies on an axis is a(n) _____ angle.

**29.** Angles in standard position that have the same initial side and same terminal side are called _____ angles.

**30.** A degree is divided into 60 equal parts called _____.

**31.** A minute is divided into 60 equal parts called _____.

**32.** A circle with radius one is called a(n) _____ circle.

**33.** The relationship between degree and radian measure of angles is 180 degrees equals _____ radians.

**34.** The length of the arc intercepted by a central angle of $\alpha$ radians on a circle of radius $r$ is _____.

# 4 Solving Conditional Trigonometric Equations

**4.1** The Inverse Trigonometric Functions

**4.2** Basic Sine, Cosine, and Tangent Equations

**4.3** Equations Involving Compositions

**4.4** Trigonometric Equations of Quadratic Type

Millions of baseball fans enjoy watching a game in which a pitcher throws a ball at 90 miles per hour at a 23-inch by 17-inch target in such a way that the batter cannot hit the ball. Most fans are aware of the important role of statistics in everything from ranking the teams to determining a player's worth. But mathematics can be used to analyze every aspect of the game.

If the batter is successful in hitting the ball, then a player tries to retrieve the ball and throw it to first base before the batter gets there. Players have developed different strategies for minimizing the time of throwing a ball to first base.

WHAT YOU WILL **LEARN**

**In this chapter we will see how to use trigonometry to analyze different strategies for throwing a baseball and see that saving even a very small amount of time is important in this high-stakes sport.**

# 4.1 The Inverse Trigonometric Functions

We introduced the inverses of the sine, cosine, and tangent functions in Section 1.5. A discussion of inverse functions in general can be found in Section P.4. In this section we study those functions in more detail and define the inverses of the other three trigonometric functions.

## The Inverse Sine Function

An inverse of a function is a function that reverses what the function does. If there is a one-to-one correspondence between the elements of the domain and the elements of the range of a function, then the function is invertible. Since $y = \sin x$ with domain $(-\infty, \infty)$ is a periodic function, it is certainly not one-to-one. However, if we restrict the domain to the interval $[-\pi/2, \pi/2]$, then the restricted function is one-to-one and invertible. Other intervals could be used, but this interval is chosen to keep the inverse function as simple as possible.

The graph of the sine function with domain $[-\pi/2, \pi/2]$ is shown in Fig. 4.1(a). Its range is $[-1, 1]$. The inverse of this restricted sine function is denoted as $f^{-1}(x) = \sin^{-1}(x)$ (read "inverse sine of $x$") or $f^{-1}(x) = \arcsin(x)$ (read "arc sine of $x$").

**Definition: The Inverse Sine Function**

For $-1 \leq x \leq 1$

$$y = \sin^{-1}(x) \text{ provided } \sin(y) = x \text{ and } -\pi/2 \leq y \leq \pi/2.$$

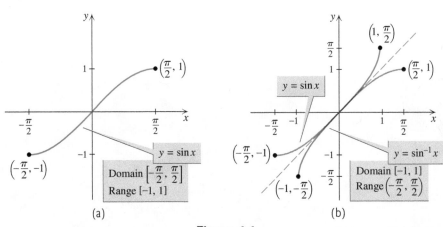

**Figure 4.1**

The domain of $y = \sin^{-1}(x)$ is $[-1, 1]$ and its range is $[-\pi/2, \pi/2]$. The graph of $y = \sin^{-1}(x)$ is a reflection about the line $y = x$ of the graph of $y = \sin(x)$ on $[-\pi/2, \pi/2]$ as shown in Fig. 4.1(b).

Depending on the context, $\sin^{-1} x$ might be an angle, a measure of an angle in degrees or radians, the length of an arc of the unit circle, or simply a real number. The expression $\sin^{-1} x$ can be read as "the angle whose sine is $x$" or "the arc length whose sine is $x$." The notation $y = \arcsin x$ reminds us that $y$ is the arc length whose sine is $x$. For example, $\arcsin(1)$ is the arc length in $[-\pi/2, \pi/2]$ whose sine is 1. Since we know that $\sin(\pi/2) = 1$, we have $\arcsin(1) = \pi/2$. We will assume that $\sin^{-1} x$ is a real number unless indicated otherwise.

Note that the $n$th power of the sine function is usually written as $\sin^n(x)$ as a shorthand notation for $(\sin x)^n$, provided $n \neq -1$. The $-1$ used in $\sin^{-1} x$ indicates the inverse function and does *not* mean reciprocal. To write $1/\sin x$ using exponents, we must write $(\sin x)^{-1}$.

**EXAMPLE 1**  Evaluating the inverse sine function

Find the exact value of each expression without using a table or a calculator.

**a.** $\sin^{-1}(1/2)$   **b.** $\arcsin\left(-\sqrt{3}/2\right)$

**Solution**

**a.** The value of $\sin^{-1}(1/2)$ is the number $\alpha$ in the interval $[-\pi/2, \pi/2]$ such that $\sin(\alpha) = 1/2$. We recall that $\sin(\pi/6) = 1/2$, and so $\sin^{-1}(1/2) = \pi/6$. Note that $\pi/6$ is the only value of $\alpha$ in $[-\pi/2, \pi/2]$ for which $\sin(\alpha) = 1/2$.

**b.** The value of $\arcsin\left(-\sqrt{3}/2\right)$ is the number $\alpha$ in the interval $[-\pi/2, \pi/2]$ such that $\sin(\alpha) = -\sqrt{3}/2$. Since $\sin(-\pi/3) = -\sqrt{3}/2$, we have $\arcsin\left(-\sqrt{3}/2\right) = -\pi/3$. Note that $-\pi/3$ is the only value of $\alpha$ in $[-\pi/2, \pi/2]$ for which $\sin(\alpha) = -\sqrt{3}/2$.

**TRY THIS.** Find the exact value of $\sin^{-1}\left(-\sqrt{2}/2\right)$. ■

**EXAMPLE 2**  Evaluating the inverse sine function

Find the exact value of each expression in degrees without using a table or a calculator.

**a.** $\sin^{-1}\left(\sqrt{2}/2\right)$   **b.** $\arcsin(0)$

**Solution**

**a.** The value of $\sin^{-1}\left(\sqrt{2}/2\right)$ in degrees is the angle $\alpha$ in the interval $[-90°, 90°]$ such that $\sin(\alpha) = \sqrt{2}/2$. We recall that $\sin(45°) = \sqrt{2}/2$, and therefore $\sin^{-1}\left(\sqrt{2}/2\right) = 45°$.

**b.** The value of $\arcsin(0)$ in degrees is the angle $\alpha$ in the interval $[-90°, 90°]$ for which $\sin(\alpha) = 0$. Since $\sin(0°) = 0$, we have $\arcsin(0) = 0°$.

**TRY THIS.** Find the exact value in degrees of $\arcsin(-1/2)$. ■

In the next example we use a calculator to find the degree measure of an angle whose sine is given. To obtain degree measure make sure the calculator is in degree mode. Scientific calculators usually have a key labeled $\sin^{-1}$ that gives values for the inverse sine function.

**EXAMPLE 3**  Finding an angle given its sine

Suppose that $\alpha$ is an angle such that $-90° < \alpha < 90°$. In each case, find $\alpha$ to the nearest tenth of a degree.

**a.** $\sin \alpha = 0.88$   **b.** $\sin \alpha = -0.27$

**Solution**

**a.** The value of $\sin^{-1}(0.88)$ is the only angle in $[-90°, 90°]$ with a sine of 0.88. Use a calculator in degree mode to get $\alpha = \sin^{-1}(0.88) \approx 61.6°$.

**b.** Use a calculator to get $\alpha = \sin^{-1}(-0.27) \approx -15.7°$.

 Figure 4.2 shows how to find the angle in part (a) on a graphing calculator and how to check. Make sure that the mode is degrees.

**TRY THIS.** Find $\alpha$ to the nearest tenth of a degree if $-90° < \alpha < 90°$ and $\sin \alpha = 0.257$. ■

**Figure 4.2**

## The Inverse Cosine Function

Since $f(x) = \cos(x)$ is not one-to-one on $(-\infty, \infty)$, we restrict the domain to $[0, \pi]$, where it is one-to-one and invertible. The graph of $f(x) = \cos(x)$ with this restricted

domain is shown in Fig. 4.3(a). Note that the range of the restricted function is $[-1, 1]$. The inverse of $f(x) = \cos x$ for $x$ in $[0, \pi]$ is denoted as $f^{-1}(x) = \cos^{-1}(x)$ or $f^{-1}(x) = \arccos(x)$. If $y = \cos^{-1}(x)$, then $y$ is the real number in $[0, \pi]$ such that $\cos(y) = x$.

**Definition: The Inverse Cosine Function**

For $-1 \le x \le 1$,

$$y = \cos^{-1}(x) \text{ provided } \cos(y) = x \text{ and } 0 \le y \le \pi.$$

The domain of $y = \cos^{-1} x$ is $[-1, 1]$ and its range is $[0, \pi]$. The expression $\cos^{-1} x$ can be read as "the angle whose cosine is $x$" or "the arc length whose cosine is $x$." The graph of $y = \cos^{-1} x$ shown in Fig. 4.3(b) is obtained by reflecting the graph of $y = \cos x$ (restricted to $[0, \pi]$) about the line $y = x$. We will assume that $\cos^{-1} x$ is a real number unless indicated otherwise.

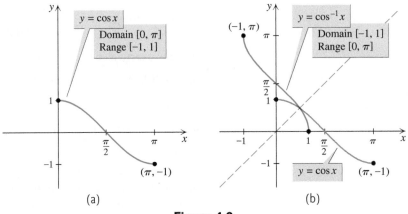

**Figure 4.3**

**EXAMPLE 4** **Evaluating the inverse cosine function**

Find the exact value of each expression without using a table or a calculator.

**a.** $\cos^{-1}(-1)$  **b.** $\arccos(-1/2)$  **c.** $\cos^{-1}\left(\sqrt{2}/2\right)$

**Solution**

**a.** The value of $\cos^{-1}(-1)$ is the number $\alpha$ in $[0, \pi]$ such that $\cos(\alpha) = -1$. We recall that $\cos(\pi) = -1$, and so $\cos^{-1}(-1) = \pi$.

**b.** The value of $\arccos(-1/2)$ is the number $\alpha$ in $[0, \pi]$ such that $\cos(\alpha) = -1/2$. We recall that $\cos(2\pi/3) = -1/2$, and so $\arccos(-1/2) = 2\pi/3$.

**c.** Since $\cos(\pi/4) = \sqrt{2}/2$, we have $\cos^{-1}\left(\sqrt{2}/2\right) = \pi/4$.

**TRY THIS.** Find the exact value of $\arccos(0)$. ∎

In the next example we use a calculator to find the degree measure of an angle whose cosine is given. Most scientific calculators have a key labeled $\cos^{-1}$ that gives values for the inverse cosine function. To get the degree measure, make sure the calculator is in degree mode.

**EXAMPLE 5** **Finding an angle given its cosine**

In each case find the angle $\alpha$ to the nearest tenth of a degree, given that $0° \le \alpha \le 180°$.

**a.** $\cos \alpha = 0.23$  **b.** $\cos \alpha = -0.82$

**Figure 4.4**

**Solution**

**a.** Use a calculator to get $\alpha = \cos^{-1}(0.23) \approx 76.7°$.
**b.** Use a calculator to get $\alpha = \cos^{-1}(-0.82) \approx 145.1°$.

Figure 4.4 shows how to find $\cos^{-1}(0.23)$ on a graphing calculator and how to check. Make sure that the mode is degrees.

**TRY THIS.** Find $\alpha$ to the nearest tenth of a degree if $0° \le \alpha \le 180°$ and $\cos \alpha = -0.361$. ∎

## The Inverse Tangent Function

Because $y = \tan(x)$ is a periodic function, its domain must be restricted before we can define the inverse function. The restricted domain for $y = \tan(x)$ is $(-\pi/2, \pi/2)$. The graph of $f(x) = \tan(x)$ with this restricted domain is shown in Fig. 4.5(a). Note that the range of the restricted function is $(-\infty, \infty)$. The inverse of $f(x) = \tan(x)$ for $x$ in $(-\pi/2, \pi/2)$ is denoted as $f^{-1}(x) = \tan^{-1}(x)$ or $f^{-1}(x) = \arctan(x)$. If $y = \tan^{-1}(x)$, then $y$ is the real number in $(-\pi/2, \pi/2)$ such that $\tan(y) = x$.

**Definition: The Inverse Tangent Function**

For any real number $x$,

$$y = \tan^{-1}(x) \text{ provided } \tan(y) = x \text{ and } -\pi/2 < y < \pi/2.$$

The domain of $y = \tan^{-1} x$ is $(-\infty, \infty)$ and its range is $(-\pi/2, \pi/2)$. The expression $\tan^{-1} x$ can be read as "the angle whose tangent is $x$" or "the arc length whose tangent is $x$." The graph of $y = \tan^{-1} x$ shown in Fig. 4.5(b) is obtained by reflecting the graph of $y = \tan x$ (restricted to $(-\pi/2, \pi/2)$) about the line $y = x$. We will assume that $\tan^{-1} x$ is a real number unless indicated otherwise.

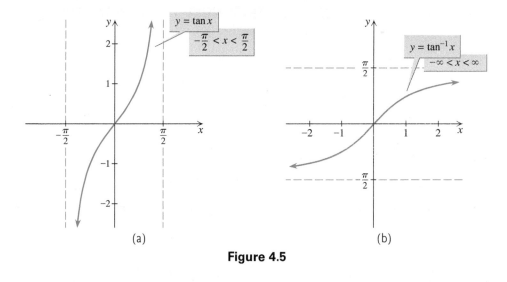

(a)                    (b)

**Figure 4.5**

**EXAMPLE 6** Evaluating the inverse tangent function

Find the exact value of each expression when possible.

**a.** $\tan^{-1}(1)$    **b.** $\arctan\left(\sqrt{3}\right)$    **c.** $\tan^{-1}(3)$

**Solution**

**a.** The value of $\tan^{-1}(1)$ is the number $\alpha$ in $(-\pi/2, \pi/2)$ such that $\tan(\alpha) = 1$. We recall that $\tan(\pi/4) = 1$, and so $\tan^{-1}(1) = \pi/4$.

**b.** The value of $\arctan\left(\sqrt{3}\right)$ is the number $\alpha$ in $(-\pi/2, \pi/2)$ such that $\tan(\alpha) = \sqrt{3}$. We recall that $\tan(\pi/3) = \sqrt{3}$, and so $\arctan\left(\sqrt{3}\right) = \pi/3$.

**c.** Use a calculator to get $\tan^{-1}(3) \approx 1.249$.

**TRY THIS.** Find the exact value of $\tan^{-1}\left(-\sqrt{3}\right)$.

## Inverses of Cosecant, Secant, and Cotangent (Optional)

The most common and useful inverse trigonometric functions are the arcsin, arccos, and arctan. These are the only ones that are needed in the remainder of this text. However, to complete our discussion of the inverse trigonometric functions, we now define inverses for the cosecant, secant, and cotangent. These functions are defined and used occasionally in calculus.

As with any periodic function, the domain must be restricted to a single cycle to get a one-to-one function that has an inverse. The restrictions of the domains for the sine, cosine, and tangent are well accepted, but there is some disagreement on restricting the domains of the cosecant, secant, and cotangent. The restrictions most commonly used in calculus texts will be given here.

The function $y = \csc(x)$ is one-to-one on the domain $[-\pi/2, 0) \cup (0, \pi/2]$, and the notation $\csc^{-1}$ or arccsc is used for the inverse function. The function $y = \sec(x)$ is one-to-one on the domain $[0, \pi/2) \cup (\pi/2, \pi]$, and the notation $\sec^{-1}$ or arcsec is used for the inverse function. The function $y = \cot(x)$ is one-to-one on the interval $(0, \pi)$, and the notation $\cot^{-1}$ or arccot is used for the inverse function. Note that the domains of these functions are the ranges for the inverse functions. The graphs of all six inverse functions, with their domains and ranges, are shown in the following Function Gallery.

### INVERSE TRIGONOMETRIC FUNCTIONS

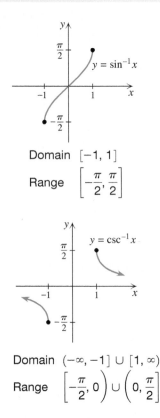

Domain  $[-1, 1]$

Range  $\left[-\dfrac{\pi}{2}, \dfrac{\pi}{2}\right]$

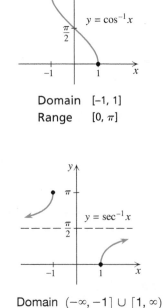

Domain  $[-1, 1]$

Range  $[0, \pi]$

Domain  $(-\infty, \infty)$

Range  $\left(-\dfrac{\pi}{2}, \dfrac{\pi}{2}\right)$

Domain  $(-\infty, -1] \cup [1, \infty)$

Range  $\left[-\dfrac{\pi}{2}, 0\right) \cup \left(0, \dfrac{\pi}{2}\right]$

Domain  $(-\infty, -1] \cup [1, \infty)$

Range  $\left[0, \dfrac{\pi}{2}\right) \cup \left(\dfrac{\pi}{2}, \pi\right]$

Domain  $(-\infty, \infty)$

Range  $(0, \pi)$

When studying inverse trigonometric functions, you should first learn to evaluate $\sin^{-1}$, $\cos^{-1}$, and $\tan^{-1}$. Those values can be used along with the identities $\csc\alpha = 1/\sin\alpha$, $\sec\alpha = 1/\cos\alpha$, and $\cot\alpha = 1/\tan\alpha$ to evaluate $\csc^{-1}$, $\sec^{-1}$, and $\cot^{-1}$. For example, $\sin(\pi/6) = 1/2$ and $\csc(\pi/6) = 2$. So the angle whose cosecant is 2 is the same as the angle whose sine is $1/2$. In symbols,

$$\csc^{-1}(2) = \sin^{-1}\left(\frac{1}{2}\right) = \frac{\pi}{6}.$$

In general, $\csc^{-1}x = \sin^{-1}(1/x)$. Likewise, $\sec^{-1}x = \cos^{-1}(1/x)$. For the inverse cotangent, $\cot^{-1}x = \tan^{-1}(1/x)$ only for positive values of $x$, because of the choice of $(0, \pi)$ as the range of the inverse cotangent. We have $\cot^{-1}(0) = \pi/2$ and if $x$ is negative $\cot^{-1}(x) = \tan^{-1}(1/x) + \pi$.

Another relationship between $\cot^{-1}$ and $\tan^{-1}$ can be obtained from their graphs in the Function Gallery on page 218. Since the graph of $y = \cot^{-1}(x)$ can be obtained from the graph of $y = \tan^{-1}(x)$ by reflecting with respect to the $x$-axis and translating upward a distance of $\pi/2$, we have $\cot^{-1}(x) = -\tan^{-1}(x) + \pi/2$ or $\cot^{-1}(x) = \pi/2 - \tan^{-1}(x)$. These identities are summarized below.

**Identities for $\csc^{-1}$, $\sec^{-1}$, and $\cot^{-1}$**

$$\csc^{-1}(x) = \sin^{-1}(1/x) \text{ for } |x| \geq 1$$

$$\sec^{-1}(x) = \cos^{-1}(1/x) \text{ for } |x| \geq 1$$

$$\cot^{-1}(x) = \begin{cases} \tan^{-1}(1/x) & \text{for } x > 0 \\ \tan^{-1}(1/x) + \pi & \text{for } x < 0 \\ \pi/2 & \text{for } x = 0 \end{cases}$$

$$\cot^{-1}(x) = \frac{\pi}{2} - \tan^{-1}(x)$$

**EXAMPLE 7** **Evaluating the inverse functions**

Find the exact value of each expression without using a table or a calculator.

**a.** $\operatorname{arcsec}(-2)$  **b.** $\csc^{-1}\left(\sqrt{2}\right)$  **c.** $\operatorname{arccot}\left(-1/\sqrt{3}\right)$

**Solution**

**a.** To evaluate the inverse secant we use the identity $\sec^{-1}(x) = \cos^{-1}(1/x)$. In this case, the arc whose secant is $-2$ is the same as the arc whose cosine is $-1/2$. So we must find $\cos^{-1}(-1/2)$. Since $\cos(2\pi/3) = -1/2$ and since $2\pi/3$ is in the range of arccos, we have $\operatorname{arccos}(-1/2) = 2\pi/3$. So

$$\operatorname{arcsec}(-2) = \operatorname{arccos}(-1/2) = \frac{2\pi}{3}.$$

**b.** To evaluate $\csc^{-1}\left(\sqrt{2}\right)$, we use the identity $\csc^{-1}(x) = \sin^{-1}(1/x)$ with $x = \sqrt{2}$. So we must find $\sin^{-1}\left(1/\sqrt{2}\right)$. Since $\sin(\pi/4) = 1/\sqrt{2}$ and since $\pi/4$ is in the range of $\csc^{-1}$, we have

$$\csc^{-1}\left(\sqrt{2}\right) = \sin^{-1}\left(1/\sqrt{2}\right) = \frac{\pi}{4}.$$

**c.** If $x$ is negative we use the identity $\cot^{-1}(x) = \tan^{-1}(1/x) + \pi$. Since $x = -1/\sqrt{3}$, we must find $\tan^{-1}\left(-\sqrt{3}\right)$. Since $\tan^{-1}\left(-\sqrt{3}\right) = -\pi/3$, we get

$$\operatorname{arccot}\left(-1/\sqrt{3}\right) = \tan^{-1}\left(-\sqrt{3}\right) + \pi = -\frac{\pi}{3} + \pi = \frac{2\pi}{3}.$$

Note that $\cot(-\pi/3) = \cot(2\pi/3) = -1/\sqrt{3}$, but $\operatorname{arccot}\left(-1/\sqrt{3}\right) = 2\pi/3$ because $2\pi/3$ is in the range of the function arccot.

**TRY THIS.** Find the exact value of $\csc^{-1}(2)$. ■

The domains and ranges for $\sin^{-1}$, $\cos^{-1}$, and $\tan^{-1}$ on a calculator agree with what we have defined. The functions $\csc^{-1}$, $\sec^{-1}$, and $\cot^{-1}$ are generally not available on a calculator, and expressions involving these functions must be written in terms of $\sin^{-1}$, $\cos^{-1}$, and $\tan^{-1}$ using the identities. The calculator values of the inverse functions are given in degrees or radians, depending on the mode setting. We will assume that the values of the inverse functions are to be in radians unless indicated otherwise.

**EXAMPLE 8** Evaluating $\sec^{-1}$, $\csc^{-1}$, and $\cot^{-1}$ with a calculator

Find the approximate value of each expression rounded to four decimal places.

**a.** $\sec^{-1}(6)$ **b.** $\csc^{-1}(4)$ **c.** $\operatorname{arccot}(2.4)$ **d.** $\cot^{-1}(0)$

**Solution**

All of these function values are found using the identities for the inverse functions.

**a.** $\sec^{-1}(6) = \cos^{-1}(1/6) \approx 1.4033$
**b.** $\csc^{-1}(4) = \sin^{-1}(1/4) \approx 0.2527$
**c.** Because $2.4 > 0$ we use the identity $\cot^{-1}(x) = \tan^{-1}(1/x)$:

$$\operatorname{arccot}(2.4) = \tan^{-1}(1/2.4) \approx 0.3948$$

**d.** The value of $\cot^{-1}(0)$ cannot be found on a calculator. But according to the identities for the inverse functions, $\cot^{-1}(0) = \pi/2$.

**TRY THIS.** Find $\sec^{-1}(4.328)$ rounded to four decimal places. ■

## Compositions of Functions

One trigonometric function can be followed by another to form a composition of functions. We can evaluate an expression such as $\sin(\sin(\alpha))$ because the sine of the real number $\sin(\alpha)$ is defined. However, it is more common to have a composition of a trigonometric function and an inverse trigonometric function. For example, $\tan^{-1}(\alpha)$ is the angle whose tangent is $\alpha$, and so $\sin(\tan^{-1}(\alpha))$ is the sine of the angle whose tangent is $\alpha$.

The composition of $y = \sin x$ restricted to $[-\pi/2, \pi/2]$ and $y = \sin^{-1} x$ is the identity function, because the functions are inverses. So

$$\sin^{-1}(\sin x) = x \qquad \text{for } x \text{ in } [-\pi/2, \pi/2]$$

and

$$\sin(\sin^{-1} x) = x \qquad \text{for } x \text{ in } [-1, 1].$$

Note that if $x$ is not in $[-\pi/2, \pi/2]$, then the sine function followed by the inverse sine function is not the identity function.

**EXAMPLE 9** Evaluating compositions of functions

Find the exact value of each composition without using a table or a calculator.

**a.** $\sin(\tan^{-1}(0))$ **b.** $\arcsin(\cos(\pi/6))$
**c.** $\tan\left(\sec^{-1}\left(\sqrt{2}\right)\right)$ **d.** $\sin^{-1}(\sin(2\pi/3))$
**e.** $\cos(\cos^{-1}(-1/2))$ **f.** $\cos^{-1}(\cos(-\pi/3))$

**Solution**

**a.** Since $\tan(0) = 0$, $\tan^{-1}(0) = 0$. Therefore,

$$\sin(\tan^{-1}(0)) = \sin(0) = 0.$$

**b.** Since $\cos(\pi/6) = \sqrt{3}/2$, we have

$$\arcsin(\cos(\pi/6)) = \arcsin\left(\sqrt{3}/2\right) = \pi/3.$$

**c.** To find $\sec^{-1}\left(\sqrt{2}\right)$, we use the identity $\sec^{-1}(x) = \cos^{-1}(1/x)$. Since $\cos(\pi/4) = 1/\sqrt{2}$ we have $\cos^{-1}\left(1/\sqrt{2}\right) = \pi/4$ and $\sec^{-1}\left(\sqrt{2}\right) = \pi/4$. Therefore,

$$\tan\left(\sec^{-1}\left(\sqrt{2}\right)\right) = \tan(\pi/4) = 1.$$

**d.** Since $\sin(2\pi/3) = \sin(\pi/3) = \sqrt{3}/2$,

$$\sin^{-1}(\sin(2\pi/3)) = \sin^{-1}\left(\sqrt{3}/2\right) = \pi/3.$$

Note that $\sin^{-1}(\sin(2\pi/3)) \neq 2\pi/3$ because $2\pi/3$ is not in the interval $[-\pi/2, \pi/2]$.

**e.** $\cos(\cos^{-1}(-1/2)) = \cos(2\pi/3) = -1/2$

**f.** $\cos^{-1}(\cos(-\pi/3)) = \cos^{-1}(1/2) = \pi/3$

📈 You can check parts (a), (b), and (c) with a graphing calculator as shown in Fig. 4.6. Note that $\pi/3 \approx 1.047$.

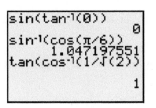

**Figure 4.6**

**TRY THIS.** Find the exact value of $\sin^{-1}(\sin(5\pi/6))$. ∎

In the next example we use the Pythagorean identities to rewrite a composition of trigonometric functions as an algebraic function.

**EXAMPLE 10** Converting compositions to algebraic functions

Find an equivalent algebraic expression for $\sin(\arctan(x))$.

**Solution**

Let $\theta = \arctan(x)$. Since $\tan(\theta) = x$ and $\tan^2(\theta) + 1 = \sec^2(\theta)$, we have $\sec^2(\theta) = x^2 + 1$ and

$$\cos^2(\theta) = \frac{1}{x^2 + 1}.$$

Since $\sin^2(\theta) = 1 - \cos^2(\theta)$, we have

$$\sin^2(\theta) = 1 - \frac{1}{x^2 + 1} = \frac{x^2 + 1}{x^2 + 1} - \frac{1}{x^2 + 1} = \frac{x^2}{x^2 + 1}$$

and

$$\sin(\theta) = \pm\sqrt{\frac{x^2}{x^2 + 1}} = \frac{x}{\sqrt{x^2 + 1}}.$$

We eliminated the radical and the $\pm$ sign in the last equation because the sign of $x$ is the same as the sign of $\sin(\theta)$. Specifically, if $x > 0$, then $0 < \theta < \frac{\pi}{2}$ and $\sin(\theta) > 0$. If $x < 0$, then $-\frac{\pi}{2} < \theta < 0$ and $\sin(\theta) < 0$. So

$$\sin(\arctan(x)) = \frac{x}{\sqrt{x^2 + 1}}.$$

**TRY THIS.** Find an equivalent algebraic expression for $\sin(\text{arccot}(x))$. ∎

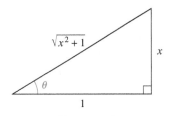

**Figure 4.7**

Example 10 can also be solved with right triangle trigonometry. Let $\theta = \arctan x$ and assume $0 < \theta < \pi/2$. Since $\tan(\theta) = x/1$, we can draw a right triangle with angle $\theta$, for which the opposite side is $x$ and the adjacent side is 1, as in Fig. 4.7. By the Pythagorean theorem, the hypotenuse is $\sqrt{x^2 + 1}$. Now use the right triangle to find

$$\sin(\arctan(x)) = \sin(\theta) = \frac{\text{opp}}{\text{hyp}} = \frac{x}{\sqrt{x^2 + 1}}.$$

## FOR THOUGHT... True or False? Explain.

**1.** $\sin^{-1}(0) = \sin(0)$

**2.** $\sin(3\pi/4) = 1/\sqrt{2}$

**3.** $\cos^{-1}(0) = 1$

**4.** $\sin^{-1}\left(\sqrt{2}/2\right) = 135°$

**5.** $\cot^{-1}(5) = \dfrac{1}{\tan^{-1}(5)}$

**6.** $\sec^{-1}(5) = \cos^{-1}(0.2)$

**7.** $\sin\left(\cos^{-1}\left(\sqrt{2}/2\right)\right) = 1/\sqrt{2}$

**8.** $\sec(\sec^{-1}(2)) = 2$

**9.** The functions $f(x) = \sin^{-1} x$ and $f^{-1}(x) = \sin x$ are inverse functions.

**10.** The secant and cosecant functions are inverses of each other.

## 4.1 EXERCISES

### CONCEPTS

*Fill in the blank.*

**1.** The _____ of $y = \arcsin(x)$ is $[-1, 1]$.

**2.** The _____ of $y = \arccos(x)$ is $[0, \pi]$.

**3.** The _____ of $y = \arctan(x)$ is $(-\infty, \infty)$.

**4.** The _____ of $y = \tan^{-1}(x)$ is $(-\pi/2, \pi/2)$.

**5.** If $y = \sin(\alpha)$, where $-\pi/2 \le \alpha \le \pi/2$, then $\alpha =$ _____.

**6.** If $y = \cos(\alpha)$, where $0 \le \alpha \le \pi$, then $\alpha =$ _____.

### SKILLS

*Find the exact value of each expression without using a calculator or table.*

**7.** $\sin^{-1}\left(1/\sqrt{2}\right)$

**8.** $\sin^{-1}(-1/2)$

**9.** $\arcsin\left(\sqrt{3}/2\right)$

**10.** $\arcsin(1/2)$

**11.** $\sin^{-1}\left(-\sqrt{2}/2\right)$

**12.** $\sin^{-1}(-1)$

**13.** $\arcsin(0)$

**14.** $\arcsin(1)$

*Find the exact value of each expression in degrees without using a calculator or table.*

**15.** $\sin^{-1}\left(-1/\sqrt{2}\right)$

**16.** $\sin^{-1}(1/2)$

**17.** $\arcsin(1/2)$

**18.** $\arcsin\left(\sqrt{2}/2\right)$

**19.** $\sin^{-1}(0)$

**20.** $\sin^{-1}(1)$

**21.** $\arcsin(-1)$

**22.** $\arcsin\left(-\sqrt{3}/2\right)$

*Find $\alpha$ to the nearest tenth of a degree, where $-90° \le \alpha \le 90°$.*

**23.** $\sin \alpha = 0.557$

**24.** $\sin \alpha = 0.93$

**25.** $\sin \alpha = -1/3$

**26.** $\sin \alpha = -1/4$

**27.** $\sin \alpha = 0.01$

**28.** $\sin \alpha = 0.9999$

*Find the exact value of each expression without using a calculator or table.*

**29.** $\cos^{-1}\left(1/\sqrt{2}\right)$

**30.** $\cos^{-1}(-1/2)$

**31.** $\arccos\left(\sqrt{3}/2\right)$

**32.** $\arccos(1/2)$

**33.** $\cos^{-1}\left(-\sqrt{2}/2\right)$

**34.** $\cos^{-1}\left(-\sqrt{3}/2\right)$

**35.** $\arccos(0)$

**36.** $\arccos(-1)$

*Find α to the nearest tenth of a degree, where $0° \leq \alpha \leq 180°$.*

**37.** $\cos \alpha = 0.06$

**38.** $\cos \alpha = 0.71$

**39.** $\cos \alpha = -1/5$

**40.** $\cos \alpha = -1/9$

**41.** $\cos \alpha = 0.01$

**42.** $\cos \alpha = -0.01$

*Find the exact value of each expression when possible. Round approximate answers to three decimal places.*

**43.** $\tan^{-1}(-1)$

**44.** $\tan^{-1}(0)$

**45.** $\arctan\left(1/\sqrt{3}\right)$

**46.** $\arctan(1)$

**47.** $\tan^{-1}\left(\sqrt{3}\right)$

**48.** $\tan^{-1}\left(-\sqrt{3}\right)$

**49.** $\arctan(5788)$

**50.** $\arctan(-5788)$

*Find the exact value of each expression without using a calculator or table.*

**51.** $\cot^{-1}\left(1/\sqrt{3}\right)$

**52.** $\cot^{-1}\left(\sqrt{3}\right)$

**53.** $\sec^{-1}(2)$

**54.** $\text{arcsec}\left(\sqrt{2}\right)$

**55.** $\csc^{-1}\left(2/\sqrt{3}\right)$

**56.** $\text{arccsc}(-2)$

**57.** $\text{arccot}\left(-\sqrt{3}\right)$

**58.** $\cot^{-1}(1)$

**59.** $\text{arcsec}(1)$

**60.** $\sec^{-1}\left(-\sqrt{2}\right)$

**61.** $\csc^{-1}(1)$

**62.** $\csc^{-1}(-1)$

**63.** $\cot^{-1}(-1)$

**64.** $\cot^{-1}(0)$

*Find the approximate value of each expression rounded to two decimal places.*

**65.** $\sec^{-1}(-3.44)$

**66.** $\text{arcsec}\left(\sqrt{6}\right)$

**67.** $\csc^{-1}(6.8212)$

**68.** $\text{arccsc}\left(-2\sqrt{2}\right)$

**69.** $\text{arccot}\left(-\sqrt{5}\right)$

**70.** $\text{arccot}(0.001)$

**71.** $\cot^{-1}(-12)$

**72.** $\cot^{-1}(15.6)$

**73.** $\cot^{-1}(-1.01)$

**74.** $\text{arccsc}(-2)$

*Find the exact value of each composition without using a calculator or table.*

**75.** $\tan(\arccos(1/2))$

**76.** $\sec\left(\arcsin\left(1/\sqrt{2}\right)\right)$

**77.** $\sin^{-1}(\cos(2\pi/3))$

**78.** $\tan^{-1}(\sin(\pi/2))$

**79.** $\cot^{-1}(\cot(\pi/6))$

**80.** $\sec^{-1}(\sec(\pi/3))$

**81.** $\arcsin(\sin(3\pi/4))$

**82.** $\arccos(\cos(-\pi/3))$

**83.** $\tan(\arctan(1))$

**84.** $\cot(\text{arccot}(0))$

**85.** $\cos^{-1}(\cos(3\pi/2))$

**86.** $\sin(\csc^{-1}(-2))$

**87.** $\sin^{-1}(\sin(3\pi/4))$

**88.** $\sin(\sin^{-1}(1))$

**89.** $\cos\left(\cos^{-1}\left(-\sqrt{3}/2\right)\right)$

**90.** $\cos^{-1}(\cos(-\pi))$

**91.** $\tan^{-1}(\tan(\pi))$

**92.** $\tan(\tan^{-1}(1))$

*Use a calculator to find the approximate value of each composition. Round answers to four decimal places. Some of these expressions are undefined.*

**93.** $\sin(\cos^{-1}(0.45))$

**94.** $\cos(\tan^{-1}(44.33))$

**95.** $\sec^{-1}(\cos(\pi/9))$

**96.** $\csc^{-1}(\sec(\pi/8))$

**97.** $\tan(\sin^{-1}(-0.7))$

**98.** $\csc^{-1}(\sin(3\pi/7))$

**99.** $\cot\left(\cos^{-1}\left(-1/\sqrt{7}\right)\right)$

**100.** $\sin^{-1}(\csc(1.08))$

**101.** $\cot(\csc^{-1}(3.6))$

**102.** $\csc(\sec^{-1}(-2.4))$

**103.** $\sec(\cot^{-1}(5.2))$

**104.** $\csc(\cot^{-1}(-3.1))$

*Find an equivalent algebraic expression for each composition.*

**105.** $\sin(\arccos(x))$

**106.** $\cos(\arcsin(x))$

**107.** $\cos(\arctan(x))$

**108.** $\tan(\arccos(x))$

**109.** $\cos(\text{arccot}(x))$

**110.** $\tan(\text{arccot}(x))$

**111.** $\tan(\arcsin(x))$

**112.** $\sec(\arcsin(x))$

**113.** $\sec(\arctan(x))$

**114.** $\csc(\arcsin(x))$

## MODELING

*In a circle with radius r, a central angle θ intercepts a chord of length c, where $\theta = \cos^{-1}\left(1 - \dfrac{c^2}{2r^2}\right)$. Use this formula to solve problems 115 and 116.*

**115.** An airplane at 2000 ft flies directly over a gun that has a range of 2400 ft as shown in the figure. What is the measure in degrees of the angle for which the airplane is within range of the gun?

2400 ft    2400 ft

θ

**Figure for Exercise 115**

**116.** A triangle has two sides that are both 5.2 m long and one side with a length of 1.3 m. Find the measure in degrees for the smallest angle of the triangle.

**117.** *Eclipse Function* The portion of the sun that is blocked out by the moon at time $t$ in a total eclipse of the sun is given by the function

$$P = \frac{\cos^{-1}(u) - \sin(\cos^{-1}(u))}{\pi}$$

where $u = 1 - 8t + 8t^2$. The eclipse begins at time $t = 0$ and ends at time $t = 1$. What percent of the sun is covered at time $t = 0.25$?

**118.** *Maximum Coverage* Using a graphing calculator, graph the function from the previous exercise. Use the maximum feature to find the maximum value of $P$ for $t$ in the interval $[0, 1]$.

**Figure for Exercises 117 and 118**

## WRITING/DISCUSSION

**119.** Graph the function $y = \sin(\sin^{-1} x)$ for $-2\pi \le x \le 2\pi$ and explain your result.

**120.** Graph the function $y = \sin^{-1}(\sin x)$ for $-2\pi \le x \le 2\pi$ and explain your result.

**121.** Graph $y = \sin^{-1}(1/x)$ and explain why the graph looks like the graph of $y = \csc^{-1} x$ shown in the Function Gallery on page 218.

**122.** Graph $y = \tan^{-1}(1/x)$ and explain why the graph does *not* look like the graph of $y = \cot^{-1} x$ shown in the Function Gallery on page 218.

## REVIEW

**123.** The shortest side of a right triangle is 7 cm, and one of the acute angles is 64°. Find the length of the hypotenuse and the length of the longer leg. Round to the nearest tenth of a centimeter.

**124.** Suppose that $\alpha$ is an angle in standard position with its terminal side in quadrant III such that $\sin \alpha = -5/6$. Find exact values for $\cos \alpha$ and $\tan \alpha$.

**125.** Determine the amplitude, period, phase shift, and range for the function $y = 5 \sin(4x - \pi) - 3$.

**126.** Determine the period, asymptotes, and range for the function $y = 3 \csc(\pi x - \pi) + 2$.

**127.** Simplify the expression $3 - \dfrac{3}{\csc^2(x)}$.

**128.** Simplify the expression $(\sin x + \cos x)^2 - \sin(2x)$.

## OUTSIDE THE BOX

**129.** *Lucky Lucy* Ms. Willis asked Lucy to come to the board to find the mean of a pair of one-digit positive integers. Lucy slowly wrote the numbers on the board. While trying to think of what to do next, she rested the chalk between the numbers to make a mark that looked like a decimal point to Ms. Willis. Ms. Willis said "correct" and asked her to find the mean for a pair of two-digit positive integers. Being a quick learner, Lucy again wrote the numbers on the board, rested the chalk between the numbers, and again Ms. Willis said "correct." Lucy had to demonstrate her ability to find the mean for a pair of three-digit and a pair of four-digit positive integers before Ms. Willis was satisfied that she understood the concept. What four pairs of integers did Ms. Willis give to Lucy? Explain why Lucy's method will not work for any other pairs of one-, two-, three-, or four-digit positive integers.

**130.** *Sum and Difference* Find the exact value of

$$\frac{\cos 15° + \sin 15°}{\cos 15° - \sin 15°}.$$

**4.1 POP QUIZ**

*Find the exact value.*

**1.** $\sin^{-1}(-1)$

**2.** $\sin^{-1}(1/2)$

**3.** $\arccos(-1)$

**4.** $\arctan(-1)$

**5.** $\tan(\arcsin(1/2))$

**6.** $\sin^{-1}(\sin(3\pi/4))$

*Find the exact value for x in the interval $[0, \pi/2]$ that satisfies each equation.*

**7.** $\sin(x) = \sqrt{2}/2$

**8.** $\cos(x) = 0$

**9.** $\tan(x) = 1$

## LINKING concepts...

### Maximizing the Viewing Angle

*A billboard that is 30 ft wide is placed 60 ft from a highway as shown in the figure. The billboard is easiest to read from the highway when the viewing angle $\alpha$ is larger than 7°.*

**a)** Show that
$$\alpha = \tan^{-1}(90/x) - \tan^{-1}(60/x).$$

**b)** Graph the function in part (a).

**c)** For what approximate values of $x$ is $\alpha$ greater than 7°?

**d)** For approximately how long does a motorist traveling at 35 mph have a viewing angle larger than 7°?

---

## 4.2 Basic Sine, Cosine, and Tangent Equations

In this section we will solve the most basic sine, cosine, and tangent equations.

### Basic Cosine Equations

An identity is satisfied by *all* values of the variable for which both sides are defined. A **conditional equation** is an equation that has at least one solution but is not an identity.

The most basic conditional equation involving cosine is of the form $\cos x = a$ where $a$ is a number in the interval $[-1, 1]$. If $a$ is not in $[-1, 1]$, $\cos x = a$ has no solution. The domain of $\cos^{-1}$ is $[-1, 1]$ and its range is $[0, \pi]$. So for $a$ in $[-1, 1]$, the equation $x = \cos^{-1} a$ provides one solution in the interval $[0, \pi]$. From that solution we can determine all of the solutions because of the periodic nature of the cosine function.

> **EXAMPLE 1** Solving $\cos(x) = a$, where $a$ is 1, 0, or −1

Find all real numbers that satisfy each equation.

**a.** $\cos x = 1$   **b.** $\cos x = 0$   **c.** $\cos x = -1$

**Solution**

**a.** Use the inverse cosine function to find one solution to $\cos x = 1$:
$$x = \cos^{-1}(1) = 0$$

Since the period of cosine is $2\pi$, any integral multiple of $2\pi$ can be added to this solution to get additional solutions. So the equation is satisfied by 0, $\pm 2\pi$, $\pm 4\pi$, and so on. Since $\cos x$ is defined as the first coordinate where an arc of length $x$ terminates on the unit circle, $\cos x = 1$ is satisfied only if the arc of length $x$ terminates at $(1, 0)$, as shown in Fig. 4.8. So there are no more solutions. The solution set is written as
$$\{x \mid x = 2k\pi\}$$

where $k$ is any integer.

**Figure 4.8**

**Figure 4.9**

**Figure 4.10**

**Figure 4.11**

Use your calculator to graph $y_1 = \cos(x)$ and $y_2 = 1$ as shown in Fig. 4.9. Use the intersect feature to find some of the intersections of the two graphs. The intersections all occur at multiples of $2\pi$, which supports our conclusion. □

**b.** Use the inverse cosine function to find one solution to $\cos x = 0$:

$$x = \cos^{-1}(0) = \frac{\pi}{2}$$

The terminal point for the arc of length $\pi/2$ is $(0, 1)$. However, an arc of length $x$ that terminates at $(0, -1)$ also satisfies $\cos x = 0$, as shown in Fig. 4.10, because $\cos x$ is the first coordinate of the terminal point of the arc. Since the distance between $(0, 1)$ and $(0, -1)$ on the unit circle is $\pi$, all arcs that terminate at these points are of the form $\pi/2 + k\pi$. So the solution set is

$$\left\{ x \mid x = \frac{\pi}{2} + k\pi \right\}$$

where $k$ is any integer.

Use your calculator to graph $y = \cos(x)$ as shown in Fig. 4.11. The solutions to $\cos(x) = 0$ correspond to the $x$-intercepts on the graph of $y = \cos(x)$ shown in Fig. 4.11. Use the zero feature to find some of the $x$-intercepts and verify that they have the form $(\pi/2 + k\pi, 0)$. □

**c.** Use $\cos^{-1}$ to find one solution to $\cos x = -1$:

$$x = \cos^{-1}(-1) = \pi$$

An arc of length $\pi$ on the unit circle terminates at $(-1, 0)$ as shown in Fig. 4.12. Since the period of cosine is $2\pi$, all arcs of length $\pi + 2k\pi$ (where $k$ is any integer) satisfy the equation. Since $(-1, 0)$ is the only point on the unit circle with first coordinate $-1$, there are no other solutions.

Graph $y_1 = \cos(x)$ and $y_2 = -1$ as shown in Fig. 4.13. Use the intersect or trace feature to verify that they intersect when $x$ has the form $\pi + 2k\pi$.

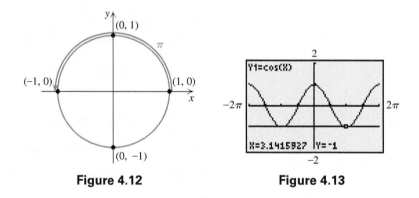

**Figure 4.12**          **Figure 4.13**

**TRY THIS.** Find the exact values of $\cos^{-1}(-1)$, $\cos^{-1}(0)$, and $\cos^{-1}(1)$. ■

Since the cosine of an arc is the first coordinate of its terminal point on the unit circle, arcs that terminate at opposite ends of a vertical chord in the unit circle have the same cosine. So, if the solution found using the inverse cosine is in the interval $(0, \pi)$ then there is another solution in the interval $(\pi, 2\pi)$.

**EXAMPLE** **2**    Solving $\cos(x) = a$ for nonzero $a$ in $(-1, 1)$

Find all real numbers that satisfy each equation.

**a.** $\cos x = -1/2$     **b.** $\cos x = 0.3$

Terminal point of $\frac{2\pi}{3}$

$\frac{2\pi}{3}$

$2\pi - \frac{2\pi}{3}$

or $\frac{4\pi}{3}$

$-\frac{1}{2}$

Terminal point of $\frac{4\pi}{3}$

**Figure 4.14**

**Figure 4.15**

**Solution**

**a.** Use $\cos^{-1}$ to find one solution to $\cos x = -1/2$:

$$x = \cos^{-1}\left(-\frac{1}{2}\right) = \frac{2\pi}{3}$$

Since the period of cosine is $2\pi$, all arcs of length $2\pi/3 + 2k\pi$ (where $k$ is any integer) satisfy the equation. The arc of length $4\pi/3$ also terminates at a point with $x$-coordinate $-1/2$ as shown in Fig. 4.14. So $4\pi/3$ is also a solution to $\cos x = -1/2$, but it is not included in the form $2\pi/3 + 2k\pi$. So the solution set to $\cos x = -1/2$ is

$$\left\{x \,\middle|\, x = \frac{2\pi}{3} + 2k\pi \text{ or } x = \frac{4\pi}{3} + 2k\pi\right\}$$

where $k$ is any integer.

Note that the calculator graph of $y_1 = \cos(x)$ in Fig. 4.15 crosses the horizontal line $y_2 = -1/2$ twice in the interval $[0, 2\pi]$, at $2\pi/3$ and $4\pi/3$. □

**b.** Use $\cos^{-1}$ on a calculator to find one solution to $\cos x = 0.3$:

$$x = \cos^{-1}(0.3) \approx 1.266$$

Since the period of cosine is $2\pi$, all arcs of length $1.266 + 2k\pi$ (where $k$ is any integer) satisfy the equation. The arc of length $2\pi - 1.266$ or $5.017$ also terminates at a point with $x$-coordinate $0.3$ as shown in Fig. 4.16. So $5.017$ is also a solution to $\cos x = 0.3$. So the solution set to $\cos x = 0.3$ is

$$\{x \mid x \approx 1.266 + 2k\pi \text{ or } x \approx 5.017 + 2k\pi\}$$

where $k$ is any integer.

The points of intersection of $y_1 = \cos(x)$ and $y_2 = 0.3$ in Fig. 4.17 correspond to the solutions to the equation $\cos(x) = 0.3$.

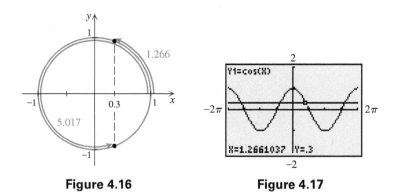

**Figure 4.16**          **Figure 4.17**

**TRY THIS.** Find all real numbers that satisfy $\cos x = \sqrt{3}/2$. ∎

The procedure used in Example 2(b) can be used to solve $\cos x = a$ for any nonzero $a$ between $-1$ and $1$. First find two values of $x$ between $0$ and $2\pi$ that satisfy the equation. (In general, $s = \cos^{-1}(a)$ and $2\pi - s$ are the values.) Next write all solutions by adding $2k\pi$ to the first two solutions. The cases $a = -1, 0$, and $1$ are handled separately because in these cases the solution set can be written as a single solution plus multiples of $\pi$ or $2\pi$.

All of the cases for solving $\cos x = a$ are summarized in the box on the next page, where $k$ is used to represent any integer. Do not try to simply memorize this summary. Practice solving $\cos x = a$ until you can do it for any value of $a$ using the unit circle as a guide.

> **SUMMARY**
>
> ### Solving cos $x = a$
>
> 1. If $-1 < a < 1$ and $a \neq 0$, then the solution set to $\cos x = a$ is
>
>    $$\{x \mid x = s + 2k\pi \text{ or } x = 2\pi - s + 2k\pi\},$$
>
>    where $s = \cos^{-1} a$ and $k$ is any integer.
> 2. The solution set to $\cos x = 1$ is $\{x \mid x = 2k\pi\}$.
> 3. The solution set to $\cos x = 0$ is $\{x \mid x = \pi/2 + k\pi\}$.
> 4. The solution set to $\cos x = -1$ is $\{x \mid x = \pi + 2k\pi\}$.
> 5. If $|a| > 1$, then $\cos x = a$ has no solution.

## Basic Sine Equations

The most basic conditional equation involving sine is $\sin x = a$ where $a$ is a number in the interval $[-1, 1]$. If $a$ is not in $[-1, 1]$, $\sin x = a$ has no solution. For $a$ in $[-1, 1]$, the equation $x = \sin^{-1} a$ provides one solution in the interval $[-\pi/2, \pi/2]$. From that solution we can determine all of the solutions.

**EXAMPLE 3** **Solving $\sin(x) = a$ for $a = 1, 0,$ or $-1$**

Find all real numbers that satisfy each equation.

**a.** $\sin(x) = 1$  **b.** $\sin(x) = 0$  **c.** $\sin(x) = -1$

**Solution**

**a.** Use the inverse sine function to find one solution to $\sin x = 1$:

$$x = \sin^{-1}(1) = \pi/2$$

Since the period of sine is $2\pi$, any integral multiple of $2\pi$ can be added to this solution to get additional solutions. Now $\sin x = 1$ is satisfied only if the arc of length $x$ on the unit circle has terminal point $(0, 1)$, as shown in Fig. 4.18. So there are no other solutions. The solution set is written as

$$\{x \mid x = \pi/2 + 2k\pi\}$$

where $k$ is any integer.

The graphs of $y_1 = \sin(x)$ and $y_2 = 1$ intersect when $x = \pi/2 + 2k\pi$ as shown in Fig. 4.19. □

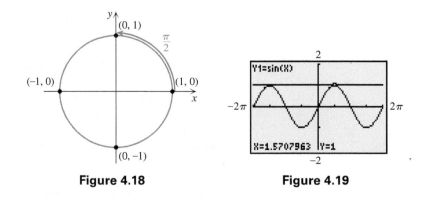

**Figure 4.18**          **Figure 4.19**

**b.** Use the inverse sine function to find one solution to $\sin x = 0$:

$$x = \sin^{-1}(0) = 0$$

The terminal point for the arc of length 0 is $(1, 0)$. However, an arc of length $x$ that terminates at $(-1, 0)$ also satisfies $\sin x = 0$, as shown in Fig. 4.20. Since the distance between $(1, 0)$ and $(-1, 0)$ on the unit circle is $\pi$, all arcs that terminate at these points are of the form $k\pi$. So the solution set is

$$\{x \mid x = k\pi\}$$

where $k$ is any integer.

The $x$-intercepts for $y_1 = \sin(x)$ occur when $x = k\pi$ as shown in Fig. 4.21. □

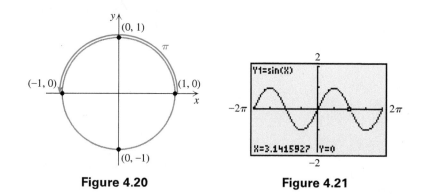

**Figure 4.20**                    **Figure 4.21**

**c.** Use $\sin^{-1}$ to find one solution to $\sin x = -1$:

$$x = \sin^{-1}(-1) = -\pi/2$$

Since the period of sine is $2\pi$ we also get $\sin x = -1$ if $x$ is $-\pi/2 + 2\pi$ or $3\pi/2$. Since there is no other terminal point for which $\sin x = -1$, the solution set is

$$\{x \mid x = 3\pi/2 + 2k\pi\}$$

where $k$ is any integer.

The graphs of $y_1 = \sin(x)$ and $y_2 = -1$ intersect when $x = 3\pi/2 + 2k\pi$ as shown in Fig. 4.22.

**TRY THIS.** Find the exact values of $\sin^{-1}(-1)$, $\sin^{-1}(0)$, and $\sin^{-1}(1)$. ■

**Figure 4.22**

In Example 3(c) we could have written the solutions in the form $-\pi/2 + 2k\pi$. However, we will always express the solutions in terms of the smallest positive angle that satisfies the equation.

Since the sine of an arc is the second coordinate of its terminal point on the unit circle, arcs that terminate at opposite ends of a horizontal chord in the unit circle have the same sine. So, if the solution to $\sin x = a$ found using the inverse sine is in $(-\pi/2, \pi/2)$ then there is another solution in $(\pi/2, 3\pi/2)$.

**EXAMPLE 4**    Solving $\sin(x) = a$ for nonzero $a$ in $(-1, 1)$

Find all real numbers that satisfy each equation.

**a.** $\sin x = -1/2$   **b.** $\sin x = 0.68$

**Solution**

**a.** Use the function $\sin^{-1}$ to find one solution to $\sin x = -1/2$:

$$x = \sin^{-1}\left(-\frac{1}{2}\right) = -\frac{\pi}{6}$$

See Fig. 4.23. The smallest positive arc with the same terminal point as $-\pi/6$ is $-\pi/6 + 2\pi$ or $11\pi/6$. Since the period of the sine function is $2\pi$, all arcs of the form $11\pi/6 + 2k\pi$ have the same terminal point and satisfy $\sin x = -1/2$. The smallest positive arc that terminates at the other end of the horizontal chord shown in Fig. 4.23 is $\pi - (-\pi/6)$ or $7\pi/6$. So $7\pi/6$ also satisfies the equation but it is not included in the form $11\pi/6 + 2k\pi$. So the solution set is

$$\left\{ x \mid x = \frac{7\pi}{6} + 2k\pi \text{ or } x = \frac{11\pi}{6} + 2k\pi \right\}$$

where $k$ is any integer.

The calculator graph of $y_1 = \sin(x)$ in Fig. 4.24 crosses the horizontal line $y_2 = -1/2$ twice in the interval $[0, 2\pi]$, at $x = 7\pi/6$ and $x = 11\pi/6$. $\square$

**Figure 4.23**  **Figure 4.24**

**Figure 4.25**

**b.** First use $\sin^{-1}$ to find one solution to $\sin x = 0.68$:

$$x = \sin^{-1}(0.68) \approx 0.748$$

Since the period is $2\pi$, all arcs of the form $0.748 + 2k\pi$ have the same terminal point and satisfy $\sin x = 0.68$. The smallest positive arc that terminates at the other end of the horizontal chord shown in Fig. 4.25 is $\pi - 0.748$ or $2.394$. So $2.394 + 2k\pi$ also satisfies $\sin x = 0.68$ and is not included in the form $0.748 + 2k\pi$. So the solution set is

$$\{ x \mid x \approx 0.748 + 2k\pi \text{ or } x \approx 2.394 + 2k\pi \}.$$

The calculator graph of $y_1 = \sin(x)$ in Fig. 4.26 crosses the horizontal line $y_2 = 0.68$ twice in the interval $[0, \pi]$, at $x \approx 0.748$ and $x \approx 2.394$.

**TRY THIS.** Find all real numbers that satisfy $\sin x = 1/2$. ∎

**Figure 4.26**

The procedure used in Example 4 can be used to solve $\sin x = a$ for any nonzero $a$ between $-1$ and $1$. First find two values of $x$ between $0$ and $2\pi$ that satisfy the equation. (In general, $s = \sin^{-1}(a)$ and $\pi - s$ work when $s$ is positive; $s + 2\pi$ and $\pi - s$ work when $s$ is negative.) Next write all solutions by adding $2k\pi$ to the first two solutions. The cases $a = -1, 0,$ and $1$ are handled separately because in these cases the solution set can be written as a single solution plus multiples of $\pi$ or $2\pi$.

All of the cases for solving $\sin x = a$ are summarized in the following box, where $k$ is used to represent any integer. Do not try to simply memorize this summary. Practice solving $\sin x = a$ until you can do it for any value of $a$ using the unit circle as a guide.

### Solving $\sin x = a$

**1.** If $-1 < a < 1$, $a \neq 0$, and $s = \sin^{-1}a$, then the solution set to $\sin x = a$ is

$$\{x \mid x = s + 2k\pi \text{ or } x = \pi - s + 2k\pi\} \qquad \text{for } s > 0.$$

and

$$\{x \mid x = s + 2\pi + 2k\pi \text{ or } x = \pi - s + 2k\pi\} \qquad \text{for } s < 0.$$

**2.** The solution set to $\sin x = 1$ is $\{x \mid x = \pi/2 + 2k\pi\}$.

**3.** The solution set to $\sin x = 0$ is $\{x \mid x = k\pi\}$.

**4.** The solution set to $\sin x = -1$ is $\{x \mid x = 3\pi/2 + 2k\pi\}$.

**5.** If $|a| > 1$, then $\sin x = a$ has no solution.

## Basic Tangent Equations

Tangent equations are a little simpler than sine and cosine equations, because the tangent function is one-to-one in its fundamental cycle, whereas sine and cosine are not one-to-one in their fundamental cycles. In the next example, we see that the solution set to $\tan x = a$ consists of any single solution plus multiples of $\pi$.

**EXAMPLE 5** Solving $\tan(x) = a$

Find all solutions, in degrees:

**a.** $\tan \alpha = 1$   **b.** $\tan \alpha = -1.34$

**Solution**

**a.** Since $\tan^{-1}(1) = 45°$ and the period of tangent is $180°$, all angles of the form $45° + k180°$ satisfy the equation. There are no additional angles that satisfy the equation. The solution set to $\tan \alpha = 1$ is

$$\{\alpha \mid \alpha = 45° + k180°\}.$$

 The calculator graphs of $y_1 = \tan(x)$ and $y_2 = 1$ in Fig. 4.27 support the conclusion that the solutions to $\tan x = 1$ are $180°$ apart. □

**b.** Since $\tan^{-1}(-1.34) \approx -53.3°$, one solution is $\alpha \approx -53.3°$. Since all solutions to $\tan \alpha = -1.34$ differ by a multiple of $180°$, $-53.3° + 180° = 126.7°$ is the smallest positive solution. So the solution set is

$$\{\alpha \mid \alpha \approx 126.7° + k180°\}.$$

 The calculator graphs of $y_1 = \tan(x)$ and $y_2 = -1.34$ in Fig. 4.28 support the conclusion that the solutions to $\tan(x) = -1.34$ are $180°$ apart.

**TRY THIS.** Find all angles $\alpha$ in degrees that satisfy $\tan \alpha = \sqrt{3}$.   ■

**Figure 4.27**

**Figure 4.28**

To solve $\tan x = a$ for any real number $a$, we first find the smallest nonnegative solution. (In general, $s = \tan^{-1}a$ works if $s > 0$ and $s + \pi$ works if $s < 0$.) Next add on all integral multiples of $\pi$. The solution to the equation $\tan x = a$ is summarized as follows.

> **SUMMARY**
>
> ### Solving $\tan x = a$
>
> If $a$ is any real number and $s = \tan^{-1} a$, then the solution set to $\tan x = a$ is
>
> $$\{x \,|\, x = s + k\pi\} \qquad \text{for } s \geq 0,$$
>
> and
>
> $$\{x \,|\, x = s + \pi + k\pi\} \qquad \text{for } s < 0.$$

In the summaries for solving sine, cosine, and tangent equations, the domain of $x$ is the set of real numbers. Similar summaries can be made if the domain of $x$ is the set of degree measures of angles. Start by finding $\sin^{-1} a$, $\cos^{-1} a$, or $\tan^{-1} a$ in degrees, then write the solution set in terms of multiples of $180°$ or $360°$.

## Isolating the Trigonometric Function

In some equations we must isolate the trigonometric function before applying the previous techniques for solving a trigonometric equation.

**EXAMPLE 6** Isolating the trigonometric function

Solve $2 \cos \alpha + 2 = 3$ for $0° \leq \alpha \leq 360°$.

**Solution**

First solve for $\cos \alpha$:

$$2 \cos \alpha + 2 = 3$$

$$2 \cos \alpha = 1$$

$$\cos \alpha = \frac{1}{2}$$

In $[0°, 360°]$, $\cos x = 1/2$ for $x = 60°$ and $x = 300°$. So the solution set is $\{60°, 300°\}$.

**TRY THIS.** Solve $2 \sin \alpha - 1 = 0$ for $0° \leq \alpha \leq 360°$. ∎

**EXAMPLE 7** Isolating the trigonometric function

Find $\alpha$ to the nearest tenth of a degree if $0° \leq \alpha \leq 180°$ and

$$\frac{\sin \alpha}{5.6} = \frac{\sin 32.6°}{12.2}.$$

**Solution**

First solve for $\sin \alpha$:

$$\frac{\sin \alpha}{5.6} = \frac{\sin 32.6°}{12.2}$$

$$\sin \alpha = \frac{5.6 \cdot \sin 32.6°}{12.2}$$

$$\sin \alpha \approx 0.2473$$

There are two angles between $0°$ and $180°$ for which $\sin \alpha \approx 0.2473$. One is $\sin^{-1}(0.2473)$ or approximately $14.3°$. The other is $180° - 14.3°$ or $165.7°$. So the solution set is $\{14.3°, 165.7°\}$.

**TRY THIS.** Find $\alpha$ to the nearest tenth of a degree if $0° \leq \alpha \leq 180°$ and $\frac{\sin 44.1°}{12.6} = \frac{\sin \alpha}{7.8}$. ∎

When performing computations with a calculator as in Example 7, always use the full accuracy of your calculator until you get to the final answer. You may write $\sin \alpha \approx 0.2473$ rounded to four places, but you should use the ten-digit value for that number shown on your calculator in the next computation to be more accurate.

## Using Inverse Functions to Solve for a Variable

Because of the definition of the inverse sine function,

$$a = \sin(b) \quad \text{is equivalent to} \quad b = \sin^{-1}(a)$$

provided $b$ is in the interval $\left[-\frac{\pi}{2}, \frac{\pi}{2}\right]$. Because of the definition of the inverse cosine function,

$$a = \cos(b) \quad \text{is equivalent to} \quad b = \cos^{-1}(a)$$

provided $b$ is in the interval $[0, \pi]$. Because of the definition of the inverse tangent function,

$$a = \tan(b) \quad \text{is equivalent to} \quad b = \tan^{-1}(a)$$

provided $b$ is in the interval $\left(-\frac{\pi}{2}, \frac{\pi}{2}\right)$. We can use these ideas to rewrite equations involving the trigonometric functions.

**EXAMPLE 8**    Solving for a certain variable

**a.** Solve $x = 5\cos(2y)$ for $y$ where $0 \le y \le \dfrac{\pi}{2}$.

**b.** Solve $w = 4\tan\left(\dfrac{a}{3}\right) + 1$ for $a$ where $-\dfrac{3\pi}{2} < a < \dfrac{3\pi}{2}$.

**Solution**

**a.** If $0 \le y \le \frac{\pi}{2}$, then $0 \le 2y \le \pi$ and we can write an equivalent equation using the inverse cosine function.

$$x = 5\cos(2y) \qquad \text{Original equation}$$

$$\frac{x}{5} = \cos(2y) \qquad \text{Isolate } \cos(2y).$$

$$2y = \cos^{-1}\left(\frac{x}{5}\right) \qquad \text{Definition of } \cos^{-1}$$

$$y = \frac{1}{2}\cos^{-1}\left(\frac{x}{5}\right)$$

**b.** If $a$ is in the interval $\left(-\frac{3\pi}{2}, \frac{3\pi}{2}\right)$, then $\frac{a}{3}$ is in the interval $\left(-\frac{\pi}{2}, \frac{\pi}{2}\right)$ and we can write an equivalent equation using the inverse tangent function:

$$w = 4\tan\left(\frac{a}{3}\right) + 1 \qquad \text{Original equation}$$

$$\frac{w-1}{4} = \tan\left(\frac{a}{3}\right) \qquad \text{Isolate } \tan\left(\frac{a}{3}\right).$$

$$\frac{a}{3} = \tan^{-1}\left(\frac{w-1}{4}\right) \qquad \text{Definition of } \tan^{-1}$$

$$a = 3\tan^{-1}\left(\frac{w-1}{4}\right)$$

**TRY THIS.** Solve $x = \frac{1}{2}\cos(3y)$ for $y$ where $0 \le y \le \frac{\pi}{3}$. ∎

# Inverses of General Trigonometric Functions

Recall that a function of the form

$$f(x) = A \sin[B(x - C)] + D,$$

$$f(x) = A \cos[B(x - C)] + D, \text{ or}$$

$$f(x) = A \tan[B(x - C)] + D$$

is in the sine, cosine, or tangent family, respectively. Each of these general trigonometric functions is a composition of a trigonometric function and several algebraic functions. Since each of the trigonometric functions has an inverse and each algebraic function has an inverse, a general trigonometric function has an inverse provided it is restricted to a suitable domain. To find an inverse we use the switch-and-solve technique from Section P.4. That is, switch or interchange $x$ and $y$, then solve for $y$ just as we did in Example 8.

**EXAMPLE 9**  **The inverse of a general sine function**

Let $f(x) = 3 \sin(2x) + 5$ where $-\frac{\pi}{4} \le x \le \frac{\pi}{4}$. Find $f^{-1}$ and determine the domain and range of $f^{-1}$.

**Solution**

Note that if $-\frac{\pi}{4} \le x \le \frac{\pi}{4}$, then $-\frac{\pi}{2} \le 2x \le \frac{\pi}{2}$. So $\sin(2x)$ is one-to-one and invertible on this domain. To find $f^{-1}$ switch or interchange $x$ and $y$, then solve for $y$:

$$f(x) = 3 \sin(2x) + 5 \qquad \text{Original function}$$

$$y = 3 \sin(2x) + 5 \qquad \text{Write } y \text{ in place of } f(x).$$

$$x = 3 \sin(2y) + 5 \qquad \text{Switch } x \text{ and } y.$$

$$x - 5 = 3 \sin(2y) \qquad \text{Solve for } y.$$

$$\frac{x - 5}{3} = \sin(2y)$$

$$\sin^{-1}\left(\frac{x - 5}{3}\right) = 2y \qquad \text{Definition of } \sin^{-1}$$

$$\frac{1}{2} \sin^{-1}\left(\frac{x - 5}{3}\right) = y$$

$$f^{-1}(x) = \frac{1}{2} \sin^{-1}\left(\frac{x - 5}{3}\right) \qquad \text{Write } f^{-1}(x) \text{ in place of } y.$$

To find the range of $f$, work with the domain of $f$ as follows:

$$-\frac{\pi}{4} \le x \le \frac{\pi}{4}$$

$$-\frac{\pi}{2} \le 2x \le \frac{\pi}{2} \qquad \text{Multiply by 2.}$$

$$-1 \le \sin(2x) \le 1 \qquad \text{Apply the sine function.}$$

$$-3 \le 3 \sin(2x) \le 3 \qquad \text{Multiply by 3.}$$

$$2 \le 3 \sin(2x) + 5 \le 8 \qquad \text{Add 5.}$$

**Figure 4.29**

The domain of $f$ is $\left[-\frac{\pi}{4}, \frac{\pi}{4}\right]$ and the range of $f$ is $[2, 8]$. So the domain of $f^{-1}$ is $[2, 8]$ and the range of $f^{-1}$ is $\left[-\frac{\pi}{4}, \frac{\pi}{4}\right]$.

Check this with a calculator by graphing $f^{-1}$ as shown in Fig. 4.29.

**TRY THIS.** Let $f(x) = 5\cos(2x) - 1$ where $0 \leq x \leq \frac{\pi}{2}$. Find $f^{-1}$ and determine the domain and range of $f^{-1}$.

## FOR THOUGHT... True or False? Explain.

1. $\sin(\pi) = 1$

2. $\cos(\pi/2) = 0$

3. $\sin(\pi/3) = 1/2$

4. $\cos(\pi/6) = \sqrt{3}/2$

5. If $x$ is in $[-1, 1]$, then $\sin^{-1}(x)$ is in $[-\pi/2, \pi/2]$.

6. If $x$ is in $[-1, 1]$, then $\cos^{-1}(x)$ is in $[0, \pi]$.

7. The only solutions to $\cos \alpha = 1/\sqrt{2}$ in $[0°, 360°)$ are $45°$ and $135°$.

8. The only solution to $\sin x = -0.55$ in $[0, \pi)$ is $\sin^{-1}(-0.55)$.

9. $\{x \mid x = -29° + k360°\} = \{x \mid x = 331° + k360°\}$ where $k$ is any integer.

10. The solution set to $\tan x = -1$ is $\{x \mid x = \frac{7\pi}{4} + k\pi\}$ where $k$ is any integer.

Scan this code to link to section lecture videos.

## 4.2 EXERCISES

### CONCEPTS

*Fill in the blank.*

1. A(n) _____ is satisfied by all real numbers for which both sides are defined.

2. A(n) _____ is an equation that has at least one solution but is not an identity.

### SKILLS

*Find all real numbers that satisfy each equation.*

3. $\cos x = 0$

4. $\cos(x) = -1$

5. $\cos(x) - 1 = 0$

6. $\cos(x) - 2 = 0$

7. $\sin(x) + 2 = 0$

8. $\sin(x) - 1 = 0$

9. $\sin(x) = -1$

10. $\sin x = 0$

11. $\tan x = 0$

12. $\tan x = 1$

13. $\tan x = -1$

14. $\tan x = \sqrt{3}$

*Find all real numbers that satisfy each equation.*

15. $\cos x = 1/2$

16. $\cos x = \sqrt{2}/2$

17. $\sin x = \sqrt{2}/2$

18. $\sin x = \sqrt{3}/2$

19. $\tan x = 1/\sqrt{3}$

20. $\tan x = \sqrt{3}/3$

21. $\cos x = -\sqrt{3}/2$

22. $\cos x = -\sqrt{2}/2$

23. $2\sin(x) + \sqrt{2} = 0$

24. $2\sin(x) + \sqrt{3} = 0$

25. $\tan(x) - \sqrt{3} = 0$

26. $\tan(x) + \sqrt{3} = 0$

*Find all angles in degrees that satisfy each equation.*

27. $\cos \alpha = 0$

28. $\cos \alpha = -1$

29. $\sin \alpha = 1$

30. $\sin \alpha = -1$

31. $\tan \alpha = 0$

32. $\tan \alpha = -1$

33. $2\cos(\alpha) - \sqrt{2} = 0$

34. $2\cos(\alpha) + 1 = 0$

35. $2\sin(\alpha) + 1 = 0$

36. $2\sin(\alpha) + \sqrt{2} = 0$

37. $\tan(\alpha) - 1 = 0$

38. $\tan(\alpha) + \sqrt{3} = 0$

*Find all angles in the interval $[0°, 360°]$ that satisfy each equation. Round approximations to the nearest tenth of a degree.*

39. $\cos \alpha = 0.873$

40. $\cos \alpha = -0.158$

41. $\sin \alpha = -0.244$

42. $\sin \alpha = 0.551$

43. $\tan \alpha = 5.42$

44. $\tan \alpha = -2.31$

45. $\cos(\alpha) - \sqrt{3} = 0$

46. $\sin(\alpha) + \sqrt{2} = 0$

*Find all real numbers in the interval* $[0, 2\pi]$ *that satisfy each equation. Round to the nearest hundredth.*

**47.** $\cos x = 0.66$

**48.** $\cos x = -0.23$

**49.** $\sin x = -1/4$

**50.** $\sin x = 4/5$

**51.** $\sqrt{6}\tan(x) - 1 = 0$

**52.** $3\tan(x) - \sqrt{7} = 0$

**53.** $\sqrt{5}\cos(x) + 2 = 0$

**54.** $7\sin(x) - \sqrt{7} = 0$

*Solve each equation.*

**55.** $2\cos(\alpha) - 2 = 0$ for $-360° \le \alpha \le 360°$

**56.** $3\sin(\beta) + 10 = 7$ for $0° \le \beta \le 360°$

**57.** $3\sin(\beta) + 6 = 5\sin(\beta) + 7$ for $0° \le \beta \le 360°$

**58.** $3\cos(\alpha) + \sqrt{12} = 5\cos(\alpha) + \sqrt{3}$ for $-360° \le \alpha \le 360°$

**59.** $\dfrac{\sin \alpha}{23.4} = \dfrac{\sin 67.2°}{25.9}$ for $0° < \alpha < 90°$

**60.** $\dfrac{\sin 9.7°}{15.4} = \dfrac{\sin \beta}{52.9}$ for $90° < \beta < 180°$

**61.** $(3.6)^2 = (5.4)^2 + (8.2)^2 - 2(5.4)(8.2)\cos \alpha$ for $0° < \alpha < 90°$

**62.** $(6.8)^2 = (3.2)^2 + (4.6)^2 - 2(3.2)(4.6)\cos \alpha$ for $90° < \alpha < 180°$

*Solve each equation for the indicated variable.*

**63.** Solve $x = 2\cos(3y)$ for $y$ where $0 \le y \le \dfrac{\pi}{3}$.

**64.** Solve $x = -4\cos\left(\dfrac{y}{2}\right)$ for $y$ where $0 \le y \le 2\pi$.

**65.** Solve $t = -6\sin(m) + 2$ for $m$ where $-\dfrac{\pi}{2} \le m \le \dfrac{\pi}{2}$.

**66.** Solve $v = 5\sin(s) - 8$ for $s$ where $-\dfrac{\pi}{2} \le s \le \dfrac{\pi}{2}$.

**67.** Solve $b = 7\tan\left(\dfrac{a}{3}\right) - d$ for $a$ where $-\dfrac{3\pi}{2} < a < \dfrac{3\pi}{2}$.

**68.** Solve $z = 9\tan(5p) + h$ for $p$ where $-\dfrac{\pi}{10} < p < \dfrac{\pi}{10}$.

**69.** Solve $q = 3\sin(\pi b - \pi)$ for $b$ where $\dfrac{1}{2} \le b \le \dfrac{3}{2}$.

**70.** Solve $a = 8\cos(2w - \pi) + 1$ for $w$ where $\dfrac{\pi}{2} \le w \le \pi$.

*Find the inverse of each function and state the domain and range of $f^{-1}$.*

**71.** $f(x) = \sin(2x)$ for $-\dfrac{\pi}{4} \le x \le \dfrac{\pi}{4}$

**72.** $f(x) = 5\sin(3x)$ for $-\dfrac{\pi}{6} \le x \le \dfrac{\pi}{6}$

**73.** $f(x) = 2\cos(3x)$ for $0 \le x \le \dfrac{\pi}{3}$

**74.** $f(x) = 3\cos(2x) + 1$ for $0 \le x \le \dfrac{\pi}{2}$

**75.** $f(x) = 3 + \tan(\pi x)$ for $-\dfrac{1}{2} < x < \dfrac{1}{2}$

**76.** $f(x) = 1 + \tan\left(\dfrac{\pi}{2}x\right)$ for $-1 < x < 1$

**77.** $f(x) = 2 - \sin(\pi x - \pi)$ for $\dfrac{1}{2} \le x \le \dfrac{3}{2}$

**78.** $f(x) = 4\cos(\pi(x - 2)) + 1$ for $2 \le x \le 3$

**79.** $f(x) = \sin^{-1}(3x)$ for $-\dfrac{1}{3} \le x \le \dfrac{1}{3}$

**80.** $f(x) = \cos^{-1}(x + 4)$ for $-5 \le x \le -3$

**81.** $f(x) = \sin^{-1}(x/2) + 3$ for $-2 \le x \le 2$

**82.** $f(x) = 2\cos^{-1}(5x) + 3$ for $-\dfrac{1}{5} \le x \le \dfrac{1}{5}$

*Find all real numbers that satisfy each equation. Round approximate answers to 2 decimal places.*

**83.** $\dfrac{\sin 33.2°}{a} = \dfrac{\sin 45.6°}{13.7}$

**84.** $\dfrac{\sin 49.6°}{55.1} = \dfrac{\sin 88.2°}{b}$

**85.** $\dfrac{\sin x}{8.5} = \dfrac{\sin(\pi/7)}{6.3}$

**86.** $\dfrac{4}{\sin(0.34)} = \dfrac{8}{\sin(y)}$

**87.** $7^2 = 5^2 + 6^2 - 2(5)(6)\cos \alpha$

**88.** $(4.1)^2 = (2.4)^2 + (5.3)^2 - 2(2.4)(5.3)\cos \alpha$

**89.** $c^2 = (3.4)^2 + (5.2)^2 - 2(3.4)(5.2)\cos(57°)$

**90.** $c^2 = (1.4)^2 + (2.6)^2 - 2(1.4)(2.6)\cos(33°)$

**91.** $3 = 5\sin(x) + 1$

**92.** $2 = 4\cos(x) + 5$

**93.** $4 = 6\cos^{-1}(x) + 1$

**94.** $3 = 4\sin^{-1}(x) + 1$

## WRITING/DISCUSSION

**95.** Use a graphing calculator to graph $y = \sin(x)$ and determine the number of solutions to $\sin(x) = 0$ in the interval $(-2\pi, 2\pi)$. What is the maximum value of this function on this interval?

**96.** Use a graphing calculator to graph $y = \cos(x)/(x - \pi/2)$ and determine the number of solutions to $\cos(x)/(x - \pi/2) = 0$ in the interval $(-2\pi, 2\pi)$. What is the minimum value of this function on this interval?

REVIEW

**97.** Find the exact value of each expression without using a calculator or table.

   **a.** $\arcsin(1/2)$         **b.** $\cos^{-1}(-1/2)$

   **c.** $\tan^{-1}(-1)$        **d.** $\sin(\pi/3)$

   **e.** $\cos(-\pi/2)$        **f.** $\sin^{-1}(-1)$

**98.** Find the exact value for each expression without using a calculator or table.

   **a.** $\tan\left(\arcsin\left(\sqrt{3}/2\right)\right)$    **b.** $\sin(\cos^{-1}(1/2))$

   **c.** $\sin(\tan^{-1}(1))$       **d.** $\cos(\sin^{-1}(1))$

**99.** Simplify $\sin(-x)\cos(-x)\tan(-x)$.

**100.** Simplify $\cos(2y)\cos(y) - \sin(2y)\sin(y)$.

**101.** Determine the amplitude, period, phase shift, and range for the function $f(x) = 3\sin(x - \pi) + 9$.

**102.** A sector of a circle has a central angle of $\pi/6$. Find the exact area of the sector if the radius of the circle is 6 inches.

OUTSIDE THE BOX

**103.** *Spiral of Triangles* The hypotenuse in the first right triangle shown in the figure has length $\sqrt{2}$. In the second the hypotenuse is $\sqrt{3}$. In the third it is $\sqrt{4}$, and so on. In this sequence of right triangles, which right triangle (after the second one) intersects the first one at more than one point?

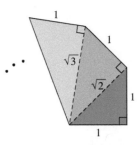

**Figure for Exercise 103**

**104.** *Natural Numbers* What is the smallest positive integer $n$ such that

$$1 + 2 + 3 + \cdots + n > 4050?$$

 **POP QUIZ**

*Find all angles $\alpha$ in degrees that satisfy each equation. Use k to represent any integer.*

**1.** $\sin\alpha = \sqrt{2}/2$       **2.** $\cos\alpha = 1/2$

**3.** $\tan\alpha = -1$         **4.** $3\sin(\alpha) - 5 = -2$

*Find $f^{-1}$ and state its domain.*

**5.** $f(x) = \cos(2x)$ for $0 \le x \le \pi/2$.

**6.** $f(x) = 5\sin(x) + 1$ for $0 \le x \le \pi/2$.

## 4.3 Equations Involving Compositions

A function such as $\sin(2x)$ is a composition of an algebraic function and a trigonometric function. To solve an equation involving $\sin(2x)$, first solve for the algebraic function $2x$, then find the value of $x$.

### Sine, Cosine, and Tangent Equations

**EXAMPLE 1**   **A sine equation with a double angle**

Find all solutions in degrees to $\sin 2\alpha = 1/\sqrt{2}$.

**Solution**

First solve for $2\alpha$ using the inverse sine function:

$$2\alpha = \sin^{-1}\left(1/\sqrt{2}\right) = 45°$$

Since $180° - 45° = 135°$, the only values for $2\alpha$ between $0°$ and $360°$ that satisfy the equation are $45°$ and $135°$. See Fig. 4.30. List all possibilities for $2\alpha$:

$$2\alpha = 45° + k360° \qquad \text{or} \qquad 2\alpha = 135° + k360°$$

$$\alpha = 22.5° + k180° \qquad \text{or} \qquad \alpha = 67.5° + k180° \qquad \text{Divide each side by 2.}$$

**Figure 4.30**

**Figure 4.31**

The solution set is

$$\{\alpha \,|\, \alpha = 22.5° + k180° \text{ or } \alpha = 67.5° + k180°\},$$

where $k$ is any integer.

The graphs of $y_1 = \sin(2x)$ and $y_2 = 1/\sqrt{2}$ in Fig. 4.31 intersect twice in the interval $(0°, 90°)$, at $x = 22.5°$ and $x = 67.5°$.

**TRY THIS.** Find all angles $\alpha$ in degrees that satisfy $\sin 2\alpha = 1$. ■

Note that in Example 1, all possible values for $2\alpha$ are found and *then* divided by 2 to get all possible values for $\alpha$. Finding $\alpha = 22.5°$ and then adding on multiples of $360°$ will not produce the same solutions. Observe the same procedure in the next example, where we wait until the final step to divide each side by 3.

**EXAMPLE 2** A tangent equation involving angle multiples

Find all solutions to $\tan 3x = -\sqrt{3}$ in the interval $(0, 2\pi)$.

**Solution**

First find the smallest positive value for $3x$ that satisfies the equation. Then form all of the solutions by adding on multiples of $\pi$. Since $\tan^{-1}(-\sqrt{3}) = -\pi/3$, the smallest positive value for $3x$ that satisfies the equation is $-\pi/3 + \pi$, or $2\pi/3$. So proceed as follows:

$$\tan 3x = -\sqrt{3}$$

$$3x = \frac{2\pi}{3} + k\pi$$

$$x = \frac{2\pi}{9} + \frac{k\pi}{3} \qquad \text{Divide each side by 3.}$$

The solutions between 0 and $2\pi$ occur if $k = 0, 1, 2, 3, 4,$ and $5$. If $k = 6$, then $x$ is greater than $2\pi$. So the solution set is

$$\left\{ \frac{2\pi}{9}, \frac{5\pi}{9}, \frac{8\pi}{9}, \frac{11\pi}{9}, \frac{14\pi}{9}, \frac{17\pi}{9} \right\}.$$

**Figure 4.32**

**Figure 4.33**

**Figure 4.34**

The graph of $y_1 = \tan(3x)$ in Fig. 4.32 intersects the graph of $y_2 = -\sqrt{3}$ six times in the interval $(0, 2\pi)$.

**TRY THIS.** Find all solutions to $\tan(4x) = 1$ in the interval $(0, \pi)$. ∎

### EXAMPLE 3   A cosine equation with a half angle

Find all solutions in degrees to $\cos(\alpha/2) = 1/2$, where $0° \le \alpha \le 1000°$.

**Solution**

First solve for $\alpha/2$ using the inverse cosine function:

$$\alpha/2 = \cos^{-1}(1/2) = 60°$$

Since $360° - 60° = 300°$, the only values for $\alpha/2$ between $0°$ and $360°$ that satisfy the equation are $60°$ and $300°$. See Fig. 4.33. List all possibilities for $\alpha/2$:

$$\frac{\alpha}{2} = 60° + k360° \qquad \text{or} \qquad \frac{\alpha}{2} = 300° + k360°$$

$$\alpha = 120° + k720° \qquad \text{or} \qquad \alpha = 600° + k720° \quad \text{Multiply each side by 2.}$$

If $k = 0$ or $k = 1$ in $\alpha = 120° + k720°$, we get $\alpha = 120°$ or $\alpha = 840°$. If $k = 0$ in $\alpha = 600° + k720°$, we get $\alpha = 600°$. Since $\alpha$ must be in $[0°, 1000°]$ the solution set is $\{120°, 600°, 840°\}$.

The graph of $y_1 = \cos(x/2)$ in Fig. 4.34 intersects the graph of $y_2 = 1/2$ three times in the interval $[0°, 1000°]$. Make sure your calculator's mode is set to degrees when you draw this graph.

**TRY THIS.** Find all solutions to $\sin(x/2) = \sqrt{3}/2$ in the interval $(0, 6\pi)$. ∎

### EXAMPLE 4   Sine of a sum

Find all solutions to $2\sin\left(x + \dfrac{\pi}{6}\right) + 1 = 0$ in the interval $[0, 2\pi)$.

**Solution**

First isolate the sine function:

$$2\sin\left(x + \frac{\pi}{6}\right) + 1 = 0$$

$$2\sin\left(x + \frac{\pi}{6}\right) = -1$$

$$\sin\left(x + \frac{\pi}{6}\right) = -\frac{1}{2}$$

On the unit circle, sine is $-1/2$ at $7\pi/6$ and $11\pi/6$.

$$x + \frac{\pi}{6} = \frac{7\pi}{6} + 2k\pi \qquad \text{or} \qquad x + \frac{\pi}{6} = \frac{11\pi}{6} + 2k\pi$$

$$x = \pi + 2k\pi \qquad \text{or} \qquad x = \frac{5\pi}{3} + 2k\pi$$

The only number of the form $\pi + 2k\pi$ in $[0, 2\pi)$ is $\pi$. The only number of the form $5\pi/3 + 2k\pi$ in $[0, 2\pi)$ is $5\pi/3$. The solution set is $\{\pi, 5\pi/3\}$.

**TRY THIS.** Find all solutions to $\sqrt{2}\sin\left(x + \dfrac{\pi}{4}\right) - 1 = 0$ in the interval $[0, 2\pi)$. ∎

## Secant, Cosecant, and Cotangent Equations

To solve equations involving secant, cosecant, or cotangent we use the definitions of these functions to write equivalent equations involving cosine, sine, or tangent.

### EXAMPLE 5  A secant equation

Find all solutions in degrees to $\sec(\alpha/3) = 2$, where $0° \le \alpha \le 360°$.

**Solution**

Rewrite the equation in terms of cosine:

$$\sec(\alpha/3) = 2$$

$$\cos(\alpha/3) = \frac{1}{2} \quad \text{Because } \cos x = 1/\sec x$$

$$\frac{\alpha}{3} = 60° + k360° \quad \text{or} \quad \frac{\alpha}{3} = 300° + k360°$$

$$\alpha = 180° + k1080° \quad \text{or} \quad \alpha = 900° + k1080°$$

The only angle in $[0°, 360°]$ that fits the above form is $180°$.

**TRY THIS.**  Find all solutions to $\csc(2x) = 2\sqrt{3}/3$ in the interval $(0, 2\pi)$.  ∎

## Equations Involving Identities

Identities can be used to simplify equations, as shown in the next example.

### EXAMPLE 6  Using an identity to simplify an equation

Find all solutions to each equation in the interval $[0, 2\pi]$.

**a.** $\sin^2(x) + \sin(2x) + \cos^2(x) = 0$  **b.** $\sin(2x) - \sin(x) = 0$

**Solution**

**a.** $\sin^2(x) + \sin(2x) + \cos^2(x) = 0$

$$\sin(2x) + 1 = 0 \qquad \sin^2(x) + \cos^2(x) = 1$$

$$\sin(2x) = -1$$

$$2x = \frac{3\pi}{2} + 2k\pi \quad 2x = \sin^{-1}(-1)$$

$$x = \frac{3\pi}{4} + k\pi$$

Now $\frac{3\pi}{4} + k\pi$ is in the interval $[0, 2\pi]$ only if $k$ is 0 or 1. So the solution set is $\left\{\frac{3\pi}{4}, \frac{7\pi}{4}\right\}$.

**b.**  $\sin(2x) - \sin(x) = 0$

$$2\sin(x)\cos(x) - \sin(x) = 0 \quad \sin(2x) = 2\sin(x)\cos(x)$$

$$\sin(x)(2\cos(x) - 1) = 0$$

$$\sin(x) = 0 \quad \text{or} \quad 2\cos(x) - 1 = 0$$

$$\sin(x) = 0 \quad \text{or} \quad \cos(x) = \frac{1}{2}$$

$$x = k\pi \quad \text{or} \qquad x = \frac{\pi}{3} + 2k\pi \quad \text{or} \quad x = \frac{5\pi}{3} + 2k\pi$$

**Figure 4.35**

Now $k\pi$ is in the interval $[0, 2\pi]$ if $k$ is 0, 1, or 2. The solutions to $\cos(x) = \dfrac{1}{2}$ in $[0, 2\pi]$ are $\dfrac{\pi}{3}$ and $\dfrac{5\pi}{3}$. So the solution set is $\left\{0, \dfrac{\pi}{3}, \pi, \dfrac{5\pi}{3}, 2\pi\right\}$.

The graph of $y = \sin(2x) - \sin(x)$ in radian mode in Fig. 4.35 intersects the $x$-axis five times in the interval $[0, 2\pi]$. So the graph supports our solution to the equation.

**TRY THIS.** Find all solutions to $\sin(2x) + \cos(-x) = 0$ in the interval $[0, 2\pi]$. ∎

## Applications

The distance $d$ (in feet) traveled by a projectile fired from the ground with an angle of elevation $\theta$ is related to the initial velocity $v_0$ (in ft/sec) by the equation $v_0^2 \sin 2\theta = 32d$. If the projectile is fired from the origin into the first quadrant, then the $x$- and $y$-coordinates (in feet) of the projectile at time $t$ (in seconds) are given by $x = v_0 t \cos \theta$ and $y = -16t^2 + v_0 t \sin \theta$.

**EXAMPLE 7** The path of a projectile

A catapult is placed 100 feet from the castle wall, which is 35 feet high. The soldier wants the burning bale of hay to clear the top of the wall and land 50 feet inside the castle wall. If the initial velocity of the bale is 70 feet per second, then at what angle should the bale of hay be launched so that it will travel 150 feet and pass over the castle wall?

**Solution**

Use the equation $v_0^2 \sin 2\theta = 32d$ with $v_0 = 70$ ft/sec and $d = 150$ ft to find $\theta$:

$$70^2 \sin 2\theta = 32(150)$$

$$\sin 2\theta = \frac{32(150)}{70^2} \approx 0.97959$$

The launch angle $\theta$ must be in the interval $(0°, 90°)$, so we look for values of $2\theta$ in the interval $(0°, 180°)$. Since $\sin^{-1}(0.97959) \approx 78.4°$, both $78.4°$ and $180° - 78.4° = 101.6°$ are possible values for $2\theta$. So possible values for $\theta$ are $39.2°$ and $50.8°$. See Fig. 4.36.

**Figure 4.36**

There are two launch angles for which the bale of hay will travel 150 feet. To determine which launch angle will cause the bale to go over the castle wall, we must find the altitude of the bale when it is 100 feet from the catapult. Use the equation $x = v_0 t \cos \theta$ to find the time at which the bale is 100 feet from the catapult (measured horizontally) by using each of the possible values for $\theta$:

$$100 = 70t \cos(39.2°) \qquad\qquad 100 = 70t \cos(50.8°)$$

$$t = \frac{100}{70 \cos(39.2°)} \approx 1.84 \text{ sec} \qquad t = \frac{100}{70 \cos(50.8°)} \approx 2.26 \text{ sec}$$

Use the equation $y = -16t^2 + v_0 t \sin \theta$ to find the altitude of the bale at time $t = 1.84$ sec and $t = 2.26$ sec:

$$y = -16(1.84)^2 + 70(1.84)\sin 39.2° \approx 27.2 \text{ ft}$$

$$y = -16(2.26)^2 + 70(2.26)\sin 50.8° \approx 40.9 \text{ ft}$$

If the burning bale is launched on a trajectory with an angle of 39.2°, then it will have an altitude of only 27.2 feet when it reaches the castle wall. If it is launched with an angle of 50.8°, then it will have an altitude of 40.9 feet when it reaches the castle wall. Since the castle wall is 35 feet tall, the 50.8° angle must be used for the bale to reach its intended target.

**TRY THIS.** Find the launch angle so that a projectile with initial velocity of 100 feet per second will travel 200 feet. ∎

## FOR THOUGHT... True or False? Explain.

**1.** If $\alpha/3 = \pi/2$, then $\alpha = \pi/6$.

**2.** If $x/2 = \pi/3 + k\pi$ for any integer $k$, then $x = 2\pi/3 + 2k\pi$.

**3.** If $2x = \pi/2 + k\pi$ for any integer $k$, then $x = \pi/4 + k\pi$.

**4.** If $\cos(x/3) = 1/2$, then $x/3 = \pi/3 + 2k\pi$ or $x/3 = 5\pi/3 + 2k\pi$.

**5.** If $\tan(5x) = 1$, then $x = \pi/4 + k\pi$.

**6.** If $\sin(6x) = \sqrt{2}/2$, then $6x = \pi/4 + 2k\pi$ or $6x = 3\pi/4 + 2k\pi$.

**7.** If $\csc(x) = 4$, then $\sin(x) = 0.25$.

**8.** If $\cot(x) = 3$, then $\tan(x) = 0.33$.

**9.** The solution set to $\cot x = 3$ for $x$ in $[0, \pi)$ is $\{\tan^{-1}(1/3)\}$.

**10.** $\left\{x \mid 3x = \frac{\pi}{2} + 2k\pi\right\} = \left\{x \mid x = \frac{\pi}{6} + 2k\pi\right\}$, where $k$ is any integer.

## 4.3 EXERCISES

### SKILLS

*Find all real numbers that satisfy each equation.*

**1.** $\cos(x/2) = 1/2$

**2.** $2\cos(2x) = -\sqrt{2}$

**3.** $\cos(3x) = 1$

**4.** $\cos(2x) = 0$

**5.** $2\sin(x/2) - 1 = 0$

**6.** $\sin(2x) = 0$

**7.** $2\sin(2x) = -\sqrt{2}$

**8.** $\sin(x/3) + 1 = 0$

**9.** $\tan(2x) = \sqrt{3}$

**10.** $\sqrt{3}\tan(3x) + 1 = 0$

**11.** $\tan(4x) = 0$

**12.** $\tan(3x) = -1$

**13.** $\sin(\pi x) = 1/2$

**14.** $\tan(\pi x/4) = 1$

**15.** $\cos(2\pi x) = 0$

**16.** $\sin(3\pi x) = 1$

*Find all solutions in $[0, 2\pi)$ for each equation.*

**17.** $\sin\left(x + \frac{\pi}{4}\right) = \frac{1}{2}$

**18.** $\cos\left(x + \frac{\pi}{2}\right) = \frac{1}{2}$

**19.** $\cos\left(x - \frac{\pi}{8}\right) = 1$

**20.** $\sin\left(x - \frac{\pi}{6}\right) = \frac{\sqrt{2}}{2}$

**21.** $2\sin(x - \pi) + 1 = 0$

**22.** $4\cos\left(x - \frac{\pi}{2}\right) - 2\sqrt{2} = 0$

**23.** $4\tan\left(x - \frac{\pi}{2}\right) - 4 = 0$

**24.** $\sqrt{3}\tan(x - \pi) - 1 = 0$

*Find all values of $\alpha$ in $[0°, 360°)$ that satisfy each equation.*

**25.** $2\sin \alpha = -\sqrt{3}$

**26.** $\tan \alpha = -\sqrt{3}$

**27.** $2\sin(2x) + \sqrt{3} = 0$

**28.** $2\cos(2x) + \sqrt{3} = 0$

**29.** $2\cos(2x) + 1 = 0$

**30.** $2\sin(2x) - 1 = 0$

**31.** $\sqrt{2}\cos(2\alpha) - 1 = 0$

**32.** $\sin(6\alpha) = 1$

**33.** $\sec(3\alpha) = -\sqrt{2}$

**34.** $\csc(5\alpha) + 2 = 0$

**35.** $\cot(\alpha/2) = \sqrt{3}$

**36.** $\sec(\alpha/2) = \sqrt{2}$

**37.** $\csc(4\alpha) = \sqrt{2}$

**38.** $\cot(4\alpha) + \sqrt{3} = 0$

*Find all real numbers in the interval $[0, 2\pi)$ that satisfy each equation.*

**39.** $2\sin(2x) - \sqrt{3} = 0$

**40.** $2\cos(2x) - \sqrt{3} = 0$

**41.** $2\cos(2x) - 1 = 0$

**42.** $2\sin(2x) + 1 = 0$

**43.** $\tan(3x) - 1 = 0$

**44.** $\cot(3x) + 1 = 0$

**45.** $\sqrt{2}\sin(x/3) - 1 = 0$

**46.** $\sqrt{2}\cos(x/2) - 1 = 0$

**47.** $\sqrt{3}\tan(x/2) - 1 = 0$

**48.** $\sqrt{3}\cot(x/3) - 3 = 0$

*Find all values of $\alpha$ in degrees that satisfy each equation. Round approximate answers to the nearest tenth of a degree.*

**49.** $\sin 3\alpha = 0.34$

**50.** $\cos 2\alpha = -0.22$

**51.** $\sin 3\alpha = -0.6$

**52.** $\tan 4\alpha = -3.2$

**53.** $\sec 2\alpha = 4.5$

**54.** $\csc 3\alpha = -1.4$

**55.** $\csc(\alpha/2) = -2.3$

**56.** $\cot(\alpha/2) = 4.7$

*Find all real numbers that satisfy each equation. Round approximate answers to the nearest hundredth.*

**57.** $3\sin(5x) - 1 = 0$

**58.** $\sqrt{3}\cos(3x) + 1 = 0$

**59.** $10\sin(x/2) + 6 = 0$

**60.** $5\tan(x/4) - \sqrt{5} = 0$

**61.** $2\sec(\pi x) - 9 = 0$

**62.** $10\csc(\pi x/2) + 14 = 0$

**63.** $3\cot(\pi x - 1) - 1 = 0$

**64.** $4 - \cot(\pi x - 2) = 0$

*Find all real numbers in the interval $[0, 2\pi]$ that satisfy each equation.*

**65.** $\cos^2(x) + \cos(2x) + \sin^2(x) = 0$

**66.** $\tan(2x) + \tan^2(x) = \sec^2(x) - 1$

**67.** $\sin(2x) - \cos(x) = 0$

**68.** $\sin(2x) + \sin\left(\dfrac{\pi}{2} - x\right) = 0$

**69.** $\sin(-x) = \tan(-x)$

**70.** $\cos(-x) = \csc\left(\dfrac{\pi}{2} - x\right)$

**71.** $\csc^2(x) + \csc(x) = 1 + \cos(-x) + \cot^2(x)$

**72.** $\tan^2(x) + \sin(-x) = \sec^2(x) + \sec(x) - 1$

**73.** $\cos(-x) = \tan\left(\dfrac{\pi}{2} - x\right)$

**74.** $\tan\left(\dfrac{x}{2}\right)\sin(x) = 0$

**75.** $\sin(x) + \sin(3x) = 0$

**76.** $\cos(3x) - \cos(5x) = 0$

**77.** $\cos(x)\cos(-2x) - \sin(x)\sin(-2x) = \dfrac{\sqrt{3}}{2}$

**78.** $\sin(x)\cos(-2x) + \cos(x)\sin(-2x) = -\dfrac{1}{2}$

**79.** $\tan(x) + \tan(2x) = 1 - \tan(x)\tan(2x)$

**80.** $\tan(x) - \tan(2x) = 1 + \tan(x)\tan(2x)$

⊿⊿ *One way to solve an equation with a graphing calculator is to rewrite the equation with 0 on the right-hand side, then graph the function that is on the left-hand side. The x-coordinate of each x-intercept of the graph is a solution to the original equation. For each equation find all real solutions (to the nearest tenth) in the interval $[0, 2\pi)$.*

**81.** $\sin(x/2) = \cos(3x)$

**82.** $2\sin(x) = \csc(x + 0.2)$

**83.** $\dfrac{x}{2} - \dfrac{\pi}{6} + \dfrac{\sqrt{3}}{2} = \sin(x)$

**84.** $x^2 = \sin(x)$

## MODELING

*Solve each problem.*

**85.** *Firing an M-16* A soldier is accused of breaking a window 3300 ft away during target practice. If the muzzle velocity for an M-16 is 325 ft/sec, then at what angle would it have to be aimed for the bullet to travel 3300 ft?

    **HINT** The distance $d$ (in feet) traveled by a projectile fired at an angle $\theta$ is related to the initial velocity $v_0$ (in feet per second) by the equation

$$v_0^2 \sin 2\theta = 32d.$$

**86.** *Firing an M-16* If you were accused of firing an M-16 into the air and breaking a window 4000 ft away, what would be your defense?

**87.** *Choosing the Right Angle* A center fielder fields a ground ball and attempts to make a 230-ft throw to home plate. If he commonly makes long throws at 90 mi/hr, find the two possible angles at which he can throw the ball to home plate. Find the time saved by choosing the smaller angle.

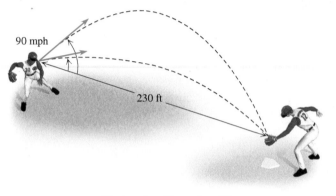

90 mph

230 ft

**Figure for Exercise 87**

**88.** *Muzzle Velocity* The 8-in. (diameter) Howitzer on the U.S. Army's M110 can propel a projectile a distance of 18,500 yd. If the angle of elevation of the barrel is 45°, then what muzzle velocity (in feet per second) is required to achieve this distance?

**89.** *Wave Action* The vertical position of a floating ball in an experimental wave tank is given by the equation $x = 2 \sin\left(\frac{\pi t}{3}\right)$, where $x$ is the number of feet above sea level and $t$ is the time in seconds. For what values of $t$ is the ball $\sqrt{3}$ ft above sea level?

**90.** *Periodic Sales* The number of car stereos sold by a national department store chain varies seasonally and is a function of the month of the year. The function

$$x = 6.2 + 3.1 \sin\left(\frac{\pi}{6}(t - 9)\right)$$

gives the anticipated sales (in thousands of units) as a function of the number of the month ($t = 1, 2, \ldots, 12$). In what month does the store anticipate selling 9300 units?

**WRITING/DISCUSSION**

**91.** Use a graphing calculator to graph $y = \sin(1/x)$ and determine the number of solutions to $\sin(1/x) = 0$ in the interval $(-\pi, \pi)$.

**92.** Use a graphing calculator to graph $y = \sin(x)/x$ and determine the number of solutions to $\sin(x)/x = 1$ in the interval $(-10, 10)$.

**REVIEW**

**93.** Find an equivalent algebraic expression for $\cos(\text{arccot}(y))$.

**94.** Find all solutions to the equation $2 \sin x = \sqrt{3}$. Use $k$ to represent any integer.

**95.** Find the exact value of each expression without using a calculator or table.
  **a.** $\arcsin(-1/2)$
  **b.** $\cos^{-1}\left(-\sqrt{2}/2\right)$
  **c.** $\tan^{-1}(1)$
  **d.** $\sin^{-1}\left(\sqrt{2}/2\right)$

**96.** Given that $\tan \alpha = -2$ and $\alpha$ is in quadrant II, find the values of the remaining five trigonometric functions at $\alpha$ by using identities.

**97.** Find the exact value of $\cos(-\pi/12)$ using a half-angle identity.

**98.** Find the period, asymptotes, and range for the function $y = \cot(2x + \pi/2)$.

**OUTSIDE THE BOX**

**99.** *Lucky Lucy* Lucy's teacher asked her to evaluate $(20 + 25)^2$. As she was trying to figure out what to do she mumbled, "twenty twenty-five." Her teacher said, "Good, 2025 is correct." Find another pair of two-digit whole numbers for which the square of their sum can be found by Lucy's method.

**100.** *Largest Integer* What is the largest integer $n$ for which $\dfrac{n^2 + 2015}{n + 1}$ is an integer?

**4.3 POP QUIZ**

*Find all real numbers in $[0, 2\pi]$ that satisfy each equation.*

**1.** $\sin(x/2) = 1/2$

**2.** $\cos(2x) = 1$

**3.** $\tan(2x) = 1$

**4.** $2 \cos(\pi x) = \sqrt{2}$

# 4.4 Trigonometric Equations of Quadratic Type

Quadratic equations can be solved by the square-root property, factoring, or the quadratic formula. We can use these same techniques on equations of quadratic type that involve the trigonometric functions. In stating formulas for solutions to equations, we will continue to use $k$ to represent any integer.

## Using the Square-Root Property

By the square-root property, $x^2 = a$ is equivalent to $x = \pm\sqrt{a}$. We can apply this property to equations involving trigonometric functions.

**EXAMPLE 1** Using the square-root property

Find all solutions to each equation in the interval $[0, 2\pi)$.

**a.** $\sin^2(x) = 1$    **b.** $\cos^2(x) = \dfrac{1}{4}$

**Solution**

**a.**
$$\sin^2(x) = 1$$
$$\sin(x) = \pm 1 \quad \text{Square-root property}$$
$$\sin(x) = 1 \qquad \text{or} \qquad \sin(x) = -1$$
$$x = \frac{\pi}{2} + 2k\pi \qquad \text{or} \qquad x = \frac{3\pi}{2} + 2k\pi$$

In the interval $[0, 2\pi)$ the solution set is $\left\{\dfrac{\pi}{2}, \dfrac{3\pi}{2}\right\}$.

**b.**
$$\cos^2(x) = \frac{1}{4}$$
$$\cos(x) = \pm\frac{1}{2} \quad \text{Square-root property}$$
$$\cos(x) = \frac{1}{2} \qquad \text{or} \qquad \cos(x) = -\frac{1}{2}$$
$$x = \frac{\pi}{3} + 2k\pi \quad \text{or} \quad x = \frac{5\pi}{3} + 2k\pi \quad \text{or} \quad x = \frac{2\pi}{3} + 2k\pi \quad \text{or} \quad x = \frac{4\pi}{3} + 2k\pi$$

In the interval $[0, 2\pi)$ the solution set is $\left\{\dfrac{\pi}{3}, \dfrac{2\pi}{3}, \dfrac{4\pi}{3}, \dfrac{5\pi}{3}\right\}$.

 The four points of intersection of $y_1 = \cos^2(x)$ and $y_2 = 1/4$ in Fig. 4.37 correspond to the four solutions to $\cos^2(x) = 1/4$.

**Figure 4.37**

**TRY THIS.** Solve $\sin^2(x) = \dfrac{1}{4}$ for $x$ in the interval $[0, 2\pi)$.  ∎

## Using Factoring

In the next two examples we solve quadratic type equations by factoring.

**EXAMPLE 2** Factoring

Find all solutions in the interval $[0, 2\pi)$ to $\tan^2(x) = \tan(x)$.

**Solution**

$$\tan^2(x) = \tan(x)$$

$$\tan^2(x) - \tan(x) = 0$$

$$\tan(x)(\tan(x) - 1) = 0 \qquad \text{Factor.}$$

$$\tan(x) = 0 \qquad \text{or} \qquad \tan(x) - 1 = 0 \qquad \text{Set each factor equal to 0.}$$

$$\tan(x) = 0 \qquad \text{or} \qquad \tan(x) = 1$$

$$x = k\pi \qquad \text{or} \qquad x = \frac{\pi}{4} + k\pi$$

In the interval $[0, 2\pi)$ the solution set is $\left\{0, \dfrac{\pi}{4}, \pi, \dfrac{5\pi}{4}\right\}$.

**TRY THIS.** Solve $\cos^2(x) + \cos(x) = 0$ in the interval $[0, 2\pi)$. ∎

In algebraic equations, we generally do not divide each side by an expression that involves a variable, and the same rule holds for trigonometric equations. In Example 2, we did not divide by $\tan(x)$ when it appeared on opposite sides of the equation. If we had, the solutions 0 and $\pi$ would have been lost.

In the next example we solve another equation by factoring. Since this equation is more complicated, we use substitution to simplify it before factoring.

**EXAMPLE 3** **Factoring using substitution**

Find all solutions to the equation

$$6 \cos^2\left(\frac{x}{2}\right) - 7 \cos\left(\frac{x}{2}\right) + 2 = 0$$

in the interval $[0, 2\pi)$. Round approximate answers to four decimal places.

**Solution**

Let $y = \cos(x/2)$ to get a quadratic equation:

$$6y^2 - 7y + 2 = 0 \qquad \text{Replace } \cos(x/2) \text{ with } y.$$

$$(2y - 1)(3y - 2) = 0 \qquad \text{Factor.}$$

$$2y - 1 = 0 \qquad \text{or} \qquad 3y - 2 = 0$$

$$y = \frac{1}{2} \qquad \text{or} \qquad y = \frac{2}{3}$$

$$\cos\frac{x}{2} = \frac{1}{2} \qquad \text{or} \qquad \cos\frac{x}{2} = \frac{2}{3} \qquad \text{Replace } y \text{ with } \cos(x/2).$$

To solve $\cos\frac{x}{2} = \frac{1}{2}$ we first find $\cos^{-1}\left(\frac{1}{2}\right) = \frac{\pi}{3}$. Since $\frac{\pi}{3}$ and $\frac{5\pi}{3}$ have the same cosine, we have

$$\frac{x}{2} = \frac{\pi}{3} + 2k\pi \qquad \text{or} \qquad \frac{x}{2} = \frac{5\pi}{3} + 2k\pi$$

$$x = \frac{2\pi}{3} + 4k\pi \qquad \text{or} \qquad x = \frac{10\pi}{3} + 4k\pi \qquad \text{Multiply by 2.}$$

To solve $\cos\frac{x}{2} = \frac{2}{3}$ we first find $\cos^{-1}\left(\frac{2}{3}\right) \approx 0.8411$. Subtract 0.8411 from $2\pi$ to get 5.4421. So we have

$$\frac{x}{2} \approx 0.8411 + 2k\pi \qquad \text{or} \qquad \frac{x}{2} \approx 5.4421 + 2k\pi$$

$$x \approx 1.6821 + 4k\pi \qquad \text{or} \qquad x \approx 10.8842 + 4k\pi \qquad \text{Multiply by 2.}$$

**Figure 4.38**

We have now found four expressions that produce infinitely many possibilities for $x$. Now check each expression for values that are in the interval $[0, 2\pi)$. We get $x = 2\pi/3$ and $x \approx 1.6821$.

 The graph of $y_1 = 6(\cos(x/2))^2 - 7\cos(x/2) + 2$ in Fig. 4.38 appears to cross the $x$-axis at $2\pi/3$ and 1.6821.

**TRY THIS.** Find all solutions to $2\sin^2(x) - \sin(x) - 1 = 0$ in the interval $(0, 2\pi)$. ∎

## Using the Quadratic Formula

For a trigonometric equation to be of quadratic type it must be written in terms of a trigonometric function and the square of that function. In the next example an identity is used to convert an equation involving sine and cosine to one with sine only. This equation is of quadratic type like Example 3, but it does not factor.

**EXAMPLE 4   An equation solved by the quadratic formula**

Find all solutions to the equation

$$\cos^2\alpha - 0.2\sin\alpha = 0.9$$

in the interval $[0°, 360°)$. Round answers to the nearest tenth of a degree.

**Solution**

First use the identity $\cos^2\alpha = 1 - \sin^2\alpha$, then use the quadratic formula on the equation of quadratic type that occurs:

$$\cos^2\alpha - 0.2\sin\alpha = 0.9$$

$$1 - \sin^2\alpha - 0.2\sin\alpha = 0.9 \quad \text{Replace } \cos^2\alpha \text{ with } 1 - \sin^2\alpha$$

$$\sin^2\alpha + 0.2\sin\alpha - 0.1 = 0$$

Now use $a = 1$, $b = 0.2$, and $c = -0.1$ in the quadratic formula:

$$\sin\alpha = \frac{-0.2 \pm \sqrt{(0.2)^2 - 4(1)(-0.1)}}{2}$$

$$\sin\alpha \approx 0.2317 \quad \text{or} \quad \sin\alpha \approx -0.4317$$

Find two positive solutions to $\sin\alpha = 0.2317$ in $[0°, 360°)$ by using a calculator to get

$$\alpha = \sin^{-1}(0.2317) \approx 13.4° \quad \text{and} \quad 180° - \alpha \approx 166.6°.$$

Now find two positive solutions to $\sin\alpha = -0.4317$ in $[0°, 360°)$. Using a calculator we get

$$\alpha = \sin^{-1}(-0.4317) \approx -25.6° \quad \text{and} \quad 180° - \alpha \approx 205.6°.$$

Since $-25.6°$ is negative, we use $-25.6° + 360°$ or $334.4°$ along with $205.6°$ as the two solutions. We list all possible solutions as

$$\alpha \approx 13.4°, 166.6°, 205.6°, \text{ or } 334.4° \ (+k360° \text{ in each case}).$$

The solutions to the equation in the interval $[0°, 360°)$ are 13.4°, 166.6°, 205.6°, and 334.4°.

**Figure 4.39**

 The graph of $y_1 = \cos^2 x - 0.2\sin x - 0.9$ in Fig. 4.39 appears to cross the $x$-axis four times in the interval $[0°, 360°)$.

**TRY THIS.** Find all solutions to $\cos\alpha - \sin^2\alpha = 0$ in the interval $[0°, 360°)$. ∎

## Squaring Each Side of an Equation

The Pythagorean identities $\sin^2 x = 1 - \cos^2 x$, $\csc^2 x = 1 + \cot^2 x$, and $\sec^2 x = 1 + \tan^2 x$ are frequently used to replace one function by the other. If an equation involves sine and cosine, cosecant and cotangent, or secant and tangent, we might be able to square each side and then use these identities.

**EXAMPLE 5    Squaring each side of the equation**

Find all values of $y$ in the interval $[0°, 360°)$ that satisfy the equation

$$\tan(3y) + 1 = \sqrt{2}\, \sec(3y).$$

**Solution**

Since the equation involves tangent and secant, we can square each side and use the identity $\sec^2 x = \tan^2 x + 1$:

$$(\tan(3y) + 1)^2 = \left(\sqrt{2}\, \sec(3y)\right)^2$$

$$\tan^2(3y) + 2\tan(3y) + 1 = 2\sec^2(3y)$$

$$\tan^2(3y) + 2\tan(3y) + 1 = 2(\tan^2(3y) + 1) \quad \text{\small Since } \sec^2 x = \tan^2 x + 1$$

$$-\tan^2(3y) + 2\tan(3y) - 1 = 0 \quad \text{\small Subtract } 2\tan^2 3y + 2 \text{ from both sides.}$$

$$\tan^2(3y) - 2\tan(3y) + 1 = 0$$

$$(\tan(3y) - 1)^2 = 0$$

$$\tan(3y) - 1 = 0$$

$$\tan(3y) = 1$$

$$3y = 45° + k180°$$

Because we squared each side, we must check for extraneous roots. First check $3y = 45°$. If $3y = 45°$, then $\tan 3y = 1$ and $\sec 3y = \sqrt{2}$. Substituting these values into $\tan(3y) + 1 = \sqrt{2}\, \sec(3y)$ gives us

$$1 + 1 = \sqrt{2} \cdot \sqrt{2}. \quad \text{\small Correct}$$

If we add any *even* multiple of 180°, or any multiple of 360°, to 45° we get the same values for $\tan 3y$ and $\sec 3y$. So for any $k$, $3y = 45° + k360°$ satisfies the original equation. Now check 45° plus *odd* multiples of 180°. If $3y = 225°(k = 1)$, then

$$\tan 3y = 1 \quad \text{and} \quad \sec 3y = -\sqrt{2}.$$

These values do not satisfy the original equation. Since $\tan 3y$ and $\sec 3y$ have these same values for $3y = 45° + k180°$ for any odd $k$, the only solutions are of the form $3y = 45° + k360°$, or

$$y = 15° + k120°.$$

The solutions in the interval $[0°, 360°)$ are 15°, 135°, and 255°.

 The graph of $y_1 = \tan(3x) + 1 - \sqrt{2}/\cos(3x)$ in Fig. 4.40 appears to touch the $x$-axis at three locations in the interval $[0°, 360°)$.

**Figure 4.40**

**TRY THIS.** Find all $\alpha$ in $[0°, 360°)$ that satisfy $\sin \alpha - \cos \alpha = \dfrac{1}{\sqrt{2}}$.    ■

There is no single method that applies to every trigonometric equation, but the following strategy will help you to solve trigonometric equations.

## Solving Trigonometric Equations

1. Know the solutions to $\sin x = a$, $\cos x = a$, and $\tan x = a$.

2. Solve an equation involving multiple angles as if the equation had a single variable.

3. Simplify complicated equations by using identities. Try to get an equation involving only one trigonometric function.

4. If possible, factor to get different trigonometric functions into separate factors.

5. For equations of quadratic type, solve by factoring or by the quadratic formula.

6. Square each side of the equation, if necessary, so that identities involving squares can be applied. (Remember to check for extraneous roots.)

## Modeling Spring Motion

In the next example we solve an equation that arises in the model for the motion of a weight on a spring.

### EXAMPLE 6    Solving a spring equation

A weight in motion attached to a spring has location $x$ given by

$$x = 2 \sin t - \cos t.$$

For what values of $t$ is the weight at position $x = 0$?

**Solution**

To solve $2 \sin t - \cos t = 0$, rewrite the equation in terms of a single trigonometric function. If we divide each side by $\cos t$, we will get $\tan t$. We can divide by $\cos t$ because the values of $t$ for which $\cos t$ is zero do not satisfy the original equation.

$$2 \sin t = \cos t$$

$$\frac{\sin t}{\cos t} = \frac{1}{2}$$

$$\tan t = \frac{1}{2}$$

Since the weight is set in motion at time $t = 0$, the values of $t$ are positive. Since $\tan^{-1}(1/2) \approx 0.46$, the weight is at position $x = 0$ for $t \approx 0.46 + k\pi$ for $k$ a non-negative integer.

 The graph of $y_1 = 2 \sin(x) - \cos(x)$ in Fig. 4.41 crosses the $x$-axis at $x = 0.46 + k\pi$.

**Figure 4.41**

**TRY THIS.** The location $x$ for a weight on a spring is given by $x = 3 \sin t - 2 \cos t$ where $t$ is time in seconds. Find all $t$ for which $x = 0$. ∎

**FOR THOUGHT...** True or False? Explain.

1. The only solutions to $\cos \alpha = 1/2$ in $[0°, 360°)$ are $60°$ and $120°$.

2. The only solution to $\sin x = -1/2$ in $[0, \pi)$ is $\sin^{-1}(-1/2)$.

3. $\{x \mid x = -30° + k360°\} = \{x \mid x = 330° + k360°\}$ where $k$ is any integer.

4. The solution set to $\tan x = 1$ is $\{x \mid x = \frac{\pi}{4} + k\pi\}$ where $k$ is any integer.

**5.** The equation $2\cos^2 x + \cos x - 1 = (2\cos x - 1)(\cos x + 1)$ is an identity.

**6.** The equation $\sin^2 x = \sin x \cos x$ is equivalent to $\sin x = \cos x$.

**7.** One solution to $\sec x = 2$ is $\dfrac{1}{\cos^{-1}(2)}$.

**8.** The solution set to $\sin x = 1/2$ for $x$ in $[0, \pi)$ is $\{\pi/6, 5\pi/6\}$.

**9.** The equation $\sin x = \cos x$ is equivalent to $\sin^2 x = \cos^2 x$.

**10.** The equation $\sin^2 x = 1$ is equivalent to $\sin x = 1$.

Scan this code
to link to section
lecture videos.

## 4.4 EXERCISES

### SKILLS

*Find all real numbers in the interval $[0, 2\pi)$ that satisfy each equation.*

**1.** $\cos^2(x) = 1$

**2.** $\tan^2(x) = 1$

**3.** $\sin^2(x) = \dfrac{1}{4}$

**4.** $\cos^2(x) = \dfrac{3}{4}$

**5.** $2\sin^2(2x) = 1$

**6.** $2\cos^2(2x) = 1$

**7.** $3\tan^2(x) - 1 = 0$

**8.** $\tan^2(x) - 3 = 0$

*Find all real numbers in the interval $[0, 2\pi)$ that satisfy each equation.*

**9.** $\sin^2(x) = \sin(x)$      **10.** $\cos(x) = \cos^2(x)$

**11.** $2\cos^2(x) + \cos(x) = 0$

**12.** $2\sin^2(x) - \sin(x) = 0$

**13.** $3\tan^3(x) - \tan(x) = 0$

**14.** $\tan^3(x) - 3\tan(x) = 0$

**15.** $\sin^2(2x) + \cos(2x) = 1$

**16.** $\tan(2x) + \sec^2(2x) = 1$

*Find all real numbers in the interval $[0, 2\pi)$ that satisfy each equation. Round approximate answers to the nearest tenth.*

**17.** $2\sin^2(x) - 3\sin(x) + 1 = 0$

**18.** $2\sin^2(x) - \sin(x) - 1 = 0$

**19.** $2\cos^2(x) + 3\cos(x) + 1 = 0$

**20.** $2\cos^2(x) + \cos(x) - 1 = 0$

**21.** $3\sin^2(x) + 2\sin(x) - 1 = 0$

**22.** $4\sin^2(x) - 5\sin(x) + 1 = 0$

**23.** $3\cos^2(2x) + 2\cos(2x) - 1 = 0$

**24.** $6\cos^2(2x) - \cos(2x) - 1 = 0$

**25.** $6\sin^2(x/2) - 5\sin(x/2) + 1 = 0$

**26.** $8\cos^2(x/2) - 2\cos(x/2) - 1 = 0$

*Find all solutions to each equation in the interval $[0°, 360°)$. Round approximate answers to the nearest tenth of a degree.*

**27.** $\sin^2(\theta) - 3\sin(\theta) - 1 = 0$

**28.** $\cos^2(\theta) - 3\cos(\theta) + 1 = 0$

**29.** $4\cos^2(\theta) - \cos(\theta) - 2 = 0$

**30.** $4\sin^2(\theta) - \sin(\theta) - 2 = 0$

**31.** $\tan^2(\theta) - 2\tan(\theta) - 1 = 0$

**32.** $\cot^2(\theta) - 4\cot(\theta) + 2 = 0$

*Find all solutions to each equation in the interval $[0°, 360°)$. Round approximate answers to the nearest tenth of a degree.*

**33.** $\tan(\theta) = \sec(\theta) - \sqrt{3}$

**34.** $\csc(\theta) - \sqrt{3} = \cot(\theta)$

**35.** $\sin(\theta) - \cos(\theta) = \dfrac{1 - \sqrt{3}}{2}$

**36.** $\sin(\theta) + \cos(\theta) = \sqrt{2}$

**37.** $\sin(\theta) - \cos(\theta) = 0.8$     **38.** $\sin(\theta) + \cos(\theta) = 1.2$

*Find all real numbers in the interval $[0, 2\pi)$ that satisfy each equation. Round approximate answers to the nearest tenth.*

**39.** $\sin(-x) = \sin(x)$      **40.** $\cos(x) + \cos(-x) = 1$

**41.** $\tan(-x) = \tan(x)$      **42.** $\sin(-x) = \cos(x)$

**43.** $3\sin^2 x = \sin x$      **44.** $2\tan^2 x = \tan x$

**45.** $2\cos^2(x) = 3\cos(x)$     **46.** $4\sin^3(x) = 5\sin(x)$

**47.** $2\cos^2 x + 3\cos x = -1$    **48.** $2\sin^2 x + \sin x = 1$

**49.** $\tan x = \sec x - \sqrt{3}$

**50.** $\csc x - \sqrt{3} = \cot x$

**51.** $\sin x + \sqrt{3} = 3\sqrt{3}\cos x$

**52.** $6\sin^2 x - 2\cos x = 5$

**53.** $5\sin^2 x - 2\sin x = \cos^2 x$

**54.** $\sin^2 x - \cos^2 x = 0$

**55.** $\tan x \sin 2x = 0$

**56.** $3\sec^2 x \tan x = 4\tan x$

**57.** $\sin 2x - \sin x \cos x = \cos x$

**58.** $2\cos^2 2x - 8\sin^2 x \cos^2 x = -1$

**59.** $\sin x \cos(\pi/4) + \cos x \sin(\pi/4) = 1/2$

**60.** $\sin(\pi/6)\cos x - \cos(\pi/6)\sin x = -1/2$

**61.** $\sin 2x \cos x - \cos 2x \sin x = -1/2$

**62.** $\cos 2x \cos x - \sin 2x \sin x = 1/2$

**63.** $4 \cdot 16^{\sin^2(x)} = 64^{\sin(x)}$

**64.** $\dfrac{1}{2} \cdot 4^{\cos^2(x)} = 2^{\cos(x)}$

*Find all values of $\theta$ in the interval of $[0°, 360°)$ that satisfy each equation. Round approximate answers to the nearest tenth of a degree.*

**65.** $\cos^2\left(\dfrac{\theta}{2}\right) = \sec\theta$

**66.** $2\sin^2\left(\dfrac{\theta}{2}\right) = \cos\theta$

**67.** $2\sin\theta = \cos\theta$

**68.** $3\sin 2\theta = \cos 2\theta$

**69.** $\sin(2\theta) = 3\sin(\theta)$

**70.** $3\cos(2\theta) = \sin^2(2\theta) - 3$

**71.** $\sin 3\theta = \csc 3\theta$

**72.** $\tan^2\theta - \cot^2\theta = 0$

**73.** $\tan^2\theta - 2\tan\theta - 1 = 0$

**74.** $\cot^2\theta - 4\cot\theta + 2 = 0$

**75.** $9\sin^2\theta + 12\sin\theta + 4 = 0$

**76.** $12\cos^2\theta + \cos\theta - 6 = 0$

**77.** $\sin x \cos x + 2\cos x = 3\sin x + 6$

**78.** $\sin^2 x + 9 = 6\sin x$

**79.** $\dfrac{\tan 3\theta - \tan\theta}{1 + \tan 3\theta \tan\theta} = \sqrt{3}$

**80.** $\dfrac{\tan 3\theta + \tan 2\theta}{1 - \tan 3\theta \tan 2\theta} = 1$

**81.** $8\cos^4\theta - 10\cos^2\theta + 3 = 0$

**82.** $4\sin^4\theta - 5\sin^2\theta + 1 = 0$

**83.** $\sec^4\theta - 5\sec^2\theta + 4 = 0$

**84.** $\cot^4\theta - 4\cot^2\theta + 3 = 0$

*Solve each problem.*

**85.** Find all points at which the graph of $y = \sin(x)$ intersects the graph of $y = \cot(x)$.

**86.** Find all points at which the graph of $y = \cos(x)$ intersects the graph of $y = \tan(x)$.

**87.** *Motion of a Spring* A block is attached to a spring and set in motion on a frictionless plane. Its location on the surface at any time $t$ in seconds is given in meters by $x = \sqrt{3}\sin 2t + \cos 2t$. For what values of $t$ is the block at its resting position $x = 0$?

**88.** *Motion of a Spring* A block is set in motion hanging from a spring and oscillates about its resting position $x = 0$ according to the function $x = -0.3\sin 3t + 0.5\cos 3t$. For what values of $t$ is the block at its resting position $x = 0$?

## WRITING/DISCUSSION

**89.** Use a calculator graph of $y = (1 - \cos x)/x$ to help you determine all of the solutions to $(1 - \cos x)/x = 0$ in the interval $(-20, 20)$. What is the maximum value of this function on this interval?

**90.** Use a calculator graph of $y = \sin(1/x^2)$ to help you determine the number of solutions to $\sin(1/x^2) = 0$ in the interval $(-\pi/100, \pi/100)$. What is the maximum value of this function on this interval?

## REVIEW

**91.** Find all solutions to $(\sin x - 1)(\sin x + 1) = 0$ in the interval $(0, 2\pi)$.

**92.** Find all solutions to $\sin(2x)\cos(2x) = 0$ in the interval $(0, 2\pi)$.

**93.** State the domain and range for each function.
   **a.** $f(x) = \sin^{-1}(x)$

   **b.** $f(x) = \arccos(x)$

   **c.** $f(x) = \tan^{-1}(x)$

**94.** Complete the sum and difference identities.
   **a.** $\sin(x + y) = $ _____

   **b.** $\sin(x - y) = $ _____

**95.** Simplify $\tan y \cos y + \csc y \sin^2 y$.

**96.** In a controlled experiment, the temperature is $0°C$ at time $t = 0$. The temperature is increased to $10°C$ at time $t = 4$ and then decreased to $-10°C$ at time $t = 12$. The temperature returns to $0°C$ at time $t = 16$. Assuming the temperature on the time interval $[0, 16]$ is a sine wave, write the temperature $y$ as a function of the time $t$.

## OUTSIDE THE BOX

**97.** *Area of a Polygon* Find the exact area of the polygon bounded by the $x$-axis, the $y$-axis, the line $2x + 3y = 7$, and the line $3x - 4y = 18$.

**98.** *Sine Cubed* If $\sin(x) + \cos(x) = 1/2$, then what is the value of $\sin^3(x) + \cos^3(x)$?

**4.4 POP QUIZ**

*Find all real numbers in* $[0, 2\pi)$ *that satisfy each equation.*

**1.** $\sin(2x) = \cos(x)$

**2.** $2\sin^2(x) + 3\sin(x) + 1 = 0$

**3.** $\sin(x) + \cos(x) = \sqrt{2}$

# LINKING concepts...

### For Individual or Group Explorations

## Modeling Baseball Strategy

*An outfielder picks up a ground ball and wants to make a quick 150-ft throw to first base. To save time, he does not throw the ball directly to the first baseman. He skips the ball off the artificial surface at 75 ft so that the ball reaches first base after a bounce, as shown in the figure. Let's examine this strategy. The equations*

$$x = v_0 t \cos \theta$$

*and*

$$y = -16t^2 + v_0 t \sin \theta + h_0$$

*give the coordinates of the ball at time t seconds for an initial angle of θ degrees, initial velocity $v_0$ ft/sec, and initial height $h_0$ ft. Assume that the ball is thrown at 130 ft/sec from a height of 5 ft and the ball is caught at a height of 5 ft.*

**a)**  At what angle from horizontal must the ball be thrown to reach first base without a skip?

**b)**  How long does it take the ball to get to first base without a skip?

**c)**  At what angle from horizontal must the ball be thrown so that it strikes the ground at 75 ft?

**d)**  How long does it take for the ball to reach the skip point at 75 ft?

**e)**  Assume that the path of the ball is symmetric with respect to the skip point and double your answer to part (d) to get the time that it takes the ball to reach first base.

**f)**  How much time is saved by skipping the ball?

**g)**  How long would it take for the ball to reach first base if it could be thrown in a straight line?

**h)**  Discuss the assumptions made for this model. Did we ignore anything that might affect our results?

# Highlights

## 4.1 The Inverse Trigonometric Functions

| | | |
|---|---|---|
| **Inverse Sine** | If $y = \sin^{-1}x$ for $x$ in $[-1, 1]$, then $y$ is the real number in $[-\pi/2, \pi/2]$ such that $\sin y = x$. | $\sin^{-1}(1) = \pi/2$ <br> $\sin^{-1}(-1/2) = -\pi/6$ |
| **Inverse Cosine** | If $y = \cos^{-1}x$ for $x$ in $[-1, 1]$, then $y$ is the real number in $[0, \pi]$ such that $\cos y = x$. | $\cos^{-1}(-1) = \pi$ <br> $\cos^{-1}(-1/2) = 2\pi/3$ |
| **Inverse Tangent** | If $y = \tan^{-1}x$ for $x$ in $(-\infty, \infty)$, then $y$ is the real number in $(-\pi/2, \pi/2)$ such that $\tan y = x$. | $\tan^{-1}(1) = \pi/4$ <br> $\tan^{-1}(-1) = -\pi/4$ |

## 4.2 Basic Sine, Cosine, and Tangent Equations

| | | |
|---|---|---|
| **Solving $\cos(x) = a$** | If $|a| \le 1$, find all solutions in $[0, 2\pi)$, then add all multiples of $2\pi$ to them and simplify. If $|a| > 1$ there are no solutions. | $\cos x = 0, x = \dfrac{\pi}{2} + k\pi$ <br> $\cos x = \dfrac{1}{2}, x = \dfrac{\pi}{3} + 2k\pi$ <br> or $x = \dfrac{5\pi}{3} + 2k\pi$ |
| **Solving $\sin(x) = a$** | If $|a| \le 1$, find all solutions in $[0, 2\pi)$, then add all multiples of $2\pi$ to them and simplify. If $|a| > 1$ there are no solutions. | $\sin x = 0, x = k\pi$ <br> $\sin x = \dfrac{1}{2}, x = \dfrac{\pi}{6} + 2k\pi$ <br> or $x = \dfrac{5\pi}{6} + 2k\pi$ |
| **Solving $\tan(x) = a$** | If $a$ is a real number, find all solutions in $(-\pi/2, \pi/2)$, then add multiples of $\pi$ to them and simplify. | $\tan x = 1, x = \dfrac{\pi}{4} + k\pi$ |

## 4.3 Equations Involving Compositions

| | | |
|---|---|---|
| **Multiple-Angle Equations** | Solve for the multiple angle and then divide or multiply to get the solution. | $\sin(2x) = 1/2$ <br> $2x = \dfrac{\pi}{6} + 2k\pi$ or $\dfrac{5\pi}{6} + 2k\pi$ <br> $x = \dfrac{\pi}{12} + k\pi$ or $\dfrac{5\pi}{12} + k\pi$ |
| **Equations Involving Identities** | Look for ways to simplify an equation by using identities. | $\sin^2 x + \sin x + \cos^2 x = 0$ <br> $\sin x + 1 = 0$ <br> $\sin x = -1$ |

## 4.4 Trigonometric Equations of Quadratic Type

| | | |
|---|---|---|
| **Quadratic Type** | Use the square-root property, factoring, or the quadratic formula to solve equations of quadratic type. | $\sin^2(x) + \sin(x) = 0$ <br> $\sin(x) = 0$ or $\sin(x) + 1 = 0$ |
| **Pythagorean Identities** | Some equations can be solved by squaring both sides and then applying a Pythagorean identity. | $\sin x + \cos x = 1$ <br> $\sin^2 x + 2\sin x \cos x + \cos^2 x = 1$ <br> $2\sin x \cos x = 0$ |

# Chapter 4 Review Exercises

*Find the exact value of each expression.*

**1.** $\sin^{-1}(-0.5)$

**2.** $\cos^{-1}(-0.5)$

**3.** $\arctan(-1)$

**4.** $\text{arccot}\left(1/\sqrt{3}\right)$

**5.** $\sec^{-1}\left(\sqrt{2}\right)$

**6.** $\csc^{-1}(2)$

**7.** $\cos(\arcsin(1/2))$

**8.** $\cos(\arcsin(-1/2))$

**9.** $\tan\left(\arccos\left(\sqrt{2}/2\right)\right)$

**10.** $\cot\left(\arcsin\left(-\sqrt{2}/2\right)\right)$

**11.** $\sin^{-1}(\sin(-\pi/4))$

**12.** $\cos^{-1}(\cos(\pi/6))$

**13.** $\sin^{-1}(\sin(3\pi/4))$

**14.** $\sin^{-1}(\sin(5\pi/6))$

**15.** $\cos^{-1}(\cos(-\pi/6))$

**16.** $\cos\left(\cos^{-1}\left(-\sqrt{3}/2\right)\right)$

**17.** $\csc^{-1}(\sec(\pi/3))$

**18.** $\sec^{-1}(\csc(\pi/4))$

*Find the exact value of each expression in degrees.*

**19.** $\sin^{-1}(1)$

**20.** $\tan^{-1}(1)$

**21.** $\arccos\left(-1/\sqrt{2}\right)$

**22.** $\text{arcsec}(2)$

**23.** $\cot^{-1}\left(\sqrt{3}\right)$

**24.** $\cot^{-1}\left(-\sqrt{3}\right)$

**25.** $\text{arccot}(0)$

**26.** $\text{arccot}\left(-\sqrt{3}/3\right)$

*Find all real numbers in $[0, 2\pi]$ that satisfy each equation.*

**27.** $\cos(x) + 1 = 0$

**28.** $\sin(x) - 1 = 0$

**29.** $2\sin(x) - 1 = 0$

**30.** $2\cos(x) + 1 = 0$

**31.** $2\tan(x) + 2 = 0$

**32.** $\sqrt{3}\tan(x) - \sqrt{3} = 0$

*Find all angles in $[0°, 360°]$ that satisfy each equation.*

**33.** $2\sin(-x) + 1 = 0$

**34.** $2\cos(-x) - 1 = 0$

**35.** $2\cos(-x) = \sqrt{2}$

**36.** $2\sin(-x) + \sqrt{3} = 0$

**37.** $\sqrt{3}\tan(-x) - 1 = 0$

**38.** $\sqrt{3}\tan(-x) + 3 = 0$

*Find all real numbers in $[0, 2\pi]$ that satisfy each equation. Round to the nearest hundredth.*

**39.** $3\sec(x) + 5 = 0$

**40.** $5\csc(x) - 9 = 0$

**41.** $2\csc(2x) = 9$

**42.** $\sec(3x) + \sqrt{5} = 0$

**43.** $\cot(x) - \sqrt{2} = 0$

**44.** $\sqrt{3}\cot(x) + 8 = 0$

*Solve each equation for the indicated variable.*

**45.** Solve $x = 3\sin(t)$ for $t$ where $-\dfrac{\pi}{2} \le t \le \dfrac{\pi}{2}$.

**46.** Solve $w = 2\tan(d)$ for $d$ where $-\dfrac{\pi}{2} < d < \dfrac{\pi}{2}$.

**47.** Solve $a = 3\sin(2y)$ for $y$ where $-\dfrac{\pi}{4} \le y \le \dfrac{\pi}{4}$.

**48.** Solve $s = -5\cos\left(\dfrac{w}{4}\right)$ for $w$ where $0 \le w \le 4\pi$.

**49.** Solve $q = -2\cos(h) + 1$ for $h$ where $0 \le h \le \pi$.

**50.** Solve $j = 4\sin(v) - 8$ for $v$ where $-\dfrac{\pi}{2} \le v \le \dfrac{\pi}{2}$.

**51.** Solve $b = 5\tan(\pi x) - 3$ for $x$ where $-\dfrac{1}{2} < x < \dfrac{1}{2}$.

**52.** Solve $p = 3\tan(\pi y - \pi) + \pi$ for $y$ where $\dfrac{1}{2} < y < \dfrac{3}{2}$.

**53.** Solve $x = a\cos^{-1}(y - 2)$ for $y$ where $1 \le y \le 3$.

**54.** Solve $x = b\sin^{-1}(y + 1) + k$ for $y$ where $-2 \le y \le 0$.

*Find the inverse of each function and state the domain and range of $f^{-1}$.*

**55.** $f(x) = \sin(x)$ for $-\dfrac{\pi}{2} \le x \le \dfrac{\pi}{2}$

**56.** $f(x) = \cos(x)$ for $0 \le x \le \pi$

**57.** $f(x) = \sin^{-1}(x)$ for $-1 \le x \le 1$

**58.** $f(x) = \cos^{-1}(x)$ for $-1 \le x \le 1$

**59.** $f(x) = \sin(3x)$ for $-\dfrac{\pi}{6} \le x \le \dfrac{\pi}{6}$

**60.** $f(x) = 4\sin(2x)$ for $-\dfrac{\pi}{4} \le x \le \dfrac{\pi}{4}$

**61.** $f(x) = 6\cos(4x)$ for $0 \le x \le \dfrac{\pi}{4}$

**62.** $f(x) = 3\cos(5x) + 2$ for $0 \le x \le \dfrac{\pi}{5}$

**63.** $f(x) = 4 + \tan\left(\dfrac{\pi x}{2}\right)$ for $-1 < x < 1$

**64.** $f(x) = 1 + \tan\left(\dfrac{\pi x}{3}\right)$ for $-\dfrac{3}{2} < x < \dfrac{3}{2}$

*Find the exact value(s) for x in the interval $\left[-\dfrac{\pi}{2}, \dfrac{\pi}{2}\right]$ that satisfy each equation. If there are no such values for x, then write "None."*

**65.** $\sin(x) = -\dfrac{1}{2}$

**66.** $\sin(x) = -\dfrac{\sqrt{3}}{2}$

**67.** $\cos(x) = -\dfrac{1}{2}$

**68.** $\cos(x) = -\dfrac{\sqrt{3}}{2}$

**69.** $\tan(x) = -1$

**70.** $\sin(x) = -\dfrac{\sqrt{2}}{2}$

**71.** $\cos(x) = \dfrac{1}{\sqrt{2}}$

**72.** $\cos(x) = 1/2$

**73.** $\tan(x) = -\dfrac{1}{\sqrt{3}}$

**74.** $\cot(x) = \sqrt{3}$

*Find all real numbers that satisfy each equation.*

**75.** $2\cos(2x) + 1 = 0$

**76.** $2\sin(2x) + \sqrt{3} = 0$

**77.** $\left(\sqrt{3}\csc(x) - 2\right)(\csc(x) - 2) = 0$

**78.** $\left(\sec(x) - \sqrt{2}\right)\left(\sqrt{3}\sec(x) + 2\right) = 0$

**79.** $2\sin^2(x) + 1 = 3\sin(x)$

**80.** $4\sin^2 x = \sin(x) + 3$

**81.** $-8\sqrt{3}\sin\dfrac{x}{2} = -12$

**82.** $-\cos\dfrac{x}{2} = \sqrt{2} + \cos\dfrac{x}{2}$

**83.** $\cos\dfrac{x}{2} - \sin x = 0$

**84.** $\sin 2x = \tan x$

**85.** $\cos 2x + \sin^2 x = 0$     **86.** $\tan \dfrac{x}{2} = \sin x$

**87.** $\sin(x)\cos(x) + \sin(x) + \cos(x) + 1 = 0$

**88.** $\sin(2x)\cos(2x) - \cos(2x) + \sin(2x) - 1 = 0$

*Find all angles* $\alpha$ *in* $[0°, 360°)$ *that satisfy each equation.*

**89.** $\sin \alpha \cos \alpha = \dfrac{1}{2}$     **90.** $\cos 2\alpha = \cos \alpha$

**91.** $\sin \alpha = \cos \alpha + 1$     **92.** $\cos \alpha \csc \alpha = \cot^2 \alpha$

**93.** $\sin^2 \alpha + \cos^2 \alpha = \dfrac{1}{2}$     **94.** $\sec^2 \alpha - \tan^2 \alpha = 0$

**95.** $4 \sin^4 2\alpha = 1$     **96.** $\sin 2\alpha = \tan \alpha$

**97.** $\tan 2\alpha = \tan \alpha$     **98.** $\tan \alpha = \cot \alpha$

**99.** $\sin(2\alpha)\cos(\alpha) + \cos(2\alpha)\sin(\alpha) = \cos(3\alpha)$

**100.** $\cos(2\alpha)\cos(\alpha) - \sin(2\alpha)\sin(\alpha) = \cot(3\alpha)$

*Solve each problem.*

**101.** Find all points where the graph of $y = \sin(x)$ intersects the graph of $y = \cos(x)$.

**102.** Find all points where the graph of $y = \sin(x)$ intersects the graph of $y = \tan(x)$.

**103.** *Motion of a Spring* A block is set in motion hanging from a spring and oscillates about its resting position $x = 0$ according to the function $x = 0.6 \sin 2t + 0.4 \cos 2t$, where $x$ is in centimeters and $t$ is in seconds. For what values of $t$ in the interval $[0, 3]$ is the block at its resting position $x = 0$?

**104.** *Battle of Gettysburg* The Confederates had at least one 24-lb mortar at the battle of Gettysburg in the Civil War. If the muzzle velocity of a projectile was 400 ft/sec, then at what angles could the cannon be aimed to hit the Union Army 3000 ft away?

> **HINT** The distance $d$ (in feet) traveled by a projectile fired at an angle $\theta$ is related to the initial velocity $v_0$ (in feet per second) by the equation $v_0^2 \sin 2\theta = 32d$.

**OUTSIDE THE BOX**

**105.** *Whole Lotta Shakin'* At the start of an economic conference between the eastern delegation and the western delegation, each delegate shook hands with every other member of his own delegation for a total of 466 handshakes. Next, each delegate shook hands with every person in the other delegation for 480 more handshakes. What was the total number of delegates at the conference?

**106.** *Minimizing* Find the minimum value of $y$ without using a calculator if

$$y = \frac{\cos^2(x) + \cos(3x)}{\cos(x)}.$$

# Chapter 4 Test

*Find the exact value of each expression.*

**1.** $\sin^{-1}(-1/2)$     **2.** $\cos^{-1}(-1/2)$

**3.** $\arctan(-1)$     **4.** $\sec^{-1}\left(2/\sqrt{3}\right)$

**5.** $\csc^{-1}\left(-\sqrt{2}\right)$     **6.** $\sin(\cos^{-1}(-1/3))$

*Find all real numbers that satisfy each equation.*

**7.** $\sin(-\theta) = 1$     **8.** $\cos 3s = \dfrac{1}{2}$

**9.** $\tan 2t = -\sqrt{3}$     **10.** $\sin 2\theta = \cos \theta$

*Find all angles* $\alpha$ *in* $[0°, 360°]$ *that satisfy each equation.*

**11.** $4 \csc(\alpha) - 8 = 0$

**12.** $\cot(\alpha/2) + 1 = 0$

**13.** $\sqrt{3} \sec(\alpha) + 1 = 3$

*Find all values of* $\alpha$ *in* $[0°, 360°)$ *that satisfy each equation. Round approximate answers to the nearest tenth of a degree.*

**14.** $\sin(\alpha)\cos(\alpha) = 0$     **15.** $\sin^2(2\alpha) - 1 = 0$

**16.** $3 \sin^2 \alpha - 4 \sin \alpha + 1 = 0$

**17.** $\dfrac{\tan 2\alpha - \tan 7\alpha}{1 + \tan 2\alpha \tan 7\alpha} = 1$

*Solve each problem.*

**18.** Solve $a = 5 \sin(2t)$ for $t$ where $-\dfrac{\pi}{4} \le t \le \dfrac{\pi}{4}$.

**19.** Find the inverse of the function

$$f(x) = \frac{1}{5}\cos\left(\frac{x}{4}\right) + \frac{1}{5}$$

where $0 \le x \le 4\pi$ and state the domain and range of $f^{-1}$.

**20.** Find all points at which the graph of $y = \cos(x)$ intersects the graph of $y = \sec(x)$.

**21.** A car with worn shock absorbers hits a pothole and oscillates about its normal riding position. At time $t$ (in seconds) the front bumper is distance $d$ (in inches) above or below its normal position, where $d = 2 \sin 3t - 4 \cos 3t$. For what values of $t$ (to the nearest tenth of a second) in the interval $[0, 4]$ is the front bumper at its normal position $d = 0$?

## TYING IT ALL TOGETHER

### Chapters P–4

*Find the exact value of each expression without using a calculator.*

**1.** $\sin(\pi/4)$

**2.** $\cos(-2\pi/3)$

**3.** $\tan(5\pi/4)$

**4.** $\sin(-7\pi/6)$

**5.** $\sin^{-1}(-1/2)$

**6.** $\cos^{-1}(-1/2)$

**7.** $\tan^{-1}(-1)$

**8.** $\tan^{-1}(0)$

**9.** $\sin(29\pi/3)$

**10.** $\cos(59\pi/6)$

**11.** $\csc(5\pi/2)$

**12.** $\sec^2(\pi)$

*Sketch at least one cycle of the graph of each function. Determine the period, amplitude, and phase shift, and find all x-intercepts.*

**13.** $y = \sin(x - \pi/6)$

**14.** $y = \sin(2x)$

**15.** $y = \cos(3x)$

**16.** $y = \cos(2x - \pi)$

**17.** $y = 3\sin(x/2 - \pi/2)$

**18.** $y = 4\cos(x + \pi/3)$

**19.** $y = -4\cos(\pi x/3)$

**20.** $y = -2\sin(\pi x - \pi)$

*Find the values of the remaining five trigonometric functions at $\alpha$ with the given information.*

**21.** $\sin(\alpha) = -1/3$ and $180° < \alpha < 270°$

**22.** $\cos(\alpha) = -1/4$ and $90° < \alpha < 180°$

**23.** $\tan(\alpha) = 3/5$ and $0° < \alpha < 90°$

**24.** $\sin(\alpha) = -4/5$ and $-90° < \alpha < 0°$

*Fill in the blanks.*

**25.** If $\omega$ is the angular velocity of a point in motion on a circle of radius $r$, then the linear velocity of the point is _____.

**26.** If $\alpha$ is an angle in standard position and $(x, y)$ is the point of intersection of the terminal side and a circle of radius $r$, then $\sin \alpha =$ _____ and $\cos \alpha =$ _____.

**27.** If $\alpha$ is an angle in standard position and $(x, y)$ is the point of intersection of the terminal side and the unit circle, then $\sin \alpha =$ _____ and $\cos \alpha =$ _____.

**28.** The identity $\sin^2 \alpha + \cos^2 \alpha = 1$ is called the _____ identity.

**29.** The value of $|A|$ is called the _____ of the function $y = A \sin x$.

**30.** For the function $y = \sin(x - C)$, $C$ is called the _____.

**31.** If the period of a sine wave is $P$, then $1/P$ is called the _____ of the wave.

**32.** The line $x = \pi/2$ is called a(n) _____ for the graph of $y = \tan x$.

**33.** If $f(-x) = f(x)$, then $f$ is called a(n) _____ function.

**34.** If $f(-x) = -f(x)$, then $f$ is called a(n) _____ function.

# 5 Applications of Trigonometry

**5.1** The Law of Sines

**5.2** The Law of Cosines

**5.3** Area of a Triangle

**5.4** Vectors

**5.5** Applications of Vectors

Mechanical devices that perform manipulative tasks under their own power (robots) are as old as recorded history. As early as 3000 B.C., Egyptians built waterclocks and articulated figures.

The first industrial robot joined GM's production line in 1961. In the 21st century, robotic "steel collar" workers will increasingly perform boring, repetitive, and dangerous jobs or jobs associated with a high degree of human error. They will assemble parts, mix chemicals, and work in areas that would be deadly to their human counterparts.

**In this chapter, we will see that trigonometry is the mathematics you need to model repetitive motion. We will learn how to use trigonometry to position a robot's arm. We'll also learn how trigonometry is used in designing machines and finding forces acting on parts.**

257

# 5.1 The Law of Sines

**Figure 5.1**

**Figure 5.2**

In Chapter 1 we were given the measures of some of the parts of a right triangle and we used the trigonometric functions to find the measures of the remaining parts. We *solved* right triangles. In this section we will solve triangles that are not right triangles.

## Oblique Triangles

Any triangle without a right angle is called an **oblique triangle.** As usual, we use $\alpha$, $\beta$, and $\gamma$ for the angles of a triangle and $a$, $b$, and $c$, respectively, for the lengths of the sides opposite those angles, as shown in Fig. 5.1. The vertices at angles $\alpha$, $\beta$, and $\gamma$ are labeled $A$, $B$, and $C$, respectively.

If we know only the angles of a triangle, then we cannot solve the triangle. For example, we cannot determine the lengths of the sides of a triangle with angles of 30°, 40°, and 110°, because there are infinitely many such triangles of different sizes. See Fig. 5.2. To solve an oblique triangle, we must know at least three parts of the triangle, at least one of which must be the length of a side. We can classify the different cases for the three known parts as follows:

1. One side and any two angles (ASA or AAS)
2. Two sides and a nonincluded angle (SSA)
3. Two sides and an included angle (SAS)
4. Three sides (SSS)

These four cases are illustrated in Fig. 5.3, with the known parts in red and the unknown parts in black.

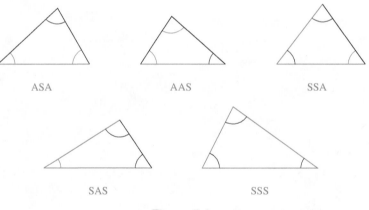

ASA          AAS          SSA

SAS          SSS

**Figure 5.3**

We can actually solve oblique triangles by dividing the triangles into right triangles and using right triangle trigonometry. Since that method is quite tedious, we develop the *law of sines* and the *law of cosines.* The first two cases listed above can be handled with the law of sines. We will discuss the last two cases in Section 5.2 when we develop the law of cosines.

## The Law of Sines

The **law of sines** gives a relationship between the sines of the angles and the sides of a triangle: *The ratio of the sine of an angle and the length of the side opposite the*

*angle is the same for each angle of a triangle.* The law of sines is stated symbolically as follows.

**Theorem: The Law of Sines**

In any triangle

$$\frac{\sin \alpha}{a} = \frac{\sin \beta}{b} = \frac{\sin \gamma}{c}.$$

**PROOF** Either the triangle is an acute triangle (all acute angles) or it is an obtuse triangle (one obtuse angle). We consider the case of the obtuse triangle here and leave the case of the acute triangle as Exercise 39. Triangle $ABC$ is shown in Fig. 5.4 below with an altitude of length $h_1$ drawn from point $C$ to the opposite side and an altitude of length $h_2$ drawn from point $B$ to the extension of the opposite side.

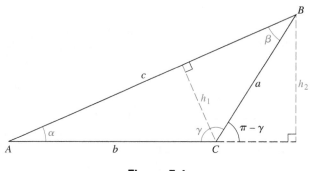

**Figure 5.4**

Since $\alpha$ and $\beta$ are now angles of right triangles, we have

$$\sin \alpha = \frac{h_1}{b} \qquad \text{or} \qquad h_1 = b \sin \alpha$$

and

$$\sin \beta = \frac{h_1}{a} \qquad \text{or} \qquad h_1 = a \sin \beta.$$

Replace $h_1$ by $b \sin \alpha$ in the equation $h_1 = a \sin \beta$ to get

$$b \sin \alpha = a \sin \beta.$$

Dividing each side of the last equation by $ab$ yields

$$\frac{\sin \alpha}{a} = \frac{\sin \beta}{b}.$$

Using the largest right triangle in Fig. 5.4, we have

$$\sin \alpha = \frac{h_2}{c} \qquad \text{or} \qquad h_2 = c \sin \alpha.$$

Note that the acute angle $\pi - \gamma$ in Fig. 5.4 is the reference angle for the obtuse angle $\gamma$. We know that sine is positive for both acute and obtuse angles, so $\sin \gamma = \sin(\pi - \gamma)$. Since $\pi - \gamma$ is an angle of a right triangle, we can write

$$\sin \gamma = \sin(\pi - \gamma) = \frac{h_2}{a} \qquad \text{or} \qquad h_2 = a \sin \gamma.$$

From $h_2 = c \sin \alpha$ and $h_2 = a \sin \gamma$ we get

$$c \sin \alpha = a \sin \gamma$$

$$\frac{c \sin \alpha}{ac} = \frac{a \sin \gamma}{ac} \qquad \text{Divide each side by } ac.$$

$$\frac{\sin \alpha}{a} = \frac{\sin \gamma}{c}. \qquad \text{Reduce.}$$

We have proved the law of sines.

The law of sines can also be written in the form

$$\frac{a}{\sin \alpha} = \frac{b}{\sin \beta} = \frac{c}{\sin \gamma}.$$

In solving triangles, it is usually simplest to use the form in which the unknown quantity appears in the numerator.

In our first example we use the law of sines to solve a triangle for which we are given two angles and an included side.

**Figure 5.5**

 **EXAMPLE 1** **Given two angles and an included side (ASA)**

Given $\beta = 34°$, $\gamma = 64°$, and $a = 5.3$, solve the triangle.

**Solution**

To sketch the triangle, first draw side $a$, then draw angles of approximately $34°$ and $64°$ on opposite ends of $a$. Label all parts as shown in Fig. 5.5. Since the sum of the three angles of a triangle is $180°$, the third angle $\alpha$ is $82°$. By the law of sines,

$$\frac{5.3}{\sin 82°} = \frac{b}{\sin 34°}$$

or

$$b = \frac{5.3 \sin 34°}{\sin 82°} \approx 3.0.$$

Again, by the law of sines,

$$\frac{c}{\sin 64°} = \frac{5.3}{\sin 82°}$$

$$c = \frac{5.3 \sin 64°}{\sin 82°} \approx 4.8.$$

So $\alpha = 82°$, $b \approx 3.0$, and $c \approx 4.8$ solves the triangle.

**TRY THIS.** Given $\alpha = 28°$, $\beta = 66°$, and $c = 8.2$, solve the triangle. ∎

The AAS case is similar to the ASA case. If two angles are known then the third can be found by using the fact that the sum of all angles is $180°$. If we know all angles and any side, then we can proceed to find the remaining sides with the law of sines as in Example 1.

## The Ambiguous Case (SSA)

In the AAS and ASA cases we are given any two angles with positive measures and the length of any side. If the total measure of the two angles is less than $180°$, then a unique triangle is determined. However, for two sides and a *nonincluded* angle

Draw $\alpha$ and $b$: find $h$

**Figure 5.6**

(SSA), there are several possibilities. So the SSA case is called the **ambiguous case.** Drawing the diagram in the proper order will help you in understanding the ambiguous case.

Suppose we are given an acute angle $\alpha$ ($0° < \alpha < 90°$) and sides $a$ and $b$. Side $a$ is opposite the angle $\alpha$ and side $b$ is adjacent to it. Draw an angle of approximate size $\alpha$ in standard position with terminal side of length $b$ as shown in Fig. 5.6. Don't draw side $a$ yet. Let $h$ be the distance from $C$ to the initial side of $\alpha$. Since $\sin \alpha = h/b$, we have $h = b \sin \alpha$. Now we are ready to draw side $a$, but there are four possibilities for its location.

1. If $a < h$, then no triangle can be formed, because $a$ cannot reach from point $C$ to the initial side of $\alpha$. This situation is shown in Fig. 5.7(a).
2. If $a = h$, then exactly one right triangle is formed, as in Fig. 5.7(b).
3. If $h < a < b$, then exactly two triangles are formed, because $a$ reaches to the initial side in two places, as in Fig. 5.7(c).
4. If $a \geq b$, then only one triangle is formed, as in Fig. 5.7(d).

| $a < h$: no triangle | $a = h$: one triangle | $h < a < b$: two triangles | $a \geq b$: one triangle |
| (a) | (b) | (c) | (d) |

**Figure 5.7**

If we start with $\alpha$, $a$, and $b$, where $90° \leq \alpha < 180°$, then there are only two possibilities for the number of triangles determined. If $a \leq b$, then no triangle is formed, as in Fig. 5.8(a). If $a > b$, one triangle is formed, as in Fig. 5.8(b).

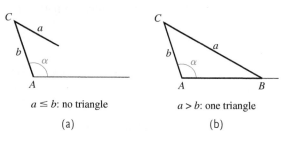

| $a \leq b$: no triangle | $a > b$: one triangle |
| (a) | (b) |

**Figure 5.8**

It is not necessary to memorize all of the SSA cases shown in Figs. 5.7 and 5.8. If you draw the triangle for a given problem in the order suggested, then it will be clear how many triangles are possible with the given parts. We must decide how many triangles are possible before we can solve the triangle(s).

**EXAMPLE 2** **SSA with no triangle**

Given $\alpha = 41°$, $a = 3.3$, and $b = 5.4$, solve the triangle.

**Solution**

**Figure 5.9**

Draw the 41° angle and label its terminal side 5.4 as shown in Fig. 5.9. Drawing an accurate 41° angle is not important. What is important is the relative positions of the given parts of the triangle. Side $a$ must be opposite the angle $\alpha$ and side $b$ must be a

side of $\alpha$. Before drawing side $a$, we check to see if it will fit by finding the length of the altitude $h$ from $C$. Using a trigonometric ratio, we get $\sin 41° = h/5.4$, or $h = 5.4 \sin 41° \approx 3.5$. Since $a$ is only 3.3, $a$ is *shorter* than the altitude $h$, and side $a$ will not reach from point $C$ to the initial side of $\alpha$. So there is no triangle with the given parts.

**TRY THIS.** Given $\beta = 38°$, $b = 2.9$, and $c = 5.9$, solve the triangle.  ■

**EXAMPLE 3** SSA with one triangle

Given $\gamma = 125°$, $b = 5.7$, and $c = 8.6$, solve the triangle.

**Solution**

Draw the 125° angle and label its terminal side 5.7. Since $8.6 > 5.7$, side $c$ will reach from point $A$ to the initial side of $\gamma$, as shown in Fig. 5.10, and form a single triangle. To solve this triangle, we use the law of sines:

$$\frac{\sin 125°}{8.6} = \frac{\sin \beta}{5.7}$$

$$\sin \beta = \frac{5.7 \sin 125°}{8.6}$$

$$\beta = \sin^{-1}\left(\frac{5.7 \sin 125°}{8.6}\right) \approx 32.9°$$

Since the sum of $\alpha$, $\beta$, and $\gamma$ is 180°, $\alpha \approx 22.1°$. Now use $\alpha$ and the law of sines to find $a$:

$$\frac{a}{\sin 22.1°} = \frac{8.6}{\sin 125°}$$

$$a = \frac{8.6 \sin 22.1°}{\sin 125°} \approx 3.9$$

So the missing parts are $a \approx 3.9$, $\alpha \approx 22.1°$, and $\beta \approx 32.9°$.

**TRY THIS.** Given $\beta = 38°$, $b = 6.4$, and $c = 5.9$, solve the triangle.  ■

**EXAMPLE 4** SSA with two triangles

Given $\beta = 56.3°$, $a = 8.3$, and $b = 7.6$, solve the triangle.

**Solution**

Draw an angle of approximately 56.3°, and label its terminal side 8.3. Side $b$ must go opposite $\beta$, but do not draw it yet. Find the length of the altitude $h$ from point $C$ to the initial side of $\beta$, as shown in Fig. 5.11(a). Since $\sin 56.3° = h/8.3$, we get $h = 8.3 \sin 56.3° \approx 6.9$. Because $b$ is longer than $h$ but shorter than $a$, there are

**Figure 5.10**

**Figure 5.11**

two triangles with the given parts, as shown in parts (b) and (c) of the figure. Imagine $b$ is attached at $C$. Side $b$ is just the right length to swing down and form a triangle in two positions along the horizontal base. In either of the two triangles we have

$$\frac{\sin \alpha}{8.3} = \frac{\sin 56.3°}{7.6}$$

$$\sin \alpha = \frac{8.3 \sin 56.3°}{7.6} \approx 0.9086.$$

This equation has two solutions in $[0°, 180°]$. For part (c) we get $\alpha = \sin^{-1}(0.9086) = 65.3°$ and for part (b) we get $\alpha = 180° - 65.3° = 114.7°$. Using the law of sines we get $\gamma = 9.0°$ and $c = 1.4$ for part (b), and we get $\gamma = 58.4°$ and $c = 7.8$ for part (c).

**TRY THIS.** Given $\beta = 38°$, $b = 4.7$, and $c = 5.9$, solve the triangle. ∎

The remaining cases for solving triangles are presented in the next section, which also contains a summary of how to proceed in each case. You may wish to refer to it now.

## Bearing

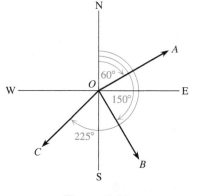

**Figure 5.12**

The measure of an angle that describes the direction of a ray is called the **bearing** of the ray. In air navigation, bearing is given as a nonnegative angle less than 360° measured in a clockwise direction from a ray pointing due north. So in Fig. 5.12, the bearing of ray $\overrightarrow{OA}$ is 60°, the bearing of ray $\overrightarrow{OB}$ is 150°, and the bearing of $\overrightarrow{OC}$ is 225°.

In marine navigation and surveying, the bearing of a ray is the acute angle the ray makes with a ray pointing due north or due south. Along with the acute angle, directions are given that indicate in which quadrant the ray lies. For example, in Fig. 5.12, $\overrightarrow{OA}$ has a bearing 60° east of north (N60°E), $\overrightarrow{OB}$ has a bearing 30° east of south (S30°E), and $\overrightarrow{OC}$ has a bearing 45° west of south (S45°W).

**EXAMPLE 5** Using bearing in solving triangles

A bush pilot left the Fairbanks Airport in a light plane and flew 100 miles toward Fort Yukon in still air on a course with a bearing of 18°. She then flew due east (bearing 90°) for some time to drop supplies to a snowbound family. After the drop, her course to return to Fairbanks had a bearing of 225°. Find her maximum distance from Fairbanks to the nearest tenth of a mile.

**Solution**

Figure 5.13 shows the course of her flight. To change course from bearing 18° to bearing 90° at point $B$, the pilot must add 72° to the bearing. So $\angle ABC$ is 108°. A bearing of 225° at point $C$ means that $\angle BCA$ is 45°. Finally, we obtain $\angle BAC = 27°$ by subtracting 18° and 45° from 90°.

**Figure 5.13**

We can find the length of $\overline{AC}$ (the maximum distance from Fairbanks) by using the law of sines:

$$\frac{b}{\sin 108°} = \frac{100}{\sin 45°}$$

$$b = \frac{100 \cdot \sin 108°}{\sin 45°}$$

$$b \approx 134.5$$

So the pilot's maximum distance from Fairbanks to the nearest tenth of a mile was 134.5 miles.

**TRY THIS.** A pilot flew north from the airport for 88 miles, then east until he headed back to the airport on a course with bearing 210°. What was his maximum distance from the airport?

## An Application Revisited

In the next example we repeat a problem that we solved in Section 1.5, Example 8. Note how much easier the solution is here, using the law of sines.

 **EXAMPLE 6** **Finding the height of an object from a distance**

The angle of elevation of the top of a water tower from point $A$ on the ground is 19.9°. From point $B$, 50.0 feet closer to the tower, the angle of elevation is 21.8°. Find the height of the tower to the nearest foot.

**Solution**

Let $y$ represent the height of the tower. All angles of triangle $ABC$ in Fig. 5.14 can be determined. Since $180° - 21.8° = 158.2°$, $\angle ABC$ is 158.2° and $\angle ACB$ is 1.9°.

**Figure 5.14**

Apply the law of sines in triangle $ABC$:

$$\frac{a}{\sin 19.9°} = \frac{50}{\sin 1.9°}$$

$$a = \frac{50 \sin 19.9°}{\sin 1.9°} \approx 513.3 \text{ ft}$$

Now using the smaller right triangle we have $\sin 21.8° = y/513.3$ or $y = 513.3 \sin 21.8° \approx 191$. So the height of the tower to the nearest foot is 191 feet.

**TRY THIS.** The angle of elevation of the top of a building from point $A$ on the ground is 24.2°. From point $B$, which is 44.5 feet closer, the angle of elevation is 38.1°. What is the height of the building?

Note that we may write $a \approx 513.3$ in Example 6, but for greater accuracy we use the calculator's value for $a$ to calculate $y$. To obtain 10-digit accuracy, a calculator actually knows more digits of a number than the 10 that it displays. So using a calculator's result from a previous computation will give more accuracy than re-entering the 10 digits for the result that appears on the display.

## FOR THOUGHT... True or False? Explain.

1. If the measures of two angles of a triangle are known, then the measure of the third angle is determined.

2. If you know the measures of all three angles of a triangle, then you can use the law of sines to find the lengths of the sides.

3. To solve a triangle, you must know the length of at least one side.

4. If $\frac{\sin 9°}{a} = \frac{\sin 17°}{88}$ then $a = \frac{\sin 17°}{88 \sin 9°}$.

5. The equation $\frac{\sin \alpha}{5} = \frac{\sin 44°}{18}$ has exactly one solution in $[0°, 180°]$.

6. One solution to $\frac{\sin \beta}{2.3} = \frac{\sin 39°}{1.6}$ is $\beta = \sin^{-1}\left(\frac{2.3 \sin 39°}{1.6}\right)$.

7. $\frac{\sin 60°}{\sqrt{3}} = \frac{\sin 30°}{1}$

8. No triangle exists with $\alpha = 60°$, $b = 10$ ft, and $a = 500$ ft.

9. There is exactly one triangle with $\beta = 30°$, $c = 20$, and $b = 10$.

10. There are two triangles with $\alpha = 135°$, $b = 17$, and $a = 19$.

Scan this code to link to section lecture videos.

## 5.1 EXERCISES

### CONCEPTS

*Fill in the blank.*

1. Any triangle without a right angle is a(n) _____ triangle.

2. To solve an oblique triangle, we must know at least _____ parts of the triangle, at least one of which must be the length of a side.

3. For the _____ case, two sides and a nonincluded angle of a triangle are known.

4. According to the _____, the ratio of the sine of an angle and the length of the side opposite the angle is the same for each angle of a triangle.

### SKILLS

*Solve each triangle. Round approximate answers to the nearest tenth.*

5.

6.

7.

8.
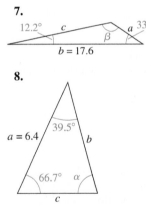

*Solve each triangle with the given parts.*

9. $\alpha = 10.3°$, $\gamma = 143.7°$, $c = 48.3$

10. $\beta = 94.7°$, $\alpha = 30.6°$, $b = 3.9$

11. $\beta = 120.7°$, $\gamma = 13.6°$, $a = 489.3$

12. $\alpha = 39.7°$, $\gamma = 91.6°$, $b = 16.4$

*Determine the number of triangles with the given parts and solve each triangle.*

13. $\alpha = 39.6°$, $c = 18.4$, $a = 3.7$

14. $\beta = 28.6°$, $a = 40.7$, $b = 52.5$

15. $\gamma = 60°$, $b = 20$, $c = 10\sqrt{3}$

16. $\alpha = 41.2°, a = 8.1, b = 10.6$

17. $\beta = 138.1°, c = 6.3, b = 15.6$

18. $\gamma = 128.6°, a = 9.6, c = 8.2$

19. $\beta = 32.7°, a = 37.5, b = 28.6$

20. $\alpha = 30°, c = 40, a = 20$

21. $\gamma = 99.6°, b = 10.3, c = 12.4$

22. $\alpha = 75.3°, a = 12.4, b = 9.8$

## MODELING

*Solve each problem.*

23. *Observing Traffic* A traffic report helicopter left the WKPR studios on a course with a bearing of 210°. After flying 12 mi to reach Interstate Highway 20, the helicopter flew due east along I-20 for some time. The helicopter headed back to WKPR on a course with a bearing of 310° and reported no accidents along I-20. For how many miles did the helicopter fly along I-20? Round to the nearest tenth of a mile.

24. *Course of a Fighter Plane* During an important NATO exercise, an F-14 Tomcat left the carrier *Nimitz* on a course with a bearing of 34° and flew 400 mi. Then the F-14 flew for some distance on a course with a bearing of 162°. Finally, the plane flew back to its starting point on a course with a bearing of 308°. What distance did the plane fly on the final leg of the journey? Round to the nearest tenth of a mile.

25. *Surveying Property* A surveyor locating the corners of a triangular piece of property started at one corner and walked 480 ft in the direction N36°W to reach the next corner. The surveyor turned and walked S21°W to get to the next corner of the property. Finally, the surveyor walked in the direction N82°E to get back to the starting point. What is the perimeter of this piece of property to the nearest foot?

26. *Sailing* Joe and Jill set sail from the same point, with Joe sailing in the direction S4°E and Jill sailing in the direction S9°W. After 4 hr, Jill was 2 mi due west of Joe. How far had Jill sailed to the nearest tenth of a mile?

27. *Wing Dimensions* The F-106 Delta Dart once held a world speed record of Mach 2.3. Its sweptback triangular wings have the dimensions given in the figure. Find the length marked $x$ in the figure to the nearest tenth of a foot.

**Figure for Exercise 27**

28. *Filtering the Sun* When the sun is directly overhead, the sun's light is filtered by approximately 10 mi of atmosphere. As the sun sets, the intensity decreases because the light must pass a greater distance through the atmosphere as shown in the figure.
   a. Find the distance $d$ in the figure when the angle of elevation of the sun is 30°. Use 3950 mi as the radius of the earth. Round to the nearest tenth of a mile.

   b. If the sun is directly overhead at 12 noon, then at what time (to the nearest tenth of a second) is the angle of elevation of the sun 30°? The distance from the earth to the sun is 93 million mi.

   c. Find the distance that the sunlight passes through the atmosphere at sunset to the nearest mile.

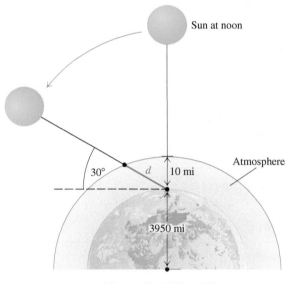

**Figure for Exercise 28**

29. *Cellular One* The angle of elevation of the top of a cellular telephone tower from point $A$ on the ground is 18.1°. From point $B$, 32.5 ft closer to the tower, the angle of elevation is 19.3°. What is the height of the tower to the nearest tenth of a foot?

30. *Moving Back* A surveyor determines that the angle of elevation of the top of a building from a point on the ground is 30.4°. He then moves back 55.4 ft and determines that the angle of elevation is 23.2°. What is the height of the building to the nearest tenth of a foot?

31. *Designing an Addition* A 40-ft-wide house has a roof with a 6-12 pitch (the roof rises 6 ft for a run of 12 ft). The owner plans a 14-ft-wide addition that will have a 3-12 pitch to its roof. See the figure. Find the lengths of $\overline{AB}$ and $\overline{BC}$ to the nearest tenth of a foot.

**Figure for Exercise 31**

**32.** *Saving an Endangered Hawk* A hill has an angle of inclination of 36° as shown in the figure. Highway engineers want to remove 400 ft of the hill, measured horizontally. The engineers plan to leave a slope alongside the highway with an angle of inclination of 62°, as shown in the figure. Located 750 ft up the hill measured from the base is a tree containing the nest of an endangered hawk. Will this tree be removed in the excavation?

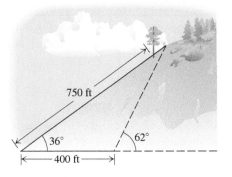

**Figure for Exercise 32**

**33.** *Making a Kite* A kite is made in the shape shown in the figure. Find length marked *x* in the figure to the nearest tenth of an inch.

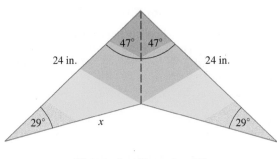

**Figure for Exercise 33**

**34.** *Flat Earth* Assume that the earth is flat with a 10-mi-thick layer of atmosphere. Find the distance that the sunlight passes through the atmosphere when the angle of elevation of the sun is 30°. Compare your answer to that obtained in part (a) of Exercise 28.

**35.** *Shot Down* A cruise missile is traveling straight across the desert at 548 mph at an altitude of 1 mi as shown in the figure. A gunner spots the missile coming in his direction and fires a projectile at the missile when the angle of elevation of the missile is 35°. If the speed of the projectile is 688 mph, then for what angle of elevation of the gun will the projectile hit the missile? Round to the nearest tenth of a degree.

**Figure for Exercise 35**

**36.** *Angle of Completion* When the ball is snapped, Jones starts running at a 50° angle to the line of scrimmage. At the moment when Jones is at a 60° angle from Smith, Jones is running at 17 ft/sec and Smith passes the ball at 60 ft/sec to Jones. However, to complete the pass, Smith must lead Jones by the angle θ shown in the figure. Find θ to the nearest tenth of a degree. (Of course, Smith found θ in his head. Note that θ can be found without knowing any distances.)

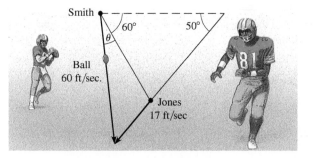

**Figure for Exercise 36**

**37.** *Rabbit and Fox* A rabbit starts running from point *A* in a straight line in the direction 30° from the north at 3.5 ft/sec, as shown in the accompanying diagram. At the same time, a fox starts running in a straight line from a position 30 ft to the west of the rabbit at 6.5 ft/sec. The fox chooses his path so that he will catch the rabbit at point *C*. In how many seconds will the fox catch the rabbit?

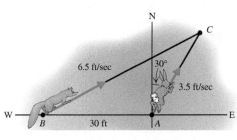

**Figure for Exercise 37**

**38.** *Head Start* How many seconds will it take for the fox in Exercise 37 to catch the rabbit if the rabbit has a 1-second head start?

## WRITING/DISCUSSION

**39.** *Law of Sines* Prove the law of sines for the case in which the triangle is an acute triangle.

**40.** *Angle Bisector Theorem* Suppose $\overline{AD}$ bisects $\angle A$ as shown in the figure on the next page. Use the law of sines to show that $AB/AC = BD/DC$. In words, the angle bisector of any angle of a triangle divides the opposite side into two segments whose lengths have the same ratio as the lengths of the adjacent sides of the angle.

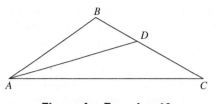

**Figure for Exercise 40**

### REVIEW

**41.** Find the exact value of each expression without using a calculator or table.

**a.** $\sin(5\pi/2)$  **b.** $\cos^{-1}\left(-\sqrt{2}/2\right)$

**c.** $\tan(5\pi/3)$  **d.** $\csc(-\pi/3)$

**e.** $\sec(-3\pi/4)$  **f.** $\sin^{-1}(-1/2)$

**42.** Given that $\alpha = 12°$, $\beta = 90°$, and $a = 3.2$ feet, solve the triangle.

**43.** Find the period of each function.

**a.** $y = 2\sin(\pi x)$  **b.** $y = -\cos(3x)$

**c.** $y = 3\tan(2\pi x)$  **d.** $y = 4\csc(2x)$

**44.** Find the values of $\sin \alpha$ and $\cos \alpha$ given that $\tan \alpha = 7/8$ and $0 < \alpha < \pi/2$.

**45.** Determine whether each function is even or odd.

**a.** $y = \sin(2x)$  **b.** $y = \cos(2x)$

**c.** $y = x\tan(x)$  **d.** $y = x^3\csc(x)$

**46.** Determine whether the equation $\cos(2x) = 2\cos(x)$ is an identity.

### OUTSIDE THE BOX

**47.** *Solving a Triangle* In the accompanying figure $AB = BC$, $\overline{AD}$ bisects $\angle BAC$, $\angle C = 72°$, and $BD = 2$. Find the exact length of $\overline{DC}$.

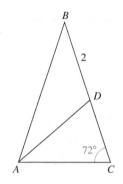

**Figure for Exercise 47**

**48.** *Costly Computers* A school district purchased $x$ computers at $y$ dollars each for a total of $640,000. Both $x$ and $y$ are whole numbers and neither is a multiple of 10. Find the absolute value of the difference between $x$ and $y$.

## 5.1 POP QUIZ

**1.** If $\alpha = 8°$, $\beta = 121°$, and $c = 12$, then what is $\gamma$?

**2.** If $\alpha = 20.4°$, $\beta = 27.3°$, and $c = 38.5$, then what is $a$ to the nearest tenth?

**3.** If $\alpha = 33.5°$, $a = 7.4$, and $b = 10.6$, then what is $\beta$ to the nearest tenth of a degree?

**4.** From one point on the ground, the angle of elevation of the top of a tree is 20°. From a point on the ground that is 50 ft closer to the tree, the angle of elevation is 25°. What is the height of the tree to the nearest foot?

## 5.2 The Law of Cosines

In Section 5.1 we discussed the AAS, ASA, and SSA cases for solving triangles using the law of sines. For the SAS and SSS cases, which cannot be handled with the law of sines, we develop the law of cosines. This law is a generalization of the Pythagorean theorem and applies to any triangle. However, it is usually stated and used only for triangles with no right angle (oblique triangles) because it is not needed for right triangles.

### The Law of Cosines

The **law of cosines** gives a formula for the square of any side of an oblique triangle in terms of the other two sides and their included angle. The law of cosines is stated symbolically for all three sides of a triangle as follows.

**Theorem: Law of Cosines**

If triangle $ABC$ is an oblique triangle with sides $a$, $b$, and $c$ and angles $\alpha$, $\beta$, and $\gamma$, then

$$a^2 = b^2 + c^2 - 2bc \cos \alpha$$

$$b^2 = a^2 + c^2 - 2ac \cos \beta$$

$$c^2 = a^2 + b^2 - 2ab \cos \gamma.$$

**PROOF**  Given triangle $ABC$, position the triangle as shown in Fig. 5.15.

**Figure 5.15**

The vertex $C$ is in the first quadrant if $\alpha$ is acute and in the second if $\alpha$ is obtuse. Both cases are shown in Fig. 5.15. In either case, the $x$-coordinate of $C$ is $b \cos \alpha$, and the $y$-coordinate of $C$ is $b \sin \alpha$. The distance from $C$ to $B$ is $a$, but we can also find that distance by using the distance formula:

$$a = \sqrt{(x - c)^2 + (y - 0)^2}$$

$$a = \sqrt{(b \cos \alpha - c)^2 + (b \sin \alpha - 0)^2} \qquad \text{Because } x = b \cos \alpha \text{ and } y = b \sin \alpha$$

$$a^2 = (b \cos \alpha - c)^2 + (b \sin \alpha)^2 \qquad \text{Square both sides.}$$

$$a^2 = b^2 \cos^2 \alpha - 2bc \cos \alpha + c^2 + b^2 \sin^2 \alpha \qquad \text{Simplify.}$$

$$a^2 = b^2(\cos^2 \alpha + \sin^2 \alpha) + c^2 - 2bc \cos \alpha \qquad \text{Factor.}$$

Using $\cos^2 x + \sin^2 x = 1$, we get the first equation of the theorem:

$$a^2 = b^2 + c^2 - 2bc \cos \alpha$$

Similar arguments with $B$ and $C$ at $(0, 0)$ produce the other two equations.  ∎

In any triangle with unequal sides, the largest angle is opposite the largest side and the smallest angle is opposite the smallest side. We need this fact in the first example.

**EXAMPLE 1**  **Given three sides of a triangle (SSS)**

Given $a = 8.2$, $b = 3.7$, and $c = 10.8$, solve the triangle. Find the angles to the nearest tenth of a degree.

**Solution**

Draw the triangle and label it as in Fig. 5.16. Since $c$ is the longest side, use

$$c^2 = a^2 + b^2 - 2ab \cos \gamma$$

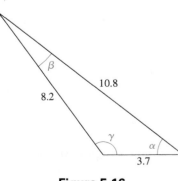

**Figure 5.16**

to find the largest angle $\gamma$:

$$-2ab \cos \gamma = c^2 - a^2 - b^2$$

$$\cos \gamma = \frac{c^2 - a^2 - b^2}{-2ab}$$

$$= \frac{(10.8)^2 - (8.2)^2 - (3.7)^2}{-2(8.2)(3.7)} \approx -0.5885$$

$$\gamma = \cos^{-1}(-0.5885) \approx 126.1°$$

We could finish with the law of cosines, but it is simpler to use the law of sines:

$$\frac{\sin \beta}{b} = \frac{\sin \gamma}{c}$$

$$\frac{\sin \beta}{3.7} = \frac{\sin 126.1°}{10.8}$$

$$\sin \beta \approx 0.2768$$

There are two solutions to $\sin \beta = 0.2768$ in $[0°, 180°]$. However, $\beta$ must be less than 90°, because $\gamma$ is 126.1°. So

$$\beta = \sin^{-1}(0.2768) \approx 16.1°.$$

Finally, $\alpha = 180° - 126.1° - 16.1° = 37.8°$. The three angles are $\alpha \approx 37.8°$, $\beta \approx 16.1°$, and $\gamma \approx 126.1°$.

**TRY THIS.** Given $a = 3.8$, $b = 9.6$, and $c = 7.7$, solve the triangle. ■

In solving the SSS case, we always *find the largest angle first* using the law of cosines. The remaining two angles must be acute angles. So when using the law of sines to find another angle, we need only find an acute solution to the equation.

In Section 5.1 we learned that two adjacent sides and a nonincluded angle (SSA) might not determine a triangle. To determine a triangle with three given sides (SSS), the sum of the lengths of any two must be greater than the length of the third side. This fact is called the **triangle inequality.** To understand the triangle inequality, try to draw a triangle with sides of lengths 1 inch, 2 inches, and 5 inches. Two sides and the included angle (SAS) will determine a triangle provided the angle is between 0° and 180°.

**EXAMPLE** **2**  **Given two sides and an included angle (SAS)**

The wing of the F-106 Delta Dart is triangular in shape with the dimensions given in Fig. 5.17. Find the length of the side labeled $c$ in Fig. 5.17.

**Solution**

Using the law of cosines, we find $c$ as follows:

$$c^2 = (19.2)^2 + (37.6)^2 - 2(19.2)(37.6) \cos 68° \approx 1241.5$$

$$c \approx \sqrt{1241.5} \approx 35.2 \text{ ft}$$

**TRY THIS.** Given $b = 5.8$, $c = 3.6$, and $\alpha = 39.5°$, find $a$. ■

The following procedure should help you in solving triangles. Any problem in which you are given enough information to determine a triangle (or possibly two) fits into one of the four listed categories. Do not attempt to simply memorize this procedure. Use it as a guide, always draw a diagram, and you will be successful at solving triangles.

37.6 ft

$c$

68°

19.2 ft

**Figure 5.17**

**PROCEDURE**

## Solving Triangles (In all cases, draw pictures.)

**ASA**    (For example $\alpha$, $c$, $\beta$)

1. Find $\gamma$ using $\gamma = 180° - \alpha - \beta$.

2. Find $a$ and $c$ using the law of sines.

**SSA**    (For example $a$, $b$, $\alpha$)

1. If $a \geq b$ there is only one triangle and $\beta$ is acute ($\alpha$ or $\gamma$ might be obtuse). Find $\beta$ using the law of sines. Then find $\gamma$ using $\gamma = 180° - \alpha - \beta$. Find $c$ using the law of sines.

2. If $a < b$, then find $h$ using $h = b \sin \alpha$.

   **a.** If $h > a$, then there is no triangle.

   **b.** If $h = a$, then there is one right triangle ($\beta = 90°$ and $b$ is the hypotenuse). Solve it using right triangle trigonometry.

   **c.** If $h < a$ there are two triangles, one with $\beta$ acute and one with $\beta$ obtuse. Find the acute $\beta$ using the law of sines. Subtract it from $180°$ to get the obtuse $\beta$. In each of the two triangles, find $\gamma$ using $\gamma = 180° - \alpha - \beta$ and $c$ using the law of sines.

**SSS**    (For example $a$, $b$, $c$)

1. Find the largest angle using the law of cosines. The largest angle is opposite the largest side.

2. Find another angle using the law of sines.

3. Find the third angle by subtracting the first two from $180°$.

**SAS**    (For example $b$, $\alpha$, $c$)

1. Find $a$ using the law of cosines.

2. Use the law of sines to find $\beta$ if $b < c$ or $\gamma$ if $c < b$. If $b = c$, then find either $\beta$ or $\gamma$.

3. Find the last angle by subtracting the first two from $180°$.

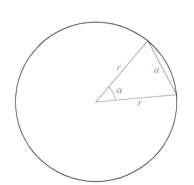

**Figure 5.18**

## Length of a Chord

A central angle $\alpha$ in a circle of radius $r$ intercepts a chord of length $a$ as shown in Fig. 5.18. Apply the law of cosines and simplify as follows.

$$a^2 = r^2 + r^2 - 2 \cdot r \cdot r \cdot \cos \alpha$$

$$a^2 = 2r^2 - 2r^2 \cos \alpha$$

$$a^2 = r^2(2 - 2 \cos \alpha)$$

Taking the square root of each side yields a formula for the length of the chord in terms of $\alpha$ and $r$. $\left(\text{Since } a > 0 \text{ and } r > 0, \sqrt{a^2} = a \text{ and } \sqrt{r^2} = r.\right)$

**Length of a Chord**

If a chord of length $a$ subtends a central angle $\alpha$ in a circle of radius $r$, then

$$a = r\sqrt{2 - 2 \cos \alpha}.$$

**EXAMPLE 3**    Finding the length of a side of a pentagon

A regular pentagon is inscribed in a circle of radius 6 in. What is the length of the side of the pentagon to the nearest tenth of an inch?

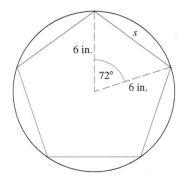

**Figure 5.19**

### Solution

The side of the pentagon $s$ is a chord in a circle of radius 6 in. that subtends a central angle of $72°$ ($360° \div 5$) as shown in Fig. 5.19. So

$$s = r\sqrt{2 - 2\cos\alpha} = 6\sqrt{2 - 2\cos 72°} \approx 7.1 \text{ in.}$$

**TRY THIS.** Find the length of the chord intercepted by a central angle of $33.8°$ in a circle of radius 22.4 feet.    ■

## Modeling Arm Motion

When you reach for a light switch, your brain controls the angles at your elbow and your shoulder that put your hand at the location of the switch. An arm on a robot works in much the same manner. In the next example we see how the law of sines and the law of cosines are used to determine the proper angles at the joints so that a robot's hand can be moved to a position given in the coordinates of the workspace.

**EXAMPLE 4**    **Positioning a robotic arm**

A robotic arm with a 0.5-meter segment and a 0.3-meter segment is attached at the origin as shown in Fig. 5.20.

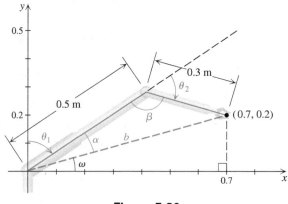

**Figure 5.20**

The computer-controlled arm is positioned by rotating each segment through angles $\theta_1$ and $\theta_2$ as shown in Fig. 5.20. Given that we want the end of the arm at the point $(0.7, 0.2)$, find $\theta_1$ and $\theta_2$ to the nearest tenth of a degree.

### Solution

In the right triangle shown in Fig. 5.20, we have $\tan\omega = 0.2/0.7$ and

$$\omega = \tan^{-1}\left(\frac{0.2}{0.7}\right) \approx 15.95°.$$

To get the angles to the nearest tenth, we use more accuracy along the way and round to the nearest tenth only on the final answer. Find $b$ using the Pythagorean theorem:

$$b = \sqrt{0.7^2 + 0.2^2} \approx 0.73$$

Find $\beta$ using the law of cosines:

$$(0.73)^2 = 0.5^2 + 0.3^2 - 2(0.5)(0.3)\cos\beta$$

$$\cos\beta = \frac{0.73^2 - 0.5^2 - 0.3^2}{-2(0.5)(0.3)} \approx -0.643$$

$$\beta = \cos^{-1}(-0.643) \approx 130.02°$$

Find $\alpha$ using the law of sines:

$$\frac{\sin \alpha}{0.3} = \frac{\sin 130.02°}{0.73}$$

$$\sin \alpha = \frac{0.3 \cdot \sin 130.02°}{0.73} \approx 0.3147$$

$$\alpha = \sin^{-1}(0.3147) \approx 18.34°$$

Since $\theta_1 = 90° - \alpha - \omega$,

$$\theta_1 = 90° - 18.34° - 15.95° = 55.71°.$$

Since $\beta$ and $\theta_2$ are supplementary,

$$\theta_2 = 180° - 130.02° = 49.98°.$$

So the longer segment of the arm is rotated approximately 55.7° and the shorter segment is rotated approximately 50.0°.

**TRY THIS.** Solve the triangle that has vertices $A\,(0, 0)$, $B\,(1, 5)$, and $C\,(8, 3)$. ∎

Since the angles in Example 4 describe a clockwise rotation, the angles could be given negative signs to indicate the direction of rotation. In robotics, the direction of rotation is important, because there may be more than one way to position a robotic arm at a desired location. In fact, Example 4 has infinitely many solutions. See if you can find another one.

## FOR THOUGHT... True or False? Explain.

1. If $\gamma = 90°$ in triangle $ABC$, then $c^2 = a^2 + b^2$.

2. If $a, b,$ and $c$ are the sides of a triangle, then $a = \sqrt{c^2 + b^2 - 2bc \cos \gamma}$.

3. If $a, b,$ and $c$ are the sides of any triangle, then $c^2 = a^2 + b^2$.

4. The smallest angle of a triangle lies opposite the shortest side.

5. The equation $\cos \alpha = -0.3421$ has two solutions in $[0°, 180°]$.

6. If the largest angle of a triangle is obtuse, then the other two are acute.

7. The equation $\sin \beta = 0.1235$ has two solutions in $[0°, 180°]$.

8. In the SSS case of solving a triangle it is best to find the largest angle first.

9. There is no triangle with sides $a = 3.4$, $b = 4.2$, and $c = 8.1$.

10. There is no triangle with $\alpha = 179.9°$, $b = 1$ inch, and $c = 1$ mile.

Scan this code to link to section lecture videos.

**5.2** EXERCISES

### CONCEPTS

*Fill in the blank.*

1. The _____ gives a formula for the square of any side of an oblique triangle in terms of the other two sides and their included angle.

2. According to the _____, the sum of the lengths of any two sides of a triangle is greater than the length of the third side.

3. To find an angle of a triangle when three sides are known, we use the law of _____.

4. The largest angle of a triangle is opposite the _____ side.

## SKILLS

*Solve each triangle. Round to the nearest tenth.*

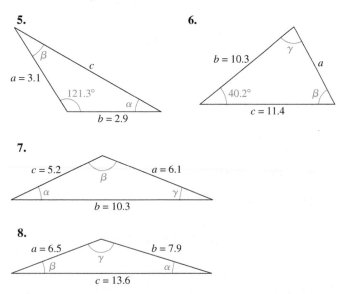

**5.**

**6.**

**7.**

**8.**

*Solve each triangle with the given information. Round to the nearest tenth.*

**9.** $a = 6.8, c = 2.4, \beta = 10.5°$

**10.** $a = 1.3, b = 14.9, \gamma = 9.8°$

**11.** $a = 18.5, b = 12.2, c = 8.1$

**12.** $a = 30.4, b = 28.9, c = 31.6$

**13.** $b = 9.3, c = 12.2, \alpha = 30°$

**14.** $a = 10.3, c = 8.4, \beta = 88°$

**15.** $a = 6.3, b = 7.1, c = 6.8$

**16.** $a = 4.1, b = 9.8, c = 6.2$

**17.** $a = 7.2, \beta = 25°, \gamma = 35°$

**18.** $b = 12.3, \alpha = 20°, \gamma = 120°$

*Determine the number of triangles with the given parts.*

**19.** $a = 3, b = 4, c = 7$

**20.** $a = 2, b = 9, c = 5$

**21.** $a = 10, b = 5, c = 8$

**22.** $a = 3, b = 15, c = 16$

**23.** $c = 10, \alpha = 40°, \beta = 60°, \gamma = 90°$

**24.** $b = 6, \alpha = 62°, \gamma = 120°$

**25.** $b = 10, c = 1, \alpha = 179°$

**26.** $a = 10, c = 4, \beta = 2°$

**27.** $b = 8, c = 2, \gamma = 45°$

**28.** $a = \sqrt{3}/2, b = 1, \alpha = 60°$

## MODELING

*Solve each problem.*

**29.** *Length of a Chord* What is the length of the chord intercepted by a central angle of 19° in a circle of radius 30 ft? Round to the nearest hundredth of a foot.

**30.** *Boating* The boat shown in the accompanying figure is 3 mi from both lighthouses and the angle between the line of sight to the lighthouses is 20°. Find the distance between the lighthouses to the nearest hundredth of a mile.
  **HINT** Use the formula for the length of a chord.

**Figure for Exercise 30**

**31.** *The Pentagon* The Pentagon in Washington D.C., as shown in the accompanying figure, is 921 ft on each side. What is the distance from a vertex to the center of the Pentagon to the nearest hundredth of a foot?

**Figure for Exercise 31**

**32.** *A Hexagon* If the length of each side of a regular hexagon is 10 ft, then what is the distance from a vertex to the center?

**33.** *Hiking* Jan and Dean started hiking from the same location at the same time. Jan hiked at 4 mph with bearing N12°E, and Dean hiked at 5 mph with bearing N31°W. How far apart were they after 6 hr? Round to the nearest tenth of a mile.

**34.** *Flying* Andrea and Carlos left the airport at the same time. Andrea flew at 180 mph on a course with bearing 80°, and Carlos flew at 240 mph on a course with bearing 210°. How far apart were they after 3 hr? Round to the nearest tenth of a mile.

**35.** *Positioning a Solar Panel* A solar panel with a width of 1.2 m is positioned on a flat roof, as shown in the figure. What is the angle of elevation $\alpha$ of the solar panel? Round to the nearest tenth of a degree.

**Figure for Exercise 35**

**36.** *Installing an Antenna* A 6-ft antenna is installed at the top of a roof as shown in the figure. A guy wire is to be attached to the top of the antenna and to a point 10 ft down the roof. If the angle of elevation of the roof is 28°, then what length guy wire is needed? Round to the nearest tenth of a foot.

**Figure for Exercise 36**

**37.** *Adjacent Pipes* An engineer wants to position three pipes at the vertices of a triangle as shown in the figure. If the pipes have radii 2 in., 3 in., and 4 in., then what are the measures of the angles of triangle *ABC* to the nearest tenth of a degree?

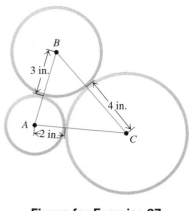

**Figure for Exercise 37**

**38.** *Firing a Torpedo* A submarine sights a moving target at a distance of 820 m. A torpedo is fired 9° ahead of the target as shown in the drawing and travels 924 m in a straight line to hit the target. How far has the target moved from the time the torpedo is fired to the time of the hit? Round to the nearest tenth of a meter.

**Figure for Exercise 38**

**39.** *Planning a Tunnel* A tunnel is planned through a mountain to connect points *A* and *B* on two existing roads as shown in the figure. If the angle between the roads at point *C* is 28°, what is the distance from point *A* to *B* to the nearest tenth of a mile? Find ∠*CBA* and ∠*CAB* to the nearest tenth of a degree.

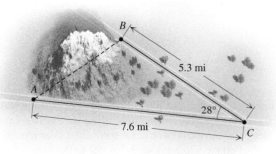

**Figure for Exercise 39**

**40.** *Scattering Angle* On June 30, 1861, Comet Tebbutt, one of the greatest comets, was visible even before sunset (*Sky & Telescope,* April 1997). One of the factors that cause a comet to be extra bright is a small *scattering angle* $\theta$ shown in the accompanying figure. When Comet Tebbutt was at its brightest, it was 0.133 a.u. from the earth and 0.894 a.u. from the sun. The earth was 1.017 a.u. from the sun. Find the *phase angle* $\alpha$ and the scattering angle $\theta$ for Comet Tebbutt on June 30, 1861. (One astronomical unit (a.u.) is the average distance between the earth and the sun.) Round to the nearest whole degree.

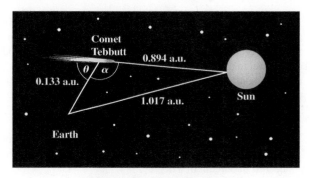

**Figure for Exercise 40**

**41.** *Side of a Pentagon* A regular pentagon is inscribed in a circle of radius 10 m. Find the length of a side of the pentagon to the nearest hundredth of a meter.

**42.** *Central Angle* A central angle $\alpha$ in a circle of radius 5 m intercepts a chord of length 1 m. What is the measure of $\alpha$ to the nearest tenth of a degree?

**43.** *Positioning a Human Arm* A human arm consists of an upper arm of 30 cm and a lower arm of 30 cm as shown in the figure. To move the hand to the point $(36, 8)$ the human brain chooses angle $\theta_1$ and $\theta_2$ as shown in the figure. Find $\theta_1$ and $\theta_2$ to the nearest tenth of a degree.

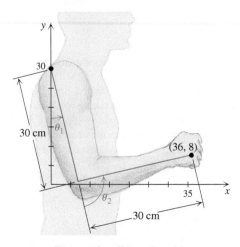

**Figure for Exercise 43**

**44.** *Making a Shaft* The end of a steel shaft for an electric motor is to be machined so that it has three flat sides of equal widths as shown in the figure. If the radius of the shaft is 10 mm and the length of the arc between each pair of flat sides must be 2.5 mm, find the width $s$ of each flat side to the nearest hundredth of a millimeter.

**Figure for Exercise 44**

**45.** *Total Eclipse* Astronomers refer to the angles alpha and beta ($\alpha$ and $\beta$) in the figure as the "diameters" of the sun and the moon, respectively. These angles/diameters vary as the sun, moon, and earth travel in their orbits. When the moon moves between the earth and sun, a total eclipse occurs provided the moon's apparent diameter is larger than that of the sun, or if beta > alpha. The actual diameters of the sun and moon are 865,000 mi and 2163 mi, respectively, and these distances do not vary.
  **a.** The distance from the earth to the sun varies from 91,400,000 mi to 94,500,000 mi. Find the minimum and maximum $\alpha$ to the nearest hundredth of a degree.
  HINT Use the law of cosines.

**b.** The distance from the earth to the moon varies from 225,800 mi to 252,000 mi. Find the minimum and maximum $\beta$ to the nearest hundredth of a degree.
  HINT Use the law of cosines.

**c.** Is it possible for the earth, the moon, and the sun to be in perfect alignment without a total eclipse occurring?

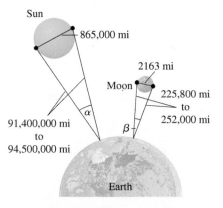

**Figure of Exercise 45**

**46.** *Total Eclipse on Jupiter* The distance from Jupiter to the sun varies from $7.406 \times 10^8$ km to $8.160 \times 10^8$ km. The diameter of the sun is $1.39 \times 10^6$ km.
  **a.** Determine the minimum and maximum diameter of the sun to the nearest hundredth of a degree as seen from Jupiter.
  HINT See the previous exercise and use the law of cosines.

  **b.** The distance from Jupiter to its moon Callisto is $1.884 \times 10^6$ km and Callisto's diameter is 2420 km. Determine the diameter of Callisto as seen from Jupiter to the nearest hundredth of a degree.
  HINT Use the law of cosines.

  **c.** Is it possible for Callisto to produce a total eclipse of the sun on Jupiter?

**47.** *Attack of the Grizzly* A forest ranger is 150 ft above the ground in a fire tower when she spots an angry grizzly bear east of the tower with an angle of depression of 10° as shown in the figure. Southeast of the tower she spots a hiker with an angle of depression of 15°. Find the distance between the hiker and the angry bear to the nearest foot.

**Figure for Exercise 47**

**48.** *Smuggler's Blues*  A smuggler's boat sets out at midnight at 20 mph on a course that makes a 40° angle with the shore as shown in the figure. One hour later a DEA boat sets out also at 20 mph from a dock 80 mi up the coast to intercept the smuggler's boat. Find the angle $\theta$ (in the figure) to the nearest tenth of a degree for which the DEA boat will intercept the smuggler's boat. At what time to the nearest minute will the interception occur?

**Figure for Exercise 48**

### WRITING/DISCUSSION

**49.** A central angle $\theta$ in a circle of radius $r$ intercepts a chord of length $a$, where $0° \leq \theta \leq 180°$. Show that $a = 2r \sin(\theta/2)$.

**50.** Explain why the second largest side in a triangle with unequal sides is opposite an acute angle.

### REVIEW

**51.** Solve the triangle with $\alpha = 108.1°$, $\beta = 18.6°$, and $c = 28.6$.

**52.** Determine the number of triangles with $\alpha = 22.5°$, $a = 5.1$, and $b = 12.6$.

**53.** The five key points on one cycle of a sine wave are $(\pi/4, 3)$, $(\pi/2, 5)$, $(3\pi/4, 3)$, $(\pi, 1)$, and $(5\pi/4, 3)$. Find the equation of the wave in the form $y = A \sin(B(x - C)) + D$.

**54.** The area of a sector of a circle with central angle $\pi/16$ is $64\pi$ in.². Find the exact radius of the circle.

**55.** Convert each degree measure to radian measure.
   **a.** 270°                **b.** 315°

   **c.** −210°            **d.** 120°

**56.** A propeller with a diameter of 6 feet is rotating at 3200 rev/min. What is the velocity in miles per hour for a point on the tip of the propeller?

### OUTSIDE THE BOX

**57.** *Moving a Refrigerator*  A box containing a refrigerator is 3 ft wide, 3 ft deep, and 6 ft high. To move it, Wally lays it on its side, then on its top, then on its other side, and finally stands it upright as shown in the figure. Exactly how far has point $A$ traveled in going from its initial location to its final location?

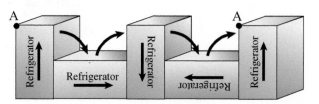

**Figure for Exercise 57**

**58.** *Arc Length*  In a circle of radius 6 the length of the chord with endpoints $A$ and $B$ is $6\sqrt{3}$. What is the length of the shorter arc with endpoints $A$ and $B$?

## 5.2 POP QUIZ

**1.** If $\alpha = 12.3°$, $b = 10.4$, and $c = 8.1$, then what is $a$?

**2.** If $a = 6$, $b = 7$, and $c = 12$, then what is $\gamma$?

**3.** How many triangles are possible with $a = 6$, $b = 12$, and $c = 5$?

**4.** What is the length of the chord (to the nearest tenth of a foot) intercepted by a central angle of 42.1° in a circle of radius 8.7 ft?

## LINKING concepts...

**For Individual or Group Explorations**

### Modeling the Best View

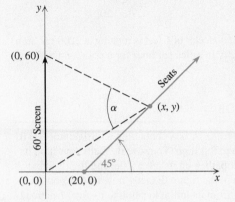

*A large-screen theater has a screen that is 60 feet high as shown in the figure. The seats are placed on a 45° incline so that the seats are close to the screen.*

**a)** Find the exact distance to the top of the screen and to the bottom of the screen for a person sitting at coordinates (30, 10).

**b)** Find the viewing angle $\alpha$ for a person sitting at (30, 10) to the nearest tenth of a degree.

**c)** Write the viewing angle $\alpha$ as a function of $x$ and graph the function using a graphing calculator.

**d)** If the good seats are those for which the viewing angle is greater than 60°, then what are the $x$-coordinates of the good seats to the nearest tenth?

**e)** If the best seat has the largest viewing angle, then what are the coordinates of the best seat?

## **5.3** Area of a Triangle

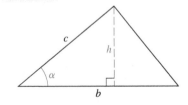

**Figure 5.21**

The well-known formula $A = \frac{1}{2}bh$ gives the area of a triangle in terms of a side and the altitude to that side. In this section we will see two new formulas for the area of a triangle.

### Area of a Triangle

Consider the triangle shown in Fig. 5.21, in which $\alpha$ is an acute angle and $h$ is inside the triangle. Since $\sin \alpha = h/c$, we get $h = c \sin \alpha$. Substitute $c \sin \alpha$ for $h$ in the formula, to get

$$A = \frac{1}{2}bc \sin \alpha.$$

This formula is also correct in all other possible cases: $\alpha$ is acute and $h$ is the side of the triangle opposite $\alpha$, $\alpha$ is acute and $h$ lies outside the triangle, $\alpha$ is a right angle, and $\alpha$ is obtuse. (See Exercise 38.) Likewise, the formula can be written using the angles $\beta$ or $\gamma$. We have the following theorem.

**Theorem: Area of a Triangle**

The area of a triangle is given by

$$A = \frac{1}{2}bc \sin \alpha, \qquad A = \frac{1}{2}ac \sin \beta, \qquad \text{and} \qquad A = \frac{1}{2}ab \sin \gamma.$$

So the area of a triangle is one-half the product of any two sides and the sine of the angle between the sides.

Note that these new formulas for the area of triangle can be used to prove the law of sines. Since the area of a triangle is the same no matter which of the three formulas is used, we can write

$$\frac{1}{2}bc \sin \alpha = \frac{1}{2}ac \sin \beta = \frac{1}{2}ab \sin \gamma.$$

Multiplying by 2 and then dividing by $abc$ yields the law of sines,

$$\frac{\sin \alpha}{a} = \frac{\sin \beta}{b} = \frac{\sin \gamma}{c}.$$

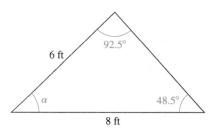

**Figure 5.22**

**EXAMPLE 1**  Area of a triangle

Find the area of the triangle shown in Fig. 5.22.

**Solution**

Since the angles of any triangle have a sum of 180°, we find that $\alpha = 39°$. Using the new area formula we get

$$A = \frac{1}{2} \cdot 8 \cdot 6 \cdot \sin(39°) \approx 15.1.$$

So the area is approximately 15.1 ft$^2$.

**TRY THIS.** Given $a = 5$ meters, $b = 11$ meters, and $\gamma = 37°$, find the area of the triangle. ∎

Surveyors describe property in terms of the lengths of the sides and the angles between those sides. In the next example we find the area of a four-sided piece of property by dividing it into triangles.

**Figure 5.23**

**EXAMPLE 2**  Area of a quadrilateral

The town of Hammond is considering the purchase of a four-sided piece of property to be used for a playground. The dimensions of the property are given in Fig. 5.23. Find the area of the property in square feet.

**Solution**

Divide the property into two triangles as shown in Fig. 5.23, and find the area of each:

$$A_1 = \frac{1}{2}(148.7)(93.5) \sin 91.5° \approx 6949.3 \text{ ft}^2$$

$$A_2 = \frac{1}{2}(100.5)(155.4) \sin 87.2° \approx 7799.5 \text{ ft}^2$$

Adding the areas of the triangles gives a total area of approximately 14,748.8 square feet. Since the property is almost a rectangle, an estimate of the area could be obtained by multiplying one of the lengths by one of the widths. However, such an estimate would be about 900 square feet in error, which is a rather large error if the property is priced per square foot.

**TRY THIS.** Two sides of a triangular piece of property are 244 feet and 206 feet, and the angle between these sides is 87.4°. Find the area to the nearest square foot. ∎

## Area of a Triangle by Heron's Formula

We now have the formulas $A = \frac{1}{2}bh$ and $A = \frac{1}{2}bc \sin \alpha$ for the area of a triangle. For one formula we must know a side and the altitude to that side, and for the other we need two sides and the measure of their included angle. Using the law of cosines, we can get a formula for the area of a triangle that involves only the lengths of the sides. This formula is known as **Heron's area formula,** named after Heron of Alexandria, who is believed to have discovered it in about A.D. 75.

**Heron's Area Formula**

The area of a triangle with sides $a$, $b$, and $c$ is given by the formula

$$A = \sqrt{S(S - a)(S - b)(S - c)}$$

where $S = (a + b + c)/2$.

**PROOF**   First rewrite the equation $a^2 = b^2 + c^2 - 2bc \cos \alpha$ as follows:

$$a^2 + 2bc \cos \alpha = b^2 + c^2$$

$$2bc \cos \alpha = b^2 + c^2 - a^2$$

$$4b^2c^2 \cos^2 \alpha = (b^2 + c^2 - a^2)^2 \quad \text{Square both sides.}$$

Now write the area formula $A = \frac{1}{2}bc \sin \alpha$ in terms of $4b^2c^2 \cos^2 \alpha$:

$$A = \frac{1}{2}bc \sin \alpha$$

$$4A = 2bc \sin \alpha \qquad\qquad \text{Multiply by 4.}$$

$$16A^2 = 4b^2c^2 \sin^2 \alpha \qquad\quad \text{Square both sides.}$$

$$16A^2 = 4b^2c^2(1 - \cos^2 \alpha) \qquad \text{Pythagorean identity}$$

$$16A^2 = 4b^2c^2 - 4b^2c^2 \cos^2 \alpha \qquad \text{Distributive property}$$

$$16A^2 = 4b^2c^2 - (b^2 + c^2 - a^2)^2 \quad \text{Substitution}$$

The last equation could be solved for $A$ in terms of the lengths of the three sides, but it would not be Heron's formula. To get Heron's formula, let $S = (a + b + c)/2$ or $2S = a + b + c$. It can be shown (see Exercise 48) that

$$4b^2c^2 - (b^2 + c^2 - a^2)^2 = 2S(2S - 2a)(2S - 2b)(2S - 2c).$$

Using this fact in the above equation yields Heron's formula:

$$16A^2 = 16S(S - a)(S - b)(S - c)$$

$$A^2 = S(S - a)(S - b)(S - c)$$

$$A = \sqrt{S(S - a)(S - b)(S - c)} \qquad\qquad ■$$

**EXAMPLE 3**   **Area of a triangle using only the sides**

A piece of property in downtown Houston is advertised for sale at $45 per square foot. If the lengths of the sides of the triangular lot are 220 feet, 234 feet, and 160 feet, then what is the asking price for the lot?

**Solution**

To use Heron's formula we first find $S$:

$$S = \frac{220 + 234 + 160}{2} = 307$$

Now use 220, 234, and 160 along with $S = 307$ in Heron's formula:

$$A = \sqrt{307(307 - 220)(307 - 234)(307 - 160)} \approx 16{,}929.69$$

At \$45 per square foot the 16,929.69 square foot lot costs \$761,836.

**TRY THIS.** Given $a = 12, b = 8$, and $c = 6$, find the area of the triangle.  ▪

## FOR THOUGHT... True or False? Explain.

1. The area of a triangle is one-half the product of the lengths of any two sides.

2. An altitude in a triangle can be inside, outside, or co-incide with a side of the triangle.

3. The area of a triangle is one-half of the product of the lengths of two sides and the measure of the included angle.

4. The area of a right triangle is one-half the product of the lengths of its legs.

5. The area of a triangle can be found if you know the lengths of the three sides.

Scan this code to link to section lecture videos.

## 5.3 EXERCISES

### CONCEPTS

*Fill in the blank.*

1. The area of a triangle with side $b$ and altitude to the side $h$ is
   _____ .

2. _____ area formula gives the area of a triangle in terms of the lengths of its sides.

### SKILLS

*Find the area of each triangle with the given parts. Round to the nearest tenth.*

3. $a = 12.9, b = 6.4, \gamma = 13.7°$

4. $a = 42.7, c = 64.1, \beta = 74.2°$

5. $\alpha = 39.4°, b = 12.6, a = 13.7$

6. $\beta = 74.2°, c = 19.7, b = 23.5$

7. $\alpha = 42.3°, \beta = 62.1°, c = 14.7$

8. $\gamma = 98.6°, \beta = 32.4°, a = 24.2$

9. $\alpha = 56.3°, \beta = 41.2°, a = 9.8$

10. $\beta = 25.6°, \gamma = 74.3°, b = 17.3$

*Find the area of each region to the nearest whole number of square units.*

11.

12.

**13.**

**14.**

**15.** $a = 16, b = 9, c = 10$

*Find the area of each triangle using Heron's formula. Round to the nearest tenth.*

**15.** $a = 16, b = 9, c = 10$   **16.** $a = 12, b = 8, c = 17$

**17.** $a = 3.6, b = 9.8, c = 8.1$

**18.** $a = 5.4, b = 8.2, c = 12.0$

**19.** $a = 346, b = 234, c = 422$

**20.** $a = 124.8, b = 86.4, c = 154.2$

*Use the most appropriate formula for the area of a triangle to find the area of each triangle. Round to the nearest tenth.*

**21.**

**22.**

**23.**

**24.**

**25.**

**26.**

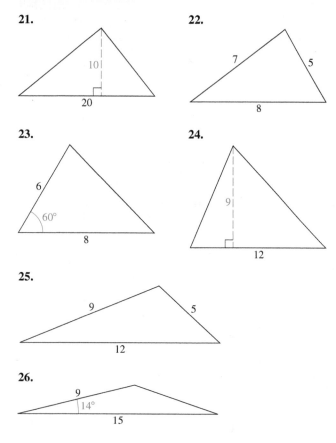

**MODELING**

*Solve each problem.*

**27.** *Making a Kite* A kite is made in the shape shown in the figure. Find the surface area of the kite in square inches. Round to the nearest tenth.
**HINT** The triangles are not right triangles.

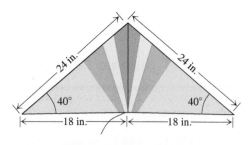

**Figure for Exercise 27**

**28.** *Area of a Wing* The F-106 Delta Dart once held a world speed record of Mach 2.3. Its sweptback triangular wings have the dimensions given in the figure. Find the area of one wing in square feet. Round to the nearest tenth.

**Figure for Exercise 28**

**29.** *Triangular Glazing* A triangular piece of glass has sides of lengths 13 in., 8 in., and 9 in. Find the area of the triangle without using Heron's formula. Round to the nearest tenth.

**30.** *Triangular Flashing* A sheet metal worker constructed a triangular metal flashing with sides of 6 ft, 5 ft, and 3 ft. If he charges $2 per square foot for this type of work, then what should he charge? Do not use Heron's formula.

**31.** *Maximum Triangle I* Find the maximum possible area for a triangle that has two vertices on a circle of radius 2 feet and the third vertex at the center of the circle.

**32.** *Maximum Triangle II* Find the maximum possible area for a triangle for which one side is the diameter of a circle of radius 2 feet and the third vertex is on the circle.

**33.** *Areas* Let $\alpha$ be the central angle (in radians) in a circle of radius $r$, as shown in the accompanying figure. The chord shown in the figure divides the sector of the circle into a triangle and a lens-shaped region. Find formulas for the following.
   **a.** The area of the triangle $A_T$ as a function of $r$ and $\alpha$

**b.** The area of the sector $A_S$ as a function of $r$ and $\alpha$

**c.** The area of the lens-shaped region $A_L$ as a function of $r$ and $\alpha$

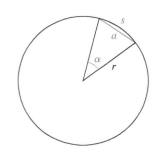

**Figure for Exercises 33 and 34**

**34.** *Arcs, Chords, and Angles* Let $s$ be the length of the arc intercepted by a central angle of $\alpha$ radians in a circle of radius $r$ and $a$ be the length of the corresponding chord. See the accompanying figure. Find formulas for the following.
**a.** The length of the arc $s$ as a function of $\alpha$ and $r$

**b.** The length of the chord $a$ as a function of $r$ and $\alpha$

**c.** The length of the chord $a$ as a function of $s$ and $\alpha$

**d.** The length of the arc $s$ as a function of $a$ and $\alpha$

**35.** *Surveying Triangular Property* A surveyor locating the corners of a triangular piece of property started at one corner and walked 480 ft in the direction N36°W to reach the next corner. The surveyor turned and walked S21°W to get to the next corner of the property. Finally, the surveyor walked in the direction N82°E to get back to the starting point. What is the area of the property in square feet? Round to the nearest tenth.

**36.** *Surveying Triangular Property* A surveyor locating the corners of a triangular piece of property started at one corner and walked 200 ft in the direction S10°E to reach the next corner. The surveyor turned and walked N80°W to get to the next corner of the property. Finally, the surveyor walked in the direction N85°E to get back to the starting point. What is the area of the property in square feet? Round to the nearest tenth.

**37.** *Surveying a Quadrilateral* A surveyor locating the corners of a four-sided piece of property started at one corner and walked 200 ft in the direction N80°E to reach the next corner. He turned and walked due north 150 ft to the next corner of the property. He then turned and walked due west to get to the fourth corner of the property. Finally, he walked in the direction S15°E to get back to the starting point. What is the area of the property in square feet? Round to the nearest tenth.

**38.** *Surveying a Quadrilateral* A surveyor locating the corners of a four-sided piece of property started at one corner and walked 400 ft in the direction S10°W to reach the second corner. She turned and walked 340 ft in the direction N86°E to the third

corner of the property. She turned and walked in the direction N5°W to get to the fourth corner of the property. Finally, the surveyor walked in the direction S75°W to get back to the starting point. What is the area of the property in square feet? Round to the nearest tenth.

**39.** *Cul-de-sac Lot* Claudia owns a lot on a cul-de-sac, as shown in the accompanying figure. The sides of the lot are parallel and are perpendicular to the back of the lot. The front of the lot is a circular arc of length 88.1 feet on a circle of radius 80 feet. If property tax is $0.08 per square foot, find the amount of property tax to the nearest dollar.

**Figure for Exercise 39**

**40.** *Lot on a Curve* Two sides of Manny's lot are parallel and perpendicular to the back boundary, as shown in the accompanying figure. The front of the lot is a circular arc of length 152.9 feet on a circle of radius 300 feet. Find the area of the lot to the nearest square foot.

**Figure for Exercise 40**

**WRITING/DISCUSSION**

**41.** *Area of a Triangle* Find the area of each triangle in the figure on the next page in square centimeters using the three different formulas that were discussed in the text. Measure the sides with a ruler and the angles with a protractor. Compare your answers. Which answer do you think is closest to the actual area of the triangle?

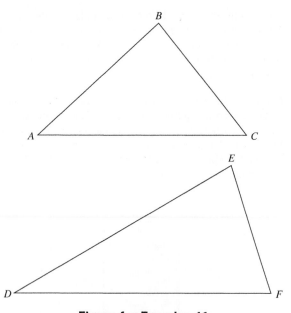

**Figure for Exercise 41**

**42.** Prove the trigonometric formula for the area of a triangle in the cases mentioned in the text but not proved in the text.

**43.** Find the area of the triangle with sides 37, 48, and 86 using Heron's formula. Explain your result.

**44.** Find the area of the triangle with sides 31, 87, and 56 using Heron's formula. Explain your result.

**45.** Find the area of the triangle with sides of length 6 ft, 9 ft, and 13 ft to the nearest tenth by using the formula

$$A = \frac{1}{4}\sqrt{4b^2c^2 - (b^2 + c^2 - a^2)^2},$$

and check your result using a different formula for the area of a triangle. Prove that this formula gives the area of any triangle with sides $a$, $b$, and $c$.

**46.** Find the area of the triangle with sides of length 11 ft, 12 ft, and 18 ft to the nearest tenth by using the formula

$$A = \frac{1}{2}bc\sin\left(\cos^{-1}\left(\frac{b^2 + c^2 - a^2}{2bc}\right)\right),$$

and check your result using a different formula for the area of a triangle. Prove that this formula gives the area of any triangle with sides $a$, $b$, and $c$.

**47.** *Cooperative Learning* Find a map or a description of a piece of property in terms of sides and angles. Show it to your class and explain how to find the area of the property in square feet.

**48.** *Heron's Formula* In the proof of Heron's formula it was stated that

$$4b^2c^2 - (b^2 + c^2 - a^2)^2 =$$
$$2S(2S - 2a)(2S - 2b)(2S - 2c).$$

Show that this equation is correct.

## REVIEW

**49.** Solve the triangle with $\beta = 122.1°$, $a = 19.4$, and $b = 22.6$.

**50.** Solve the triangle with $\alpha = 22.1°$, $a = 19.4$, and $c = 144.2$.

**51.** Solve the triangle with $\alpha = 33.2°$, $b = 9.4$, and $c = 4.3$.

**52.** Find all solutions to the equation $2\cos(x) + 1 = 0$. Use $k$ to represent any integer.

**53.** Find all solutions to the equation $4\sin^2(3x) - 3 = 0$ in the interval $(0, \pi)$.

**54.** Find the period, asymptotes, and range for the function $y = 5\tan(\pi x + \pi)$.

## OUTSIDE THE BOX

**55.** *Watering the Lawn* Josie places her lawn sprinklers at the vertices of a triangle that has sides of 9 m, 10 m, and 11 m. The sprinklers water in circular patterns with radii of 4, 5, and 6 meters. No area is watered by more than one sprinkler. What amount of area inside the triangle is not watered by any of the three sprinklers? Give your answer to the nearest thousandth of a square meter.

**56.** *Equal Areas* In the accompanying figure, $P$ is an arbitrary point inside a regular hexagon. Show that the total area of the three shaded triangles is equal to the total area of the three unshaded triangles.

**Figure for Exercise 56**

## 5.3 POP QUIZ

**1.** Find the area of the triangle in which $a = 6$ ft, $b = 8$ ft, and $c = 10$ ft.

**2.** Find the area of the triangle in which $a = 6$ ft, $b = 15$ ft, and $\gamma = 66.7°$. Round to the nearest tenth of a square foot.

**3.** Find the exact area of the triangle with sides of lengths 7, 8, and 9 using Heron's formula.

## Discovering Area and Circumference Formulas

*In this exercise you will use the formula for the area of a triangle,*

$$A = \frac{1}{2}bc\sin\alpha,$$

*to develop the formulas for the area and circumference of a circle. As shown in the figure, you will be using polygons that circumscribe a circle of radius r.*

**a)** Suppose a regular pentagon circumscribes a circle of radius $r$ as shown in the figure. Show that the area of the pentagon is $5r^2\tan(36°)$.

**b)** Now suppose that a regular polygon of $n$ sides circumscribes a circle of radius $r$. Show that the area of the $n$-gon is given by $A = nr^2\tan(180°/n)$.

**c)** Does the area of the polygon in part (b) vary directly or inversely as the square of the radius?

**d)** Find the constant of proportionality for a decagon, kilogon, and megagon.

**e)** What happens to the shape of the $n$-gon as $n$ increases? So what can you conclude is the formula for the area of a circle of radius $r$?

**f)** Use trigonometry to find a formula for the perimeter of an $n$-gon that circumscribes a circle of radius $r$.

**g)** Use the result of part (f) to find a formula for the circumference of a circle of radius $r$.

**h)** Graph the function $y = x\tan(\pi/x)$ where $x$ is a real number (radian mode). Identify all vertical and horizontal asymptotes.

# 5.4 Vectors

By using the law of sines and the law of cosines we can find all of the missing parts of any triangle for which we have enough information to determine its shape. In this section we will use these and many other tools of trigonometry in the study of vectors.

## Definitions

Quantities such as length, area, volume, temperature, and time have magnitude only and are completely characterized by a single real number with appropriate units (such as feet, degrees, or hours). Such quantities are called **scalar quantities,** and the corresponding real numbers are **scalars.** Quantities that involve both a magnitude and direction, such as velocity, acceleration, and force, are **vector quantities,** and they can be represented by **directed line segments.** These directed line segments are called **vectors.**

For an example of vectors, consider two baseballs hit into the air as shown in Fig. 5.24. One is hit with an initial velocity of 30 ft/sec at an angle of 20° from the horizontal and the other is hit with an initial velocity of 60 ft/sec at an angle of 45° from the horizontal. The length of a vector represents the **magnitude** of the vector quantity. So the vector representing the faster baseball is drawn twice as long as the other vector. The **direction** is indicated by the position of the vector and the arrowhead at one end.

**Figure 5.24**

We have used the notation $\overline{AB}$ to name a line segment with endpoints $A$ and $B$ and $\overrightarrow{AB}$ to name a ray with initial point $A$ and passing through $B$. When studying vectors, the notation $\overrightarrow{AB}$ is used to name a vector with **initial point** $A$ and **terminal point** $B$. The vector $\overrightarrow{AB}$ terminates at $B$ while the ray $\overrightarrow{AB}$ goes beyond $B$. Since the direction of the vector is the same as the order of the letters, we will omit the arrow and use boldface type to indicate a vector in print. In handwritten work an arrow over the letter or letters indicates a vector. Thus **AB** and $\overrightarrow{AB}$ represent the same vector. The vector **BA** is a vector with initial point $B$ and terminal point $A$, and is not the same as **AB.** A vector whose endpoints are not specified is named by a single uppercase or lowercase letter. For example, $\mathbf{b}, \mathbf{B}, \vec{b},$ and $\vec{B}$ are names of vectors. The magnitude of vector **A** is written as $|\mathbf{A}|$.

Two vectors are **equal** if they have the same magnitude and the same direction. Equal vectors may be in different locations. They do not necessarily coincide. In Fig. 5.25, $\mathbf{A} = \mathbf{B}$ because they have the same direction and the same magnitude. Vector **B** is *not* equal to vector **C** because they have the same magnitude but opposite direction. $\mathbf{C} \neq \mathbf{D}$ because they have different magnitudes and different directions. The **zero vector** is a vector that has no magnitude and no direction. It is denoted by a boldface zero, **0.**

**Figure 5.25**

## Scalar Multiplication and Addition

Suppose two tugboats are pulling on a barge that has run aground in the Mississippi River as shown in Fig. 5.26.

**Figure 5.26**

The yellow boat exerts a force of 2000 pounds in an easterly direction and its force is represented by the vector **A.** If the red boat exerts a force of 4000 pounds in the same direction, then its force can be represented by the vector 2**A.** This example illustrates the operation of **scalar multiplication.**

**Definition:**
**Scalar Multiplication**

For any scalar $k$ and vector **A,** $k$**A** is a vector with magnitude $|k|$ times the magnitude of **A.**

If $k > 0$, then the direction of $k$**A** is the same as the direction of **A.**
If $k < 0$, the direction of $k$**A** is opposite to the direction of **A.**
If $k = 0$, then $k$**A** = **0.**

The vector 2**A** has twice the magnitude of **A,** and $-\frac{2}{3}$**A** has two-thirds of the magnitude. For $-1$**A** we write $-$**A.** Some examples of scalar multiplication are shown in Fig. 5.27.

Suppose two draft horses are pulling on a tree stump with forces of 200 pounds and 300 pounds as shown in Fig. 5.28, with an angle of 65° between the forces.

**Figure 5.27**

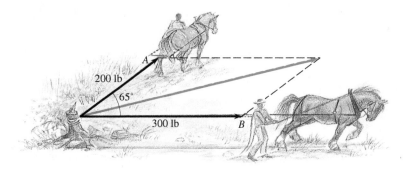

**Figure 5.28**

The two forces are represented by vectors **A** and **B.** If **A** and **B** had the same direction, then there would be a total force of 500 pounds acting on the stump. However, a total force of 500 pounds is not achieved because of the angle between the forces. It can be shown by experimentation that one force acting along the diagonal of the parallelogram shown in Fig. 5.28, with a magnitude equal to the length of the diagonal, has the same effect on the stump as the two forces **A** and **B.** In physics, this result is known as the **parallelogram law.** The single force **A** + **B** acting along the diagonal is called the **sum** or **resultant** of **A** and **B.** The parallelogram law motivates our definition of vector addition.

**Definition: Vector Addition**

To find the resultant or sum **A** + **B** of any vectors **A** and **B,** position **B** (without changing its magnitude or direction) so that the initial point of **B** coincides with the terminal point of **A,** as in Fig. 5.29. The vector that begins at the initial point of **A** and terminates at the terminal point of **B** is the vector **A** + **B.**

Note that the vector **A** + **B** in Fig. 5.29 coincides with the diagonal of a parallelogram whose adjacent sides are **A** and **B.** We can see from Fig. 5.29 that **A** + **B** = **B** + **A,** and vector addition is commutative. If **A** and **B** have the same direction or opposite directions, then no parallelogram is formed, but **A** + **B** can be found by the procedure given in the definition. Each vector in the sum **A** + **B** is called a **component** of the sum.

**Figure 5.29**

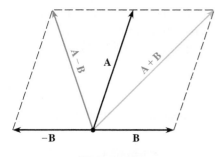

**Figure 5.30**

For every vector **A** there is a vector −**A,** the **opposite** of **A,** having the same magnitude as **A** but opposite direction. The sum of a vector and its opposite is the zero vector, **A** + (−**A**) = **0.** Vector subtraction is defined like subtraction of real numbers. For any two vectors **A** and **B, A** − **B** = **A** + (−**B**). Figure 5.30 shows **A** − **B.**

The law of sines and the law of cosines can be used to find the magnitude and direction of a resultant vector. Remember that in any parallelogram the opposite sides are equal and parallel, and adjacent angles are supplementary. The diagonals of a parallelogram do not bisect the angles of a parallelogram unless the adjacent sides of the parallelogram are equal in length.

## Horizontal and Vertical Components

When a peregrine falcon dives at its prey with a velocity of 120 ft/sec at an angle of 40° from the horizontal, its location is changing both horizontally and vertically. So its velocity vector can be thought of as a sum of a horizontal velocity vector and a vertical velocity vector. In fact, any nonzero vector **w** is the sum of a **vertical component** and a **horizontal component** as shown in Fig. 5.31. The horizontal component is denoted $\mathbf{w_x}$ and the vertical component is denoted $\mathbf{w_y}.$ The vector **w** is the diagonal of the rectangle formed by the vertical and horizontal components.

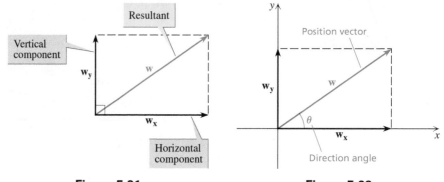

**Figure 5.31**         **Figure 5.32**

If a vector **w** is placed in a rectangular coordinate system so that its initial point is the origin (as in Fig. 5.32), then **w** is called a **position vector** or **radius vector. A** position vector is a convenient representative to focus on when considering all vectors that are equal to a certain vector. The angle $\theta$ ($0° \le \theta < 360°$) formed by the positive $x$-axis and a position vector is the **direction angle** for the position vector (or any other vector that is equal to the position vector).

If the vector **w** has magnitude $r$, direction angle $\theta$, horizontal component $\mathbf{w_x},$ and vertical component $\mathbf{w_y},$ as shown in Fig. 5.32, then by using trigonometric ratios we get

$$\cos \theta = \frac{|\mathbf{w_x}|}{r} \qquad \text{and} \qquad \sin \theta = \frac{|\mathbf{w_y}|}{r}$$

or

$$|\mathbf{w_x}| = r \cos \theta \qquad \text{and} \qquad |\mathbf{w_y}| = r \sin \theta.$$

If the direction of **w** is such that $\sin \theta$ or $\cos \theta$ is negative, then we can write

$$|\mathbf{w_x}| = |r \cos \theta| \qquad \text{and} \qquad |\mathbf{w_y}| = |r \sin \theta|.$$

**EXAMPLE 1**   **Finding horizontal and vertical components**

Find the magnitude of the horizontal and vertical components for a vector **v** with magnitude 8.3 and direction angle 121.3°.

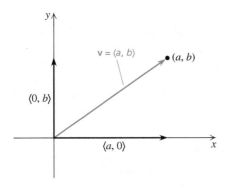

**Figure 5.33**

**Solution**

The vector **v** and its horizontal and vertical components $\mathbf{v_x}$ and $\mathbf{v_y}$ are shown in Fig. 5.33. The magnitudes of $\mathbf{v_x}$ and $\mathbf{v_y}$ are found as follows:

$$|\mathbf{v_x}| = |8.3 \cos 121.3°| \approx 4.3$$

$$|\mathbf{v_y}| = |8.3 \sin 121.3°| \approx 7.1$$

The direction angle for $\mathbf{v_x}$ is 180°, and the direction angle for $\mathbf{v_y}$ is 90°.

**TRY THIS.** Find the magnitude of the horizontal and vertical components for a vector **v** with magnitude 5.6 and direction angle 22°. ∎

## Component Form of a Vector

Any vector is the resultant of its horizontal and vertical components. Since the horizontal and vertical components of a vector determine the vector, it is convenient to use a notation for vectors that involves them. The notation $\langle a, b \rangle$ is used for the position vector with terminal point $(a, b)$ as shown in Fig. 5.34. The form $\langle a, b \rangle$ is called **component form** because its horizontal component is $\langle a, 0 \rangle$ and its vertical component is $\langle 0, b \rangle$. Since the vector $\mathbf{v} = \langle a, b \rangle$ extends from $(0, 0)$ to $(a, b)$, its magnitude is the distance between these points:

$$|\mathbf{v}| = \sqrt{a^2 + b^2}$$

**Figure 5.34**

In Example 2 we find the magnitude and direction for a vector given in component form and in Example 3 we find the component form for a vector given as a directed line segment.

EXAMPLE **2** **Finding magnitude and direction from component form**

Find the magnitude and direction angle of the vector $\mathbf{v} = \langle -2, -5 \rangle$.

**Solution**

We find the magnitude of **v** shown in Fig. 5.35 using $|\mathbf{v}| = \sqrt{a^2 + b^2}$:

$$|\mathbf{v}| = \sqrt{(-2)^2 + (-5)^2} = \sqrt{29}$$

Using trigonometric ratios we get $\sin \theta = -5/\sqrt{29}$. Since

$$\sin^{-1}\left(-5/\sqrt{29}\right) \approx -68.2°,$$

we have $\theta = 180° - (-68.2°) = 248.2°$.

**TRY THIS.** Find the magnitude and direction angle for $\mathbf{v} = \langle 2, -6 \rangle$. ∎

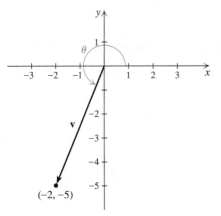

**Figure 5.35**

EXAMPLE **3** **Finding the component form given magnitude and direction**

Find the component form for a vector of magnitude 40 mph with direction angle 330°.

**Solution**

To find the component form, we need the terminal point $(a, b)$ for a vector of magnitude 40 mph positioned as shown in Fig. 5.36. Since $a = r \cos \theta$ and $b = r \sin \theta$, we have

$$a = 40 \cos 330° = 40 \frac{\sqrt{3}}{2} = 20\sqrt{3}$$

**Figure 5.36**

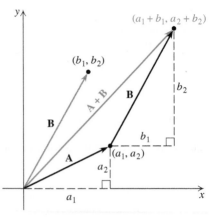

**Figure 5.37**

and

$$b = 40 \sin 330° = 40\left(-\frac{1}{2}\right) = -20.$$

So the component form of the vector is $\langle 20\sqrt{3}, -20 \rangle$.

**TRY THIS.** Find the component form for a vector of magnitude 50 mph with direction angle 120°. ∎

We originally defined vectors as directed line segments and performed operations with these directed line segments. When vectors are written in component form, operations with vectors are easier to perform. Figure 5.37 illustrates the addition of $\mathbf{A} = \langle a_1, a_2 \rangle$ and $\mathbf{B} = \langle b_1, b_2 \rangle$ where $(a_1, a_2)$ and $(b_1, b_2)$ are in the first quadrant. It is easy to see that the endpoint of $\mathbf{A} + \mathbf{B}$ is $(a_1 + b_1, a_2 + b_2)$ and so

$$\mathbf{A} + \mathbf{B} = \langle a_1 + b_1, a_2 + b_2 \rangle.$$

So the sum of two vectors can be found in component form by adding the components instead of drawing directed line segments. The component forms for the three operations that we have already studied and a new operation, *dot product,* are given in the following box. The **dot product** of two vectors is a scalar found by multiplying the corresponding components and adding the results. (When multiplying matrices in algebra you find dot products of row vectors and column vectors.)

**Rules for Scalar Product, Vector Sum and Difference, and Dot Product**

If $\mathbf{A} = \langle a_1, a_2 \rangle$, $\mathbf{B} = \langle b_1, b_2 \rangle$, and $k$ is a scalar, then

1. $k\mathbf{A} = \langle ka_1, ka_2 \rangle$                      **Scalar product**
2. $\mathbf{A} + \mathbf{B} = \langle a_1 + b_1, a_2 + b_2 \rangle$      **Vector sum**
3. $\mathbf{A} - \mathbf{B} = \langle a_1 - b_1, a_2 - b_2 \rangle$      **Vector difference**
4. $\mathbf{A} \cdot \mathbf{B} = a_1 b_1 + a_2 b_2$                  **Dot product**

**EXAMPLE 4**    Operations with vectors in component form

Let $\mathbf{w} = \langle -3, 2 \rangle$ and $\mathbf{z} = \langle 5, -1 \rangle$. Perform the operations indicated.

**a.** $\mathbf{w} - \mathbf{z}$    **b.** $-8\mathbf{z}$    **c.** $3\mathbf{w} + 4\mathbf{z}$    **d.** $\mathbf{w} \cdot \mathbf{z}$

**Solution**

**a.** $\mathbf{w} - \mathbf{z} = \langle -3, 2 \rangle - \langle 5, -1 \rangle = \langle -8, 3 \rangle$
**b.** $-8\mathbf{z} = -8\langle 5, -1 \rangle = \langle -40, 8 \rangle$
**c.** $3\mathbf{w} + 4\mathbf{z} = 3\langle -3, 2 \rangle + 4\langle 5, -1 \rangle = \langle -9, 6 \rangle + \langle 20, -4 \rangle = \langle 11, 2 \rangle$
**d.** $\mathbf{w} \cdot \mathbf{z} = \langle -3, 2 \rangle \cdot \langle 5, -1 \rangle = (-3)(5) + (2)(-1) = -17$

**TRY THIS.** Let $\mathbf{u} = \langle -1, -3 \rangle$ and $\mathbf{v} = \langle 3, -4 \rangle$. Find $\mathbf{u} + 3\mathbf{v}$ and $\mathbf{u} \cdot \mathbf{v}$. ∎

## The Angle between Two Vectors

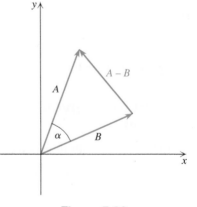

**Figure 5.38**

If $\mathbf{A} = k\mathbf{B}$ for a nonzero scalar $k$, then $\mathbf{A}$ and $\mathbf{B}$ are **parallel** vectors. If $\mathbf{A}$ and $\mathbf{B}$ have the same direction ($k > 0$), the angle between $\mathbf{A}$ and $\mathbf{B}$ is 0°. If they have opposite directions ($k < 0$), the angle between them is 180°. If $\mathbf{A}$ and $\mathbf{B}$ are nonparallel vectors with the same initial point, then the vectors $\mathbf{A}$, $\mathbf{B}$, and $\mathbf{A} - \mathbf{B}$ form a triangle as in Fig. 5.38. The **angle between the vectors** $\mathbf{A}$ and $\mathbf{B}$ is the angle $\alpha$ shown in Fig. 5.38. The angle between two vectors is in the interval $[0°, 180°]$. If the angle between $\mathbf{A}$ and $\mathbf{B}$ is 90°, then the vectors are **perpendicular** or **orthogonal.**

The following theorem relates the angle between two vectors, the magnitudes of the vectors, and the dot product of the vectors. The theorem is simply a vector version of the law of cosines.

**Theorem: The Angle between Two Vectors and the Dot Product**

If **A** and **B** are nonzero vectors and $\alpha$ is the smallest positive angle between them, then

$$\cos \alpha = \frac{\mathbf{A} \cdot \mathbf{B}}{|\mathbf{A}|\,|\mathbf{B}|}.$$

**PROOF** Let $\mathbf{A} = \langle a_1, a_2 \rangle$, $\mathbf{B} = \langle b_1, b_2 \rangle$, and $\mathbf{A} - \mathbf{B} = \langle a_1 - b_1, a_2 - b_2 \rangle$. The vectors **A**, **B**, and **A** − **B** form a triangle as shown in Fig. 5.38. Apply the law of cosines to this triangle and simplify as follows.

$$|\mathbf{A}|^2 + |\mathbf{B}|^2 - 2|\mathbf{A}|\,|\mathbf{B}|\cos \alpha = |\mathbf{A} - \mathbf{B}|^2$$

$$(a_1)^2 + (a_2)^2 + (b_1)^2 + (b_2)^2 - 2|\mathbf{A}|\,|\mathbf{B}|\cos \alpha = (a_1 - b_1)^2 + (a_2 - b_2)^2$$

$$-2|\mathbf{A}|\,|\mathbf{B}|\cos \alpha = -2a_1b_1 - 2a_2b_2$$

$$|\mathbf{A}|\,|\mathbf{B}|\cos \alpha = a_1b_1 + a_2b_2$$

$$|\mathbf{A}|\,|\mathbf{B}|\cos \alpha = \mathbf{A} \cdot \mathbf{B}$$

$$\cos \alpha = \frac{\mathbf{A} \cdot \mathbf{B}}{|\mathbf{A}|\,|\mathbf{B}|}$$ ∎

Note that $\cos \alpha = 0$ if and only if $\mathbf{A} \cdot \mathbf{B} = 0$. So two vectors are perpendicular if and only if their dot product is zero. Two vectors are parallel if and only if $\cos \alpha = \pm 1$.

**EXAMPLE 5** **Angle between two vectors**

Find the smallest positive angle between each pair of vectors.

**a.** $\langle -3, 2 \rangle, \langle 4, 5 \rangle$   **b.** $\langle -5, 9 \rangle, \langle 9, 5 \rangle$

**Solution**

**a.** Use the dot product theorem to find the cosine of the angle and then a calculator to find the angle:

$$\cos \alpha = \frac{\langle -3, 2 \rangle \cdot \langle 4, 5 \rangle}{|\langle -3, 2 \rangle|\,|\langle 4, 5 \rangle|} = \frac{-2}{\sqrt{13}\sqrt{41}}$$

$$\alpha = \cos^{-1}\left(-2/\left(\sqrt{13}\sqrt{41}\right)\right) \approx 94.97°$$

**b.** Because $\langle -5, 9 \rangle \cdot \langle 9, 5 \rangle = 0$, the vectors are perpendicular and the angle between them is 90°.

**TRY THIS.** Find the smallest positive angle between $\langle 1, 3 \rangle$ and $\langle 5, 2 \rangle$. ∎

**Figure 5.39**

## Unit Vectors

The vectors $\mathbf{i} = \langle 1, 0 \rangle$ and $\mathbf{j} = \langle 0, 1 \rangle$ are called **unit vectors** because each has magnitude one. For any vector $\langle a_1, a_2 \rangle$ as shown in Fig. 5.39 we have

$$\langle a_1, a_2 \rangle = a_1 \langle 1, 0 \rangle + a_2 \langle 0, 1 \rangle = a_1\mathbf{i} + a_2\mathbf{j}.$$

The form $a_1\mathbf{i} + a_2\mathbf{j}$ is called a **linear combination** of the vectors **i** and **j.** These unit vectors are thought of as fundamental vectors, because any vector can be expressed as a linear combination of them.

### EXAMPLE 6 Unit vectors

Write each vector as a linear combination of the unit vectors **i** and **j**.

**a.** $A = \langle -2, 6 \rangle$  **b.** $B = \langle -4, -1 \rangle$

**Solution**

**a.** $A = \langle -2, 6 \rangle = -2\mathbf{i} + 6\mathbf{j}$
**b.** $B = \langle -4, -1 \rangle = -4\mathbf{i} - \mathbf{j}$

**TRY THIS.** Write $\langle -1, 7 \rangle$ as a linear combination of **i** and **j**. ∎

## FOR THOUGHT... True or False? Explain.

1. The vector 2**v** has the same direction as **v** but twice the magnitude of **v**.

2. The magnitude of **A** + **B** is the sum of the magnitudes of **A** and **B**.

3. The magnitude of −**A** is equal to the magnitude of **A**.

4. For any vector **A**, **A** + (−**A**) = **0**.

5. The parallelogram law says that the opposite sides of a parallelogram are equal in length.

6. In the coordinate plane, the direction angle is the angle formed by the *y*-axis and a vector with initial point at the origin.

7. If **v** has magnitude *r* and direction angle θ, then the horizontal component of **v** is a vector with direction angle 0° and magnitude *r* cos θ.

8. The magnitude of $\langle 3, -4 \rangle$ is 5.

9. The vectors $\langle -2, 3 \rangle$ and $\langle -6, 9 \rangle$ have the same direction angle.

10. The direction angle of $\langle -2, 2 \rangle$ is $\cos^{-1}\left(-2/\sqrt{8}\right)$.

## 5.4 EXERCISES

### CONCEPTS

*Fill in the blank.*

1. A(n) _____ quantity involves both magnitude and direction.

2. Two vectors with the same magnitude and direction are _____ vectors.

3. The length of a vector represents the _____ of the vector quantity.

4. If **A** and **B** are vectors, then **A** + **B** is the _____ or _____ of **A** and **B**.

5. The _____ indicates that the resultant of two vectors lies along the diagonal of the parallelogram formed by the two vectors.

6. The angle formed by the positive *x*-axis and a position vector is the _____ angle.

7. The form $\langle a, b \rangle$ is the _____ form of a vector.

8. If the angle between two vectors is 90°, then the vectors are _____ or _____.

### SKILLS

*For each given pair of vectors* **A** *and* **B** *draw the vectors* **A** + **B** *and* **A** − **B**.

9.

10.

**11.**    **12.**

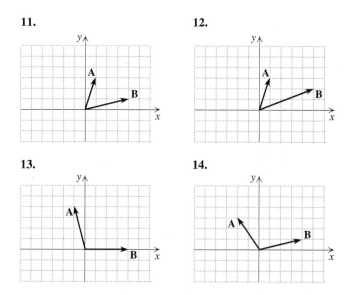

**13.**    **14.**

*Determine which of the position vectors shown in the following figure has the given magnitude and direction angle.*

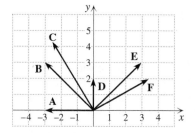

**15.** Magnitude 2 and direction angle 90°

**16.** Magnitude 3 and direction angle 180°

**17.** Magnitude $3\sqrt{2}$ and direction angle 45°

**18.** Magnitude 4 and direction angle 30°

**19.** Magnitude $3\sqrt{2}$ and direction angle 135°

**20.** Magnitude 5 and direction angle 120°

*Find the magnitude of the horizontal and vertical components for each vector* **v** *with the given magnitude and given direction angle θ. Round to the nearest tenth.*

**21.** $|\mathbf{v}| = 4.5, \theta = 65.2°$    **22.** $|\mathbf{v}| = 6000, \theta = 13.1°$

**23.** $|\mathbf{v}| = 8000, \theta = 155.1°$    **24.** $|\mathbf{v}| = 445, \theta = 211.1°$

**25.** $|\mathbf{v}| = 234, \theta = 248°$    **26.** $|\mathbf{v}| = 48.3, \theta = 349°$

*Find the exact magnitude and direction angle to the nearest tenth of a degree of each vector.*

**27.** $\langle \sqrt{3}, 1 \rangle$    **28.** $\langle -1, \sqrt{3} \rangle$

**29.** $\langle -\sqrt{2}, \sqrt{2} \rangle$    **30.** $\langle \sqrt{2}, -\sqrt{2} \rangle$

**31.** $\langle 8, -8\sqrt{3} \rangle$    **32.** $\langle -1/2, -\sqrt{3}/2 \rangle$

**33.** $\langle 5, 0 \rangle$    **34.** $\langle 0, -6 \rangle$

**35.** $\langle -3, 2 \rangle$    **36.** $\langle -4, -2 \rangle$

**37.** $\langle 3, -1 \rangle$    **38.** $\langle 2, -6 \rangle$

*Find the component form for each vector* **v** *with the given magnitude and direction angle θ. Give exact values using radicals when possible. Otherwise round to the nearest tenth.*

**39.** $|\mathbf{v}| = 8, \theta = 45°$    **40.** $|\mathbf{v}| = 12, \theta = 120°$

**41.** $|\mathbf{v}| = 290, \theta = 145°$    **42.** $|\mathbf{v}| = 5.3, \theta = 321°$

**43.** $|\mathbf{v}| = 18, \theta = 347°$    **44.** $|\mathbf{v}| = 3000, \theta = 209.1°$

*Let* $\mathbf{r} = \langle 3, -2 \rangle, \mathbf{s} = \langle -1, 5 \rangle,$ *and* $\mathbf{t} = \langle 4, -6 \rangle.$ *Perform the operations indicated. Write the vector answers in the form* $\langle a, b \rangle.$

**45.** **5r**    **46.** **−4s**

**47.** **2r + 3t**    **48.** **r − t**

**49.** **s + 3t**    **50.** $\dfrac{\mathbf{r} + \mathbf{s}}{2}$

**51.** **r − (s + t)**    **52.** **r − s − t**

**53.** **r · s**    **54.** **s · t**

*Find the smallest positive angle to the nearest tenth of a degree between each given pair of vectors.*

**55.** $\langle 2, 1 \rangle, \langle 3, 5 \rangle$

**56.** $\langle 2, 3 \rangle, \langle 1, 5 \rangle$

**57.** $\langle -1, 5 \rangle, \langle 2, 7 \rangle$

**58.** $\langle -2, -5 \rangle, \langle 1, -9 \rangle$

**59.** $\langle -6, 5 \rangle, \langle 5, 6 \rangle$

**60.** $\langle 2, 7 \rangle, \langle 7, -2 \rangle$

*Determine whether each pair of vectors is parallel, perpendicular, or neither.*

**61.** $\langle -2, 3 \rangle, \langle 6, 4 \rangle$    **62.** $\langle 2, 3 \rangle, \langle 8, 12 \rangle$

**63.** $\langle 1, 7 \rangle, \langle -2, -14 \rangle$    **64.** $\langle 2, -4 \rangle, \langle 2, 1 \rangle$

**65.** $\langle 5, 3 \rangle, \langle 2, 5 \rangle$    **66.** $\langle 2, 6 \rangle, \langle 6, 2 \rangle$

*Write each vector as a linear combination of the unit vectors* **i** *and* **j.**

**67.** $\langle 2, 1 \rangle$    **68.** $\langle 1, 5 \rangle$

**69.** $\langle -3, \sqrt{2} \rangle$    **70.** $\langle \sqrt{2}, -5 \rangle$

**71.** $\langle 0, -9 \rangle$    **72.** $\langle -1/2, 0 \rangle$

**73.** $\langle -7, -1 \rangle$    **74.** $\langle 1, 1 \rangle$

*Given that* $\mathbf{A} = \langle 3, 1 \rangle$ *and* $\mathbf{B} = \langle -2, 3 \rangle$, *find the magnitude and direction angle for each of the following vectors. Give exact answers using radicals when possible. Otherwise round to the nearest tenth.*

**75.** $\mathbf{A} + \mathbf{B}$

**76.** $\mathbf{A} - \mathbf{B}$

**77.** $-3\mathbf{A}$

**78.** $5\mathbf{B}$

**79.** $\mathbf{B} - \mathbf{A}$

**80.** $\mathbf{B} + \mathbf{A}$

**81.** $-\mathbf{A} + \dfrac{1}{2}\mathbf{B}$

**82.** $\dfrac{1}{2}\mathbf{A} - 2\mathbf{B}$

*For each given pair of vectors, find the magnitude and direction angle of the resultant. Give exact answers using radicals when possible. Otherwise round to the nearest tenth.*

**83.**

**84.**

**85.**

**86.**

**87.**

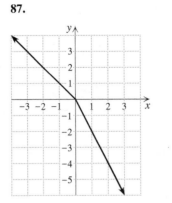

**88.**

*Solve each problem.*

**89.** *Gaining Altitude* An airplane with an air speed of 520 mph is climbing at an angle of 30° from the horizontal. What are the magnitudes of the horizontal and vertical components of the speed vector? Round to the nearest tenth.

**90.** *Acceleration of a Missile* A missile is fired with an angle of elevation of 22°, and an acceleration of 30 m/sec². What are the magnitudes of the horizontal and vertical components of the acceleration vector? Round to the nearest tenth.

## WRITING/DISCUSSION

**91.** *Distributive* Prove that scalar multiplication is distributive over vector addition, first using the component form and then using a geometric argument.

**92.** *Associative* Prove that vector addition is associative, first using the component form and then using a geometric argument.

## REVIEW

**93.** Find the exact area of the triangle whose sides are 3, 3, and 1.

**94.** Find the area of the triangle with $a = 3.6$, $b = 4.5$, and $\gamma = 37.1°$.

**95.** Complete the sum and difference identities.
   **a.** $\tan(x + y) = \underline{\hspace{2cm}}$

   **b.** $\tan(x - y) = \underline{\hspace{2cm}}$

**96.** State the three Pythagorean identities.

**97.** The length of the hypotenuse of a right triangle is 66 feet and one of the acute angles is 33°. Find the other acute angle and the lengths of the legs.

**98.** Suppose that $\alpha$ is an angle in standard position whose terminal side contains the point $(-3, 5)$. Find $\sin \alpha$, $\cos \alpha$, and $\tan \alpha$.

**OUTSIDE THE BOX**

**99.** *Packing Billiard Balls* There are several ways to tightly pack nine billiard balls each with radius 1 into a rectangular box. Find the volume of the box in each of the following cases and determine which box has the least volume.
  **a.** Four balls are placed so that they just fit into the bottom of the box, then another layer of four, then one ball in the middle tangent to all four in the second layer, as shown from the side in the accompanying figure.

**Figure for Exercise 99(a)**

**b.** Four balls are placed so that they just fit into the bottom of the box as in (a), then one is placed in the middle on top of the first four. Finally, four more are placed so that they just fit at the top of the box.

**c.** The box is packed with layers of four, one, and four as in (b), but the box is required to be cubic. In this case, the four balls in the bottom layer will not touch each other and the four balls in the top layer will not touch each other. The ball in the middle will be tangent to all of the other eight balls.

**100.** *Big Trig Equation* How many solutions are there with $0 < x \le \pi$ to the equation

$$\cos x \cos 2x \cos 3x \cos 4x \cos 5x \cos 6x = 0?$$

**5.4 POP QUIZ**

**1.** Find $|\mathbf{v_x}|$ and $|\mathbf{v_y}|$ if $|\mathbf{v}| = 5.6$ and the direction angle of $\mathbf{v}$ is $33.9°$. Round to the nearest tenth.

**2.** Find the exact magnitude and direction angle to the nearest tenth of the vector $\mathbf{v} = \langle -2, 6 \rangle$.

**3.** Find $\mathbf{v} - \mathbf{w}$, $3\mathbf{v}$, and $\mathbf{v} \cdot \mathbf{w}$, if $\mathbf{v} = \langle -1, 3 \rangle$ and $\mathbf{w} = \langle 2, 6 \rangle$.

**4.** Find the smallest positive angle to the nearest tenth of a degree between $\langle 1, 4 \rangle$ and $\langle 2, 6 \rangle$.

# 5.5 Applications of Vectors

In this section we use the operations with vectors from Section 5.4 to solve problems involving forces.

## Basic Resultant Problems

When modeling forces that are acting on an object, we usually assume that all forces are acting on a single point. Since the resultant of two forces is the diagonal of a parallelogram, you will need to recall some facts about parallelograms. In a parallelogram, opposite sides are parallel and equal, and opposite angles are equal. Adjacent angles are supplementary, which means that they have a sum of $180°$.

**EXAMPLE 1**  **Magnitude and direction of a resultant**

Two draft horses are pulling on a tree stump with forces of 200 pounds and 300 pounds, as shown in Fig. 5.40 on the next page. If the angle between the forces

is 65°, then what is the magnitude of the resultant force? What is the angle between the resultant and the 300-pound force?

**Figure 5.40**

### Solution

The resultant coincides with the diagonal of the parallelogram shown in Fig. 5.40. Each acute angle of this parallelogram is 65°. Since adjacent angles of a parallelogram are supplementary, each obtuse angle is 115°. So the resultant is a side of a triangle in which the other two sides are 200 and 300, and their included angle is 115°. The law of cosines can be used to find the magnitude of the resultant vector **v:**

$$|\mathbf{v}|^2 = 300^2 + 200^2 - 2(300)(200) \cos 115° \approx 180,714.19$$

$$|\mathbf{v}| \approx 425.1$$

We can use the law of sines to find the angle $\alpha$ between the resultant and the 300-pound force.

$$\frac{\sin \alpha}{200} = \frac{\sin 115°}{425.1}$$

$$\sin \alpha = \frac{200 \sin 115°}{425.1} \approx 0.4264$$

Since $\sin^{-1}(0.4264) \approx 25.2°$, both 25.2° and 154.8° are solutions to $\sin \alpha = 0.4264$. Since $\alpha$ must be smaller than 65°, the angle between the resultant and the 300-pound force is approximately 25.2°. So one horse pulling in the direction of the resultant with a force of approximately 425.1 pounds would have the same effect on the stump as the two horses pulling at an angle of 65°.

**TRY THIS.** Forces of 100 pounds and 200 pounds are acting on a point. If the angle between the forces is 30°, then what is the magnitude of the resultant? ∎

**Figure 5.41**

Note that magnitude and direction of a resultant can be found using component form if the vectors are in a coordinate system. Figure 5.41 shows the vectors of Example 1 in a coordinate system. For simplicity we place one of the vectors on the positive $x$-axis. The component forms are $\langle 300, 0 \rangle$ and $\langle 200 \cos(65°), 200 \sin(65°) \rangle$. The component form of the resultant is

$$\langle 300 + 200 \cos(65°), 0 + 200 \sin(65°) \rangle$$

or approximately $\langle 384.5, 181.3 \rangle$. The magnitude of the resultant is $\sqrt{384.5^2 + 181.3^2}$ or approximately 425.1 pounds. The angle between the resultant and the 300-pound force is $\tan^{-1}(181.3/384.5)$ or approximately 25.2°.

## Inclined-Plane Problems

Anyone who has ever moved a refrigerator knows that it is easier to roll it up a ramp than to lift it into a truck. However, the steeper the ramp, the more force it takes. The amount of force depends on the weight of the refrigerator, the amount of incline in the ramp, and friction. Since friction can be minimized by using a well-oiled refrigerator dolly, we will always assume that there is no friction in our models of inclined-plane problems.

The weight of an object on an inclined plane is always modeled as a vertical vector and its length is fixed regardless of the incline of the plane. See Fig. 5.42. The force required to move the object is always modeled as a vector parallel to the inclined plane and its length increases as the incline increases as shown in Fig. 5.42. The resultant of these two forces is a vector perpendicular to the plane. It is what a bathroom scale would read if it were trapped between the object and the plane. The length of the resultant decreases as the incline increases.

**Figure 5.42**

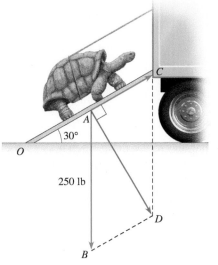

**Figure 5.43**

**EXAMPLE 2** **Finding the force to move an object**

Workers at the Audubon Zoo must move a giant tortoise to his new home. Find the amount of force required to pull a 250-pound tortoise up a ramp leading into a truck. The angle of elevation of the ramp is 30°. See Fig. 5.43.

**Solution**

The weight of the tortoise is a 250-pound force in a downward direction, shown as vector **AB** in Fig. 5.43. The tortoise exerts a force against the ramp at a 90° angle with the ramp, shown as vector **AD**. The magnitude of **AD** is less than the magnitude of **AB** because of the incline of the ramp. The force required to pull the tortoise up the ramp is vector **AC** in Fig. 5.43. The force against the ramp, **AD**, is the resultant of **AB** and **AC**. Since $\angle O = 30°$, we have $\angle OAB = 60°$ and $\angle BAD = 30°$. Because $\angle CAD = 90°$, we have $\angle ADB = 90°$. We can find any of the missing parts to the right triangle $ABD$. Since

$$\sin 30° = \frac{\text{opposite}}{\text{hypotenuse}} = \frac{|\mathbf{BD}|}{|\mathbf{AB}|},$$

we have

$$|\mathbf{BD}| = |\mathbf{AB}| \sin 30° = 250 \cdot \frac{1}{2} = 125 \text{ lb.}$$

Since the opposite sides of a parallelogram are equal, the magnitude of **AC** is also 125 pounds.

**TRY THIS.** Find the amount of force required to push an 800-pound block of ice up a ramp that is inclined 10°. ■

**EXAMPLE 3** **Finding the weight of an object**

If a force of 86 pounds is required to push a safe up a ramp that is inclined 10°, then what is the weight of the safe?

**Solution**

The weight of the safe is a force in a downward direction, shown as vector **AB** in Fig. 5.44.

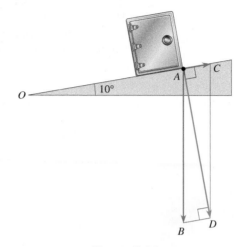

**Figure 5.44**

The force required to move it up the ramp is parallel to the ramp, shown as vector **AC.** The resultant of these two vectors is vector **AD,** which is perpendicular to the ramp. Since $\angle O = 10°$, we have $\angle OAB = 80°$ and $\angle BAD = 10°$. Because $\angle CAD = 90°$, we have $\angle ADB = 90°$ and we can find any of the missing parts to right triangle $ADB$. Since the opposite sides of a parallelogram are equal in length, $|\mathbf{AC}| = |\mathbf{BD}|$.

$$\sin 10° = \frac{86}{|\mathbf{AB}|}$$

$$|\mathbf{AB}| = \frac{86}{\sin 10°} \approx 495.3$$

So the weight of the safe is approximately 495.3 pounds.

**TRY THIS.** A landscaper uses 100 pounds of force to pull a cart full of rocks up a driveway that is inclined 15°. What is the weight of the cart? ■

In the next example we find the amount of incline from knowing the amount of force required to keep an object in position on an inclined plane. Since there is no friction involved, the force required to move the object is the same as the force required just to keep the object in a fixed position.

**EXAMPLE 4** **Finding the amount of incline**

A motorist exerts 150 pounds of force to keep his 3000-pound car from rolling down a hill. Find the angle of inclination of the hill to the nearest tenth of a degree.

**Solution**

The angle of inclination of the hill, $\angle A$ in Fig. 5.45, is the same as $\angle BCD$, the angle between the resultant and the 3000-pound vertical force. Since $\angle BCD$ is an acute angle of a right triangle, we can use right triangle trigonometry.

$$\sin(\angle BCD) = \frac{\text{opposite}}{\text{hypotenuse}} = \frac{150}{3000} = 0.05$$

$$\angle BCD = \sin^{-1}(0.05) \approx 2.9°$$

So the angle of inclination of the hill is about 2.9°.

**Figure 5.45**

**TRY THIS.** If 300 pounds of force is required to push a 1000-pound safe up a ramp, then what is the angle of inclination of the ramp?

## Navigation Problems

If an airplane heads directly against the wind or with the wind, the wind speed is subtracted from or added to the air speed of the plane to get the ground speed of the plane. When the wind is at some other angle to the direction of the plane, some portion of the wind speed will be subtracted from or added to the air speed to determine the ground speed. In addition, the wind causes the plane to travel on a course different from where it is headed. We can use the vector $\mathbf{v_1}$ to represent the heading and air speed of the plane, as shown in Fig. 5.46.

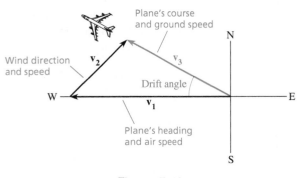

**Figure 5.46**

The vector $\mathbf{v_2}$ represents the wind direction and speed. The resultant of $\mathbf{v_1}$ and $\mathbf{v_2}$ is the vector $\mathbf{v_3}$, where $\mathbf{v_3}$ represents the course and ground speed of the plane. The angle between the heading and the course is the **drift angle.** Recall that the bearing of a vector used to describe direction in air navigation is a nonnegative angle smaller than 360° measured in a clockwise direction from due north. For example, the wind in Fig. 5.46 is out of the southwest and has a bearing of 45°. The vector $\mathbf{v_1}$ describing the heading of the plane points due west and has a bearing of 270°.

**EXAMPLE 5** Finding the course and ground speed

An airplane is headed due west (bearing 270°) with an air speed of 250 mph. The wind is from the north (bearing 180°) at 30 mph. Find the drift angle, the ground speed, and the course of the airplane. Round to one decimal place.

**Solution**

Figure 5.47 shows the vectors for the heading $\mathbf{v_1}$, the wind $\mathbf{v_2}$, and the course $\mathbf{v_3}$.

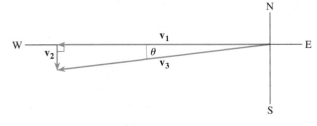

**Figure 5.47**

Since the wind direction is perpendicular to the heading, we can solve this problem with right triangle trigonometry.

$$\theta = \tan^{-1}\left(\frac{30}{250}\right) \approx 6.8°$$

The drift angle is 6.8°. The ground speed is the length of $\mathbf{v_3}$, which is the hypotenuse of a right triangle:

$$\left|\mathbf{v_3}\right| = \sqrt{250^2 + 30^2} \approx 251.8$$

So the ground speed is approximately 251.8 mph. The airplane is headed due west, but is blown 6.8° to the south. So its course has bearing 270° − 6.8° or 263.2°.

**TRY THIS.** An airplane is headed due east with an air speed of 200 mph. The wind is out of the south (bearing 0°) at 40 mph. Find the drift angle, the ground speed, and the course of the airplane. ∎

**EXAMPLE 6** Finding the course and ground speed

An airplane is headed due east with an air speed of 300 mph. The wind is out of the northeast (bearing 225°) at 50 mph. Find the drift angle, the ground speed, and the course of the airplane.

**Solution**

Figure 5.48 shows vectors for the heading $\mathbf{v_1}$, the wind $\mathbf{v_2}$, and the course $\mathbf{v_3}$.

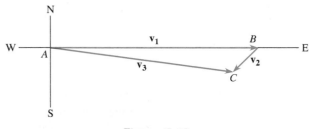

**Figure 5.48**

Since the bearing of the wind is 225°, $\angle ABC = 45°$. Use the law of cosines to find $\left|\mathbf{v_3}\right|$:

$$\left|\mathbf{v_3}\right|^2 = 300^2 + 50^2 - 2(300)(50)\cos 45° \approx 71{,}286.8$$

$$\left|\mathbf{v_3}\right| \approx 267.0$$

Use the law of sines to find the drift angle:

$$\frac{\sin \angle BAC}{50} = \frac{\sin 45°}{267.0}$$

$$\sin \angle BAC = \frac{50 \sin 45°}{267.0} \approx 0.1324$$

$$\angle BAC \approx 7.6°$$

So the ground speed is 267.0 mph, the drift angle is 7.6°, and the bearing of the course is 97.6°.

**TRY THIS.** An airplane is headed due west with an air speed of 400 mph. The wind is out of the northwest (bearing 135°) at 90 mph. Find the drift angle, the ground speed, and the course of the airplane. ∎

**EXAMPLE 7** Finding course and ground speed
of an airplane

The heading of a Lear jet has a bearing of 320°. The wind is 70 mph with a bearing of 190°. If the air speed of the plane is 400 mph, then find the drift angle, the ground speed, and the course of the airplane.

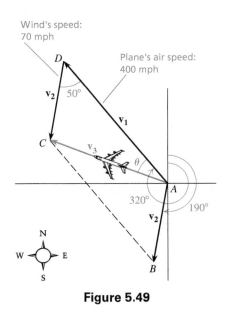

Wind's speed:
70 mph

Plane's air speed:
400 mph

**Figure 5.49**

N
W   E
S

**Solution**

Figure 5.49 shows vectors for the heading $\mathbf{v_1}$, the wind $\mathbf{v_2}$, and the course $\mathbf{v_3}$. Subtract the bearing of the wind from that of the heading to get

$$\angle DAB = 320° - 190° = 130°.$$

Since $ABCD$ is a parallelogram, its adjacent angles are supplementary and $\angle CDA = 50°$. Apply the law of cosines to triangle $ACD$ to find the length of $\mathbf{v_3}$:

$$|\mathbf{v_3}|^2 = 70^2 + 400^2 - 2(70)(400)\cos 50° \approx 128{,}903.9$$

$$|\mathbf{v_3}| \approx 359.0$$

The ground speed is approximately 359.0 mph. The drift angle $\theta$ is found by using the law of sines:

$$\frac{\sin \theta}{70} = \frac{\sin 50°}{359.0}$$

$$\sin \theta = \frac{70 \sin 50°}{359.0}$$

$$\approx 0.1494$$

Since $\theta$ is an acute angle, $\theta = \sin^{-1}(0.1494) \approx 8.6°$. The course $\mathbf{v_3}$ has a bearing of $320° - 8.6°$ or approximately $311.4°$.

**TRY THIS.** A jet is headed northwest with an air speed of 500 mph. The wind is 100 mph with a bearing of 200°. Find the drift angle, the ground speed, and the course of the jet. ∎

**FOR THOUGHT...** True or False? Explain.

1. The force required to push a 99-kg block of ice up a frictionless ramp inclined at 88° is nearly 99 kg.

2. The force required to move a 99-kg block of ice on a frictionless horizontal surface is 0 kg.

3. The weight of an object is always modeled by a horizontal vector.

4. The force required to move an object on an inclined plane is modeled by a vector parallel to the plane.

5. The resultant of the weight and the force required to move an object up an inclined plane is a vector perpendicular to the plane.

6. If the course of an airplane is northeast, then the bearing for the course vector is 45°.

7. If the wind is blowing from the southwest, then the bearing for the wind vector is 225°.

8. If a plane is heading east with air speed 400 mph and the wind is from the northeast, then the ground speed is greater than 400 mph.

9. If a plane is heading south with wind from the northeast and the drift angle is 5°, then the bearing for the course is 175°.

10. If a plane is heading southeast with wind from the north and a drift angle of 3°, then the bearing for the course is 138°.

 **EXERCISES**

**SKILLS**

*In each case, find the magnitude of the resultant force and the angle between the resultant and each force. Round to the nearest tenth.*

1. Forces of 3 lb and 8 lb act at an angle of 90° to each other.

2. Forces of 2 lb and 12 lb act at an angle of 60° to each other.

3. Forces of 4.2 newtons (a unit of force from physics) and 10.3 newtons act at an angle of 130° to each other.

4. Forces of 34 newtons and 23 newtons act at an angle of 100° to each other.

## MODELING

*Solve each problem. Round answers to the nearest tenth.*

**5.** *Magnitude of a Force* The resultant of a 10-lb force and an-other force has a magnitude of 12.3 lb at an angle of 23.4° with the 10-lb force. Find the magnitude of the other force and the angle between the two forces.

**6.** *Magnitude of a Force* The resultant of a 15-lb force and an-other force has a magnitude of 9.8 lb at an angle of 31° with the 15-lb force. Find the magnitude of the other force and the angle between the other force and the resultant.

**7.** *Moving a Donkey* Two prospectors are pulling on ropes at-tached around the neck of a donkey that does not want to move. One prospector pulls with a force of 55 lb, and the other pulls with a force of 75 lb. If the angle between the ropes is 25°, as shown in the figure, then how much force must the donkey use in order to stay put? (The donkey knows the proper direction in which to apply his force.)

55 lb

25°

75 lb

**Figure for Exercise 7**

**8.** *Pushing a Shopping Cart* Ronnie, Phyllis, and Ted are con-ducting a vector experiment in a Walmart parking lot. Ronnie is pushing a cart containing Phyllis to the east at 5 mph while Ted is pushing it to the north at 3 mph. What is Phyllis's speed and in what direction (measured from north) is she moving?

*Solve each problem.*

**9.** *Winch Force* Find the amount of force required for a winch to pull a 3000-lb car up a ramp that is inclined 20°.

**10.** *Rock and Roll* In Roman mythology, Sisyphus, king of Corinth, revealed a secret of Zeus and thus incurred the god's wrath. As punishment, Zeus banished him to Hades, where he was doomed for eternity to roll a rock uphill, only to have it roll back on him. If Sisyphus stands in front of a 4000-lb spherical rock on a 20° incline, as shown in the figure, then what force applied in the direction of the incline would keep the rock from rolling down the incline?

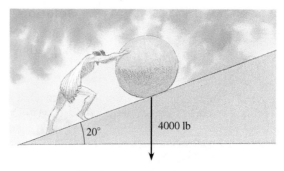

20°    4000 lb

**Figure for Exercise 10**

**11.** *Ice Force* If the amount of force required to push a block of ice up an ice-covered driveway that is inclined at 25° is 100 pounds, then what is the weight of the block?

**12.** *Weight of a Ball* A solid steel ball is placed on a 10° incline. If a force of 3.2 lb in the direction of the incline is required to keep the ball in place, then what is the weight of the ball?

**13.** *Super Force* If Superman exerts 1000 pounds of force to pre-vent a 5000-lb boulder from rolling down a hill and crushing a bus full of children, then what is the angle of inclination of the hill?

**14.** *Sisy's Slope* If Sisyphus exerts a 500-lb force in rolling his 4000-lb spherical boulder uphill, then what is the angle of in-clination of the hill?

**15.** *Due East* A plane is headed due east with an air speed of 240 mph. The wind is from the north at 30 mph. Find the bear-ing for the course and the ground speed of the plane.

**16.** *Due West* A plane is headed due west with an air speed of 300 mph. The wind is from the north at 80 mph. Find the bearing for the course and the ground speed of the plane.

**17.** *Ultralight* An ultralight is flying northeast at 50 mph. The wind is from the north at 20 mph. Find the bearing for the course and the ground speed of the ultralight.

**18.** *Superlight* A superlight is flying northwest at 75 mph. The wind is from the south at 40 mph. Find the bearing for the course and the ground speed of the superlight.

**19.** *Course of an Airplane* An airplane is heading on a bearing of 102° with an air speed of 480 mph. If the wind is out of the northeast (bearing 225°) at 58 mph, then what are the bearing of the course and the ground speed of the airplane?

**20.** *Course of a Helicopter* The heading of a helicopter has a bearing of 240°. If the 70-mph wind has a bearing of 185° and the air speed of the helicopter is 195 mph, then what are the bearing of the course and the ground speed of the helicopter?

**21.** *Going with the Flow* A river is 2000 ft wide and flowing at 6 mph from north to south. A woman in a canoe starts on the eastern shore and heads west at her normal paddling speed of 2 mph. In what direction (measured clockwise from north) will she actually be traveling? How far downstream from a point directly across the river will she land?

**22.** *Crossing a River* If the woman in Exercise 21 wants to go directly across the river and she paddles at 8 mph, then in what direction (measured clockwise from north) must she aim her canoe? How long will it take her to go directly across the river?

**23.** *Distance and Rate* A group of trigonometry students wants to cross a river that is 0.2 mi wide and has a current of 1 mph. Their boat goes 3 mph in still water.
  **a.** Write the distance the boat travels as a function of the angle $\beta$ shown in the figure.

  **b.** Write the actual speed of the boat as a function of $\alpha$ and $\beta$.

**Figure for Exercises 23 and 24**

**24.** *Minimizing the Time* Write the time for the trip in the previous exercise as a function of $\alpha$. Find the angle $\alpha$ for which they will cross the river in the shortest amount of time.

**25.** *My Three Elephants* Amal uses three elephants to pull a very large log out of the jungle. The papa elephant pulls with 800 lb of force, the mama elephant pulls with 500 lb of force, and the baby elephant pulls with 200 lb of force. The angles between the forces are shown in the figure. What is the magnitude of the resultant of all three forces?

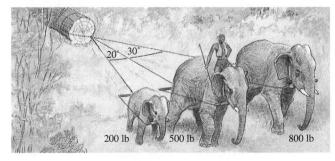

**Figure for Exercise 25**

**26.** *Mama's Direction* If mama is pulling due east in Exercise 25, then in what direction will the log move?

## REVIEW

**27.** Solve the triangle in which $a = 5$, $b = 7$, and $c = 10$.

**28.** Find the dot product of the vectors $\langle -2, 6 \rangle$ and $\langle 3, 5 \rangle$.

**29.** Find the smallest positive angle between the vectors $\langle -3, 5 \rangle$ and $\langle 1, 6 \rangle$.

**30.** Find the area of the triangle in which $\alpha = 10.6°$, $b = 5.7$ feet, and $c = 12.2$ feet.

**31.** A tall building casts a shadow of length 230 feet when the angle of elevation of the sun is 48°. Find the height of the building.

**32.** Simplify the expression $\dfrac{1 - \sin^2 x \csc^2 x + \sin^2 x}{\cos^2 x}$.

## OUTSIDE THE BOX

**33.** *Disappearing Dogs* Someone opened up the cages at Pet Depot and more than 100 puppies got away! There were exactly 300 puppies to begin with. The Daily Mixup reported: "Of the pups that remained, a third were Dobermans, a quarter were schnauzers, a fifth were beagles, a seventh were poodles, and a ninth were dachshunds. The original number of beagles was three times the number of dachshunds that stayed." The Daily Mixup got just one of the fractions wrong. How many beagles escaped?

**34.** *Fifteen Degrees* Show that $\sin(\pi/12) + \cos(\pi/12) = \sqrt{6}/2$.

## 5.5 POP QUIZ

**1.** Forces of 50 lb and 100 lb are acting on a point. If the angle between the forces is 80°, then what is the magnitude of the resultant (to the nearest tenth of a pound)?

**2.** If a force of 100 lb is required to push a motorcycle up a ramp inclined at 12°, then what is the weight of the motorcycle (to the nearest pound)?

**3.** An airplane is headed due east with an air speed of 300 mph. The wind is from the north at 60 mph. What is the ground speed and heading of the airplane? Round to the nearest tenth.

# LINKING
## concepts...   For Individual or Group Explorations

### Minimizing the Total Time

*Suk wants to go from point A to point B as shown in the figure. Point B is 1 mile upstream and on the north side of the 0.4-mi-wide river. The speed of the current is 0.5 mph. Her boat will travel 4 mph in still water. When she gets to the north side she will travel by bicycle at 6 mph. Let $\alpha$ represent the bearing of her course and $\beta$ represent the drift angle as shown in the figure.*

a)  When $\alpha = 12°$ find $\beta$ and the actual speed of the boat. Round to hundredths.

b)  Find the total time for the trip to the nearest tenth of a minute when $\alpha = 12°$.

c)  Write the total time for the trip as a function of $\alpha$.

d)  Graph the function that you found in part (c).

e)  Find the approximate angle $\alpha$ that minimizes the total time for the trip and find the corresponding $\beta$. Round to the nearest tenth of a degree.

# Highlights

## 5.1  The Law of Sines

| Law of Sines | In any triangle $\frac{\sin \alpha}{a} = \frac{\sin \beta}{b} = \frac{\sin \gamma}{c}$. | $a = \sqrt{3}, b = 1, c = 2$ $\alpha = \pi/3, \beta = \pi/6, \gamma = \pi/2$ $\dfrac{\sin\left(\frac{\pi}{3}\right)}{\sqrt{3}} = \dfrac{\sin\left(\frac{\pi}{6}\right)}{1} = \dfrac{\sin\left(\frac{\pi}{2}\right)}{2}$ |
|---|---|---|

## 5.2  The Law of Cosines

| Law of Cosines | If $a$ is the side opposite angle $\alpha$ in any triangle, then $a^2 = b^2 + c^2 - 2bc \cos \alpha$. | $a = 4, b = 5, c = 6$ $4^2 = 5^2 + 6^2 - 60 \cos \alpha$ |
|---|---|---|
| Length of a Chord | A central angle $\alpha$ in a circle of radius $r$ intercepts a chord of length $a$ where $a = r\sqrt{2 - 2 \cos \alpha}$. | $\alpha = \pi/4, r = 2$ $a = 2\sqrt{2 - 2 \cos(\pi/4)}$ |

## 5.3 Area of a Triangle

| | | |
|---|---|---|
| **Area of a Triangle** | The area of a triangle is one-half the product of any two sides and the sine of the angle between them: $A = \frac{1}{2} bc \sin \alpha$. | Equilateral triangle with sides of length 4: $A = \frac{1}{2} \cdot 4 \cdot 4 \cdot \sin(60°) = 4\sqrt{3}$ |
| **Heron's Area Formula** | If $a$, $b$, and $c$ are the sides of a triangle and $S = (a + b + c)/2$, then $A = \sqrt{S(S - a)(S - b)(S - c)}$. | $S = (4 + 6 + 8)/2 = 9$ $A = \sqrt{9 \cdot 5 \cdot 3 \cdot 1} = \sqrt{135}$ |

## 5.4 Vectors

| | | |
|---|---|---|
| **Component Form** | The vector with initial point $(0, 0)$ and terminal point $(a, b)$ is denoted $\langle a, b \rangle$. | $\langle 2, 3 \rangle$ has initial point $(0, 0)$ and terminal point $(2, 3)$. |
| **Magnitude** | $\lvert \langle a, b \rangle \rvert = \sqrt{a^2 + b^2}$ | $\lvert \langle 2, 3 \rangle \rvert = \sqrt{13}$ |
| **Scalar Product** | If $k$ is a scalar and $\mathbf{A}$ is a vector, then $\lvert k\mathbf{A} \rvert = \lvert k \rvert \cdot \lvert \mathbf{A} \rvert$. If $k > 0$, $k\mathbf{A}$ has same direction as $\mathbf{A}$. If $k < 0$, $k\mathbf{A}$ has opposite direction to that of $\mathbf{A}$. | $4\langle 1, 2 \rangle = \langle 4, 8 \rangle$ $-4\langle 1, 2 \rangle = \langle -4, -8 \rangle$ |
| **Resultant of A and B** | Place $\mathbf{B}$ so that its initial point coincides with the terminal point of $\mathbf{A}$; then $\mathbf{A} + \mathbf{B}$ is the vector from the initial point of $\mathbf{A}$ to the terminal point of $\mathbf{B}$. | $\mathbf{A} = \langle 1, 4 \rangle, \mathbf{B} = \langle 2, 5 \rangle$ $\mathbf{A} + \mathbf{B} = \langle 3, 9 \rangle$ |
| **Dot Product** | $\langle a, b \rangle \cdot \langle c, d \rangle = ac + bd$ | $\langle 1, 4 \rangle \cdot \langle 2, 5 \rangle = 22$ |
| **Angle between Two Vectors** | If $\alpha$ is the angle between nonzero vectors $\mathbf{A}$ and $\mathbf{B}$, then $\cos \alpha = \dfrac{\mathbf{A} \cdot \mathbf{B}}{\lvert \mathbf{A} \rvert \lvert \mathbf{B} \rvert}$. | $\mathbf{A} = \langle 1, 4 \rangle, \mathbf{B} = \langle 2, 5 \rangle$ $\cos \alpha = \dfrac{22}{\sqrt{17}\sqrt{29}}$ |

## 5.5 Applications of Vectors

| | |
|---|---|
| **Basic Resultant Problems** | All vectors act on a single point. The resultant of two vectors acting on a point is the diagonal of the parallelogram determined by the two vectors. |
| **Inclined-Plane Problems** | The weight vector always points downward. The force vector to move the object is parallel to the inclined plane. The resultant is perpendicular to the inclined plane. |
| **Navigation Problems** | Bearing is measured in degrees from north. The ground speed vector is the resultant of the air speed vector and wind speed vector. |

# Chapter 5 Review Exercises

*Solve each triangle that exists with the given parts. If there is more than one triangle with the given parts, then solve each one. Round to the nearest tenth.*

1. $\gamma = 48°, a = 3.4, b = 2.6$
2. $a = 6, b = 8, c = 10$
3. $\alpha = 13°, \beta = 64°, c = 20$
4. $\alpha = 50°, a = 3.2, b = 8.4$
5. $a = 3.6, b = 10.2, c = 5.9$
6. $\beta = 36.2°, \gamma = 48.1°, a = 10.6$
7. $a = 30.6, b = 12.9, c = 24.1$
8. $\alpha = 30°, a = \sqrt{3}, b = 2\sqrt{3}$
9. $\beta = 22°, c = 4.9, b = 2.5$
10. $\beta = 121°, a = 5.2, c = 7.1$

*Find the area of each triangle. Round to the nearest tenth.*

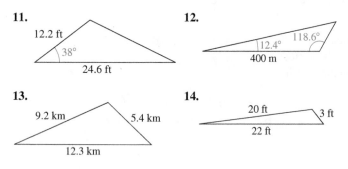

11. 12.2 ft, 38°, 24.6 ft

12. 12.4°, 118.6°, 400 m

13. 9.2 km, 5.4 km, 12.3 km

14. 20 ft, 3 ft, 22 ft

*Find the magnitude of the horizontal and vertical components for each vector **v** with the given magnitude and given direction angle θ. Round to the nearest tenth.*

15. $|\mathbf{v}| = 6, \theta = 23.3°$
16. $|\mathbf{v}| = 4.5, \theta = 156°$
17. $|\mathbf{v}| = 3.2, \theta = 231.4°$
18. $|\mathbf{v}| = 7.3, \theta = 344°$

*Find the magnitude and direction for each vector. Round to the nearest tenth.*

19. $\langle 2, 3 \rangle$
20. $\langle -4, 3 \rangle$
21. $\langle -3.2, -5.1 \rangle$
22. $\langle 2.1, -3.8 \rangle$

*Find the component form for each vector **v** with the given magnitude and direction angle. Round to the nearest tenth.*

23. $|\mathbf{v}| = \sqrt{2}, \theta = 45°$
24. $|\mathbf{v}| = 6, \theta = 60°$
25. $|\mathbf{v}| = 9.1, \theta = 109.3°$
26. $|\mathbf{v}| = 5.5, \theta = 344.6°$

*Perform the vector operations. Write your answer in the form $\langle a, b \rangle$ if the answer is a vector.*

27. $2\langle -3, 4 \rangle$
28. $-3\langle 4, -1 \rangle$
29. $\langle 2, -5 \rangle - 2\langle 1, 6 \rangle$
30. $3\langle 1, 2 \rangle + 4\langle -1, -2 \rangle$
31. $\langle -1, 5 \rangle \cdot \langle 4, 2 \rangle$
32. $\langle -4, 7 \rangle \cdot \langle 7, 4 \rangle$

*Rewrite each vector **v** in the form $a_1\mathbf{i} + a_2\mathbf{j}$ where $\mathbf{i} = \langle 1, 0 \rangle$ and $\mathbf{j} = \langle 0, 1 \rangle$. Give exact answers.*

33. In component form, $\mathbf{v} = \langle -4, 8 \rangle$.
34. In component form, $\mathbf{v} = \langle 3.2, -4.1 \rangle$.
35. The direction angle for **v** is 30° and its magnitude is 7.2.
36. The magnitude of **v** is 6 and it has the same direction as the vector $\langle 2, 5 \rangle$.

*Determine whether the given vectors are parallel or perpendicular.*

37. $\langle 2, 6 \rangle, \langle 4, 12 \rangle$
38. $\langle -3, 7 \rangle, \langle 3, -7 \rangle$
39. $\langle -6, 2 \rangle, \langle 2, 6 \rangle$
40. $\langle -7, 1 \rangle, \langle 1, 7 \rangle$
41. $\langle -3, 8 \rangle, \langle 9, -24 \rangle$
42. $\langle -2, 5 \rangle, \langle -4, 10 \rangle$

*Solve each problem. Round to the nearest tenth.*

43. *Resultant Force* Forces of 12 lb and 7 lb act at a 30° angle to each other. Find the magnitude of the resultant force and the angle that the resultant makes with each force.

44. *Pushing a Cycle* How much force does it take to push an 800-lb motorcycle up a ramp that is inclined at a 25° angle?

45. *Course of a Cessna* A twin-engine Cessna is heading on a bearing of 35° with an air speed of 180 mph. If the wind is out of the west (bearing 90°) at 40 mph, then what is the bearing of its course? What is the ground speed of the airplane?

46. *Course of an Ultralight* An ultralight airplane has a heading of 340° with an air speed of 25 mph. If the wind is out of the northeast (bearing 225°) at 10 mph, then what is the bearing of its course? What is the ground speed of the airplane?

47. *Dividing Property* Mrs. White Eagle gave each of her children approximately half of her four-sided lot in Gallup by dividing it on a diagonal. If Susan's piece is 482 ft by 364 ft by 241 ft and Seth's piece is 482 ft by 369 ft by 238 ft, then which child got the larger piece?

48. *Area, Area, Area* A surveyor found that two sides of a triangular lot were 135.4 ft and 164.1 ft, with an included angle of 86.4°. Find the area of this lot using each of the three area formulas.

49. *Pipeline Detour* A pipeline was planned to go from point $A$ to $B$ as shown in the figure. However, Mr. Smith would not give permission for the pipeline to cross his property. The pipeline was laid 431 ft from $A$ to $C$ and then 562 ft from $C$ to $B$. If $\angle C$ is 122° and the cost of the pipeline was $21.60/ft, then how much extra was spent (to the nearest dollar) to go around Mr. Smith's property?

**Figure for Exercise 49**

**50.** *In the Wrong Place* In a lawsuit filed against a crane operator, a pedestrian of average height claims that he was struck by a wrecking ball. At the time of the accident, the operator had the ball extended 40 ft from the end of the 60-ft boom as shown in the figure, and the angle of elevation of the boom was 53°. How far from the crane would the pedestrian have to stand to be struck by this wrecking ball?

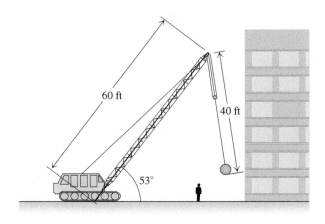

**Figure for Exercise 50**

**51.** *Shady Triangle* Find the exact area of the shaded triangle shown in the accompanying figure.

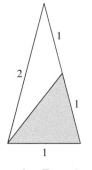

**Figure for Exercise 51**

**52.** *Detroit Pistons* The pistons in a gasoline engine are connected to a crankshaft as shown in the figure. If the length of the connecting rod is $c$ and the radius of revolution around the center

of the crankshaft is $r$, then the distance from the center of the crankshaft to the center of the piston $a$ varies from $c + r$ to $c - r$.

**a.** Show that $a = \sqrt{c^2 - r^2 \sin^2 \theta} + r \cos \theta$.

**b.** Suppose that the crankshaft is rotating at 426 rpm and time $t = 0$ min corresponds to $\theta = 0$ and $a = c + r$. If $c = 12$ in., $r = 2$ in., and $t = 0.1$ min, then what is $a$?

**Figure for Exercise 52**

**53.** *Total Eclipse* Since the sun and moon appear to be about the same size from earth, a total eclipse occurs when the earth, moon, and sun are in line. Assume the sun and moon both have apparent radius $r$. What percent of the sun is blocked when the length of the chord $a$ shown in the accompanying figure is 80% of the apparent diameter of the sun?

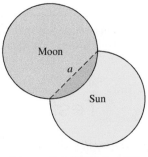

**Figure for Exercise 53**

**54.** *Tank Stand* An engineer is designing a concrete stand that will hold a cylindrical tank of radius 5 feet as shown in the accompanying figure. The stand is to be 8 feet wide, 3 feet high, and 1 foot thick. The top will be an arc of a circle of radius 5 feet. Find the volume of concrete (to the nearest tenth of a cubic foot) that is needed to make one stand.

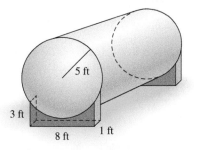

**Figure for Exercise 54**

**55.** *Total Eclipse Function* When the sun and moon have the same apparent radius in the sky, the moon can pass in front of the sun and totally block out the sun. Assume that the moon moves at a constant rate, and that the eclipse begins at time $t = 0$ and ends at time $t = 1$. Verify that the function

$$P = \frac{\cos^{-1}(u) - \sin(\cos^{-1}(u))}{\pi},$$

where $u = 1 - 8t + 8t^2$, gives the portion of the sun that is blocked by the moon at time $t$.

**56.** *Partial Eclipse* Because the distance from the earth to the moon varies, the apparent diameter of the moon could be 95% of the apparent diameter of the sun when an eclipse occurs. In this case, what is the maximum percentage of the sun that can be blocked by the moon?

**Figure for Exercises 55 and 56**

OUTSIDE THE BOX

**57.** *Lakefront Property* A man-made lake in the shape of a triangle is bounded on each of its sides by a square lot as shown in

the figure. The square lots are 8, 13, and 17 acres, respectively. What is the area of the lake in square feet?

**Figure for Exercise 57**

**58.** *More Circles* A circular arc centered at $A$ is drawn from $B$ to $D$ inside the square $ABCD$, as shown in the accompanying figure. The length of $AB$ is 1. The small circle is tangent to the sides of the square and to the circular arc. Find the radius of the small circle.

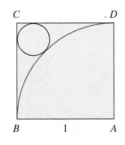

**Figure for Exercise 58**

# Chapter 5 Test

*Determine the number of triangles with the given parts and solve each triangle. Round to the nearest tenth.*

**1.** $\alpha = 30°, b = 4, a = 2$

**2.** $\alpha = 60°, b = 4.2, a = 3.9$

**3.** $a = 3.6, \alpha = 20.3°, \beta = 14.1°$

**4.** $a = 2.8, b = 3.9, \gamma = 17°$

**5.** $a = 4.1, b = 8.6, c = 7.3$

*Given* $\mathbf{A} = \langle -3, 2 \rangle$ *and* $\mathbf{B} = \langle 1, 4 \rangle$, *find the exact magnitude and direction angle (to the nearest tenth) for each of the following vectors.*

**6.** $\mathbf{A} + \mathbf{B}$       **7.** $\mathbf{A} - \mathbf{B}$       **8.** $3\mathbf{B}$

*Solve each problem. Round to the nearest tenth.*

**9.** Find the area of the triangle in which $\alpha = 22°, b = 12$ ft, and $c = 10$ ft.

**10.** Find the area of the triangle in which $a = 4.1$ m, $b = 6.8$ m, and $c = 9.5$ m.

**11.** A vector $\mathbf{v}$ in the coordinate plane has direction angle $\theta = 37.2°$ and $|\mathbf{v}| = 4.6$. Find real numbers $a_1$ and $a_2$ such that $\mathbf{v} = a_1\mathbf{i} + a_2\mathbf{j}$ where $\mathbf{i} = \langle 1, 0 \rangle$ and $\mathbf{j} = \langle 0, 1 \rangle$.

**12.** Determine whether the vectors $\langle -3, 5 \rangle$ and $\langle 5, 3 \rangle$ are perpendicular.

**13.** Determine the amount of force required to push a 1000-lb riding lawnmower up a ramp that is inclined at a 40° angle.

**14.** The bearing of an airplane is 40° with an air speed of 240 mph. If the wind is out of the northwest (bearing 135°) at 30 mph, then what are the bearing of the course and the ground speed of the airplane?

# TYING IT ALL TOGETHER

## Chapters P–5

*Determine the exact values of* $\sin\theta$, $\cos\theta$, $\tan\theta$, $\csc\theta$, $\sec\theta$, *and* $\cot\theta$ *for each given angle, without using a calculator.*

**1.** $\theta = \pi/6$       **2.** $\theta = \pi/4$       **3.** $\theta = \pi/3$       **4.** $\theta = \pi/2$

*Determine the exact value (in radians) of each expression without using a calculator.*

**5.** $\sin^{-1}(1)$       **6.** $\sin^{-1}(-1)$       **7.** $\sin^{-1}(-1/2)$       **8.** $\sin^{-1}(1/2)$

**9.** $\cos^{-1}(-1)$       **10.** $\cos^{-1}(1)$       **11.** $\cos^{-1}\left(-\sqrt{3}/2\right)$       **12.** $\cos^{-1}\left(\sqrt{3}/2\right)$

**13.** $\tan^{-1}(0)$       **14.** $\tan^{-1}(1)$       **15.** $\tan^{-1}(-1)$       **16.** $\tan^{-1}\left(1/\sqrt{3}\right)$

*Find all real solutions to each equation.*

**17.** $\sin x = 0$       **18.** $\sin^2 x - \sin x = 0$       **19.** $\sin^2 x - \sin x = 2$

**20.** $2\sin 2x - 2\cos x + 2\sin x = 1$       **21.** $4x\sin x + 2\sin x - 2x = 1$       **22.** $\sin(2x) = 1/2$

**23.** $\tan(4x) = 1/\sqrt{3}$       **24.** $\sin^2 x + \cos^2(x) = 1$

*Sketch one cycle of the graph of each function. State the amplitude, period, phase shift, domain, and range.*

**25.** $y = \sin(3x)$       **26.** $y = 3\sin(2x)$       **27.** $y = 2\cos(\pi x - \pi)$

**28.** $y = \cos(2x - \pi/2) + 1$       **29.** $y = \tan(x - \pi/2)$       **30.** $y = \tan(\pi x) - 3$

*Fill in the blanks.*

**31.** If $\alpha$ is an acute angle in a right triangle, then $\sin(\alpha)$ is the length of the side _____ $\alpha$ divided by the length of the _____.

**32.** If $\alpha$ is an acute angle in a right triangle, then $\cos(\alpha)$ is the length of the side _____ to $\alpha$ divided by the length of the _____.

**33.** A central angle of one radian in a unit circle intercepts an arc of length _____.

**34.** The _____ for the function $y = \sin(B(x - C))$ is $2\pi/B$.

**35.** The identities $\sin^2 x + \cos^2 x = 1$, $1 + \cot^2 x = \csc^2 x$, and $\tan^2 x + 1 = \sec^2 x$ are called the _____ identities.

**36.** Cosine and secant are _____ functions, whereas the other four trigonometric functions are _____.

**37.** A triangle without a right angle is called a(n) _____ triangle.

**38.** The _____ indicates that the ratio of the sine of an angle and the side opposite the angle is the same for each angle of a triangle.

**39.** The _____ gives a formula for the square of any side of an oblique triangle in terms of the other two sides and their included angle.

**40.** The _____ indicates that the sum of the lengths of any two sides of a triangle is greater than the length of the third side.

# 6

# Complex Numbers, Polar Coordinates, and Parametric Equations

**6.1** Complex Numbers

**6.2** Trigonometric Form of Complex Numbers

**6.3** Powers and Roots of Complex Numbers

**6.4** Polar Equations

**6.5** Parametric Equations

Designing and building custom motorcycle wheels has never been easier. Thanks to computer-aided design (CAD) programs and computerized numerical control (CNC) machines, unique wheels are being manufactured for custom bikes. There is even a company that allows you to design your own custom wheels on its Web site.

The process starts with an idea and a solid block (or billet) of aluminum. The design is drawn with a CAD program, and then the CNC machine carves the wheel to the proper specifications. The process is slow, wasteful, and costly, but for the enthusiast the results are spectacular. The computing cost is insignificant compared to the cost of running the CNC machine, which is a complex robot that can cut and measure with an accuracy of a ten thousandth of an inch.

WHAT YOU WILL **LEARN**

**In this chapter we will study polar coordinates and see how CAD programs use them to make drawings.**

# 6.1 Complex Numbers

Our system of numbers developed as the need arose. Numbers were first used for counting. As society advanced, the rational numbers were formed to express fractional parts and ratios. Negative numbers were invented to express losses or debts. When it was discovered that the exact size of some very real objects could not be expressed with rational numbers, the irrational numbers were added to the system, forming the set of real numbers. In this section we introduce the final expansion of the number system, the complex numbers.

## Definitions

The simple quadratic equation $x^2 = 4$ has two solutions (2 and $-2$). However, $x^2 = -1$ has no real solution because the square of every real number is nonnegative. So that equations such as $x^2 = -1$ will have solutions, imaginary numbers are defined. They are combined with the set of real numbers to form the set of complex numbers.

The imaginary numbers are based on the solution of the equation $x^2 = -1$. Since no real number solves this equation, a solution is called an *imaginary number*. The imaginary number $i$ is defined to be a solution to this equation.

**Definition: Imaginary Number *i***

> The imaginary number $i$ is defined as
> $$i = \sqrt{-1} \quad \text{and} \quad i^2 = -1.$$

A complex number is formed as the sum of a real number and a real multiple of $i$.

**Definition: Complex Numbers**

> The set of **complex numbers** is the set of all numbers of the form $a + bi$, where $a$ and $b$ are real numbers.

In the complex number $a + bi$, $a$ is called the **real part** and $bi$ is called the **imaginary part.** If $b \neq 0$, then $a + bi$ is called an **imaginary number.** Two complex numbers $a + bi$ and $c + di$ are **equal** if and only if $a = c$ and $b = d$.

The form $a + bi$ is called the **standard form** of a complex number, but for convenience we use a few variations of that form. If $a = 0$, then $a$ is omitted and only the imaginary part is written. If $b = 0$, then only $a$ is written and the complex number is a real number. If $b$ is a radical, then $i$ is usually written before $b$. For example, we write $2 + i\sqrt{3}$ rather than $2 + \sqrt{3}i$, which could be confused with $2 + \sqrt{3i}$ where $i$ is inside the radical. If $b$ is negative, we use the minus symbol instead of the plus symbol. For example, instead of $3 + (-2)i$ we write $3 - 2i$. If both $a$ and $b$ are zero, the complex number $0 + 0i$ is the real number 0. For a complex number involving fractions, such as $\frac{1}{3} - \frac{2}{3}i$, we may write $\frac{1 - 2i}{3}$.

**EXAMPLE 1** **Standard form of a complex number**

Determine whether each complex number is real or imaginary and write it in the standard form $a + bi$.

**a.** $3i$ **b.** $87$ **c.** $4 - 5i$ **d.** $0$ **e.** $\dfrac{1 + \pi i}{2}$

**Solution**

**a.** The complex number $3i$ is imaginary, and $3i = 0 + 3i$.
**b.** The complex number 87 is a real number, and $87 = 87 + 0i$.

**c.** The complex number $4 - 5i$ is imaginary, and $4 - 5i = 4 + (-5)i$.

**d.** The complex number $0$ is real, and $0 = 0 + 0i$.

**e.** The complex number $\dfrac{1 + \pi i}{2}$ is imaginary, and $\dfrac{1 + \pi i}{2} = \dfrac{1}{2} + \dfrac{\pi}{2}i$.

**TRY THIS.** Determine whether $i - 5$ is real or imaginary and write it in standard form. ∎

Just as there are two types of real numbers (rational and irrational), there are two types of complex numbers, the real numbers and the imaginary numbers. The relationship between these sets of numbers is shown in Fig. 6.1.

**Complex Numbers**

| Real Numbers | Imaginary Numbers |
|---|---|
| $3, -\frac{1}{2}, \pi, \sqrt{5}$, etc. | $i, \frac{1}{2}i, -2 + 3i, \sqrt{2} - i$, etc. |

**Figure 6.1**

## Addition, Subtraction, and Multiplication

Writing and thinking of complex numbers as binomials make it easy to perform operations with complex numbers. The sum of $4 + 5i$ and $1 - 7i$ is found as if we were adding binomials with $i$ being the variable:

$$(4 + 5i) + (1 - 7i) = 5 - 2i$$

To find the sum, we add the real parts and add the imaginary parts. Subtraction is done similarly:

$$(9 - 2i) - (6 + 5i) = 3 - 7i$$

To multiply complex numbers, we use the FOIL method for multiplying binomials:

$$(2 + 3i)(5 - 4i) = 10 - 8i + 15i - 12i^2$$
$$= 10 + 7i - 12(-1) \qquad \text{Replace } i^2 \text{ with } -1.$$
$$= 22 + 7i$$

In the following box we give the definitions of addition, subtraction, and multiplication of complex numbers.

**Definition: Addition, Subtraction, and Multiplication**

If $a + bi$ and $c + di$ are complex numbers, we define their sum, difference, and product as follows:

$$(a + bi) + (c + di) = (a + c) + (b + d)i$$
$$(a + bi) - (c + di) = (a - c) + (b - d)i$$
$$(a + bi)(c + di) = (ac - bd) + (bc + ad)i$$

Note that it is not necessary to memorize these definitions to perform these operations with complex numbers. We get the same results by working with complex numbers as if they were binomials in which $i$ is the variable, replacing $i^2$ by $-1$ wherever it occurs.

EXAMPLE **2** **Operations with complex numbers**

Perform the indicated operations with the complex numbers.

**a.** $(-2 + 3i) + (-4 - 9i)$ **b.** $(-1 - 5i) - (3 - 2i)$
**c.** $2i(3 + i)$ **d.** $(3i)^2$
**e.** $(-3i)^2$ **f.** $(5 - 2i)(5 + 2i)$

**Solution**

**a.** $(-2 + 3i) + (-4 - 9i) = -6 - 6i$
**b.** $(-1 - 5i) - (3 - 2i) = -4 - 3i$
**c.** $2i(3 + i) = 6i + 2i^2$
$\quad\quad\quad\quad = 6i + 2(-1)$
$\quad\quad\quad\quad = -2 + 6i$
**d.** $(3i)^2 = 3^2i^2 = 9(-1) = -9$
**e.** $(-3i)^2 = (-3)^2i^2$
$\quad\quad\quad\quad = 9(-1) = -9$
**f.** $(5 - 2i)(5 + 2i) = 25 - 4i^2$
$\quad\quad\quad\quad\quad\quad\quad = 25 - 4(-1) = 29$

**TRY THIS.** Find the product $(4 - 3i)(1 + 2i)$. ▪

## Powers of *i*

Since $i^2 = -1$, we have $i^3 = i^2 \cdot i = -1 \cdot i = -i$. Since $i^3 = -i$, we have

$$i^4 = i^3 \cdot i = -i \cdot i = -i^2 = -(-1) = 1.$$

The first eight powers of *i* are listed here:

$$i^1 = i \quad\quad\quad\quad i^5 = i$$
$$i^2 = -1 \quad\quad\quad\quad i^6 = -1$$
$$i^3 = -i \quad\quad\quad\quad i^7 = -i$$
$$i^4 = 1 \quad\quad\quad\quad i^8 = 1$$

This list could be continued in this pattern, but any other whole-number power of *i* can be obtained from knowing the first four powers. We can simplify a power of *i* by using the fact that $i^4 = 1$ and $(i^4)^n = 1$ for any integer *n*.

EXAMPLE **3** **Simplifying a power of *i***

Simplify $i^{83}$.

**Solution**

Divide 83 by 4 and write $83 = 4 \cdot 20 + 3$. So

$$i^{83} = (i^4)^{20} \cdot i^3 = 1 \cdot i^3 = -i.$$

**TRY THIS.** Simplify $i^{35}$. ▪

## Division of Complex Numbers

For real numbers, *c* is the quotient of *a* and *b* provided that $c \cdot b = a$. The same definition applies to complex numbers. For example, $m + ni$ is the quotient of $2 + 3i$ and $4 - 5i$ provided that $(m + ni)(4 - 5i) = 2 + 3i$. However, it is not so easy to find $m + ni$ because of the way these complex numbers are multiplied. Before we learn a process for finding the quotient of two complex numbers, we need to

understand complex conjugates. The complex numbers $a + bi$ and $a - bi$ are called **complex conjugates** of each other. The product of a complex number and its conjugate is a real number.

 **EXAMPLE 4** **Products of complex conjugates**

Find the product of the given complex number and its conjugate.

**a.** $3 - i$    **b.** $4 + 2i$    **c.** $-i$

**Solution**

**a.** The conjugate of $3 - i$ is $3 + i$, and

$$(3 - i)(3 + i) = 9 - i^2 = 10.$$

**b.** The conjugate of $4 + 2i$ is $4 - 2i$, and

$$(4 + 2i)(4 - 2i) = 16 - 4i^2 = 20.$$

**c.** The conjugate of $-i$ is $i$, and

$$-i \cdot i = -i^2 = -(-1) = 1.$$

**TRY THIS.** Find the product of $3 - 5i$ and its conjugate.    ∎

In general we have the following theorem about complex conjugates.

**Theorem:**
**Complex Conjugates**

If $a$ and $b$ are real numbers, then the product of $a + bi$ and its conjugate $a - bi$ is the real number $a^2 + b^2$. In symbols,

$$(a + bi)(a - bi) = a^2 + b^2.$$

We use complex conjugates to find the quotient of complex numbers in a process that is similar to rationalizing the denominator.

**PROCEDURE**

**Dividing Complex Numbers**

The quotient of $a + bi$ and $c + di$ is the complex number that is obtained by multiplying the numerator and denominator of $\frac{a + bi}{c + di}$ by the conjugate of the denominator.

**EXAMPLE 5** **Dividing imaginary numbers**

Write each quotient in the form $a + bi$.

**a.** $\dfrac{8 - i}{2 + i}$    **b.** $\dfrac{1}{5 - 4i}$    **c.** $\dfrac{3 - 2i}{i}$

**Solution**

**a.** Multiply the numerator and denominator by $2 - i$, the conjugate of $2 + i$:

$$\frac{8 - i}{2 + i} = \frac{(8 - i)(2 - i)}{(2 + i)(2 - i)}$$

$$= \frac{16 - 10i + i^2}{4 - i^2} = \frac{15 - 10i}{5} = 3 - 2i$$

Check division using multiplication: $(3 - 2i)(2 + i) = 8 - i$.

**b.** $\dfrac{1}{5-4i} = \dfrac{1(5+4i)}{(5-4i)(5+4i)}$

$\qquad = \dfrac{5+4i}{25+16} = \dfrac{5+4i}{41} = \dfrac{5}{41} + \dfrac{4}{41}i$

Check:

$$\left(\dfrac{5}{41} + \dfrac{4}{41}i\right)(5-4i) = \dfrac{25}{41} - \dfrac{20}{41}i + \dfrac{20}{41}i - \dfrac{16}{41}i^2$$

$$= \dfrac{25}{41} + \dfrac{16}{41} = 1$$

**c.** $\dfrac{3-2i}{i} = \dfrac{(3-2i)(-i)}{i(-i)}$

$\qquad = \dfrac{-3i + 2i^2}{-i^2} = \dfrac{-2-3i}{1} = -2 - 3i$

Check: $(-2-3i)(i) = -3i^2 - 2i = 3 - 2i$

**TRY THIS.** Write $\dfrac{4}{1+i}$ in the form $a + bi$. ∎

## Roots of Negative Numbers

In Examples 2(d) and 2(e), we saw that both $(3i)^2 = -9$ and $(-3i)^2 = -9$. This means that in the complex number system there are two square roots of $-9$, $3i$ and $-3i$. For any positive real number $b$, we have $\left(i\sqrt{b}\right)^2 = -b$ and $\left(-i\sqrt{b}\right)^2 = -b$. So there are two square roots of $-b$, $i\sqrt{b}$ and $-i\sqrt{b}$. We call $i\sqrt{b}$ the **principal square root** of $-b$ and make the following definition.

**Definition: Square Root of a Negative Number**

For any positive real number $b$,

$$\sqrt{-b} = i\sqrt{b}.$$

In the real number system, $\sqrt{-2}$ and $\sqrt{-8}$ are undefined, but in the complex number system, they are defined as $\sqrt{-2} = i\sqrt{2}$ and $\sqrt{-8} = i\sqrt{8}$. Even though we now have meaning for a symbol such as $\sqrt{-2}$, *all operations with complex numbers must be performed after converting to the $a + bi$ form.* If we perform operations with roots of negative numbers using properties of the real numbers, we can get contradictory results:

$$\sqrt{-2} \cdot \sqrt{-8} = \sqrt{(-2)(-8)} = \sqrt{16} = 4 \qquad \text{Incorrect}$$
$$\sqrt{-2} \cdot \sqrt{-8} = i\sqrt{2} \cdot i\sqrt{8} = i^2 \cdot \sqrt{16} = -1 \cdot 4 = -4 \qquad \text{Correct}$$

The product rule $\sqrt{a} \cdot \sqrt{b} = \sqrt{ab}$ applies only to nonnegative numbers $a$ and $b$.

**EXAMPLE 6** **Square roots of negative numbers**

Write each expression in the form $a + bi$, where $a$ and $b$ are real numbers.

**a.** $\sqrt{-8} + \sqrt{-18}$  **b.** $\dfrac{-4 + \sqrt{-50}}{4}$  **c.** $\sqrt{-27}\left(\sqrt{9} - \sqrt{-2}\right)$

**Solution**

The first step in each case is to write the principal square roots of the negative numbers.

a. $\sqrt{-8} + \sqrt{-18} = i\sqrt{8} + i\sqrt{18}$

$= 2i\sqrt{2} + 3i\sqrt{2} = 5i\sqrt{2}$

b. $\dfrac{-4 + \sqrt{-50}}{4} = \dfrac{-4 + i\sqrt{50}}{4}$

$= \dfrac{-4 + 5i\sqrt{2}}{4} = -1 + \dfrac{5}{4}i\sqrt{2}$ or $-1 + \dfrac{5\sqrt{2}}{4}i$

c. $\sqrt{-27}\left(\sqrt{9} - \sqrt{-2}\right) = 3i\sqrt{3}\left(3 - i\sqrt{2}\right)$

$= 9i\sqrt{3} - 3i^2\sqrt{6} = 3\sqrt{6} + 9i\sqrt{3}$

**TRY THIS.** Write $\dfrac{2 - \sqrt{-12}}{2}$ in the form $a + bi$.

## Complex Solutions to Equations

We can use the operations for complex numbers defined in this section to determine whether a given complex number satisfies an equation.

**EXAMPLE 7**  Complex number solutions to equations

Determine whether the complex number $3 + i\sqrt{2}$ satisfies $x^2 - 6x + 11 = 0$.

**Solution**

Replace $x$ by $3 + i\sqrt{2}$ in the polynomial $x^2 - 6x + 11$:

$\left(3 + i\sqrt{2}\right)^2 - 6\left(3 + i\sqrt{2}\right) + 11$

$= 3^2 + 6i\sqrt{2} + \left(i\sqrt{2}\right)^2 - 18 - 6i\sqrt{2} + 11$

$= 9 + 6i\sqrt{2} - 2 - 18 - 6i\sqrt{2} + 11$

$= 0$

Since the value of the polynomial is zero for $x = 3 + i\sqrt{2}$, the complex number $3 + i\sqrt{2}$ satisfies the equation $x^2 - 6x + 11 = 0$.

**TRY THIS.** Determine whether $1 + i\sqrt{3}$ satisfies $x^2 - 2x + 4 = 0$.

**FOR THOUGHT...** True or False? Explain.

1. The multiplicative inverse of $i$ is $-i$.

2. The conjugate of $i$ is $-i$.

3. The set of complex numbers is a subset of the set of real numbers.

4. $\left(\sqrt{3} - i\sqrt{2}\right)\left(\sqrt{3} + i\sqrt{2}\right) = 5$

5. $(2 - 5i)(2 + 5i) = 4 + 25$

6. $5 - \sqrt{-9} = 5 - 9i$

7. If $P(x) = x^2 + 9$, then $P(3i) = 0$.

8. The imaginary number $-3i$ is a root to the equation $x^2 + 9 = 0$.

9. $i^4 = 1$

10. $i^{18} = 1$

Scan this code
to link to section
lecture videos.

## 6.1 EXERCISES

### CONCEPTS

*Fill in the blank.*

1. Numbers of the form $a + bi$ where $a$ and $b$ are real numbers are _____.

2. In $a + bi$, $a$ is the _____ and $b$ is the _____.

3. A number of the form $a + bi$ with $b \neq 0$ is a(n) _____.

4. The _____ square root of $-b$ where $b$ is a positive real number is $i\sqrt{b}$.

### SKILLS

*Determine whether each complex number is real or imaginary and write it in the standard form $a + bi$.*

5. $6i$

6. $-3i + \sqrt{6}$

7. $\dfrac{1 + i}{3}$

8. $-72$

9. $\sqrt{7}$

10. $-i\sqrt{5}$

11. $\dfrac{\pi}{2}$

12. $0$

*Perform the indicated operations and write your answers in the form $a + bi$, where $a$ and $b$ are real numbers.*

13. $(3 - 3i) + (4 + 5i)$

14. $(-3 + 2i) + (5 - 6i)$

15. $(1 - i) - (3 + 2i)$

16. $(6 - 7i) - (3 - 4i)$

17. $-6i(3 - 2i)$

18. $-3i(5 + 2i)$

19. $(2 - 3i)(4 + 6i)$

20. $(3 - i)(5 - 2i)$

21. $(5 - 2i)(5 + 2i)$

22. $(4 + 3i)(4 - 3i)$

23. $\left(\sqrt{3} - i\right)\left(\sqrt{3} + i\right)$

24. $\left(\sqrt{2} + i\sqrt{3}\right)\left(\sqrt{2} - i\sqrt{3}\right)$

25. $(3 + 4i)^2$

26. $(-6 - 2i)^2$

27. $\left(\sqrt{5} - 2i\right)^2$

28. $\left(\sqrt{6} + i\sqrt{3}\right)^2$

29. $i^{17}$

30. $i^{24}$

31. $i^{98}$

32. $i^{19}$

33. $i^{-4}$

34. $i^{-13}$

35. $i^{-1}$

36. $i^{-27}$

*Find the product of the given complex number and its conjugate.*

37. $3 - 9i$

38. $4 + 3i$

39. $\dfrac{1}{2} + 2i$

40. $\dfrac{1}{3} - i$

41. $i$

42. $-i\sqrt{5}$

43. $3 - i\sqrt{3}$

44. $\dfrac{5}{2} + i\dfrac{\sqrt{2}}{2}$

*Write each quotient in the form $a + bi$.*

45. $\dfrac{1}{2 - i}$

46. $\dfrac{1}{5 + 2i}$

47. $\dfrac{-3i}{1 - i}$

48. $\dfrac{3i}{-2 + i}$

49. $\dfrac{-2 + 6i}{2}$

50. $\dfrac{-6 - 9i}{-3}$

51. $\dfrac{-3 + 3i}{i}$

52. $\dfrac{-2 - 4i}{-i}$

53. $\dfrac{1 - i}{3 + 2i}$

54. $\dfrac{4 + 2i}{2 - 3i}$

55. $\dfrac{\sqrt{2} - i\sqrt{3}}{\sqrt{3} + i\sqrt{2}}$

56. $\dfrac{\sqrt{3} - i\sqrt{2}}{\sqrt{2} + i\sqrt{3}}$

*Write each expression in the form $a + bi$, where $a$ and $b$ are real numbers.*

57. $\sqrt{-4} - \sqrt{-9}$

58. $\sqrt{-16} + \sqrt{-25}$

59. $\sqrt{-4} - \sqrt{16}$

60. $\sqrt{-3} \cdot \sqrt{-3}$

61. $\left(\sqrt{-6}\right)^2$

62. $\left(\sqrt{-5}\right)^3$

63. $\sqrt{-2} \cdot \sqrt{-50}$

64. $\dfrac{-6 + \sqrt{-3}}{3}$

65. $\dfrac{-2 + \sqrt{-20}}{2}$

66. $\dfrac{9 - \sqrt{-18}}{-6}$

67. $-3 + \sqrt{3^2 - 4(1)(5)}$

68. $1 - \sqrt{(-1)^2 - 4(1)(1)}$

69. $\sqrt{-8}\left(\sqrt{-2} + \sqrt{8}\right)$

70. $\sqrt{-6}\left(\sqrt{2} - \sqrt{-3}\right)$

71. $\left(\sqrt{-4} + \sqrt{8}\right)\left(\sqrt{-1} - \sqrt{2}\right)$

72. $\left(\sqrt{-54} - \sqrt{2}\right)^2$

*Let $P(x) = x^2 + 4x + 5$, $T(x) = 2x^2 + 1$, and $W(x) = x^2 - 6x + 14$. Find each of the following.*

73. $P(-2 + i)$

74. $T\left(\dfrac{i\sqrt{2}}{2}\right)$

**75.** $P(-2 - i)$

**76.** $T\left(\dfrac{-i\sqrt{2}}{2}\right)$

**77.** $P(1 + i)$

**78.** $T(3 - i)$

**79.** $W\left(3 + i\sqrt{5}\right)$

**80.** $W\left(3 - i\sqrt{5}\right)$

**81.** $W(-1)$

**82.** $W(2 - i)$

*Determine whether the given complex number satisfies the equation following it.*

**83.** $2i, x^2 + 4 = 0$

**84.** $-4i, 2x^2 + 32 = 0$

**85.** $1 - i, x^2 + 2x - 2 = 0$

**86.** $1 + i, x^2 - 2x + 2 = 0$

**87.** $3 - 2i, x^2 - 6x + 13 = 0$

**88.** $2 - i, x^2 - 4x - 6 = 0$

**89.** $\dfrac{i\sqrt{3}}{3}, 3x^2 + 1 = 0$

**90.** $\dfrac{-i\sqrt{3}}{3}, 3x^2 - 1 = 0$

**91.** $2 + i\sqrt{3}, x^2 - 4x + 7 = 0$

**92.** $\sqrt{2} + i\sqrt{3}, x^2 - 2\sqrt{2}x + 5 = 0$

## WRITING/DISCUSSION

**93.** Explain in detail how to find $i^n$ for any positive integer $n$.

**94.** Find a complex number $a + bi$ such that $a^2 + b^2$ is irrational.

**95.** Is it true that the product of a complex number and its conjugate is a real number? Explain.

**96.** Show that the reciprocal of $a + bi$ for $a$ and $b$ not both zero is
$$\dfrac{a}{a^2 + b^2} - \dfrac{b}{a^2 + b^2}i.$$

## REVIEW

**97.** Find the amount of force required to push a 600-pound motorcycle up a ramp that is inclined at $20°$.

**98.** Find the magnitude and direction angle for the vector $\langle -3, -9 \rangle$.

**99.** Solve the triangle in which $a = 5$, $b = 9$, and $c = 12$.

**100.** Find all solutions in radians to $\sin x = 0.88$. Round to the nearest hundredth.

**101.** Simplify $\dfrac{\cos(2x)}{\sin^2(x)} - \csc^2(x)$.

**102.** Find a function of the form $y = A\cos(B(x - C)) + D$ that has the same graph as $y = \sin(2x - \pi/4)$.

## OUTSIDE THE BOX

**103.** *Leaning Ladder* A 7-ft ladder is leaning against a vertical wall. There is a point near the bottom of the ladder that is 1 ft from the ground and 1 ft from the wall. Find the exact or approximate distance from the top of the ladder to the ground.

1 ft

1 ft

**Figure for Exercise 103**

**104.** *Twin Squares* Find two different numbers such that each of the numbers is the square of the other.

## 6.1 POP QUIZ

**1.** Find the sum of $3 + 2i$ and $4 - i$.

**2.** Find the product of $4 - 3i$ and $2 + i$.

**3.** Find the product of $2 - 3i$ and its conjugate.

**4.** Write $\dfrac{5}{2 - 3i}$ in the form $a + bi$.

**5.** Determine whether $4 + 2i$ satisfies $x^2 - 8x + 20 = 0$.

# 6.2 Trigonometric Form of Complex Numbers

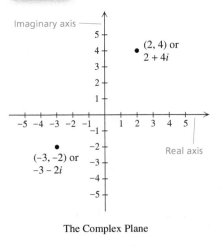

The Complex Plane

**Figure 6.2**

Until now we have concentrated on applying the trigonometric functions to real-life situations. So the trigonometric form of complex numbers, introduced in this section, may be somewhat surprising. However, with this form we can expand our knowledge of complex numbers and find powers and roots of complex numbers that we could not find without it.

## The Complex Plane

A complex number $a + bi$ can be thought of as an ordered pair $(a, b)$, where $a$ is the real part and $b$ is the imaginary part of the complex number. Ordered pairs representing complex numbers are found in a coordinate system just like ordered pairs of real numbers. The horizontal axis is called the **real axis,** and the vertical axis is called the **imaginary axis,** as shown in Fig. 6.2. This coordinate system for complex numbers is called the **complex plane.** The complex plane provides an order to the complex number system and allows us to treat complex numbers like vectors.

We use the same symbols to indicate the absolute value of a real number or a complex number. The absolute value $a + bi$ is written as $|a + bi|$ and it is defined as follows:

**Definition: Absolute Value or Modulus of $a + bi$**

> The **absolute value** or **modulus** of the complex number $a + bi$ is defined by
>
> $$|a + bi| = \sqrt{a^2 + b^2}.$$

The absolute value of the complex number $a + bi$ is the distance between $a + bi$ and the origin in the complex plane. The absolute value of a real number is the distance on the number line between the real number and origin. Note that the absolute value of $a + bi$ is the same as the magnitude of the vector $\langle a, b \rangle$, which we studied in the last chapter.

**Figure 6.3**

**EXAMPLE 1** **Graphing complex numbers**

Graph each complex number and find its absolute value.

**a.** $3 + 4i$   **b.** $-2 - i$

**Solution**

**a.** The complex number $3 + 4i$ is located in the first quadrant, as shown in Fig. 6.3, 3 units to the right of the origin and 4 units upward. Using the definition of absolute value of a complex number, we have

$$|3 + 4i| = \sqrt{3^2 + 4^2} = 5.$$

**b.** The complex number $-2 - i$ is located in the third quadrant, as shown in Fig. 6.3, and

$$|-2 - i| = \sqrt{(-2)^2 + (-1)^2} = \sqrt{5}.$$

**TRY THIS.** Find the absolute value of $5 - i$.   ∎

## Trigonometric Form of a Complex Number

Multiplication and division of complex numbers in the standard form $a + bi$ are rather complicated, as are finding powers and roots, but in trigonometric form these operations are simpler and more natural. A complex number is written in

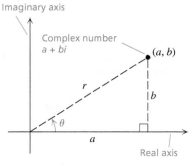

**Figure 6.4**

**Definition: Trigonometric Form of a Complex Number**

**trigonometric form** using a process similar to finding the horizontal and vertical components of a vector.

Consider the complex number $a + bi$ shown in Fig. 6.4, where

$$r = |a + bi| = \sqrt{a^2 + b^2}.$$

Since $(a, b)$ is a point on the terminal side of the angle $\theta$ in standard position, we have

$$a = r \cos \theta \quad \text{and} \quad b = r \sin \theta.$$

Substituting for $a$ and $b$, we get

$$a + bi = (r \cos \theta) + (r \sin \theta)i = r(\cos \theta + i \sin \theta).$$

> If $z = a + bi$ is a complex number, then the **trigonometric form** of $z$ is
>
> $$z = r(\cos \theta + i \sin \theta)$$
>
> where $r = \sqrt{a^2 + b^2}$ and $\theta$ is an angle in standard position whose terminal side contains the point $(a, b)$. An abbreviation for $r(\cos \theta + i \sin \theta)$ is $r \operatorname{cis} \theta$.

The number $r$ is the **modulus** of $a + bi$, and $\theta$ is the **argument** of $a + bi$. Since there are infinitely many angles whose terminal side contains $(a, b)$, the trigonometric form of a complex number is not unique. However, we usually use the smallest nonnegative value for $\theta$ that satisfies

$$a = r \cos \theta \quad \text{or} \quad b = r \sin \theta.$$

In the next two examples, we convert from standard to trigonometric form and vice versa.

**Figure 6.5**

**Figure 6.6**

**EXAMPLE 2** **Writing a complex number in trigonometric form**

Write each complex number in trigonometric form.

**a.** $-2\sqrt{3} + 2i$  **b.** $3 - 2i$

**Solution**

**a.** First we locate $-2\sqrt{3} + 2i$ in the complex plane, as shown in Fig. 6.5. Find $r$ using $a = -2\sqrt{3}$, $b = 2$, and $r = \sqrt{a^2 + b^2}$.

$$r = \sqrt{\left(-2\sqrt{3}\right)^2 + 2^2} = 4$$

Since $a = r \cos \theta$, we have $-2\sqrt{3} = 4 \cos \theta$ or

$$\cos \theta = -\frac{\sqrt{3}}{2}.$$

The smallest nonnegative solution to this equation is $\theta = 5\pi/6$. Since the terminal side of $5\pi/6$ goes through $\left(-2\sqrt{3}, 2\right)$, we get

$$-2\sqrt{3} + 2i = 4(\cos 5\pi/6 + i \sin 5\pi/6).$$

**b.** Locate $3 - 2i$ in the complex plane, as shown in Fig. 6.6. Next, find $r$ by using $a = 3$, $b = -2$, and $r = \sqrt{a^2 + b^2}$:

$$r = \sqrt{(3)^2 + (-2)^2} = \sqrt{13}$$

Since $a = r \cos \theta$, we have $3 = \sqrt{13} \cos \theta$ or

$$\cos \theta = \frac{3}{\sqrt{13}}.$$

Use a calculator to get

$$\cos^{-1}\left(\frac{3}{\sqrt{13}}\right) \approx 33.7°.$$

Since $\theta$ is an angle whose terminal side contains $(3, -2)$, choose $\theta = 360° - 33.7° = 326.3°$. So

$$3 - 2i \approx \sqrt{13}(\cos 326.3° + i \sin 326.3°).$$

**TRY THIS.** Write $1 + 2i$ in trigonometric form using degree measure.    ■

**EXAMPLE 3** **Writing a complex number in standard form**

Write each complex number in the form $a + bi$.

**a.** $\sqrt{2}(\cos(\pi/4) + i \sin(\pi/4))$    **b.** $3.6(\cos 143° + i \sin 143°)$

**Solution**

**a.** Since $\cos(\pi/4) = \sqrt{2}/2$ and $\sin(\pi/4) = \sqrt{2}/2$, we have

$$\sqrt{2}(\cos(\pi/4) + i \sin(\pi/4)) = \sqrt{2}\left(\frac{\sqrt{2}}{2} + i\frac{\sqrt{2}}{2}\right)$$

$$= 1 + i.$$

**b.** Use a calculator to find $\cos 143°$ and $\sin 143°$.

$$3.6(\cos 143° + i \sin 143°) \approx -2.875 + 2.167i$$

**TRY THIS.** Write $12\left(\cos\frac{\pi}{6} + i \sin\frac{\pi}{6}\right)$ in standard form.    ■

## Products and Quotients in Trigonometric Form

In standard form, multiplication and division of complex numbers are rather complicated, whereas addition and subtraction are simple. In trigonometric form, addition and subtraction are complicated, whereas multiplication and division are simple. We multiply the complex numbers

$$z_1 = r_1(\cos \theta_1 + i \sin \theta_1) \quad \text{and} \quad z_2 = r_2(\cos \theta_2 + i \sin \theta_2)$$

as follows:

$$z_1 z_2 = r_1(\cos \theta_1 + i \sin \theta_1) \cdot r_2(\cos \theta_2 + i \sin \theta_2)$$

$$= r_1 r_2(\cos \theta_1 + i \sin \theta_1)(\cos \theta_2 + i \sin \theta_2)$$

$$= r_1 r_2[(\cos \theta_1 \cos \theta_2 - \sin \theta_1 \sin \theta_2) + i(\sin \theta_1 \cos \theta_2 + \cos \theta_1 \sin \theta_2)]$$

Using the identities for cosine and sine of a sum, we get

$$z_1 z_2 = r_1 r_2[\cos(\theta_1 + \theta_2) + i \sin(\theta_1 + \theta_2)].$$

The last equation gives us a rule for finding the product of the complex numbers $z_1$ and $z_2$ in trigonometric form: *To find the product $z_1 z_2$, multiply the moduli $r_1$ and $r_2$ and add the arguments $\theta_1$ and $\theta_2$.*

We find the quotient of $z_1$ and $z_2$ as follows:

$$\frac{z_1}{z_2} = \frac{r_1(\cos\theta_1 + i\sin\theta_1)}{r_2(\cos\theta_2 + i\sin\theta_2)}$$

$$= \frac{r_1(\cos\theta_1 + i\sin\theta_1)(\cos\theta_2 - i\sin\theta_2)}{r_2(\cos\theta_2 + i\sin\theta_2)(\cos\theta_2 - i\sin\theta_2)}$$

$$= \frac{r_1}{r_2}[(\cos\theta_1\cos\theta_2 + \sin\theta_1\sin\theta_2) + i(\sin\theta_1\cos\theta_2 - \cos\theta_1\sin\theta_2)]$$

Using the identities for cosine and sine of a difference, we get

$$\frac{z_1}{z_2} = \frac{r_1}{r_2}[\cos(\theta_1 - \theta_2) + i\sin(\theta_1 - \theta_2)].$$

The last equation gives us a rule for finding the quotient of the complex numbers $z_1$ and $z_2$ in trigonometric form: *To find the quotient $z_1/z_2$, divide the moduli $r_1$ and $r_2$ and subtract the arguments $\theta_1$ and $\theta_2$.* The rules for multiplying and dividing complex numbers in trigonometric form are stated precisely in the following theorem.

**Theorem: The Product and Quotient of Complex Numbers**

If $z_1 = r_1(\cos\theta_1 + i\sin\theta_1)$ and $z_2 = r_2(\cos\theta_2 + i\sin\theta_2)$, then

$$z_1 z_2 = r_1 r_2[\cos(\theta_1 + \theta_2) + i\sin(\theta_1 + \theta_2)]$$

and

$$\frac{z_1}{z_2} = \frac{r_1}{r_2}[\cos(\theta_1 - \theta_2) + i\sin(\theta_1 - \theta_2)].$$

**EXAMPLE 4** **A product and quotient using trigonometric form**

Find $z_1 z_2$ and $z_1/z_2$ for

$$z_1 = 6(\cos 2.4 + i\sin 2.4) \quad \text{and} \quad z_2 = 2(\cos 1.8 + i\sin 1.8).$$

Express the answers in the form $a + bi$.

**Solution**

Find the product by multiplying the moduli and adding the arguments:

$$z_1 z_2 = 6 \cdot 2(\cos(2.4 + 1.8) + i\sin(2.4 + 1.8))$$

$$= 12(\cos 4.2 + i\sin 4.2)$$

$$= 12\cos 4.2 + 12i\sin 4.2$$

$$\approx -5.88 - 10.46i \qquad \text{Radian mode}$$

Find the quotient by dividing the moduli and subtracting the arguments:

$$\frac{z_1}{z_2} = \frac{6}{2}(\cos(2.4 - 1.8) + i\sin(2.4 - 1.8))$$

$$= 3(\cos(0.6) + i\sin(0.6))$$

$$= 3\cos 0.6 + 3i\sin 0.6$$

$$\approx 2.48 + 1.69i$$

**TRY THIS.** Find $z_1 z_2$ for $z_1 = 4\left(\cos\frac{\pi}{12} + i\sin\frac{\pi}{12}\right)$ and $z_2 = 8\left(\cos\frac{\pi}{12} + i\sin\frac{\pi}{12}\right)$. ■

EXAMPLE **5**    **A product and quotient using trigonometric form**

Use trigonometric form to find $z_1 z_2$ and $z_1/z_2$ if

$$z_1 = -2 + 2i\sqrt{3} \quad \text{and} \quad z_2 = \sqrt{3} + i.$$

**Solution**

First find the modulus of $z_1$ and $z_2$:

$$|z_1| = \sqrt{(-2)^2 + (2\sqrt{3})^2} = 4 \quad \text{and} \quad |z_2| = \sqrt{(\sqrt{3})^2 + 1^2} = 2$$

Now $120°$ is a positive angle whose terminal side contains $(-2, 2\sqrt{3})$ and $30°$ is a positive angle whose terminal side contains $(\sqrt{3}, 1)$. So

$$z_1 = 4(\cos 120° + i \sin 120°) \quad \text{and} \quad z_2 = 2(\cos 30° + i \sin 30°).$$

To find the product we multiply the moduli and add the arguments:

$$z_1 z_2 = 4 \cdot 2[\cos(120° + 30°) + i \sin(120° + 30°)]$$

$$= 8(\cos 150° + i \sin 150°)$$

Using $\cos 150° = -\sqrt{3}/2$ and $\sin 150° = 1/2$, we get

$$z_1 z_2 = 8\left(-\frac{\sqrt{3}}{2} + i\frac{1}{2}\right) = -4\sqrt{3} + 4i.$$

To find the quotient, we divide the moduli and subtract the arguments:

$$\frac{z_1}{z_2} = \frac{4}{2}[\cos(120° - 30°) + i \sin(120° - 30°)]$$

$$= 2(\cos 90° + i \sin 90°)$$

$$= 2(0 + i)$$

$$= 2i$$

Check these results by performing the operations with $z_1$ and $z_2$ in standard form.

**TRY THIS.** Find $z_1 z_2$ using trigonometric form for $z_1 = 2\sqrt{3} + 2i$ and $z_2 = 3 + 3i\sqrt{3}$. ∎

## Complex Conjugates

The notation $\bar{z}$ (read "z bar") is used for the complex conjugate of $z$. If $z = a + bi$ and $\bar{z} = a - bi$, then $z\bar{z} = a^2 + b^2$, which is a real number. In trigonometric form

$$z = a + bi = r(\cos \theta + i \sin \theta),$$

and we could write $\bar{z} = a - bi = r(\cos \theta - i \sin \theta)$. However, $r(\cos \theta - i \sin \theta)$ is not written in trigonometric form, which is $r(\cos \theta + i \sin \theta)$. (Trigonometric form is necessary for multiplying and dividing by the rules that we just learned.) Because $\cos(-\theta) = \cos(\theta)$ and $\sin(-\theta) = -\sin(\theta)$, we have

$$\bar{z} = a - bi = r(\cos \theta - i \sin \theta) = r(\cos(-\theta) + i \sin(-\theta)).$$

**Theorem:**
**Complex Conjugates**

The conjugate of the complex number $r(\cos \theta + i \sin \theta)$ is

$$r(\cos(-\theta) + i \sin(-\theta)).$$

The product of $z$ and $\bar{z}$ is a real number:

$$z \cdot \bar{z} = (r(\cos\theta + i\sin\theta))(r(\cos(-\theta) + i\sin(-\theta)))$$
$$= r^2(\cos 0 + i\sin 0)$$
$$= r^2$$

So the product of $z$ and its conjugate $\bar{z}$ is the real number $r^2$.

**EXAMPLE 6**    **Complex conjugates in trigonometric form**

Find the product of $2(\cos(\pi/3) + i\sin(\pi/3))$ and its conjugate using multiplication in trigonometric form.

**Solution**

The conjugate is $2(\cos(-\pi/3) + i\sin(-\pi/3))$. We find the product by multiplying the moduli and adding the arguments:

$$2\left(\cos\frac{\pi}{3} + i\sin\frac{\pi}{3}\right) \cdot 2\left(\cos\left(-\frac{\pi}{3}\right) + i\sin\left(-\frac{\pi}{3}\right)\right) = 4(\cos 0 + i\sin 0) = 4$$

**TRY THIS.** Find the product of $6(\cos 30° + i\sin 30°)$ and its conjugate using multiplication in trigonometric form.    ∎

## FOR THOUGHT...   True or False? Explain.

1. The complex number $3 - 4i$ lies in quadrant IV of the complex plane.

2. The absolute value of $-2 - 5i$ is $2 + 5i$.

3. If $\theta$ is an angle whose terminal side contains $(1, -3)$, then $\cos\theta = 1/\sqrt{10}$.

4. If $\theta$ is an angle whose terminal side contains $(-2, -3)$, then $\tan\theta = 2/3$.

5. $i = 1(\cos 0° + i\sin 0°)$

6. The smallest positive solution to $\sqrt{3} = 2\cos\theta$ is $\theta = 30°$.

7. The modulus of $2 - 5i$ is $\sqrt{29}$.

8. An argument for $2 - 4i$ is $\theta = 360° - \cos^{-1}(1/\sqrt{5})$.

9. $3(\cos\pi/4 + i\sin\pi/4) \cdot 2(\cos\pi/2 + i\sin\pi/2) = 6(\cos 3\pi/4 + i\sin 3\pi/4)$

10. $\dfrac{3(\cos\pi/4 + i\sin\pi/4)}{2(\cos\pi/2 + i\sin\pi/2)} = 1.5(\cos\pi/4 + i\sin\pi/4)$

Scan this code to link to section lecture videos.

## 6.2  EXERCISES

### CONCEPTS

*Fill in the blank.*

1. Complex numbers are graphed in the _____ plane.

2. In the complex plane, the horizontal axis is the _____ axis and the vertical axis is the _____ axis.

3. The value of $\sqrt{a^2 + b^2}$ is the _____ of the complex number $a + bi$.

4. Complex numbers can be expressed in standard form or _____ form.

5. In the trigonometric form of a complex number $r(\cos\theta + i\sin\theta)$, $r$ is the _____ and $\theta$ is the _____.

6. The product of two complex numbers in trigonometric form is found by _____ their moduli and _____ their arguments.

7. The _____ of two complex numbers in trigonometric form is found by dividing their moduli and subtracting their arguments.

8. The _____ of the complex number $r(\cos\theta + i\sin\theta)$ is $r(\cos(-\theta) + i\sin(-\theta))$.

### SKILLS

*Graph each complex number, and find its absolute value.*

9. $2 - 6i$

10. $-1 - i$

11. $-2 + 2i\sqrt{3}$

12. $-\sqrt{3} - i$

**13.** $8i$

**14.** $-3i$

**15.** $-9$

**16.** $-\sqrt{6}$

**17.** $\dfrac{1}{\sqrt{2}} - \dfrac{i}{\sqrt{2}}$

**18.** $\dfrac{\sqrt{3}}{2} + \dfrac{i}{2}$

**19.** $3 + 3i$

**20.** $-4 - 4i$

*Write each complex number in trigonometric form, using degree measure for the argument.*

**21.** $-3 + 3i$

**22.** $4 - 4i$

**23.** $-\dfrac{3}{\sqrt{2}} + \dfrac{3i}{\sqrt{2}}$

**24.** $\dfrac{\sqrt{3}}{6} + \dfrac{i}{6}$

**25.** $8$

**26.** $-7$

**27.** $i\sqrt{3}$

**28.** $-5i$

**29.** $-\sqrt{3} + i$

**30.** $-2 - 2i\sqrt{3}$

**31.** $3 + 4i$

**32.** $-2 + i$

**33.** $-3 + 5i$

**34.** $-2 - 4i$

**35.** $3 - 6i$

**36.** $5 - 10i$

*Write each complex number in the form $a + bi$.*

**37.** $\sqrt{2}(\cos 45° + i \sin 45°)$    **38.** $6(\cos 30° + i \sin 30°)$

**39.** $\dfrac{\sqrt{3}}{2}(\cos 150° + i \sin 150°)$

**40.** $12(\cos(\pi/10) + i \sin(\pi/10))$

**41.** $\dfrac{1}{2}(\cos 3.7 + i \sin 3.7)$

**42.** $4.3(\cos(\pi/9) + i \sin(\pi/9))$

**43.** $3(\cos 90° + i \sin 90°)$

**44.** $4(\cos 180° + i \sin 180°)$

**45.** $\sqrt{3}(\cos(3\pi/2) + i \sin(3\pi/2))$

**46.** $8.1(\cos \pi + i \sin \pi)$

**47.** $\sqrt{6}(\cos 60° + i \sin 60°)$

**48.** $0.5(\cos(5\pi/6) + i \sin(5\pi/6))$

*Perform the indicated operations. Write the answer in the form $a + bi$.*

**49.** $2(\cos 150° + i \sin 150°) \cdot 3(\cos 300° + i \sin 300°)$

**50.** $\sqrt{3}(\cos 45° + i \sin 45°) \cdot \sqrt{2}(\cos 315° + i \sin 315°)$

**51.** $\sqrt{3}(\cos 10° + i \sin 10°) \cdot \sqrt{2}(\cos 20° + i \sin 20°)$

**52.** $8(\cos 100° + i \sin 100°) \cdot 3(\cos 35° + i \sin 35°)$

**53.** $[3(\cos 45° + i \sin 45°)]^2$

**54.** $\left[ \sqrt{5}(\cos(\pi/12) + i \sin(\pi/12)) \right]^2$

**55.** $\dfrac{4(\cos(\pi/3) + i \sin(\pi/3))}{2(\cos(\pi/6) + i \sin(\pi/6))}$

**56.** $\dfrac{9(\cos(\pi/4) + i \sin(\pi/4))}{3(\cos(5\pi/4) + i \sin(5\pi/4))}$

**57.** $\dfrac{4.1(\cos 36.7° + i \sin 36.7°)}{8.2(\cos 84.2° + i \sin 84.2°)}$

**58.** $\dfrac{18(\cos 121.9° + i \sin 121.9°)}{2(\cos 325.6° + i \sin 325.6°)}$

*Find $z_1 z_2$ and $z_1/z_2$ for each pair of complex numbers, using trigonometric form. Write the answer in the form $a + bi$.*

**59.** $z_1 = 4 + 4i,\ z_2 = -5 - 5i$

**60.** $z_1 = -3 + 3i,\ z_2 = -2 - 2i$

**61.** $z_1 = \sqrt{3} + i,\ z_2 = 2 + 2i\sqrt{3}$

**62.** $z_1 = -\sqrt{3} + i,\ z_2 = 4\sqrt{3} - 4i$

**63.** $z_1 = 2 + 2i,\ z_2 = \sqrt{2} - i\sqrt{2}$

**64.** $z_1 = \sqrt{5} + i\sqrt{5},\ z_2 = -\sqrt{6} - i\sqrt{6}$

**65.** $z_1 = 3 + 4i,\ z_2 = -5 - 2i$

**66.** $z_1 = 3 - 4i,\ z_2 = -1 + 3i$

**67.** $z_1 = 2 - 6i,\ z_2 = -3 - 2i$

**68.** $z_1 = 1 + 4i,\ z_2 = -4 - 2i$

**69.** $z_1 = 3i,\ z_2 = 1 + i$

**70.** $z_1 = 4,\ z_2 = -3 + i$

*Find the product of the given complex number and its complex conjugate in trigonometric form.*

**71.** $3\left(\cos \dfrac{\pi}{6} + i \sin \dfrac{\pi}{6}\right)$    **72.** $5\left(\cos \dfrac{\pi}{7} + i \sin \dfrac{\pi}{7}\right)$

**73.** $2(\cos 7° + i \sin 7°)$    **74.** $6(\cos 5° + i \sin 5°)$

*Solve each problem.*

**75.** If $z = 3 + 3i$, find $z^3$ by writing $z$ in trigonometric form and computing the product $z \cdot z \cdot z$.

**76.** If $z = \sqrt{3} + i$, find $z^4$ by writing $z$ in trigonometric form and computing the product $z \cdot z \cdot z \cdot z$.

## WRITING/DISCUSSION

**77.** Show that the reciprocal of $z = r(\cos \theta + i \sin \theta)$ is $z^{-1} = r^{-1}(\cos \theta - i \sin \theta)$ provided $r \neq 0$.

**78.** If $z = r(\cos \theta + i \sin \theta)$, then find $z^2$ and $z^{-2}$ provided $r \neq 0$.

**79.** Find the sum of $6(\cos 9° + i \sin 9°)$ and $3(\cos 5° + i \sin 5°)$. Find the sum of $1 + 3i$ and $5 - 7i$. Is it easier to add complex numbers in trigonometric form or standard form?

**80.** Find the quotient when $6(\cos 9° + i \sin 9°)$ is divided by $3(\cos 5° + i \sin 5°)$. Find the quotient when $1 + 3i$ is divided by $5 - 7i$. Is it easier to divide complex numbers in trigonometric form or standard form?

## REVIEW

**81.** Perform the indicated operations and write your answers in $a + bi$ form.
  **a.** $(3 + 5i)^2$

  **b.** $\dfrac{1 + i}{2 + i}$

  **c.** $\sqrt{-8} + \sqrt{-50}$

**82.** Find all solutions in radians to $\cos(2x) = 0.89$. Round to the nearest hundredth.

**83.** The five key points on one cycle of a sine wave are $(\pi/3, -1)$, $(2\pi/3, -2)$, $(\pi, -3)$, $(4\pi/3, -2)$, and $(5\pi/3, -1)$. Find the equation of the wave in the form $y = A \cos(B(x - C)) + D$.

**84.** Convert each degree measure to the exact radian measure.
  **a.** $180°$                     **b.** $-150°$

  **c.** $120°$                     **d.** $-225°$

**85.** Find the period of each function.
  **a.** $y = 3 \sin(2x) + 5$

  **b.** $y = \dfrac{1}{2} \cos(2\pi x - 7) - 12$

  **c.** $f(x) = \tan(4x - 4) + 3$

  **d.** $f(x) = \dfrac{3}{4} \sec(3x + 1) - 9$

**86.** Simplify each expression.
  **a.** $\sin^2(\beta) + \cos^2(\beta)$

  **b.** $\csc^2(2y) - \cot^2(2y)$

  **c.** $\tan^2(9w) - \sec^2(9w)$

## OUTSIDE THE BOX

**87.** *Filling a Triangle*  Fiber-optic cables just fit inside a triangular pipe as shown in the figure. The cables have circular cross sections and the cross section of the pipe is an equilateral triangle with sides of length 1. Suppose that there are $n$ cables of the same size in the bottom row, $n - 1$ of that size in the next row, and so on.
  **a.** Write the total cross-sectional area of the cables as a function of $n$.

  **b.** As $n$ approaches infinity, will the triangular pipe get totally filled with cables? Explain.

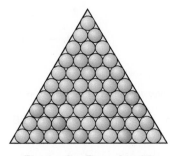

**Figure for Exercise 87**

**88.** *Regular Polygons*  An equilateral triangle and a regular hexagon have equal perimeters. What is the ratio of the area of the triangle to the area of the hexagon?

## 6.2 POP QUIZ

**1.** Find the absolute value of $3 - i$.

**2.** Write $5 + 5i$ in trigonometric form.

**3.** Write $2\sqrt{3}(\cos(5\pi/6) + i \sin(5\pi/6))$ in standard form.

**4.** Find $zw$ if $z = \sqrt{2}(\cos(\pi/12) + i \sin(\pi/12))$ and $w = 3\sqrt{2}(\cos(\pi/12) + i \sin(\pi/12))$.

**5.** Find the product of $\sqrt{2}(\cos(\pi/4) + i \sin(\pi/4))$ and its complex conjugate.

# 6.3  Powers and Roots of Complex Numbers

We know that the square root of a negative number is an imaginary number, but we have not yet encountered roots of imaginary numbers. In this section we will use the trigonometric form of a complex number to find any power or root of a complex number.

# De Moivre's Theorem

Powers of a complex number can be found by repeated multiplication. Repeated multiplication can be done with standard form or trigonometric form. However, in trigonometric form a product is found by simply multiplying the moduli and adding the arguments. So, if

$$z = r(\cos \theta + i \sin \theta),$$

then

$$z^2 = r(\cos \theta + i \sin \theta) \cdot r(\cos \theta + i \sin \theta)$$

$$= r^2(\cos 2\theta + i \sin 2\theta). \qquad \text{Since } r \cdot r = r^2 \text{ and } \theta + \theta = 2\theta$$

We can now find $z^3$ because $z^3 = z^2 \cdot z$:

$$z^3 = r^2(\cos 2\theta + i \sin 2\theta) \cdot r(\cos \theta + i \sin \theta)$$

$$= r^3(\cos 3\theta + i \sin 3\theta)$$

Multiplying $z^3$ by $z$ gives

$$z^4 = r^4(\cos 4\theta + i \sin 4\theta).$$

If you examine the results of $z^2, z^3$, and $z^4$ you will see a pattern, which can be stated as follows.

**De Moivre's Theorem**

> If $z = r(\cos \theta + i \sin \theta)$ is a complex number and $n$ is any positive integer, then
>
> $$z^n = r^n(\cos n\theta + i \sin n\theta).$$

De Moivre's theorem is named after the French mathematician Abraham De Moivre (1667–1754).

**EXAMPLE 1** **A power of a complex number**

Use De Moivre's theorem to simplify $\left(-\sqrt{3} + i\right)^8$. Write the answer in the form $a + bi$.

**Solution**

First write $-\sqrt{3} + i$ in trigonometric form. The modulus is 2, and the argument is 150°. So

$$-\sqrt{3} + i = 2(\cos 150° + i \sin 150°).$$

Use De Moivre's theorem to find the eighth power:

$$\left(-\sqrt{3} + i\right)^8 = [2(\cos 150° + i \sin 150°)]^8$$

$$= 2^8[\cos(8 \cdot 150°) + i \sin(8 \cdot 150°)]$$

$$= 256[\cos 1200° + i \sin 1200°]$$

Since $1200° = 3 \cdot 360° + 120°$,

$$\cos 1200° = \cos 120° = -\frac{1}{2} \qquad \text{and} \qquad \sin 1200° = \sin 120° = \frac{\sqrt{3}}{2}.$$

Use these values to simplify the trigonometric form of $\left(-\sqrt{3} + i\right)^8$:

$$\left(-\sqrt{3} + i\right)^8 = 256\left(-\frac{1}{2} + i\frac{\sqrt{3}}{2}\right)$$

$$= -128 + 128i\sqrt{3}$$

**TRY THIS.** Simplify $(1 + i)^6$. ∎

We have seen how De Moivre's theorem is used to find a positive integral power of a complex number. It is also used for finding roots of complex numbers.

## Roots of a Complex Number

In the real number system, $a$ is an $n$th root of $b$ if $a^n = b$. We have a similar definition of $n$th root in the complex number system.

**Definition: $n$th Root of a Complex Number**

The complex number $w = a + bi$ is an $n$th root of the complex number $z$ if

$$(a + bi)^n = z.$$

In Example 1, we saw that $\left(-\sqrt{3} + i\right)^8 = -128 + 128i\sqrt{3}$. So $-\sqrt{3} + i$ is an eighth root of $-128 + 128i\sqrt{3}$.

The definition of $n$th root does not indicate how to find an $n$th root, but a formula for all $n$th roots can be found by using trigonometric form. Suppose that the complex number $w$ is an $n$th root of the complex number $z$, where the trigonometric forms of $w$ and $z$ are

$$w = s(\cos \alpha + i \sin \alpha) \quad \text{and} \quad z = r(\cos \theta + i \sin \theta).$$

By De Moivre's theorem, $w^n = s^n(\cos n\alpha + i \sin n\alpha)$. Since $w^n = z$, we have

$$w^n = s^n(\cos n\alpha + i \sin n\alpha) = r(\cos \theta + i \sin \theta).$$

This last equation gives two different trigonometric forms for $w^n$. The absolute value of $w^n$ in one form is $s^n$ and in the other it is $r$. So $s^n = r$, or $s = r^{1/n}$. Because the argument of a complex number is any angle in the complex plane whose terminal side goes through the point representing the complex number, any two angles used for the argument must differ by a multiple of $2\pi$. Since the argument for $w^n$ is either $n\alpha$ or $\theta$, we have $n\alpha = \theta + 2k\pi$ or

$$\alpha = \frac{\theta + 2k\pi}{n}.$$

We can get $n$ different values for $\alpha$ from this formula by using $k = 0, 1, 2, \ldots, n - 1$. These results are summarized in the following theorem.

**Theorem: $n$th Roots of a Complex Number**

For any positive integer $n$, the complex number $z = r(\cos \theta + i \sin \theta)$ has exactly $n$ distinct $n$th roots given by the expression

$$r^{1/n}\left[\cos\left(\frac{\theta + 2k\pi}{n}\right) + i \sin\left(\frac{\theta + 2k\pi}{n}\right)\right]$$

for $k = 0, 1, 2, \ldots, n - 1$.

The $n$ distinct $n$th roots of a complex number all have the same modulus, $r^{1/n}$. What distinguishes between these roots is the $n$ different arguments determined by

$$\alpha = \frac{\theta + 2k\pi}{n}$$

for $k = 0, 1, 2, \ldots, n - 1$. If $\theta$ is measured in degrees, then the $n$ different arguments are determined by

$$\alpha = \frac{\theta + k360°}{n}$$

for $k = 0, 1, 2, \ldots, n - 1$. Note that if $k \geq n$, we get new values for $\alpha$ but no new $n$th roots, because the values of $\sin \alpha$ and $\cos \alpha$ have already occurred for some

$k < n$. For example, if $k = n$, then $\alpha = \theta/n + 2\pi$, and the values for $\cos \alpha$ and $\sin \alpha$ are the same as they were for $\alpha = \theta/n$, which corresponds to $k = 0$.

**EXAMPLE 2** Finding $n$th roots

Find all of the fourth roots of the complex number $-8 + 8i\sqrt{3}$.

**Solution**

First find the modulus of $-8 + 8i\sqrt{3}$:

$$\sqrt{(-8)^2 + \left(8\sqrt{3}\right)^2} = 16$$

Since the terminal side of $\theta$ contains $\left(-8, 8\sqrt{3}\right)$, we have

$$\cos \theta = -\frac{8}{16} = -\frac{1}{2}.$$

Choose $\theta = 120°$. So the fourth roots are generated by the expression

$$16^{1/4}\left[\cos\left(\frac{120° + k360°}{4}\right) + i \sin\left(\frac{120° + k360°}{4}\right)\right].$$

Evaluating this expression for $k = 0, 1, 2,$ and $3$ (because $n = 4, n - 1 = 3$), we get

$$2[\cos 30° + i \sin 30°] = \sqrt{3} + i$$
$$2[\cos 120° + i \sin 120°] = -1 + i\sqrt{3}$$
$$2[\cos 210° + i \sin 210°] = -\sqrt{3} - i$$
$$2[\cos 300° + i \sin 300°] = 1 - i\sqrt{3}.$$

So, the fourth roots of $-8 + 8i\sqrt{3}$ are $\sqrt{3} + i$, $-1 + i\sqrt{3}$, $-\sqrt{3} - i$, and $1 - i\sqrt{3}$.

**TRY THIS.** Find all fourth roots of $-8 - 8i\sqrt{3}$. ∎

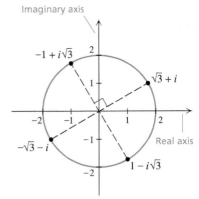

Fourth Roots of $-8 + 8i\sqrt{3}$

**Figure 6.7**

The graph of the four fourth roots found in Example 2 reveals an interesting pattern. Since the modulus of $-8 + 8i\sqrt{3}$ is $16, -8 + 8i\sqrt{3}$ lies on a circle of radius 16 centered at the origin. Since the modulus of each fourth root is $16^{1/4} = 2$, the fourth roots lie on a circle of radius 2 centered at the origin as shown in Fig. 6.7. Note how the fourth roots in Fig. 6.7 divide the circumference of the circle into four equal parts. In general, the $n$th roots of $z = r(\cos \theta + i \sin \theta)$ divide the circumference of a circle of radius $r^{1/n}$ into $n$ equal parts. This symmetric arrangement of the $n$th roots on a circle provides a simple check on whether the $n$th roots are correct.

Consider the equation $x^6 - 1 = 0$. By the fundamental theorem of algebra, this equation has six solutions in the complex number system if multiplicity is considered. Since this equation is equivalent to $x^6 = 1$, each solution is a sixth root of 1. The roots of 1 are called the **roots of unity.** The fundamental theorem of algebra does not guarantee that all of the roots are unique, but the theorem on the $n$th roots of a complex number does.

**EXAMPLE 3** The six sixth roots of unity

Solve the equation $x^6 - 1 = 0$.

**Solution**

The solutions to the equation are the sixth roots of 1. Since $1 = 1(\cos 0° + i \sin 0°)$, the expression for the sixth roots is

$$1^{1/6}\left[\cos\left(\frac{0° + k360°}{6}\right) + i \sin\left(\frac{0° + k360°}{6}\right)\right].$$

**Figure 6.8**

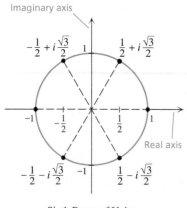

Sixth Roots of Unity

**Figure 6.9**

Evaluating this expression for $k = 0, 1, 2, 3, 4, 5$ gives the six roots:

$$\cos 0° + i \sin 0° = 1 \qquad\qquad \cos 60° + i \sin 60° = \frac{1}{2} + i\frac{\sqrt{3}}{2}$$

$$\cos 120° + i \sin 120° = -\frac{1}{2} + i\frac{\sqrt{3}}{2} \qquad \cos 180° + i \sin 180° = -1$$

$$\cos 240° + i \sin 240° = -\frac{1}{2} - i\frac{\sqrt{3}}{2} \qquad \cos 300° + i \sin 300° = \frac{1}{2} - i\frac{\sqrt{3}}{2}$$

You can check the six roots using a graphing calculator that handles complex numbers as shown in Fig. 6.8. Note that the calculator does not always get the exact answer.

**TRY THIS.** Find all complex solutions to $x^3 - 27 = 0$. ∎

A graph of the sixth roots of unity found in Example 3 is shown in Fig. 6.9. Notice that they divide the circle of radius 1 into six equal parts. According to the conjugate pairs theorem in algebra, the complex solutions to a polynomial equation with real coefficients occur in conjugate pairs. (Conjugate pairs are $a + bi$ and $a - bi$.) In Example 3, the complex solutions of $x^6 - 1 = 0$ did occur in conjugate pairs. This fact gives us another method of checking whether the six roots are correct. In general, the $n$th roots of any real number occur in conjugate pairs. Note that the fourth roots of the imaginary number in Example 2 did not occur in conjugate pairs.

## FOR THOUGHT... True or False? Explain.

1. $(2 + 3i)^2 = 4 + 9i^2$

2. If $z = 2(\cos 120° + i \sin 120°)$, then $z^3 = 8i$.

3. If $z$ is a complex number with modulus $r$, then the modulus of $z^4$ is $r^4$.

4. If the argument of $z$ is $\theta$, then the argument of $z^4$ is $\cos 4\theta$.

5. $[\cos(\pi/3) + i \sin(\pi/6)]^2 = \cos(2\pi/3) + i \sin(\pi/3)$

6. If $\cos \alpha = \cos \beta$, then $\alpha = \beta + 2k\pi$ for some integer $k$.

7. One of the fifth roots of 32 is $2(\cos 72° + i \sin 72°)$.

8. All solutions to $x^8 = 1$ in the complex plane lie on the unit circle.

9. All solutions to $x^8 = 1$ lie on the axes or the lines $y = \pm x$.

10. The equation $x^4 + 81 = 0$ has two real and two imaginary solutions.

## 6.3 EXERCISES

### CONCEPTS

*Fill in the blank.*

1. Powers of a complex number can be found using _____ theorem.

2. The roots of 1 are called the roots of _____.

### SKILLS

*Use De Moivre's theorem to simplify each expression. Write the answer in the form $a + bi$.*

3. $[3(\cos 30° + i \sin 30°)]^3$

4. $[2(\cos 45° + i \sin 45°)]^5$

5. $\left[\sqrt{2}(\cos 120° + i \sin 120°)\right]^4$

6. $\left[\sqrt{3}(\cos 150° + i \sin 150°)\right]^3$

7. $[\cos(\pi/12) + i \sin(\pi/12)]^8$

8. $[\cos(\pi/6) + i \sin(\pi/6)]^9$

9. $\left[\sqrt{6}(\cos(2\pi/3) + i \sin(2\pi/3))\right]^4$

10. $\left[\sqrt{18}(\cos(5\pi/6) + i \sin(5\pi/6))\right]^3$

11. $[4.3(\cos 12.3° + i \sin 12.3°)]^5$

12. $[4.9(\cos 37.4° + i \sin 37.4°)]^6$

*Simplify each expression, by using trigonometric form and De Moivre's theorem.*

**13.** $(2 + 2i)^3$

**14.** $(1 - i)^3$

**15.** $\left(\sqrt{3} - i\right)^4$

**16.** $\left(-2 + 2i\sqrt{3}\right)^4$

**17.** $\left(-3 - 3i\sqrt{3}\right)^5$

**18.** $\left(2\sqrt{3} - 2i\right)^5$

**19.** $(2 + 3i)^4$

**20.** $(4 - i)^5$

**21.** $(2 - i)^4$

**22.** $(-1 - 2i)^6$

**23.** $(1.2 + 3.6i)^3$

**24.** $(-2.3 - i)^3$

*Find the indicated roots. Express answers in trigonometric form.*

**25.** The square roots of $4(\cos 90° + i \sin 90°)$

**26.** The cube roots of $8(\cos 30° + i \sin 30°)$

**27.** The fourth roots of $\cos 120° + i \sin 120°$

**28.** The fifth roots of $32(\cos 300° + i \sin 300°)$

**29.** The sixth roots of $64(\cos \pi + i \sin \pi)$

**30.** The fourth roots of $16[\cos(3\pi/2) + i \sin(3\pi/2)]$

*Find the indicated roots in the form $a + bi$. Check by graphing the roots in the complex plane.*

**31.** The cube roots of 1

**32.** The cube roots of 8

**33.** The fourth roots of 16

**34.** The fourth roots of 1

**35.** The fourth roots of $-1$

**36.** The fourth roots of $-16$

**37.** The cube roots of $i$

**38.** The cube roots of $-8i$

**39.** The square roots of $-2 + 2i\sqrt{3}$

**40.** The square roots of $-4i$

**41.** The square roots of $1 + 2i$

**42.** The cube roots of $-1 + 3i$

*Find all complex solutions to each equation. Express answers in the form $a + bi$.*

**43.** $x^3 + 1 = 0$

**44.** $x^3 + 125 = 0$

**45.** $x^4 - 81 = 0$

**46.** $x^4 + 81 = 0$

**47.** $x^2 + 2i = 0$

**48.** $ix^2 + 3 = 0$

**49.** $x^7 - 64x = 0$

**50.** $x^9 - x = 0$

*Find all complex solutions to each equation. Express answers in trigonometric form.*

**51.** $x^5 - 2 = 0$

**52.** $x^5 + 3 = 0$

**53.** $x^4 + 3 - i = 0$

**54.** $ix^3 + 2 - i = 0$

*Solve each problem.*

**55.** Write the expression $[\cos(\pi/3) + i \sin(\pi/6)]^3$ in the form $a + bi$.

**56.** Solve the equation $x^6 - 1 = 0$ by factoring and the quadratic formula. Compare your answers to Example 3 of this section.

**57.** Use the quadratic formula and De Moivre's theorem to solve
$$x^2 + (-1 + i)x - i = 0.$$

**58.** Use the quadratic formula and De Moivre's theorem to solve
$$x^2 + (-1 - 3i)x + (-2 + 2i) = 0.$$

## WRITING/DISCUSSION

**59.** Explain why $x^6 - 2x^3 + 1 = 0$ has three distinct solutions, $x^6 - 2x^3 = 0$ has four distinct solutions, and $x^6 - 2x = 0$ has six distinct solutions.

**60.** Find all real numbers $a$ and $b$ for which it is true that $\sqrt{a + bi} = \sqrt{a} + i\sqrt{b}$.

## REVIEW

**61.** Find the absolute value of the complex number $3 + 5i$.

**62.** Write $4 - 4i$ in trigonometric form, using degree measure for the argument.

**63.** Perform the indicated operation:
$$2(\cos(\pi/6) + i \sin(\pi/6)) \cdot 3(\cos(\pi/3) + i \sin(\pi/3))$$
Write the result in $a + bi$ form.

**64.** At one point on the ground, the angle of elevation of the line of sight to the top of a building is $20°$. At a point that is 100 feet closer to the building, the angle of elevation is $30°$. Find the height of the building to the nearest foot.

**65.** Find all solutions to the equation $2 \sin^2(2\pi x) - 1 = 0$ in the interval $(0, \pi/2)$.

**66.** Find the area of the triangle whose sides are 10 feet, 14 feet, and 18 feet.

## OUTSIDE THE BOX

**67.** *Related Angles*
   **a.** Find the triangle with the smallest perimeter and whole-number sides such that the measure of one of the angles is twice the measure of another angle.

   **b.** Find general expressions that could be used to determine infinitely many such triangles and find the next two larger such triangles.

**68.** *Swim Meet* Two swimmers of unequal ability are at opposite ends of a pool. They simultaneously dive in and swim the length of the pool and back at a constant rate. They pass for the first time 75 feet from one end of the pool and for the second time 25 feet from the other end. What is the length of the pool?

**6.3** POP QUIZ

1. Use De Moivre's theorem to simplify $(1 + i)^8$.

2. Find all fourth roots of $-16$.

# **6.4** Polar Equations

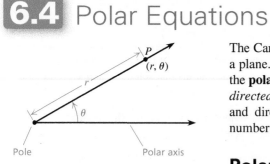

**Figure 6.10**

The Cartesian coordinate system is a system for describing the location of points in a plane. It is not the only coordinate system available. In this section we will study the **polar coordinate system.** In this coordinate system, a point is located by using a *directed distance* and an *angle* in a manner that will remind you of the magnitude and direction angle of a vector and the modulus and argument for a complex number.

## Polar Coordinates

In the rectangular coordinate system, points are named according to their position with respect to an $x$-axis and a $y$-axis. In the polar coordinate system, we have a fixed point called the **pole** and a fixed ray called the **polar axis.** A point $P$ has coordinates $(r, \theta)$ where $r$ is the directed distance from the pole to $P$. Note that $r$ can be positive or negative. The angle $\theta$ is an angle whose initial side is the polar axis and whose terminal side contains the point $P$. See Fig. 6.10.

Since we are so familiar with rectangular coordinates, we retain the $x$- and $y$-axes when using polar coordinates. The pole is placed at the origin, and the polar axis is placed along the positive $x$-axis. The angle $\theta$ is any angle (in degrees or radians) in standard position whose terminal side contains the point. As usual, $\theta$ is positive for a counterclockwise rotation and negative for a clockwise rotation. In polar coordinates, $r$ can be any real number. For example, the ordered pair $(2, \pi/4)$ represents the point that lies two units from the origin on the terminal side of the angle $\pi/4$. The point $(0, \pi/4)$ is at the origin. The point $(-2, \pi/4)$ lies two units from the origin on the line through the terminal side of $\pi/4$ but in the direction opposite to $(2, \pi/4)$. See Fig. 6.11. Any ordered pair in polar coordinates names a single point, but the coordinates of a point in polar coordinates are not unique. For example, $(-2, \pi/4)$, $(2, 5\pi/4)$, and $(2, -3\pi/4)$ all name the same point.

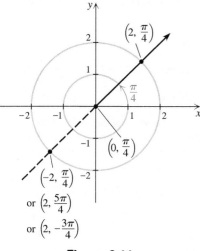

**Figure 6.11**

> **EXAMPLE 1** Plotting points in polar coordinates

Plot the points whose polar coordinates are $(2, 5\pi/6)$, $(-3, \pi)$, $(1, -\pi/2)$, and $(-1, 450°)$.

**Solution**

The terminal side of $5\pi/6$ lies in the second quadrant. The point $(2, 5\pi/6)$ is two units from the origin along this ray. The terminal side of the angle $\pi$ points in the direction of the negative $x$-axis, but since the first coordinate of $(-3, \pi)$ is negative, the point is located 3 units in the direction opposite to the direction of the ray. So $(-3, \pi)$ lies 3 units from the origin on the positive $x$-axis. The point $(-3, \pi)$ is the same point as $(3, 0)$. The terminal side of $-\pi/2$ lies on the negative $y$-axis. So $(1, -\pi/2)$ lies 1 unit from the origin on the negative $y$-axis. The terminal side of $450°$ lies on the positive $y$-axis. Since the first coordinate of $(-1, 450°)$ is negative, the point is located 1 unit in the opposite direction. The point $(-1, 450°)$ is the same point as $(1, -\pi/2)$. All of these points are shown in Fig. 6.12.

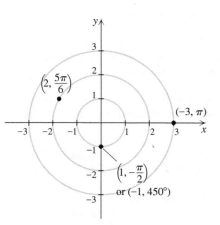

**Figure 6.12**

**TRY THIS.** Plot $(2, 3\pi/4)$ and $(-1, \pi/2)$ in polar coordinates. ∎

If you search for *polar coordinate graph paper* on the Internet, you will find sites that have free grids that you can print like the one in Fig. 6.13. Using grids like this will make your graphing in polar coordinates easier and more accurate. Or, make copies of this page and use the grid given here.

**Figure 6.13**

**Figure 6.14**

## Polar–Rectangular Conversions

To graph equations in polar coordinates, we must be able to switch the coordinates of a point in either system to the other. Suppose that $(r, \theta)$ is a point in the first quadrant with $r > 0$ and $\theta$ acute as shown in Fig. 6.14. If $(x, y)$ is the same point in rectangular coordinates, then $x$ and $y$ are the lengths of the legs of the right triangle shown in Fig. 6.14. Using the Pythagorean theorem and trigonometric ratios we have

$$x^2 + y^2 = r^2, \qquad x = r\cos\theta, \qquad y = r\sin\theta, \qquad \text{and} \qquad \tan\theta = \frac{y}{x}.$$

It can be shown that these equations hold for any point $(r, \theta)$, except that $\tan\theta$ is undefined if $x = 0$. The following rules for converting from one system to the other follow from these relationships.

**Polar–Rectangular Conversion Rules**

To convert $(r, \theta)$ to rectangular coordinates $(x, y)$, use

$$x = r\cos\theta \qquad \text{and} \qquad y = r\sin\theta.$$

To convert $(x, y)$ to polar coordinates $(r, \theta)$, use $r = \sqrt{x^2 + y^2}$ and any angle $\theta$ in standard position whose terminal side contains $(x, y)$.

There are several ways to find an angle $\theta$ whose terminal side contains $(x, y)$. One way to find $\theta$ (provided $x \neq 0$) is to find an angle that satisfies $\tan\theta = y/x$ and goes through $(x, y)$. Remember that the angle $\tan^{-1}(y/x)$ is between $-\pi/2$ and $\pi/2$ and will go through $(x, y)$ only if $x > 0$. If $x = 0$, then the point $(x, y)$ is on the $y$-axis, and you can use $\theta = \pi/2$ or $\theta = -\pi/2$ as appropriate.

**Figure 6.15**

**Figure 6.16**

**EXAMPLE 2** Polar–rectangular conversion

**a.** Convert $(6, 210°)$ to rectangular coordinates.
**b.** Convert $(-3, 6)$ to polar coordinates.

**Solution**

**a.** Use $r = 6$, $\cos 210° = -\sqrt{3}/2$, and $\sin 210° = -1/2$ in the formulas $x = r \cos \theta$ and $y = r \sin \theta$:

$$x = 6\left(-\frac{\sqrt{3}}{2}\right) = -3\sqrt{3} \quad \text{and} \quad y = 6\left(-\frac{1}{2}\right) = -3$$

So $(6, 210°)$ in polar coordinates is $\left(-3\sqrt{3}, -3\right)$ in rectangular coordinates. You can check this result with a graphing calculator as shown in Fig. 6.15. □

**b.** To convert $(-3, 6)$ to polar coordinates, first find $r$:

$$r = \sqrt{(-3)^2 + 6^2} = 3\sqrt{5}$$

Since $y/x = -2$ we can use a calculator to find that

$$\tan^{-1}(-2) \approx -63.4°.$$

Now the terminal side of $-63.4°$ does not go through $(-3, 6)$. To get an angle whose terminal side contains $(-3, 6)$, use $\theta \approx 180° - 63.4°$ or $116.6°$. So $(-3, 6)$ in polar coordinates is $\left(3\sqrt{5}, 116.6°\right)$. Since there are infinitely many representations for any point in polar coordinates, this answer is not unique. In fact, another possibility is $\left(-3\sqrt{5}, -63.4°\right)$. You can check this result with a graphing calculator as shown in Fig. 6.16.

**TRY THIS.** Convert $(3, 45°)$ to rectangular coordinates and convert $\left(-2, 2\sqrt{3}\right)$ to polar coordinates. ∎

## Polar Equations

An equation in two variables (typically $x$ and $y$) that is graphed in the rectangular coordinate system is called a **rectangular** or **Cartesian equation.** An equation in two variables (typically $r$ and $\theta$) that is graphed in the polar coordinate system is called a **polar equation.** Certain polar equations are easier to graph than the equivalent Cartesian equations. We can graph a polar equation in the same manner that we graph a rectangular equation; that is, we can simply plot enough points to get the shape of the graph. However, since most of our polar equations involve trigonometric functions, finding points on these curves can be tedious. A graphing calculator can be used to great advantage here.

**EXAMPLE 3** Graphing a polar equation

Sketch the graph of the polar equation $r = 2 \cos \theta$.

**Solution**

If $\theta = 0°$, then $r = 2 \cos 0° = 2$. So the ordered pair $(2, 0°)$ is on the graph. If $\theta = 30°$, then $r = 2 \cos 30° = \sqrt{3}$, and $\left(\sqrt{3}, 30°\right)$ is on the graph. These ordered pairs and several others that satisfy the equation are listed in the following table:

| $\theta$ | 0° | 30° | 45° | 60° | 90° | 120° | 135° | 150° | 180° |
|---|---|---|---|---|---|---|---|---|---|
| $r$ | 2 | $\sqrt{3}$ | $\sqrt{2}$ | 1 | 0 | $-1$ | $-\sqrt{2}$ | $-\sqrt{3}$ | $-2$ |

Plot these points and draw a smooth curve through them to get the graph shown in Fig. 6.17. If $\theta$ is larger than 180° or smaller than 0°, we get different ordered pairs, but they are all located on the curve drawn in Fig. 6.17. For example, $\left(-\sqrt{3}, 210°\right)$ satisfies $r = 2\cos\theta$, but it has the same location as $\left(\sqrt{3}, 30°\right)$.

**Figure 6.17**

**Figure 6.18**

 To check this graph with a calculator, set the mode to polar and enter $r = 2\cos\theta$. The calculator graph in Fig. 6.18 supports our conclusions.

**TRY THIS.** Graph $r = 4\sin\theta$ in polar coordinates. ▪

The graph $r = 2\cos\theta$ in Fig. 6.17 looks like a circle. To verify that it is a circle, we can convert the polar equation to an equivalent rectangular equation because we know the form of the equation of a circle in rectangular coordinates. This conversion is done in Example 6.

In the next example, a simple polar equation produces a curve that is not usually graphed when studying rectangular equations because the equivalent rectangular equation is quite complicated.

**EXAMPLE 4** **Graphing a four-leaf rose**

Sketch the graph of the polar equation $r = 3\sin 2\theta$.

**Solution**

The ordered pairs in the following table satisfy the equation $r = 3\sin 2\theta$. The values of $r$ are rounded to the nearest tenth.

| $\theta$ | 0° | 15° | 30° | 45° | 60° | 90° | 135° | 180° | 225° | 270° | 315° | 360° |
|---|---|---|---|---|---|---|---|---|---|---|---|---|
| $r$ | 0 | 1.5 | 2.6 | 3 | 2.6 | 0 | −3 | 0 | 3 | 0 | −3 | 0 |

As $\theta$ varies from 0° to 90°, the value of $r$ goes from 0 to 3, then back to 0, creating a loop in quadrant I. As $\theta$ varies from 90° to 180°, the value of $r$ goes from 0 to −3 then back to 0, creating a loop in quadrant IV. As $\theta$ varies from 180° to 270°, a loop in quadrant III is formed; and as $\theta$ varies from 270° to 360°, a loop in quadrant II is formed. If $\theta$ is chosen greater than 360° or less than 0°, we get the same points over

and over because the sine function is periodic. The graph of $r = 3 \sin 2\theta$ is shown in Fig. 6.19. The graph is called a **four-leaf rose.**

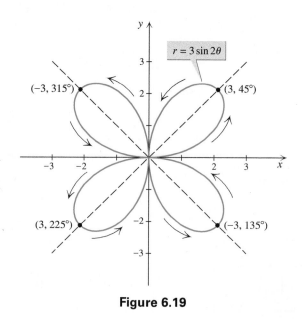

**Figure 6.19**

With a graphing calculator you can make a table of ordered pairs as shown in Fig. 6.20. Make this table yourself and scroll through the table to see how the radius oscillates between 3 and $-3$ as the angle varies. The calculator graph of $r = 3 \sin(2\theta)$ in polar mode with $\theta$ between $0°$ and $360°$ is shown in Fig. 6.21. The calculator graph supports the graph shown in Fig. 6.19.

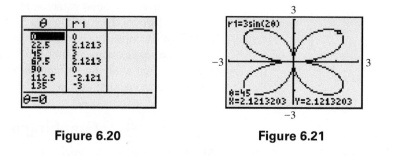

**Figure 6.20**                    **Figure 6.21**

**TRY THIS.**  Graph $r = \cos 2\theta$ in polar coordinates.                  ■

### EXAMPLE 5  Graphing the spiral of Archimedes

Sketch the graph of $r = \theta$, where $\theta$ is in radians and $\theta \geq 0$.

**Solution**

The ordered pairs in the following table satisfy $r = \theta$. The values of $r$ are rounded to the nearest tenth.

| $\theta$ | 0 | $\frac{\pi}{4}$ | $\frac{\pi}{2}$ | $\frac{3\pi}{4}$ | $\pi$ | $\frac{3\pi}{2}$ | $2\pi$ | $\frac{5\pi}{2}$ | $3\pi$ |
|---|---|---|---|---|---|---|---|---|---|
| $r$ | 0 | 0.8 | 1.6 | 2.4 | 3.1 | 4.7 | 6.3 | 7.8 | 9.4 |

The graph of $r = \theta$, called **the spiral of Archimedes,** is shown in Fig. 6.22. As the value of $\theta$ increases, the value of $r$ increases, causing the graph to spiral out from the pole. There is no repetition of points as there was in Examples 3 and 4 because no periodic function is involved.

**Figure 6.22**

**Figure 6.23**

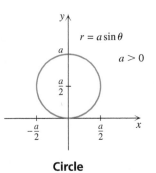

**Figure 6.24**

With a graphing calculator you can make a table of ordered pairs for $r = \theta$ as shown in Fig. 6.23. The calculator graph of $r = \theta$ in polar coordinates for $\theta$ ranging from 0 through 40 radians is shown in Fig. 6.24. This graph shows much more of the spiral than Fig. 6.22. It supports the conclusion that the graph of $r = \theta$ is a spiral.

**TRY THIS.**  Graph $r = -\theta$ for $\theta$ in radians and $\theta \geq 0$.    ■

The following Function Gallery shows the graphs of several types of polar equations. You should graph these equations on your graphing calculator.

**FUNCTION GALLERY**

## FUNCTIONS IN POLAR COORDINATES

Spiral            Circle            Circle

*(continued)*

**Three-Leaf Rose**

**Four-Leaf Rose**

**Lemniscate**

**Cardioid**

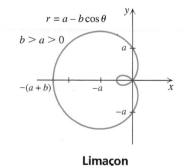

**Limaçon**

**Limaçon**

## Converting Equations

We know that certain types of rectangular equations have graphs that are particular geometric shapes such as lines, circles, and parabolas. We can use our knowledge of equations in rectangular coordinates with equations in polar coordinates (and vice versa) by converting the equations from one system to the other. For example, we can determine whether the graph of $r = 2 \cos \theta$ in Example 3 is a circle by finding the equivalent Cartesian equation and deciding whether it is the equation of a circle. When converting from one system to another, we use the relationships

$$x^2 + y^2 = r^2, \qquad x = r \cos \theta, \qquad \text{and} \qquad y = r \sin \theta.$$

**EXAMPLE 6** **Converting a polar equation to a rectangular equation**

Write an equivalent rectangular equation for the polar equation $r = 2 \cos \theta$.

**Solution**

First multiply each side of $r = 2 \cos \theta$ by $r$ to get $r^2 = 2r \cos \theta$. Now eliminate $r$ and $\theta$ by making substitutions using $r^2 = x^2 + y^2$ and $x = r \cos \theta$.

$$r^2 = 2r \cos \theta$$

$$x^2 + y^2 = 2x$$

The standard equation of a circle with center $(h, k)$ and radius $r$ is $(x - h)^2 + (y - k)^2 = r^2$. Complete the square to get $x^2 + y^2 = 2x$ into the standard form of the equation of a circle:

$$x^2 - 2x + y^2 = 0$$

$$x^2 - 2x + 1 + y^2 = 1$$

$$(x - 1)^2 + y^2 = 1$$

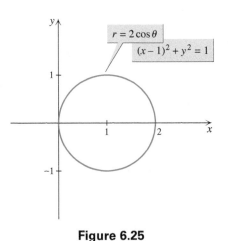

**Figure 6.25**

Since we recognize $(x - 1)^2 + y^2 = 1$ as the equation of a circle centered at $(1, 0)$ with radius 1, the graph of $r = 2 \cos \theta$ shown in Fig. 6.25 is a circle centered at $(1, 0)$ with radius 1.

**TRY THIS.** Convert $r = 3 \sin \theta$ to a rectangular equation. ∎

In the next example we convert the rectangular equation of a line and circle into polar coordinates.

**EXAMPLE 7** **Converting a rectangular equation to a polar equation**

For each rectangular equation, write an equivalent polar equation.

**a.** $y = 3x - 2$    **b.** $x^2 + y^2 = 9$

**Solution**

**a.** Substitute $x = r \cos \theta$ and $y = r \sin \theta$, and then solve for $r$:

$$y = 3x - 2$$
$$r \sin \theta = 3r \cos \theta - 2$$
$$r \sin \theta - 3r \cos \theta = -2$$
$$r(\sin \theta - 3 \cos \theta) = -2$$
$$r = \frac{-2}{\sin \theta - 3 \cos \theta}$$

**Figure 6.26**

The graph in Fig. 6.26 supports the conclusion that we have found the polar coordinate form for the line $y = 3x - 2$. □

**b.** Substitute $r^2 = x^2 + y^2$ to get polar coordinates:

$$x^2 + y^2 = 9$$
$$r^2 = 9$$
$$r = \pm 3$$

A polar equation for a circle of radius 3 centered at the origin is $r = 3$. Another equation for the same circle is $r = -3$.

**Figure 6.27**

The graph in Fig. 6.27 supports the conclusion that $r = 3$ is a circle in polar coordinates.

**TRY THIS.** Convert $y = -2x + 5$ into a polar equation. ∎

In Example 7 we saw that a straight line has a rather simple equation in rectangular coordinates but a more complicated equation in polar coordinates. A circle centered at the origin has a very simple equation in polar coordinates but a more complicated equation in rectangular coordinates. The graphs of simple polar equations are typically circular or somehow "centered" at the origin.

## Applications of Polar Coordinates

In computer-assisted design software it is often more convenient to use polar coordinates rather than rectangular coordinates when telling the computer how to draw a figure. The computer uses a coordinate system at each vertex of the figure. From each vertex, the user gives the polar coordinates of a point. The computer then draws a line segment from the vertex to the point.

EXAMPLE **8** **Drawing a triangle with polar coordinates**

List the polar coordinates that would be used at each vertex to locate the next vertex. Start at $A$ and go in clockwise manner. Use a positive $r$ and the smallest nonnegative $\theta$.

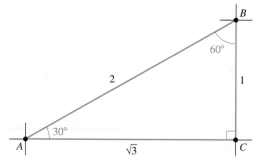

**Solution**

Starting with a coordinate system centered at $A$, point $B$ has polar coordinates $(2, 30°)$. In a coordinate system centered at $B$, point $C$ has polar coordinates $(1, 270°)$. In a coordinate system centered at $C$, point $A$ has coordinates $\left(\sqrt{3}, 180°\right)$.

**TRY THIS.** A triangle has vertices $(0, 0)$, $(5, 5)$, and $(5, 0)$. List the polar coordinates that would be used at each vertex to locate the next vertex, going clockwise from the origin. ∎

## FOR THOUGHT... True or False? Explain.

**1.** The distance of the point $(r, \theta)$ from the origin depends only on $r$.

**2.** The distance of the point $(r, \theta)$ from the origin is $r$.

**3.** The ordered pairs $(2, \pi/4)$, $(2, -3\pi/4)$, and $(-2, 5\pi/4)$ all represent the same point in polar coordinates.

**4.** The equations relating rectangular and polar coordinates are $x = r\sin\theta$, $y = r\cos\theta$, and $x^2 + y^2 = r^2$.

**5.** The point $(-4, 225°)$ in polar coordinates is $\left(2\sqrt{2}, 2\sqrt{2}\right)$ in rectangular coordinates.

**6.** The graph of $0 \cdot r + \theta = \pi/4$ in polar coordinates is a straight line.

**7.** The graphs of $r = 5$ and $r = -5$ are identical.

**8.** The ordered pairs $\left(-\sqrt{2}/2, \pi/3\right)$ and $\left(\sqrt{2}/2, \pi/3\right)$ satisfy $r^2 = \cos 2\theta$.

**9.** The graph of $r = 1/\sin\theta$ is a vertical line.

**10.** The graphs of $r = \theta$ and $r = -\theta$ are identical.

## 6.4 EXERCISES

### CONCEPTS

*Fill in the blank.*

**1.** In the polar coordinate system, the _____ axis is placed along the positive $x$-axis.

**2.** In the polar coordinate system, the _____ is placed at the origin.

**3.** The graph of the polar equation $r = 3\sin 2\theta$ is a(n) _____.

**4.** The graph of the polar equation $r = \theta$ is the _____.

**5.** To convert $(r, \theta)$ to rectangular coordinates $(x, y)$, we use $x = $ _____ and $y = $ _____.

**6.** To convert $(x, y)$ to polar coordinates $(r, \theta)$, we use $r = $ _____ and any angle $\theta$ in standard position whose terminal side contains _____.

## SKILLS

*Find polar coordinates for each given point using radian measure for the angle.*

**7.**          **8.**

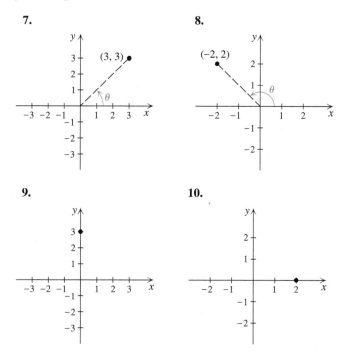

**9.**          **10.**

*Plot the points whose polar coordinates are given.*

**11.** $(3, \pi/6)$     **12.** $(2, \pi/4)$     **13.** $(-2, 2\pi/3)$

**14.** $(-1, \pi/6)$    **15.** $(2, -\pi/4)$   **16.** $(1, -2\pi/3)$

**17.** $(3, -225°)$    **18.** $(2, -180°)$   **19.** $(-2, 45°)$

**20.** $(-3, 30°)$     **21.** $(4, 390°)$    **22.** $(3, 13\pi/6)$

*Convert the polar coordinates of each point to rectangular coordinates.*

**23.** $(1, \pi/6)$              **24.** $(2, \pi/4)$

**25.** $(-3, 3\pi/2)$            **26.** $(-2, 2\pi)$

**27.** $\left(\sqrt{2}, 135°\right)$      **28.** $\left(\sqrt{3}, 150°\right)$

**29.** $\left(-\sqrt{6}, -60°\right)$     **30.** $\left(-\sqrt{2}/2, -45°\right)$

*Convert the rectangular coordinates of each point to polar coordinates. Use degrees for $\theta$.*

**31.** $\left(\sqrt{3}, 3\right)$        **32.** $(4, 4)$

**33.** $(-2, 2)$             **34.** $\left(-2, 2\sqrt{3}\right)$

**35.** $(0, 2)$              **36.** $(-2, 0)$

**37.** $(-3, -3)$            **38.** $(2, -2)$

**39.** $(1, 4)$              **40.** $(-2, 3)$

**41.** $\left(\sqrt{2}, -2\right)$       **42.** $\left(-2, -\sqrt{3}\right)$

*Sketch the graph of each polar equation.*

**43.** $r = 2 \sin \theta$            **44.** $r = 3 \cos \theta$

**45.** $r = 3 \cos 2\theta$          **46.** $r = -2 \sin 2\theta$

**47.** $r = 2\theta$ for $\theta$ in radians

**48.** $r = \theta$ for $\theta \leq 0$ and $\theta$ in radians

**49.** $r = 1 + \cos \theta$ (cardioid)

**50.** $r = 1 - \cos \theta$ (cardioid)

**51.** $r^2 = 9 \cos 2\theta$ (lemniscate)

**52.** $r^2 = 4 \sin 2\theta$ (lemniscate)

**53.** $r = 4 \cos 2\theta$ (four-leaf rose)

**54.** $r = 3 \sin 2\theta$ (four-leaf rose)

**55.** $r = 2 \sin 3\theta$ (three-leaf rose)

**56.** $r = 4 \cos 3\theta$ (three-leaf rose)

**57.** $r = 1 + 2 \cos \theta$ (limaçon)

**58.** $r = 2 + \cos \theta$ (limaçon)

**59.** $r = 3.5$

**60.** $r = -5$

**61.** $\theta = 30°$

**62.** $\theta = 3\pi/4$

*For each polar equation, write an equivalent rectangular equation.*

**63.** $r = 4 \cos \theta$           **64.** $r = 2 \sin \theta$

**65.** $r = \dfrac{3}{\sin \theta}$          **66.** $r = \dfrac{-2}{\cos \theta}$

**67.** $r = 3 \sec \theta$           **68.** $r = 2 \csc \theta$

**69.** $r = 5$                **70.** $r = -3$

**71.** $\theta = \dfrac{\pi}{4}$            **72.** $\theta = 0$

**73.** $r = \dfrac{2}{1 - \sin \theta}$      **74.** $r = \dfrac{3}{1 + \cos \theta}$

*For each rectangular equation, write an equivalent polar equation.*

**75.** $x = 4$               **76.** $y = -6$

**77.** $y = -x$             **78.** $y = x\sqrt{3}$

**79.** $x^2 = 4y$            **80.** $y^2 = 2x$

**81.** $x^2 + y^2 = 4$       **82.** $2x^2 + y^2 = 1$

**83.** $y = 2x - 1$

**84.** $y = -3x + 5$

**85.** $x^2 + (y - 1)^2 = 1$

**86.** $(x + 1)^2 + y^2 = 4$

*Graph each pair of polar equations on the same screen of your calculator and use the trace feature to estimate the polar coordinates of all points of intersection of the curves. Check your calculator manual to see how to graph polar equations on your calculator.*

**87.** $r = 1$, $r = 2 \sin 3\theta$

**88.** $r = \sin \theta$, $r = \sin 2\theta$

**89.** $r = 3 \sin 2\theta$, $r = 1 - \cos \theta$ **90.** $r = 3 \sin 4\theta$, $r = 2$

*Using a coordinate system centered at each vertex, specify the polar coordinates of the next vertex. Start at vertex A and go clockwise.*

**91.**

**92.**

**93.**

**94.**

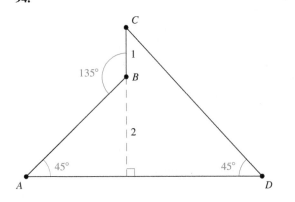

**95.** Explain why $\theta$ must be radian measure for the equation $r = 2\theta$, but $\theta$ can be in radians or degrees for the equation $r = 2 \cos \theta$.

**96.** Show that a polar equation for the straight line $y = mx$ is $\theta = \tan^{-1} m$.

**REVIEW**

**97.** Use De Moivre's theorem to simplify $(1 + i)^{12}$.

**98.** Find all fourth roots of $-8 - 8i\sqrt{3}$.

**99.** Perform the indicated operations. Write the results in $a + bi$ form.
**a.** $(4 + 3i)^2$

**b.** $\dfrac{2 + i}{4 - i}$

**c.** $(-3 + 5i) - (-4 - 2i)$

**100.** The force required to push a riding lawnmower up a ramp inclined at 12° is 130 pounds. Find the weight of the riding lawnmower to the nearest pound.

**101.** Determine the range of each function.
**a.** $f(x) = 3 \sin(2x) + 1$

**b.** $f(x) = x^2 + 1$

**c.** $f(x) = -5 \tan(3x) + 4$

**d.** $f(x) = 2 \sec(3x - \pi/4)$

**102.** A globe with a diameter of 12 inches is rotating at 3 revolutions per second. How fast in miles per hour is a point on the 45th parallel moving? Round to the nearest tenth.

**OUTSIDE THE BOX**

**103.** *Two Squares and Two Triangles* The two squares with areas of 25 and 36 shown in the accompanying figure are positioned so that $AB = 7$. Find the exact area of triangle *TSC*.

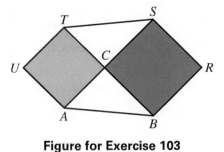

**Figure for Exercise 103**

**104.** *Summing Factorials* What is the units digit in the sum

$$1! + 2! + 3! + 4! + \cdots + 1775! + 1776!?$$

## 6.4 POP QUIZ

1. If the polar coordinates of a point are $(-1, 15\pi/4)$, then in which quadrant does the point lie?

2. Convert $(4, 150°)$ to rectangular coordinates.

3. Convert the rectangular coordinates $(-2, 2)$ to polar coordinates using radians for the angle.

4. Convert the polar equation $r = 4 \cos \theta$ into a rectangular equation.

5. Find the center and radius of the circle $r = -16 \sin \theta$.

6. Convert $y = x + 1$ into polar coordinates.

## LINKING
# concepts...
### For Individual or Group Explorations

### Curves with Strange Names

*Some of the famous curves studied by mathematicians a few centuries ago had strange names such as the cissoid, the lituus, the witch of Agnesi, and the nephroid of Freeth. With a graphing calculator it is relatively easy to see what these curves look like. The lituus is shown in the accompanying figure. Imagine how difficult it was to draw an accurate graph a few hundred years ago.*

**a)** The Cartesian equation for the *conchoid of Nichomedes* is $x^2 = (2 - x)^2(x^2 + y^2)$. Convert it to polar coordinates and graph it in polar coordinates. Who/what was Nichomedes?

**b)** The polar equation for the *witch of Agnesi* is $r^3 \cos^2 \theta \sin \theta + 4r \sin \theta - 8 = 0$. Convert it to rectangular coordinates and draw its graph. Find out Agnesi's whole name, her nationality, and when she lived.

**c)** Graph the *lituus* $r^2\theta = 1$ for $0 \le \theta \le 50$.

**d)** Graph the *nephroid of Freeth*, $r = 1 + 2 \sin(\theta/2)$. Who/what is Freeth? What is a nephroid?

**e)** The rectangular equation for the *folium of Descartes* is $x^3 + y^3 - 6xy = 0$. Convert it to polar coordinates and draw its graph. Does this graph have an asymptote?

**f)** Graph the *bifolium* $(x^2 + y^2)^2 = 4x^2y$.

**g)** Graph the *cissoid* $(4 - x)y^2 = x^3$.

# 6.5 Parametric Equations

We know how to graph points in the plane using the rectangular coordinate system and using polar coordinates. Points in the plane can also be located using parametric equations. We have already used parametric equations when we gave the $x$- and $y$-coordinates of a projectile as functions of time in Chapter 4. In this section we will study parametric equations in more detail.

## Graphs of Parametric Equations

If $f(t)$ and $g(t)$ are functions of $t$, where $t$ is in some interval of real numbers, then the equations $x = f(t)$ and $y = g(t)$ are called **parametric equations.** The variable $t$ is called the **parameter** and the graph of the parametric equations is said to be defined **parametrically.** If the parameter is thought of as time, then we know when each point of the graph is plotted. If no interval is specified for $t$, then $t$ is assumed to be any real number for which both $f(t)$ and $g(t)$ are defined.

> **EXAMPLE 1** **Graphing a line segment**
>
> Graph the parametric equations $x = 3t - 2$ and $y = t + 1$ for $t$ in the interval $[0, 3]$.

**Solution**

If $t = 0$, then $x = 3(0) - 2 = -2$ and $y = 0 + 1 = 1$. So $t = 0$ produces the point $(-2, 1)$ on the graph. Make a table of ordered pairs with some more values of $t$ in the interval $[0, 3]$:

| $t$ | $x = 3t - 2$ | $y = t + 1$ |
|---|---|---|
| 0 | $-2$ | 1 |
| 1 | 1 | 2 |
| 2 | 4 | 3 |
| 3 | 7 | 4 |

Now plot the points $(-2, 1)$, $(1, 2)$, $(4, 3)$, and $(7, 4)$. Of course, $t$ is not restricted to just the values in the table. Other values of $t$ in the interval $[0, 3]$ will produce points that fall in between these points. The graph of the parametric equations is shown in Fig. 6.28. Note that the graph is a line segment that starts at $(-2, 1)$ and ends at $(7, 4)$ because $t$ is restricted to the interval $[0, 3]$. We draw solid dots at the endpoints to show that they are included in the graph.

To check with a graphing calculator, set your calculator to parametric mode and enter the parametric equations as shown in Fig. 6.29(a). Set the limits on the viewing window and the parameter as in Fig. 6.29(b). The graph is shown in Fig. 6.29(c).

**Figure 6.28**

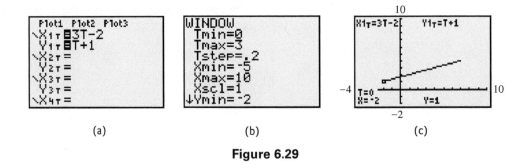

(a)  (b)  (c)

**Figure 6.29**

**TRY THIS.** Graph $x = t + 5$ and $y = 2t - 1$ for $t$ in $[0, 5]$. ∎

## Eliminating the Parameter

In rectangular coordinates we know that $y = mx + b$ is a line, $(x - h)^2 + (y - k)^2 = r^2$ is a circle, and $y = ax^2 + bx + c$ is a parabola. Since we have little experience with parametric equations, it may not be obvious when a system of

parametric equations has a familiar graph. However, it is often possible to identify a graph by eliminating the parameter and writing an equation involving only $x$ and $y$.

**EXAMPLE 2** **Eliminating the parameter**

Eliminate the parameter and identify the graph of the parametric equations.

**a.** $x = 3t - 2, y = t + 1, -\infty < t < \infty$
**b.** $x = 7 \sin t, y = 7 \cos t, -\infty < t < \infty$

**Solution**

**a.** Solve $y = t + 1$ for $t$ to get $t = y - 1$. Now replace $t$ in $x = 3t - 2$ with $y - 1$:

$$x = 3(y - 1) - 2$$

$$x = 3y - 5$$

$$3y = x + 5$$

$$y = \frac{1}{3}x + \frac{5}{3}$$

Because the equation has the form $y = mx + b$, we know that its graph is a line with slope $1/3$ and $y$-intercept $(0, 5/3)$. Because there is no restriction on $t$, the graph is the entire line. This conclusion is consistent with Example 1, where the same parametric equations with a different interval for $t$ produced a segment of this line.

**b.** The simplest way to eliminate the parameter in this case is to use the trigonometric identity $\sin^2 \theta + \cos^2 \theta = 1$. Because $\sin t = x/7$ and $\cos t = y/7$ we have $(x/7)^2 + (y/7)^2 = 1$ or $x^2 + y^2 = 49$. So the graph is a circle centered at the origin with radius 7.

 You can check this conclusion with a graphing calculator as shown in Fig. 6.30.

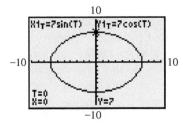

**Figure 6.30**

**TRY THIS.** Eliminate the parameter and identify the graph $x = 4t - 9$ and $y = -t + 1$ for $t$ in $(-\infty, \infty)$.  ∎

## Writing Parametric Equations

Because a nonvertical straight line has a unique slope and $y$-intercept, it has a unique equation in slope-intercept form. However, a polar equation for a curve is not unique and neither are parametric equations for a curve. For example, consider the line $y = 2x + 1$. For parametric equations we could let $x = t$ and $y = 2t + 1$. We could also let $x = 4t$ and $y = 8t + 1$. We could even write $x = t^3 + 7$ and $y = 2t^3 + 15$. Each of these pairs of parametric equations produces the line $y = 2x + 1$.

**EXAMPLE 3** **Writing parametric equations for a line segment**

Write parametric equations for the line segment between $(1, 3)$ and $(5, 8)$ for $t$ in the interval $[0, 2]$ with $t = 0$ corresponding to $(1, 3)$ and $t = 2$ corresponding to $(5, 8)$.

**Solution**

We can make both parametric equations linear functions of $t$. If $x = mt + b$ and $t = 0$ corresponds to $x = 1$, then $1 = m \cdot 0 + b$ and $b = 1$. So $x = mt + 1$. If $t = 2$ corresponds to $x = 5$ then $5 = m \cdot 2 + 1$ or $m = 2$. So we have

$$x = 2t + 1.$$

Now we use the same reasoning to get the equation for $y$. When $t = 0$, $y$ should be 3. So $3 = m \cdot 0 + b$, or $b = 3$. When $t = 2$, $y$ should be 8. So $8 = m \cdot 2 + 3$, or $m = 2.5$. So

$$y = 2.5t + 3.$$

You can use a graphing calculator to check as shown in Fig. 6.31.

**Figure 6.31**

**TRY THIS.** Write parametric equations for the line segment between $(1, 2)$ and $(8, 10)$ with $t$ in the interval $[0, 1]$, where $t = 0$ corresponds to $(1, 2)$ and $t = 1$ corresponds to $(8, 10)$. ∎

While it may not seem obvious how to write parametric equations for a particular rectangular equation, there is a simple way to find parametric equations for a polar equation $r = f(\theta)$. Because $x = r \cos \theta$ and $y = r \sin \theta$ we can substitute $f(\theta)$ for $r$ and write $x = f(\theta) \cos \theta$ and $y = f(\theta) \sin \theta$. In this case the parameter is $\theta$ and we have parametric equations for the polar curve. This method of graphing polar equations can be used on calculators that are capable of handling parametric equations but not polar equations.

**EXAMPLE 4** **Converting a polar equation to parametric equations**

Write parametric equations for the polar equation $r = 1 - \cos \theta$ and graph the parametric equations.

**Solution**

Replace $r$ by $1 - \cos \theta$ in the equations $x = r \cos \theta$ and $y = r \sin \theta$ to get $x = (1 - \cos \theta) \cos \theta$ and $y = (1 - \cos \theta) \sin \theta$. We know from Section 6.4 that the graph of $r = 1 - \cos \theta$ is the cardioid shown in Fig. 6.32. So the graph of these parametric equations is the same cardioid.

**Figure 6.32**

**Figure 6.33**

⊿⊿ The calculator graph of the parametric equations in Fig. 6.33 appears to be a cardioid and supports the graph shown in Fig. 6.32.

**TRY THIS.** Write parametric equations for the polar equation $r = 3 \cos \theta$.  ∎

## Shooting Baskets with Parametric Equations

With $x$ and $y$ in feet and $t$ in seconds, the parametric equations for the path of a projectile are

$$x = v_0(\cos \theta)t \quad \text{and} \quad y = -16t^2 + v_0(\sin \theta)t + h_0$$

where $\theta$ is the angle of inclination in degrees, $v_0$ is the initial velocity in feet per second, and $h_0$ is the initial height in feet. With these equations and a graphing calculator you can illustrate the flight of a basketball.

To enter these equations into a graphing calculator, first set the MODE as in Fig. 6.34. We will throw the ball from the location $(0, 6)$ to a basket at $(20, 10)$. To get these locations to appear on the screen we use STAT PLOT. Use STAT EDIT to enter these ordered pairs in L1 and L2 as shown in Fig. 6.35. Then use STAT PLOT to turn on the feature that plots a square mark at these locations as in Fig. 6.36.

**Figure 6.34**          **Figure 6.35**          **Figure 6.36**

Using $A$ for the initial velocity, $B$ for the angle of inclination of the throw, and $h_0 = 6$ feet, enter the parametric equations using $Y =$ as in Fig. 6.37. To get the "ball," move the cursor to the left of $X_{1T}$ and press enter until a small circle appears to the left of $X_{1T}$ as in Fig. 6.37. Set the window so that the time $T$ ranges from 0 to 2 seconds in steps of 0.1 second, as in Fig. 6.38. Set the window so that $-2 \leq x \leq 25$ and $0 \leq y \leq 20$. Press QUIT to enter the initial values for $A$ and $B$ on the home screen using the STO button as in Fig. 6.39. Our first guess is $A = 25$ ft/sec and $B = 60°$.

**Figure 6.37**          **Figure 6.38**          **Figure 6.39**

Now press GRAPH to see the ball tossed toward the basket. After the toss, the path of the ball will appear as in Fig. 6.40.

With the initial velocity of 25 feet per second and the angle 60°, the ball falls short of the basket as seen in Fig. 6.40. Enter a new angle and velocity on the home screen until you find a combination for which the path of the ball goes through the basket at $(20, 10)$. Before pressing GRAPH, press DRAW and ClrDraw to clear the old path from the graph screen.

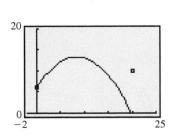

**Figure 6.40**

## FOR THOUGHT... True or False? Explain.

1. If $x = 3t + 1$ and $y = 4t - 2$, then $t$ is the variable and $x$ and $y$ are the parameters.

2. Parametric equations are graphed in the rectangular coordinate system.

3. The graph of $x = 0.5t$ and $y = 2t + 1$ is a straight line with slope 4.

4. The graph of $x = \cos t$ and $y = \sin t$ is a sine wave.

5. The graph of $x = \tan(t)$ and $y = \tan^2(t)$ for $-\pi/2 < t < \pi/2$ is a parabola.

6. The graph of $x = 3t + 1$ and $y = 6t - 1$ for $0 \le t \le 3$ includes the point $(2, 1)$.

7. The graph of $x = w^2 - 3$ and $y = w + 5$ for $-2 < w < 2$ includes the point $(1, 7)$.

8. The graph of $x = \tan(t)$ and $y = 2 \tan(t)$ lies entirely within the first quadrant.

9. The graph of $x = -\sin t$ and $y = \cos t$ for $0 < t < \pi/2$ lies entirely within the second quadrant.

10. The polar equation $r = \cos \theta$ can be graphed using the parametric equations $x = \cos^2 \theta$ and $y = \cos \theta \sin \theta$.

Scan this code
to link to section
lecture videos.

## 6.5 EXERCISES

### CONCEPTS

*Fill in the blank.*

1. If both $x$ and $y$ are functions of $t$, then $x = f(t)$ and $y = g(t)$ are _____ equations.

2. If $x = f(t)$ and $y = g(t)$, then $t$ is the _____.

### SKILLS

*Complete the table that accompanies each pair of parametric equations.*

3. $x = 4t + 1$,
   $y = t - 2$,
   for $0 \le t \le 3$

| $t$ | $x$ | $y$ |
|-----|-----|-----|
| 0 |   |   |
| 1 |   |   |
|   | 7 |   |
|   |   | 1 |

4. $x = 3 - t$,
   $y = 2t + 5$,
   for $2 \le t \le 7$

| $t$ | $x$ | $y$ |
|-----|-----|-----|
| 2 |   |   |
| 3 |   |   |
|   | -2 |   |
|   |   | 19 |

5. $x = t^2$,
   $y = 3t - 1$,
   for $1 \le t \le 5$

| $t$ | $x$ | $y$ |
|-----|-----|-----|
| 1 |   |   |
| 2.5 |   |   |
|   | 5 |   |
|   |   | 11 |
|   | 25 |   |

6. $x = \sqrt{t}$,
   $y = t + 4$,
   for $0 \le t \le 9$

| $t$ | $x$ | $y$ |
|-----|-----|-----|
| 0 |   |   |
| 2 |   |   |
| 4 |   |   |
|   |   | 12 |
|   | 3 |   |

*Graph each pair of parametric equations in the rectangular coordinate system.*

7. $x = 3t - 2$, $y = t + 3$, for $0 \le t \le 4$

8. $x = 4 - 3t$, $y = 3 - t$, for $1 \le t \le 3$

9. $x = t - 1$, $y = t^2$, for $t$ in $(-\infty, \infty)$

10. $x = t - 3$, $y = 1/t$, for $t$ in $(-\infty, \infty)$

11. $x = \sqrt{w}$, $y = \sqrt{1 - w}$, for $0 < w < 1$

12. $x = t - 2$, $y = \sqrt{t + 2}$, for $-2 \le t \le 7$

13. $x = \cos t$, $y = \sin t$        14. $x = 0.5t$, $y = \sin t$

*Eliminate the parameter and identify the graph of each pair of parametric equations.*

15. $x = 4t - 5$, $y = 3 - 4t$

16. $x = 5t - 1$, $y = 4t + 6$

17. $x = -4 \sin 3t$, $y = 4 \cos 3t$

18. $x = 2 \sin t \cos t$, $y = 3 \sin 2t$

19. $x = t + 4$, $y = \sqrt{t - 5}$

20. $x = t - 5$, $y = t^2 - 10t + 25$

21. $x = \tan t$, $y = 2 \tan t + 3$

22. $x = \tan t$, $y = -\tan^2 t + 3$

*Write a pair of parametric equations that will produce the indicated graph. Answers may vary.*

**23.** The line segment starting at $(2, 3)$ with $t = 0$ and ending at $(5, 9)$ with $t = 2$

**24.** The line segment starting at $(-2, 4)$ with $t = 3$ and ending at $(5, -9)$ with $t = 7$

**25.** That portion of the circle $x^2 + y^2 = 4$ that lies in the third quadrant

**26.** That portion of the circle $x^2 + y^2 = 9$ that lies below the x-axis

**27.** The vertical line through $(3, 1)$

**28.** The horizontal line through $(5, 2)$

**29.** The circle whose polar equation is $r = 2 \sin \theta$

**30.** The four-leaf rose whose polar equation is $r = 5 \sin(2\theta)$

*Graph the following pairs of parametric equations with the aid of a graphing calculator. These are uncommon curves that would be difficult to describe in rectangular or polar coordinates.*

**31.** $x = \cos 3t, y = \sin t$       **32.** $x = \sin t, y = t^2$

**33.** $x = t - \sin t, y = 1 - \cos t$ (cycloid)

**34.** $x = t - \sin t, y = -1 + \cos t$ (inverted cycloid)

**35.** $x = 4 \cos t - \cos 4t, y = 4 \sin t - \sin 4t$ (epicycloid)

**36.** $x = \sin^3 t, y = \cos^3 t$ (hypocycloid)

## MODELING

*The following problems involve the parametric equations for the path of a projectile.*

$$x = v_0(\cos \theta)t \quad and \quad y = -16t^2 + v_0(\sin \theta)t + h_0,$$

*where $\theta$ is the angle of inclination of the projectile at the launch, $v_0$ is the initial velocity of the projectile in feet per second, and $h_0$ is the initial height of the projectile in feet.*

**37.** An archer shoots an arrow from a height of 5 ft at an angle of inclination of 30° with a velocity of 300 ft/sec. Write the parametric equations for the path of the projectile and sketch the graph of the parametric equations.

**38.** If the arrow of Exercise 37 strikes a target at a height of 5 ft, then how far is the target from the archer?

**39.** For how many seconds is the arrow of Exercise 37 in flight?

**40.** What is the maximum height reached by the arrow in Exercise 37?

## REVIEW

**41.** Write a rectangular equation that is equivalent to the polar equation $r = 8 \cos \theta$.

**42.** Find all complex solutions to the equation $x^4 + 1 = 0$.

**43.** Find the trigonometric form for the complex number $3 - 3i\sqrt{3}$. Use radian measure for the argument.

**44.** Solve the equation $w = \dfrac{1}{3} \cos(4b)$ for $b$ where $0 \le b \le \dfrac{\pi}{4}$.

**45.** Find all solutions to $2 \cos^2(x) + \cos(x) - 1 = 0$ on the interval $[0, 2\pi]$.

**46.** Find the exact value of $\tan \alpha$ if $\sin \alpha = -7/8$ and $\pi < \alpha < 3\pi/2$.

## OUTSIDE THE BOX

**47.** *Double Boxed* A rectangular box contains a delicate statue. The shipping department places the box containing the statue inside a 3-ft by 4-ft rectangular box as shown from above in the accompanying figure. If the box containing the statue is 1 ft wide, then what is its length? Find a four-decimal place approximation.

**Figure for Exercise 47**

**48.** *Quadratic Roots* The two roots to the quadratic equation $x^2 - 56x + c = 0$ are prime numbers. What is the value of $c$?

## 6.5 POP QUIZ

**1.** The graph of $x = 2t + 5$ and $y = 3t - 7$ for $t$ in $[3, 5]$ is a line segment. What are the endpoints?

**2.** Eliminate the parameter and identify the graph of $x = 3 \cos t$ and $y = 3 \sin t$ for $-\infty < t < \infty$.

**3.** Write parametric equations for the line segment between $(0, 1)$ and $(3, 5)$ where $t = 0$ corresponds to $(0, 1)$ and $t = 4$ corresponds to $(3, 5)$.

# Highlights

## 6.1 Complex Numbers

| | | |
|---|---|---|
| **Standard Form** | Numbers of the form $a + bi$ where $a$ and $b$ are real numbers, $i = \sqrt{-1}$, and $i^2 = -1$ | $2 + 3i, -\pi + i\sqrt{2},$ <br> $6, 0, \frac{1}{2}i$ |
| **Add, Subtract, Multiply** | Add, subtract, and multiply like binomials with variable $i$, using $i^2 = -1$ to simplify. | $(3 - 2i)(4 + 5i)$ <br> $= 12 + 7i - 10i^2$ <br> $= 22 + 7i$ |
| **Divide** | Divide by multiplying the numerator and denominator by the complex conjugate of the denominator. | $6/(1 + i) =$ <br> $\dfrac{6(1 - i)}{(1 + i)(1 - i)} = 3 - 3i$ |
| **Square Roots of Negative Numbers** | Square roots of negative numbers must be converted to standard form using $\sqrt{-b} = i\sqrt{b}$ for $b > 0$ before doing computations. | $\sqrt{-4} \cdot \sqrt{-9} = 2i \cdot 3i = -6$ |

## 6.2 Trigonometric Form of Complex Numbers

| | | |
|---|---|---|
| **Absolute Value** | $\left|a + bi\right| = \sqrt{a^2 + b^2}$ | $\left|2 + 3i\right| = \sqrt{13}$ |
| **Trigonometric Form** | $z = a + bi = r(\cos\theta + i\sin\theta)$, where $a = r\cos\theta, b = r\sin\theta, r = \sqrt{a^2 + b^2}$, and $\theta$ is an angle in standard position whose terminal side contains $(a, b)$ | $z = \sqrt{3} + i$ <br> $= 2\left(\cos\frac{\pi}{6} + i\sin\frac{\pi}{6}\right)$ |
| **Multiplying and Dividing** | $z_1 = r_1(\cos\theta_1 + i\sin\theta_1)$ <br> $z_2 = r_2(\cos\theta_2 + i\sin\theta_2)$ <br> $z_1 z_2 = r_1 r_2\left[\cos(\theta_1 + \theta_2) + i\sin(\theta_1 + \theta_2)\right]$ <br> $\frac{z_1}{z_2} = \frac{r_1}{r_2}\left[\cos(\theta_1 - \theta_2) + i\sin(\theta_1 - \theta_2)\right]$ | $z_1 = 8\left(\cos\frac{\pi}{3} + i\sin\frac{\pi}{3}\right)$ <br> $z_2 = 2\left(\cos\frac{\pi}{6} + i\sin\frac{\pi}{6}\right)$ <br> $z_1 z_2 = 16\left(\cos\frac{\pi}{2} + i\sin\frac{\pi}{2}\right)$ <br> $\frac{z_1}{z_2} = 4\left(\cos\frac{\pi}{6} + i\sin\frac{\pi}{6}\right)$ |

## 6.3 Powers and Roots of Complex Numbers

| | | |
|---|---|---|
| **De Moivre's Theorem** | If $z = r(\cos\theta + i\sin\theta)$ and $n$ is a positive integer, then $z^n = r^n(\cos n\theta + i\sin n\theta)$. | $z = 2\left(\cos\frac{\pi}{6} + i\sin\frac{\pi}{6}\right)$ <br> $z^3 = 8\left(\cos\frac{\pi}{2} + i\sin\frac{\pi}{2}\right)$ |
| **Roots** | The $n$ distinct $n$th roots of $r(\cos\theta + i\sin\theta)$ are <br> $r^{1/n}\left[\cos\left(\frac{\theta + 2k\pi}{n}\right) + i\sin\left(\frac{\theta + 2k\pi}{n}\right)\right]$ <br> for $k = 0, 1, 2, \ldots, n - 1$. | Square roots of $z$: <br> $\sqrt{2}\left(\cos\frac{\pi}{12} + i\sin\frac{\pi}{12}\right)$ <br> $\sqrt{2}\left(\cos\frac{13\pi}{12} + i\sin\frac{13\pi}{12}\right)$ |

## 6.4 Polar Equations

| | | |
|---|---|---|
| **Polar Coordinates** | If $r > 0$, then $(r, \theta)$ is $r$ units from the origin on the terminal side of $\theta$ in standard position. <br> If $r < 0$, then $(r, \theta)$ is $\left|r\right|$ units from the origin on the extension of the terminal side of $\theta$. | $(2, \pi/4)$ <br><br> $(-2, 5\pi/4)$ |
| **Converting** | Polar to rectangular: $x = r\cos\theta, y = r\sin\theta$ <br> Rectangular to polar: $r = \sqrt{x^2 + y^2}$ and the terminal side of $\theta$ contains $(x, y)$. | Polar: $(2, \pi/4)$ <br> Rectangular: $\left(\sqrt{2}, \sqrt{2}\right)$ |

## 6.5 Parametric Equations

| Parametric Equations | $x = f(t)$ and $y = g(t)$ where $t$ is a parameter in some interval of real numbers | $x = 2t, y = t^2$ for $t$ in $[0, 5]$ |
| --- | --- | --- |
| Converting to Rectangular | Eliminate the parameter. | $y = (x/2)^2$ for $x$ in $[0, 10]$ |

# Chapter 6 Review Exercises

*Write each expression in the form a + bi where a and b are real numbers.*

**1.** $(3 - 7i) + (-4 + 6i)$

**2.** $(-6 - 3i) - (3 - 2i)$

**3.** $(4 - 5i)^2$

**4.** $7 - i(2 - 3i)^2$

**5.** $(1 - 3i)(2 - 6i)$

**6.** $(0.3 + 2i)(0.3 - 2i)$

**7.** $(2 - 3i) \div i$

**8.** $(-2 + 4i) \div (-i)$

**9.** $\dfrac{1 + i}{2 - 3i}$

**10.** $\dfrac{3 - i}{4 - 3i}$

**11.** $\dfrac{6 + \sqrt{-8}}{2}$

**12.** $\dfrac{-2 - \sqrt{-18}}{2}$

**13.** $i^{34} + i^9$

**14.** $i^{55} - i^6$

*Find the absolute value of each complex number.*

**15.** $3 - 5i$

**16.** $3.6 + 4.8i$

**17.** $\sqrt{5} + i\sqrt{3}$

**18.** $-2\sqrt{2} + 3i\sqrt{5}$

*Write each complex number in trigonometric form using degree measure for the argument.*

**19.** $-4.2 + 4.2i$

**20.** $3 - i\sqrt{3}$

**21.** $-2.3 - 7.2i$

**22.** $4 + 9.2i$

*Write each complex number in the form a + bi.*

**23.** $\sqrt{3}(\cos 150° + i \sin 150°)$

**24.** $\sqrt{2}(\cos 225° + i \sin 225°)$

**25.** $6.5(\cos 33.1° + i \sin 33.1°)$

**26.** $14.9(\cos 289.4° + i \sin 289.4°)$

*Find the product and quotient of each pair of complex numbers, using trigonometric form. Write your answers in a + bi form.*

**27.** $z_1 = 2.5 + 2.5i, z_2 = -3 - 3i$

**28.** $z_1 = -\sqrt{3} + i, z_2 = -2 - 2i\sqrt{3}$

**29.** $z_1 = 2 + i, z_2 = 3 - 2i$     **30.** $z_1 = -3 + i, z_2 = 2 - i$

*Use De Moivre's theorem to simplify each expression. Write the answer in the form a + bi.*

**31.** $[2(\cos 45° + i \sin 45°)]^3$

**32.** $\left[\sqrt{3}(\cos 210° + i \sin 210°)\right]^4$

**33.** $(4 + 4i)^3$

**34.** $\left(1 - i\sqrt{3}\right)^4$

*Find the indicated roots. Express answers in the form a + bi.*

**35.** The square roots of $i$

**36.** The cube roots of $-i$

**37.** The cube roots of $\sqrt{3} + i$

**38.** The square roots of $3 + 3i$

**39.** The cube roots of $2 + i$

**40.** The cube roots of $3 - i$

**41.** The fourth roots of $625i$

**42.** The fourth roots of $-625i$

*Convert the polar coordinates of each point to rectangular coordinates.*

**43.** $(5, 60°)$

**44.** $(-4, 30°)$

**45.** $\left(\sqrt{3}, 100°\right)$

**46.** $\left(\sqrt{5}, 230°\right)$

*Convert the rectangular coordinates of each point to polar coordinates. Use radians for $\theta$.*

**47.** $\left(-2, -2\sqrt{3}\right)$

**48.** $\left(-3\sqrt{2}, 3\sqrt{2}\right)$

**49.** $(2, -3)$

**50.** $(-4, -5)$

*Sketch the graph of each polar equation.*

**51.** $r = -2 \sin \theta$

**52.** $r = 5 \sin 3\theta$

**53.** $r = 2 \cos 2\theta$

**54.** $r = 1.1 - \cos \theta$

**55.** $r = 500 + \cos \theta$

**56.** $r = 500$

**57.** $r = \dfrac{1}{\sin \theta}$

**58.** $r = \dfrac{-2}{\cos \theta}$

*For each polar equation, write an equivalent rectangular equation.*

**59.** $r = \dfrac{1}{\sin \theta + \cos \theta}$

**60.** $r = -6 \cos \theta$

**61.** $r = -5$

**62.** $r = \dfrac{1}{1 + \sin \theta}$

*For each rectangular equation, write an equivalent polar equation.*

**63.** $y = 3$

**64.** $x^2 + (y + 1)^2 = 1$

**65.** $x^2 + y^2 = 49$          **66.** $2x + 3y = 6$

*Sketch the graph of each pair of parametric equations.*

**67.** $x = 3t, y = 3 - t$, for $t$ in $(0, 1)$

**68.** $x = t - 3, y = t^2$, for $t$ in $(-\infty, \infty)$

**69.** $x = -\sin t, y = -\cos t$, for $t$ in $[0, \pi/2]$

**70.** $x = -\cos t, y = \sin t$, for $t$ in $[0, \pi]$

## OUTSIDE THE BOX

**71.** *Laying Pipe* A circular pipe with radius 1 is placed in a V-shaped trench whose sides form an angle of $\theta$ radians. In the cross section shown here, the pipe touches the sides of the trench at points $A$ and $B$.

**a.** Find the area inside the circle and below the line segment $AB$ in terms of $\theta$ (the blue area).

**b.** Find the area below the circle and inside the trench in terms of $\theta$ (the red area).

**c.** If $\theta$ is nearly equal to $\pi$, then the blue area and the red area are both very small. If $\theta$ is equal to $\pi$, then both areas are zero. As $\theta$ approaches $\pi$ there is a limit to the ratio of the blue area to the red area. Use a calculator to determine this limit.

**Figure for Exercise 71**

**72.** *Square Roots* Solve the equation

$$\sqrt{10 + 4\sqrt{x}} - \sqrt{10 - 4\sqrt{x}} = 4.$$

# Chapter **6** Test

*Perform the indicated operations with the complex numbers in standard form and write the answer in standard form $a + bi$, where $a$ and $b$ are real.*

**1.** $(4 - 3i)^2$          **2.** $\dfrac{2 - i}{3 + i}$

**3.** $i^6 - i^{35}$          **4.** $\sqrt{-8}\left(\sqrt{-2} + \sqrt{6}\right)$

*Write each complex number in trigonometric form, using degree measure for the argument.*

**5.** $3 + 3i$          **6.** $-1 + i\sqrt{3}$          **7.** $-4 - 2i$

*Perform the indicated operations. Write the answer in the form $a + bi$.*

**8.** $3(\cos 20° + i \sin 20°) \cdot 2(\cos 25° + i \sin 25°)$

**9.** $[2(\cos 10° + i \sin 10°)]^9$          **10.** $\dfrac{3(\cos 63° + i \sin 63°)}{2(\cos 18° + i \sin 18°)}$

*Give the rectangular coordinates for each of the following points in the polar coordinate system.*

**11.** $(5, 30°)$

**12.** $(-3, -\pi/4)$

**13.** $(33, 217°)$

*Sketch the graph of each equation in polar coordinates.*

**14.** $r = 5 \cos \theta$

**15.** $r = 3 \cos 2\theta$

*Sketch the graph of each pair of parametric equations.*

**16.** $x = 2t$ and $y = 4t - 6$ for $t$ in $[-1, 3]$

**17.** $x = 3 \sin t$ and $y = 3 \cos t$ for $t$ in $[0, \pi]$

*Solve each problem.*

**18.** Find all of the fourth roots of $-81$.

**19.** Write an equation equivalent to $x^2 + y^2 + 5y = 0$ in polar coordinates.

**20.** Write an equation equivalent to $r = 5 \sin 2\theta$ in rectangular coordinates.

**21.** Find a pair of parametric equations whose graph is the line segment joining the points $(-2, -3)$ and $(4, 5)$.

## TYING IT ALL TOGETHER

### Chapters P–6

*Evaluate each expression without using a calculator.*

**1.** $\sin(\pi/4)$      **2.** $\cos(\pi/3)$      **3.** $\sin^{-1}\left(1/\sqrt{2}\right)$      **4.** $\sin^{-1}\left(-\sqrt{2}/2\right)$      **5.** $\cos^{-1}(1/2)$

**6.** $\cos^{-1}(-1/2)$      **7.** $\cos(\tan^{-1}(1))$      **8.** $\sin(\tan^{-1}(-1))$      **9.** $\sin(\cos^{-1}(3/5))$      **10.** $\tan(\sin^{-1}(-3/5))$

*Find all solutions to each equation.*

**11.** $\sin x = 1$      **12.** $\cos x = 1$      **13.** $\sin^2 x = 1$

**14.** $\cos^2 x = 1$      **15.** $2 \sin x = 1$      **16.** $2 \cos x = 1$

**17.** $2 \sin x \cos x = 1$      **18.** $\sin^2 x + \cos^2 x = 1$      **19.** $x \sin(\pi/3) = 1$

**20.** $x \sin(\pi/2) + x \cos(\pi/6) = 1$      **21.** $2 \sin 2x - 2 \cos x + 2 \sin x = 1$      **22.** $4x \sin x + 2 \sin x - 2x = 1$

*Sketch the graph of each function using rectangular or polar coordinates as appropriate.*

**23.** $y = \sin x$      **24.** $y = \cos x$      **25.** $r = \sin \theta$      **26.** $r = \cos \theta$      **27.** $r = \theta$

**28.** $y = x$      **29.** $y = \pi/2$      **30.** $r = \sin(\pi/3)$      **31.** $y = \sqrt{\sin x}$      **32.** $y = \cos^2(x)$

*Fill in the blanks.*

**33.** _____ gives the area of a triangle in terms of the lengths of the sides only.

**34.** A(n) _____ is a directed line segment.

**35.** Quantities that are completely characterized by a single real number are called _____ quantities.

**36.** The length of a vector is also called its _____.

**37.** The sum of two vectors is called the _____.

**38.** The _____ of two vectors is a scalar that is found by multiplying the corresponding components and adding the results.

**39.** If the angle between two vectors is 90°, then the vectors are perpendicular or _____.

**40.** A number of the form $a + bi$, where $a$ and $b$ are real numbers, is called a(n) _____.

**41.** The numbers $a + bi$ and $a - bi$ are called _____.

**42.** _____ theorem gives a formula for finding powers of a complex number in trigonometric form.

# PHOTO CREDITS

## Chapter P

### Section P.1

**For Thought: 1.** F  **2.** F  **3.** F  **4.** F  **5.** T  **6.** F
**7.** T  **8.** T  **9.** T  **10.** F

**Exercises:**
**1.** ordered  **3.** $x$-axis  **5.** Pythagorean theorem
**7.** linear equation  **9.** $(4, 1)$, I  **11.** $(1, 0)$, $x$-axis
**13.** $(5, -1)$, IV  **15.** $(-4, -2)$, III  **17.** $(-2, 4)$, II  **19.** $c = 2$
**21.** $b = \sqrt{5}$  **23.** $a = 4$  **25.** $2\sqrt{7}$  **27.** $\sqrt{5}/3$  **29.** $\sqrt{6}/3$
**31.** $2\sqrt{15}/5$  **33.** $\sqrt{3}/3$  **35.** $\sqrt{6}/3$  **37.** $5, (2.5, 5)$

**39.** $2\sqrt{2}, (0, -1)$  **41.** $1, \left(\dfrac{\sqrt{2}}{4}, \dfrac{\sqrt{2}}{4}\right)$  **43.** $\sqrt{5}, \left(\dfrac{5\sqrt{2}}{2}, \dfrac{5\sqrt{3}}{2}\right)$

**45.** $\sqrt{74}, (-1.3, 1.3)$  **47.** $\dfrac{\sqrt{\pi^2 + 4}}{2}, \left(\dfrac{3\pi}{4}, \dfrac{1}{2}\right)$  **49.** $\pi, \left(\dfrac{3\pi}{2}, 0\right)$

**51.** $(0, 0), 4$  **53.** $(-6, 0), 6$

**55.** $(2, -2), 2\sqrt{2}$

**57.** $x^2 + y^2 = 7$  **59.** $(x + 2)^2 + (y - 5)^2 = 1/4$
**61.** $(x - 3)^2 + (y - 5)^2 = 34$  **63.** $x^2 + y^2 = 1$
**65.** $(x + 1)^2 + (y - 2)^2 = 5$  **67.** $(x - 1)^2 + (y - 3)^2 = 4$
**69.** $a = \pm 4/5$  **71.** $a = \pm\sqrt{21}/5$
**73.** $(0, -4), (4/3, 0)$  **75.** $(0, -6), (2, 0)$

**77.** $(0, 80), (80, 0)$  **79.** $(0, 30)\ (-90, 0)$

**81.** $(0, -1800), (1200, 0)$  **83.** $(0, 0.0025), (0.005, 0)$

**85.**  **87.**

**89.**  **91.**

**93.**  **95.**

**97.** 10 ft     **99. a.** 4/3 cm
**b.** $(x - 4/3)^2 + y^2 = 4/9$, $(x + 4/3)^2 + y^2 = 4/9$
**101.** $\left(x - \sqrt{2}\right)^2 + (y + 1/2)^2 = 1/4$
**103.** $(21, 23.5)$; in 1991 the median age at first marriage was 23.5.
**109.** Break into 1-, 2-, and 4-dollar pieces.

## Section P.2

**For Thought: 1.** T    **2.** F    **3.** F    **4.** T    **5.** T    **6.** T
**7.** F    **8.** T    **9.** T    **10.** T
**Exercises:**
**1.** function    **3.** domain, range    **5.** function    **7.** Both
**9.** $a$ is a function of $b$    **11.** $b$ is a function of $a$    **13.** Neither
**15.** Both    **17.** Both    **19.** $b$ is a function of $a$    **21.** $A = s^2$
**23.** $s = \dfrac{d\sqrt{2}}{2}$    **25.** $P = 4s$    **27.** $A = \dfrac{P^2}{16}$
**29.** $(-\infty, \infty)$, $(-\infty, \infty)$    **31.** $(-\infty, \infty)$, $\{5\}$

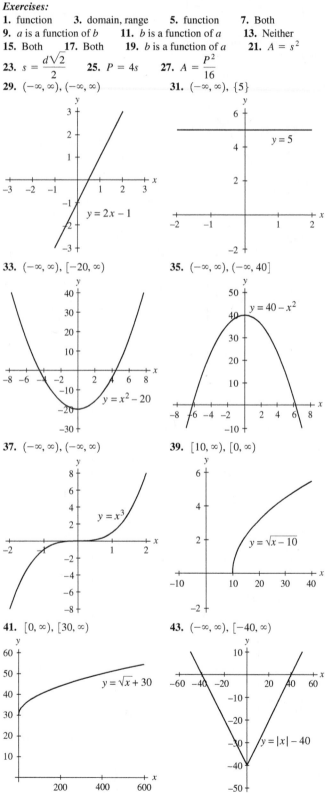

**33.** $(-\infty, \infty)$, $[-20, \infty)$    **35.** $(-\infty, \infty)$, $(-\infty, 40]$

**37.** $(-\infty, \infty)$, $(-\infty, \infty)$    **39.** $[10, \infty)$, $[0, \infty)$

**41.** $[0, \infty)$, $[30, \infty)$    **43.** $(-\infty, \infty)$, $[-40, \infty)$

**45.** $(-\infty, \infty)$, $[0, \infty)$

**47.** 10    **49.** $-6$    **51.** 8    **53.** 0.3808    **55.** $-2$    **57.** 78
**59.** $3a^2 - a$    **61.** $3x^2 + x$    **63.** 0 or 1/3    **65.** 1 or $-7$
**67.** $C = 353n$    **69.** $C = 50 + 35n$
**71.** $C \approx 1.822$

**73.** 81.7 in.$^3$    **75.** $B = 2\sqrt{\dfrac{D}{\pi S \cdot N}}$

**77.** Pythagorean, legs, hypotenuse    **79.** $\sqrt{29}$
**81.** $(10, 0)$, $(0, -20/3)$    **83.** $-94$

## Section P.3

**For Thought: 1.** F    **2.** F    **3.** F    **4.** T    **5.** T    **6.** F
**7.** T    **8.** F    **9.** T    **10.** T
**Exercises:**
**1.** rigid    **3.** reflection    **5.** right, left    **7.** odd
**9.** transformation
**11.**                                **13.**

**15.**                                **17.**

**19.**

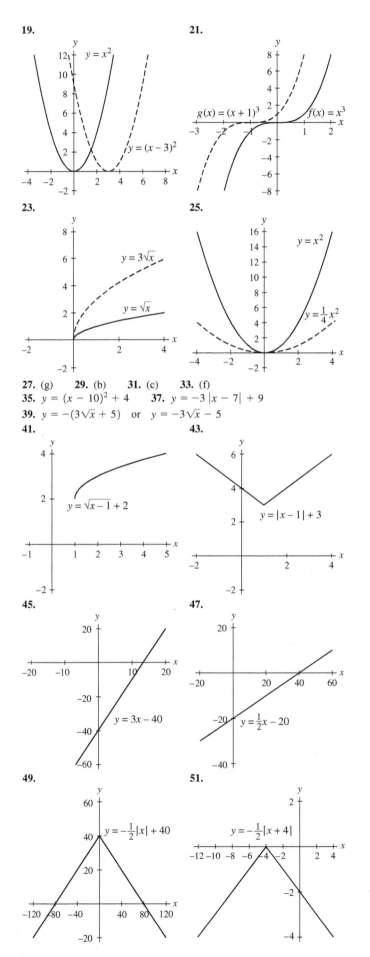

**21.**

$g(x) = (x + 1)^3$     $f(x) = x^3$

**23.**

$y = 3\sqrt{x}$

$y = \sqrt{x}$

**25.**

$y = x^2$

$y = \frac{1}{4}x^2$

**27.** (g)    **29.** (b)    **31.** (c)    **33.** (f)

**35.** $y = (x - 10)^2 + 4$    **37.** $y = -3\,|x - 7| + 9$

**39.** $y = -(3\sqrt{x} + 5)$  or  $y = -3\sqrt{x} - 5$

**41.**

$y = \sqrt{x - 1} + 2$

**43.**

$y = |x - 1| + 3$

**45.**

$y = 3x - 40$

**47.**

$y = \frac{1}{2}x - 20$

**49.**

$y = -\frac{1}{2}|x| + 40$

**51.**

$y = -\frac{1}{2}|x + 4|$

**53.**

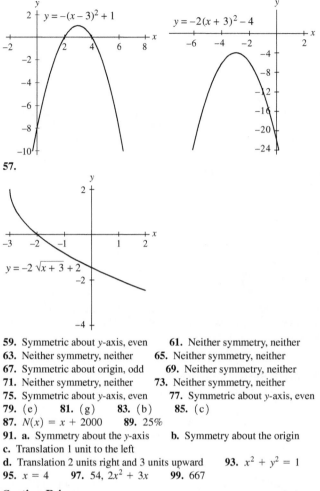

$y = -(x - 3)^2 + 1$

**55.**

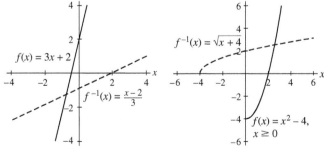

$y = -2(x + 3)^2 - 4$

**57.**

$y = -2\sqrt{x + 3} + 2$

**59.** Symmetric about $y$-axis, even    **61.** Neither symmetry, neither
**63.** Neither symmetry, neither    **65.** Neither symmetry, neither
**67.** Symmetric about origin, odd    **69.** Neither symmetry, neither
**71.** Neither symmetry, neither    **73.** Neither symmetry, neither
**75.** Symmetric about $y$-axis, even    **77.** Symmetric about $y$-axis, even
**79.** (e)    **81.** (g)    **83.** (b)    **85.** (c)
**87.** $N(x) = x + 2000$    **89.** 25%
**91. a.** Symmetry about the $y$-axis    **b.** Symmetry about the origin
**c.** Translation 1 unit to the left
**d.** Translation 2 units right and 3 units upward    **93.** $x^2 + y^2 = 1$
**95.** $x = 4$    **97.** 54, $2x^2 + 3x$    **99.** 667

**Section P.4**

*For Thought:* **1.** T    **2.** F    **3.** F    **4.** T    **5.** F    **6.** F
**7.** T    **8.** F    **9.** F    **10.** T
*Exercises:*
**1.** composition    **3.** invertible    **5.** switch-and-solve
**7.** $y = 6x - 1$    **9.** $y = x^2 + 6x + 7$    **11.** $y = x$    **13.** $y = x$
**15.** 5    **17.** 5    **19.** 50    **21.** $3x^2 + 2$    **23.** $9x^2 - 6x + 2$
**25.** $\dfrac{x^2 + 2}{3}$    **27.** $F = g \circ h$    **29.** $H = h \circ g$    **31.** $N = f \circ h$
**33.** $f^{-1}(x) = \dfrac{x + 7}{3}$    **35.** $f^{-1}(x) = (x - 2)^2 + 3$ for $x \geq 2$
**37.** $f^{-1}(x) = -x - 9$    **39.** $f^{-1}(x) = -\dfrac{1}{x}$
**41.** $f^{-1}(x) = (x - 5)^3 + 9$    **43.** $f^{-1}(x) = \sqrt{x} + 2$
**53.**

$f(x) = 3x + 2$

$f^{-1}(x) = \dfrac{x - 2}{3}$

**55.**

$f^{-1}(x) = \sqrt{x + 4}$

$f(x) = x^2 - 4,$
$x \geq 0$

**57.**

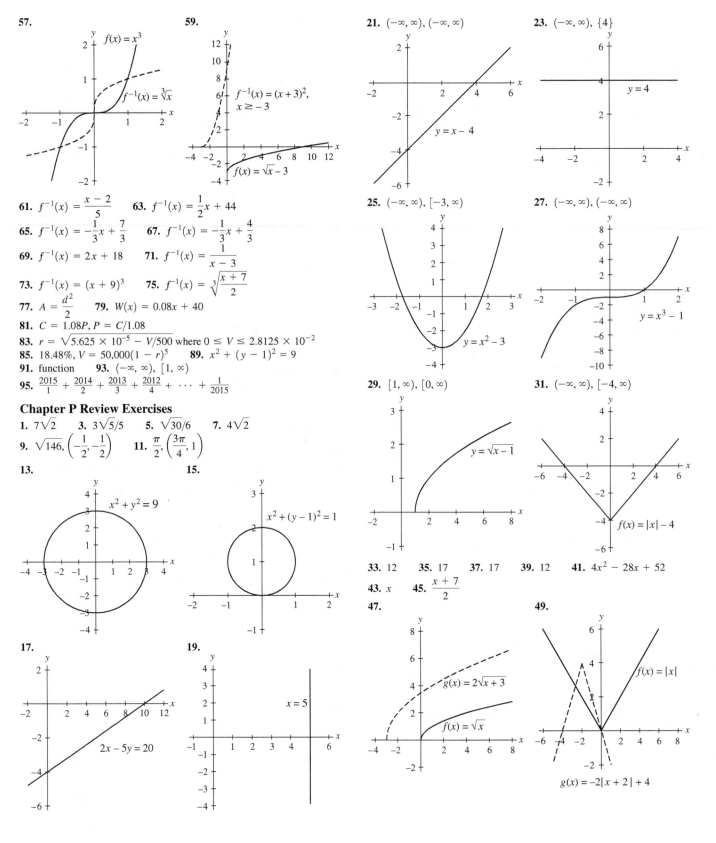

**59.**

**61.** $f^{-1}(x) = \dfrac{x-2}{5}$    **63.** $f^{-1}(x) = \dfrac{1}{2}x + 44$

**65.** $f^{-1}(x) = -\dfrac{1}{3}x + \dfrac{7}{3}$    **67.** $f^{-1}(x) = -\dfrac{1}{3}x + \dfrac{4}{3}$

**69.** $f^{-1}(x) = 2x + 18$    **71.** $f^{-1}(x) = \dfrac{1}{x-3}$

**73.** $f^{-1}(x) = (x+9)^3$    **75.** $f^{-1}(x) = \sqrt[3]{\dfrac{x+7}{2}}$

**77.** $A = \dfrac{d^2}{2}$    **79.** $W(x) = 0.08x + 40$

**81.** $C = 1.08P,\ P = C/1.08$

**83.** $r = \sqrt{5.625 \times 10^{-5} - V/500}$ where $0 \le V \le 2.8125 \times 10^{-2}$

**85.** $18.48\%,\ V = 50{,}000(1 - r)^5$    **89.** $x^2 + (y-1)^2 = 9$

**91.** function    **93.** $(-\infty, \infty),\ [1, \infty)$

**95.** $\dfrac{2015}{1} + \dfrac{2014}{2} + \dfrac{2013}{3} + \dfrac{2012}{4} + \cdots + \dfrac{1}{2015}$

### Chapter P Review Exercises

**1.** $7\sqrt{2}$    **3.** $3\sqrt{5}/5$    **5.** $\sqrt{30}/6$    **7.** $4\sqrt{2}$

**9.** $\sqrt{146},\ \left(-\dfrac{1}{2}, -\dfrac{1}{2}\right)$    **11.** $\dfrac{\pi}{2},\ \left(\dfrac{3\pi}{4}, 1\right)$

**13.**

**15.**

**17.**

**19.**

**21.** $(-\infty, \infty),\ (-\infty, \infty)$

**23.** $(-\infty, \infty),\ \{4\}$

**25.** $(-\infty, \infty),\ [-3, \infty)$

**27.** $(-\infty, \infty),\ (-\infty, \infty)$

**29.** $[1, \infty),\ [0, \infty)$

**31.** $(-\infty, \infty),\ [-4, \infty)$

**33.** 12    **35.** 17    **37.** 17    **39.** 12    **41.** $4x^2 - 28x + 52$

**43.** $x$    **45.** $\dfrac{x+7}{2}$

**47.**

**49.**

**51.**

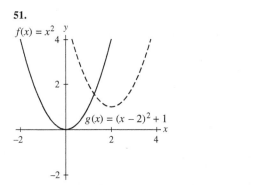

$f(x) = x^2$

$g(x) = (x - 2)^2 + 1$

**53.** $N = h \circ j$    **55.** $R = h \circ g$    **57.** $F = f \circ g$    **59.** $H = f \circ g \circ j$
**61.** $y = |x| - 3, (-\infty, \infty), [-3, \infty)$
**63.** $y = -2|x| + 4, (-\infty, \infty), (-\infty, 4]$
**65.** $y = |x + 2| + 1, (-\infty, \infty), [1, \infty)$    **67.** Symmetric about $y$-axis
**69.** Symmetric about origin    **71.** Neither symmetry
**73.** Symmetric about $y$-axis
**75.**        **77.**

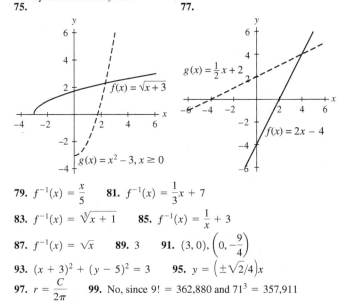

$f(x) = \sqrt{x + 3}$

$g(x) = x^2 - 3, x \geq 0$

$g(x) = \frac{1}{2}x + 2$

$f(x) = 2x - 4$

**79.** $f^{-1}(x) = \dfrac{x}{5}$    **81.** $f^{-1}(x) = \dfrac{1}{3}x + 7$

**83.** $f^{-1}(x) = \sqrt[3]{x + 1}$    **85.** $f^{-1}(x) = \dfrac{1}{x} + 3$

**87.** $f^{-1}(x) = \sqrt{x}$    **89.** 3    **91.** $(3, 0), \left(0, -\dfrac{9}{4}\right)$

**93.** $(x + 3)^2 + (y - 5)^2 = 3$    **95.** $y = \left(\pm\sqrt{2}/4\right)x$

**97.** $r = \dfrac{C}{2\pi}$    **99.** No, since $9! = 362{,}880$ and $71^3 = 357{,}911$

## Chapter P Test

**1.** $(-\infty, \infty), [0, \infty)$    **2.** $[9, \infty), [0, \infty)$    **3.** $(-\infty, \infty), (-\infty, 2]$

**4.**        **5.**

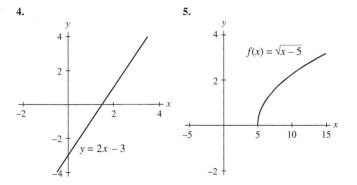

$y = 2x - 3$

$f(x) = \sqrt{x - 5}$

**6.**        **7.**

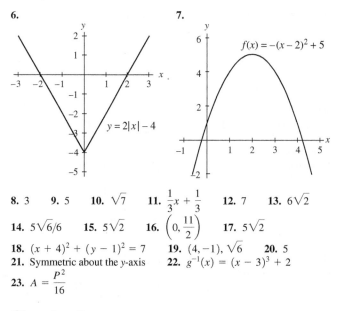

$y = 2|x| - 4$

$f(x) = -(x - 2)^2 + 5$

**8.** 3    **9.** 5    **10.** $\sqrt{7}$    **11.** $\dfrac{1}{3}x + \dfrac{1}{3}$    **12.** 7    **13.** $6\sqrt{2}$

**14.** $5\sqrt{6}/6$    **15.** $5\sqrt{2}$    **16.** $\left(0, \dfrac{11}{2}\right)$    **17.** $5\sqrt{2}$

**18.** $(x + 4)^2 + (y - 1)^2 = 7$    **19.** $(4, -1), \sqrt{6}$    **20.** 5
**21.** Symmetric about the $y$-axis    **22.** $g^{-1}(x) = (x - 3)^3 + 2$

**23.** $A = \dfrac{P^2}{16}$

## Chapter 1

### Section 1.1

***For Thought:***  **1.** T    **2.** T    **3.** T    **4.** F    **5.** F    **6.** T
**7.** T    **8.** F    **9.** T    **10.** F
***Exercises:***
**1.** angle    **3.** standard position    **5.** obtuse    **7.** coterminal
**9.** minute    **11.** $420°, 780°, -300°, -660°$
**13.** $390°, 750°, -330°, -690°$    **15.** $585°, 945°, -135°, -495°$
**17.** $315°, 675°, -405°, -765°$    **19.** $270°, 630°, -450°, -810°$
**21.** $150°, 510°, -570°, -930°$    **23.** Yes    **25.** No    **27.** Yes
**29.** Yes    **31.** No    **33.** I    **35.** III    **37.** IV    **39.** III
**41.** $-310°$    **43.** $-220°$    **45.** $-90°$    **47.** $40°$    **49.** $20°$
**51.** $340°$    **53.** $180.54°$    **55.** (c)    **57.** (e)    **59.** (h)
**61.** (g)    **63.** $13.2°$    **65.** $-8.505°$    **67.** $28.0858°$
**69.** $155.5811°$    **71.** $75°30'$    **73.** $39°24'$    **75.** $-17°19'48''$
**77.** $18°7'23''$    **79.** $58°6'$    **81.** $31°34'$    **83.** $61°7'20''$
**85.** $82°15'25''$    **87.** $60°51'13''$    **89.** $65°13'14''$    **91.** $47°41'6''$
**93.** $21°36'34''$    **95.** $4°23'23''$    **97.** $53°28'$    **99.** $41°50'54''$
**101.** $39°40'44''$    **103.** $14°54'54''$    **105.** $21°10'35.3''$
**107.** $180°, 180°$    **109.** $360°, 360°$    **111.** $20°$    **113.** $15, 7$
**115.** 5    **117.** $5\sqrt{2}$ feet    **119.** $\left(\sqrt{3} - 1\right)/4$

### Section 1.2

***For Thought:***  **1.** F    **2.** F    **3.** T    **4.** T    **5.** T    **6.** F
**7.** T    **8.** F    **9.** T    **10.** T
***Exercises:***
**1.** unit    **3.** $s = \alpha r$
**5.** $30° = \pi/6, 45° = \pi/4, 60° = \pi/3, 90° = \pi/2, 120° = 2\pi/3, 135° = 3\pi/4, 150° = 5\pi/6, 180° = \pi, 210° = 7\pi/6, 225° = 5\pi/4, 240° = 4\pi/3, 270° = 3\pi/2, 300° = 5\pi/3, 315° = 7\pi/4, 330° = 11\pi/6, 360° = 2\pi$
**7.** $\pi/4$    **9.** $\pi/2$    **11.** $2\pi/3$    **13.** $5\pi/6$    **15.** $\pi/10$
**17.** $60°$    **19.** $75°$    **21.** $135°$    **23.** $-1080°$    **25.** $136.937°$
**27.** $-7.334°$    **29.** $0.653$    **31.** $-0.241$    **33.** $0.936$
**35.** $7\pi/3, 13\pi/3, -5\pi/3, -11\pi/3$    **37.** $5\pi/2, 9\pi/2, -3\pi/2, -7\pi/2$
**39.** $8\pi/3, 14\pi/3, -4\pi/3, -10\pi/3$    **41.** $7.5, 13.8, -5.1, -11.4$
**43.** I    **45.** III    **47.** I    **49.** IV    **51.** II    **53.** IV
**55.** (g)    **57.** (b)    **59.** (h)    **61.** (d)    **63.** $\pi$    **65.** $3\pi/2$
**67.** $\pi/2$    **69.** $\pi/3$    **71.** $5\pi/3$    **73.** $2.04$    **75.** $3\pi/4$
**77.** $5\pi/6$    **79.** $3\pi/4$    **81.** $-2\pi/3$    **83.** $9.4$ ft    **85.** $209.4$ mi
**87.** $33.9$ m    **89.** 1 mi    **91.** $3.2$ km    **93.** $954.9$ ft    **95.** $3\pi$
**97.** $24\pi$    **99.** $3102$ mi    **101.** $29.0°$
**103.** Eratosthenes $41{,}143$ km, currently $40{,}074$ km
**105.** $60°$    **107.** $A = 18\theta/(2 + \theta)^2$    **109.** $5890$ ft$^2$

**111.** 33.5 in.²    **113.** $100\left(3\pi/2 + \sqrt{3}\right)$ m² or 644.4 m²

**115. a.** 22.5 in.³    **b.** $V = \dfrac{\pi(360 - \alpha)^2 \sqrt{720\alpha - \alpha^2}}{2{,}187{,}000}$

**119.** angle, rays    **121.** $x = \pi$    **123.** right    **125. a.** 10
**b.** 8, 16, 18    **c.** If $n = 2^m$ for some integer $m$, then $k = 2^m$. If
$2^m < n < 2^{m+1}$ for some integer $m$, then $k = 2(n - 2^m)$.

## Section 1.3

*For Thought:*   **1.** F    **2.** T    **3.** F    **4.** F    **5.** F    **6.** F
**7.** F    **8.** T    **9.** T    **10.** F

*Exercises:*

**1.** angular velocity    **3.** linear velocity    **5.** 5 rad/min
**7.** $8\pi$ rad/sec    **9.** 1036.7 ft/min    **11.** 95.5 rev/hr
**13.** 188.5 rad/min    **15.** 2 rev/min    **17.** 4,071,504.1 rad/hr
**19.** 44 ft/sec    **21.** 3141.6 rad/sec    **23.** 45.4 rad/sec
**25.** 3.6 rad/sec    **27.** 61.6 mph    **29.** 102.6 mph
**31.** 143.7 mph    **33.** 282.7 rad/min, 918.9 in./min
**35.** 0.8 ft/sec    **37.** 166.6 mph    **39.** 30.1 ft/sec
**41.** $\pi/12$ or 0.3 mph    **43.** 6721.7 mph    **45.** $\omega = 2/r$, decreasing
**47.** *B:* 10 ft/sec, 40 rad/sec, *C:* 10 ft/sec, 24 rad/sec
**49.** 9.6 mph    **53. a.** $13°6'36''$    **b.** $18°58'39''$    **55.** 210°
**57.** $A = \pi r^2$    **59. a.** 12:21:49.1    **b.** No    **c.** No

## Section 1.4

*For Thought:*   **1.** T    **2.** T    **3.** T    **4.** T    **5.** T    **6.** T
**7.** T    **8.** F    **9.** T    **10.** T

*Exercises:*

**1.** $\sin(\alpha), \cos(\alpha)$    **3.** reciprocals

**5.** $\dfrac{2\sqrt{5}}{5}, \dfrac{\sqrt{5}}{5}, 2, \dfrac{\sqrt{5}}{2}, \sqrt{5}, \dfrac{1}{2}$    **7.** 1, 0, undefined, 1, undefined, 0

**9.** $\dfrac{\sqrt{2}}{2}, \dfrac{\sqrt{2}}{2}, 1, \sqrt{2}, \sqrt{2}, 1$    **11.** $\dfrac{\sqrt{2}}{2}, -\dfrac{\sqrt{2}}{2}, -1, \sqrt{2}, -\sqrt{2}, -1$

**13.** $-\dfrac{3\sqrt{13}}{13}, -\dfrac{2\sqrt{13}}{13}, \dfrac{3}{2}, -\dfrac{\sqrt{13}}{3}, -\dfrac{\sqrt{13}}{2}, \dfrac{2}{3}$    **15.** 0    **17.** $-1$

**19.** 0    **21.** Undefined    **23.** $-1$    **25.** $-1$    **27.** $\dfrac{\sqrt{2}}{2}$

**29.** $\dfrac{\sqrt{2}}{2}$    **31.** $-1$    **33.** $\sqrt{2}$    **35.** $\dfrac{1}{2}$    **37.** $\dfrac{1}{2}$    **39.** $-\dfrac{\sqrt{3}}{3}$

**41.** $-2$    **43.** 2    **45.** $\sqrt{3}$    **47.** $\sqrt{3}/3$    **49.** $-1$    **51.** 1
**53.** $2 + \sqrt{3}$    **55.** $\sqrt{2}$    **57.** $\left(\sqrt{2} - \sqrt{6}\right)/4$    **59.** 1/4
**61.** $-\sqrt{3}$    **63.** 0.6820    **65.** $-0.6366$    **67.** 0.0105
**69.** 2.4780    **71.** 1.4960    **73.** 2.7475    **75.** 0.8578

**77.** 0.2679    **79.** 0.9894    **81.** 2.9992    **83.** 1    **85.** $\dfrac{1}{2}$

**87.** $\dfrac{\sqrt{2}}{2}$    **91.** $\dfrac{4}{3}$    **93.** $\dfrac{3}{10}$    **95. a.** II    **b.** IV    **c.** III
**d.** II    **97.** 4.3 in.    **99.** $160°42'16''$    **101.** 195°

**103.** 178.5 mi/hr    **105.** 6-PNC, 5-CNP, 3-NCP, 1-NPC

## Section 1.5

*For Thought:*   **1.** T    **2.** F    **3.** F    **4.** F    **5.** F    **6.** F
**7.** T    **8.** F    **9.** T    **10.** F

*Exercises:*

**1.** right    **3.** inverse sine    **5.** adjacent side, hypotenuse    **7.** 45°
**9.** 60°    **11.** 60°    **13.** 0°    **15.** 83.6°    **17.** 67.6°
**19.** 26.1°    **21.** 29.1°
**23.** $\sin \alpha = 2\sqrt{13}/13$, $\cos \alpha = 3\sqrt{13}/13$, $\tan \alpha = 2/3$, $\csc \alpha = $
$\sqrt{13}/2$, $\sec \alpha = \sqrt{13}/3$, $\cot \alpha = 3/2$
**25.** $\sqrt{5}/5, 2\sqrt{5}/5, 1/2, 2\sqrt{5}/5, \sqrt{5}/5, 2$
**27.** $3\sqrt{34}/34, 5\sqrt{34}/34, 3/5, 5\sqrt{34}/34, 3\sqrt{34}/34, 5/3$
**29.** 4/5, 3/5, 4/3, 3/5, 4/5, 3/4    **31.** $c = 10, \alpha \approx 36.9°, \beta \approx 53.1°$
**33.** $a \approx 5.7, \alpha \approx 43.7°, \beta \approx 46.3°$    **35.** $\beta = 74°, a \approx 5.5, b \approx 19.2$
**37.** $\beta = 50°51', b \approx 11.1, c \approx 14.3$    **39.** 50 ft    **41.** 1.7 mi
**43.** 43.2 m    **45.** 25.1 ft    **47.** 22 m, 57.6°    **49.** 6.5 inches
**51.** 41.3 in.    **53.** 3.08 m    **55.** 11.2 mph    **57.** 153.1 m

**59.** 4.5 km    **61.** $1.9 \times 10^{13}$ mi, 3.26 yr    **63.** 75°
**65. a.** $72\left(1 + 2\sqrt{2} - \pi\right)$ m²    **b.** 88.2%    **67.** 11,189 mi
**69.** 41.60 ft by 90.14 ft    **71.** 5.987 ft    **73.** 2.768 ft    **75.** 4798 mi
**79. a.** 3/5    **b.** 4/5    **c.** 3/4    **81. a.** $\sqrt{3}/2$    **b.** $-\sqrt{3}/2$
**c.** $-\sqrt{3}$    **83.** quadrant III    **85.** 21.216 in.

## Section 1.6

*For Thought:*   **1.** T    **2.** T    **3.** F    **4.** T    **5.** F    **6.** T
**7.** T    **8.** F    **9.** T    **10.** T

*Exercises:*

**1.** fundamental    **3.** 0    **5.** $-12/13$    **7.** $-4/5$    **9.** $2\sqrt{2}/3$
**11.** 30°, $\pi/6$    **13.** 60°, $\pi/3$    **15.** 60°, $\pi/3$    **17.** 30°, $\pi/6$
**19.** 45°, $\pi/4$    **21.** 45°, $\pi/4$    **23.** $\sqrt{2}/2$    **25.** 1/2
**27.** $-\sqrt{2}/2$    **29.** $-\sqrt{3}/2$    **31.** $-\sqrt{2}/2$    **33.** $-1/2$

**35.** $\dfrac{\sqrt{2}}{2}, -\dfrac{\sqrt{2}}{2}, -1, \sqrt{2}, -\sqrt{2}, -1$

**37.** $-\dfrac{\sqrt{3}}{2}, -\dfrac{1}{2}, \sqrt{3}, -\dfrac{2\sqrt{3}}{3}, -2, \dfrac{\sqrt{3}}{3}$

**39.** $-\dfrac{\sqrt{3}}{2}, \dfrac{1}{2}, -\sqrt{3}, -\dfrac{2\sqrt{3}}{3}, 2, -\dfrac{\sqrt{3}}{3}$

**41.** $-\dfrac{\sqrt{2}}{2}, -\dfrac{\sqrt{2}}{2}, 1, -\sqrt{2}, -\sqrt{2}, 1$

**43.** F    **45.** T    **47.** F    **49.** T    **51.** F    **53.** T
**55.** 120°F and 84°F    **57. a.** 3    **b.** $-7\sqrt{2}/2$
**59.** 1.708 in., 1.714 in.    **61.** 74.0 mph    **63. a.** 5/3
**b.** 5/4    **c.** 3/4    **65. a.** 2    **b.** $-2$    **c.** $\sqrt{3}/3$
**67.** $-30°$    **69.** 0.118°

## Chapter 1 Review Exercises

**1.** 28°    **3.** $206°45'33''$    **5.** 180°    **7.** 108°    **9.** 300°
**11.** 270°    **13.** $11\pi/6$    **15.** $-5\pi/3$
**17.**

| $\theta$ deg | 0 | 30 | 45 | 60 | 90 | 120 | 135 | 150 | 180 |
|---|---|---|---|---|---|---|---|---|---|
| $\theta$ rad | 0 | $\dfrac{\pi}{6}$ | $\dfrac{\pi}{4}$ | $\dfrac{\pi}{3}$ | $\dfrac{\pi}{2}$ | $\dfrac{2\pi}{3}$ | $\dfrac{3\pi}{4}$ | $\dfrac{5\pi}{6}$ | $\pi$ |
| $\sin \theta$ | 0 | $\dfrac{1}{2}$ | $\dfrac{\sqrt{2}}{2}$ | $\dfrac{\sqrt{3}}{2}$ | 1 | $\dfrac{\sqrt{3}}{2}$ | $\dfrac{\sqrt{2}}{2}$ | $\dfrac{1}{2}$ | 0 |
| $\cos \theta$ | 1 | $\dfrac{\sqrt{3}}{2}$ | $\dfrac{\sqrt{2}}{2}$ | $\dfrac{1}{2}$ | 0 | $-\dfrac{1}{2}$ | $-\dfrac{\sqrt{2}}{2}$ | $-\dfrac{\sqrt{3}}{2}$ | $-1$ |

**19.** $-\sqrt{2}/2$    **21.** $\sqrt{3}$    **23.** $-2\sqrt{3}/3$    **25.** 0    **27.** 0
**29.** $-1$    **31.** $\sqrt{3}/3$    **33.** $-\sqrt{2}/2$    **35.** $-2$    **37.** $-\sqrt{3}/3$
**39.** 0.6947    **41.** $-0.0923$    **43.** 0.1869    **45.** $-3.8470$
**47.** 2.9238    **49.** 1.4975    **51.** 45°    **53.** 0°    **55.** 30°
**57.** 30°    **59.** 5/13, 12/13, 5/12, 13/5, 13/12, 12/5
**61.** $c = \sqrt{13}, \alpha \approx 33.7°, \beta \approx 56.3°$
**63.** $\beta = 68.7°, c \approx 8.8, b \approx 8.2$    **65.** $-2\sqrt{6}/5$    **67.** 94.2 ft/hr
**69.** 6.9813 ft    **71.** $c = \sqrt{89}$ ft, 32.0°, 58.0°
**73.** $\beta = 70.7°, b \approx 11.3$ ft, $a \approx 4.0$ ft    **75.** 218 ft, 345 ft
**77.** 293 ft    **79.** 56.1 in.    **81.** 22.8 in.
**83. a.** $A = x^2 \cos^{-1}(x/2)$    **b.** $\sqrt{2}$ ft or $\sqrt{3}$ ft    **c.** $x \approx 1.5882$ ft

## Chapter 1 Test

**1.** 1/2    **2.** $-1/2$    **3.** $\sqrt{2}/2$    **4.** 1/2    **5.** $\sqrt{3}/3$
**6.** $\sqrt{3}$    **7.** Undefined    **8.** $-1$    **9.** Undefined
**10.** 1    **11.** $\pi/4$ or 45°    **12.** $\pi/6$ or 30°    **13.** 28.85 m
**14.** 134.07°    **15.** Coterminal    **16.** $-\sqrt{15}/4$
**17.** 647.2 rad/min    **18.** 7.97 mph
**19.** $\sin \alpha = -2\sqrt{29}/29, \cos \alpha = 5\sqrt{29}/29, \tan \alpha = -2/5,$
$\csc \alpha = -\sqrt{29}/2, \sec \alpha = \sqrt{29}/5, \cot \alpha = -5/2$
**20.** $\beta = 66°, a \approx 2.0$ ft, $b \approx 4.6$ ft    **21.** 12.2 m    **22.** 1117 ft

# Chapter 2

## Section 2.1

***For Thought:*** **1.** T   **2.** F   **3.** T   **4.** F   **5.** F   **6.** T
**7.** F   **8.** F   **9.** T   **10.** T

***Exercises:***
**1.** $\sin(\alpha)$, $\cos(\alpha)$   **3.** sine wave   **5.** fundamental cycle
**7.** phase shift   **9.** starting, maximum, inflection, minimum, ending
**11.** 0   **13.** $\sqrt{3}$   **15.** 1/2   **17.** 2   **19.** 0   **21.** 0
**23.** $-2$   **25.** $-\sqrt{2}/2$   **27.** $(\pi/4, 0)$   **29.** $(3\pi/4, 3)$
**31.** $(-\pi/4, -1)$   **33.** $(5\pi/4, 0)$   **35.** $(0, 0)$   **37.** $(2\pi/3, 1)$
**39.** $(\pi/6, -1)$   **41.** $(-4\pi/3, 1)$   **43.** $(7\pi/6, 1)$
**45.** $(2\pi/3, 2)$   **47.** $(-4\pi/3, 3)$   **49.** $(13\pi/6, -2)$
**51.** $(3\pi/2, 0)$   **53.** $(\pi/8, 2)$   **55.** $(\pi/3, 1)$
**57.** $(5\pi/12, -4)$   **59.** $(0, 0), (\pi/4, 2), (\pi/2, 0), (3\pi/4, -2)$
**61.** $(\pi/4, 0), (5\pi/8, 2), (\pi, 0), (11\pi/8, -2)$
**63.** $(0, 2), (\pi/12, 3), (\pi/6, 2), (\pi/4, 1)$   **65.** $2, 2\pi, 0, [-2, 2]$
**67.** $1, 2\pi, \pi/2, [-1, 1]$   **69.** $2, 2\pi, -\pi/3, [-2, 2]$
**71.** $1, 0, [-1, 1]$; points:
$(0, 0), (\pi/2, -1), (\pi, 0),$
$(3\pi/2, 1), (2\pi, 0)$

**73.** $3, 0, [-3, 3]$; points:
$(0, 0), (\pi/2, -3), (\pi, 0),$
$(3\pi/2, 3), (2\pi, 0)$

**75.** $1/2, 0, [-1/2, 1/2]$; points:
$(0, 1/2), (\pi/2, 0), (\pi, -1/2),$
$(3\pi/2, 0), (2\pi, 1/2)$

**77.** $1, -\pi, [-1, 1]$; points:
$(-\pi, 0), (-\pi/2, 1),$
$(0, 0), (\pi/2, -1), (\pi, 0)$

**79.** $1, \pi/3, [-1, 1]$; points:
$(\pi/3, 1), (5\pi/6, 0),$
$(4\pi/3, -1), (11\pi/6, 0),$
$(7\pi/3, 1)$

**81.** $1, 0, [1, 3]$; points:
$(0, 3), (\pi/2, 2), (\pi, 1),$
$(3\pi/2, 2), (2\pi, 3)$

**83.** $1, 0, [-2, 0]$; points:
$(0, -1), (\pi/2, -2), (\pi, -1),$
$(3\pi/2, 0), (2\pi, -1)$

**85.** $1, -\pi/4, [1, 3]$; points:
$(-\pi/4, 2), (\pi/4, 3),$
$(3\pi/4, 2), (5\pi/4, 1),$
$(7\pi/4, 2)$

**87.** $2, -\pi/6, [-1, 3]$; points:
$(-\pi/6, 3), (\pi/3, 1),$
$(5\pi/6, -1), (4\pi/3, 1),$
$(11\pi/6, 3)$

**89.** $2, \pi/3, [-1, 3]$; points:
$(\pi/3, 1), (5\pi/6, -1),$
$(4\pi/3, 1), (11\pi/6, 3),$
$(7\pi/3, 1)$

**91.** $y = 2\sin(x - \pi/2)$   **93.** $y = 2\sin(x + \pi) + 1$
**95.** $y = 2\cos(x - \pi/2)$   **97.** $y = 2\cos(x - \pi) + 1$
**99.** $y = \sin(x - \pi/4)$   **101.** $y = \sin(x + \pi/2)$
**103.** $y = -\cos(x - \pi/5)$   **105.** $y = -\cos(x - \pi/8) + 2$
**107.** $y = -3\cos(x + \pi/4) - 5$
**109.** $A$ determines stretching, shrinking, and reflection; $C$ is the phase shift; and $D$ is the vertical translation.

**111.** $\pi/6$   **113.** 67,000 mph   **115.** 21.1°   **117.** $\dfrac{2\sqrt{3} - 3}{3}$

## Section 2.2

***For Thought:*** **1.** T   **2.** F   **3.** T   **4.** T   **5.** F   **6.** F
**7.** T   **8.** T   **9.** F   **10.** T

***Exercises:***
**1.** period   **3.** amplitude   **5.** $3, \pi/2, 0$   **7.** $2, \pi, -\pi/4$
**9.** $2, 2, 1$
**11.** $2\pi/3, 0, [-1, 1]$; points:
$(0, 0), (\pi/6, 1), (\pi/3, 0),$
$(\pi/2, -1), (2\pi/3, 0)$

**13.** $\pi, 0, [-1, 1]$; points:
$(0, 0), (\pi/4, -1),$
$(\pi/2, 0), (3\pi/4, 1), (\pi, 0)$

**15.** $\pi/2, 0, [1, 3]$; points:
(0, 3), ($\pi/8$, 2), ($\pi/4$, 1),
($3\pi/8$, 2), ($\pi/2$, 3)

**17.** $8\pi, 0, [1, 3]$; points:
(0, 2), ($2\pi$, 1), ($4\pi$, 2),
($6\pi$, 3), ($8\pi$, 2)

**19.** $6, 0, [-1, 1]$; points:
(0, 0), (1.5, 1), (3, 0),
(4.5, −1), (6, 0)

**21.** $\pi, \pi/2, [-1, 1]$; points:
($\pi/2$, 0), ($3\pi/4$, 1), ($\pi$, 0),
($5\pi/4$, −1), ($3\pi/2$, 0)

**23.** $4, -3, [-1, 1]$; points:
(−3, 0), (−2, 1), (−1, 0),
(0, −1), (1, 0)

**25.** $\pi, -\pi/6, [-1, 3]$; points:
(−$\pi/6$, 3), ($\pi/12$, 1),
($\pi/3$, −1), ($7\pi/12$, 1),
($5\pi/6$, 3)

**27.** $2\pi/3, \pi/6, [-3/2, -1/2]$;
points: ($\pi/6$, −1),
($\pi/3$, −3/2), ($\pi/2$, −1),
($2\pi/3$, −1/2), ($5\pi/6$, −1)

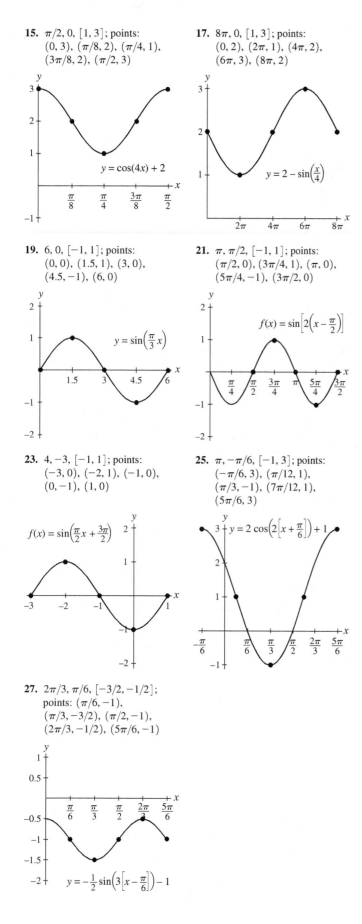

**29.** $y = 2 \sin\left[2\left(x - \dfrac{\pi}{4}\right)\right]$     **31.** $y = 3 \sin\left[\dfrac{3}{2}\left(x + \dfrac{\pi}{3}\right)\right] + 3$

**33.** 0     **35.** 0     **37.** 0     **39.** $\sin(x - \pi/4)$
**41.** $3 \sin(x - \pi/4)$     **43.** 100 cycles/sec     **45.** 40 cycles/hr
**47.** $3, \pi$

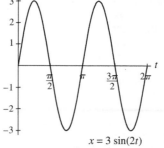

$x = 3 \sin(2t)$

**49.** 11 yr     **51. a.** 1300 cc and 500 cc     **b.** 30 breaths/min
**53.** 12, 15,000, −3; 25,000; $y = 15,000 \sin(\pi x/6 + \pi/2) + 25,000$;
$17,500     **55. a.** 40, 65, $y = 65 \sin(\pi x/20)$     **b.** 40 days
**c.** −38.2 m/sec     **d.** Planet is between the earth and Rho.
**57.** $y = \sin(\pi x/10) + 1$
**59.** $y = 50.0 \sin(0.214x - 0.615) + 48.8$; 29.4 days; 98%
**63.** 1/2, $2\pi$, $\pi/2$, [2.5, 3.5]     **65.** 289 feet     **67.** 98°44′
**69.** One possibility is WRONG = 25938 and RIGHT = 51876.

## Section 2.3

*For Thought:* **1.** F     **2.** T     **3.** T     **4.** F     **5.** T     **6.** T
**7.** F     **8.** T     **9.** T     **10.** T
*Exercises:*
**1.** domain     **3.** asymptote     **5.** 2     **7.** $-\sqrt{2}$     **9.** Undefined
**11.** Undefined     **13.** 92.6     **15.** 100.0     **17.** 627.9
**19.** −418.6     **21.** $\pi$     **23.** $4\pi/3$     **25.** 2
**27.** $(-\infty, -2] \cup [2, \infty)$     **29.** $(-\infty, -1/2] \cup [1/2, \infty)$
**31.** $(-\infty, -2] \cup [0, \infty)$
**33.** $2\pi, x = \pi/2 + k\pi$,
$(-\infty, -2] \cup [2, \infty)$

**35.** $2\pi/3, x = \pi/6 + k\pi/3$,
$(-\infty, -1] \cup [1, \infty)$

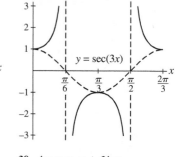

**37.** $2\pi, x = \pi/4 + k\pi$,
$(-\infty, -1] \cup [1, \infty)$

**39.** $4\pi, x = \pi + 2k\pi$,
$(-\infty, -1] \cup [1, \infty)$

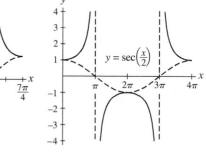

**41.** $2, x = 1/2 + k$,
$(-\infty, -2] \cup [2, \infty)$

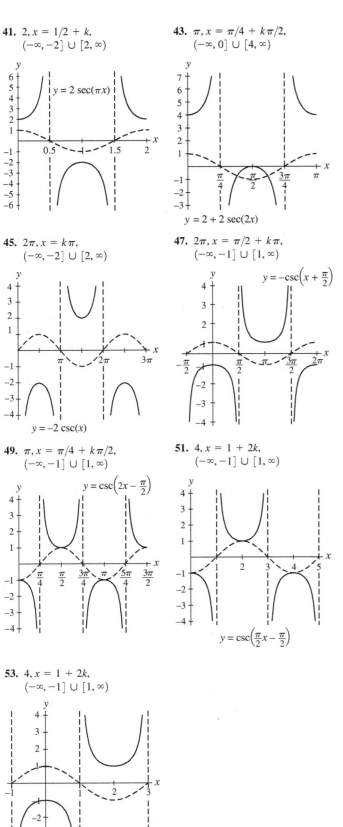

$y = 2 \sec(\pi x)$

**43.** $\pi, x = \pi/4 + k\pi/2$,
$(-\infty, 0] \cup [4, \infty)$

$y = 2 + 2\sec(2x)$

**45.** $2\pi, x = k\pi$,
$(-\infty, -2] \cup [2, \infty)$

$y = -2\csc(x)$

**47.** $2\pi, x = \pi/2 + k\pi$,
$(-\infty, -1] \cup [1, \infty)$

$y = -\csc\left(x + \dfrac{\pi}{2}\right)$

**49.** $\pi, x = \pi/4 + k\pi/2$,
$(-\infty, -1] \cup [1, \infty)$

$y = \csc\left(2x - \dfrac{\pi}{2}\right)$

**51.** $4, x = 1 + 2k$,
$(-\infty, -1] \cup [1, \infty)$

$y = \csc\left(\dfrac{\pi}{2}x - \dfrac{\pi}{2}\right)$

**53.** $4, x = 1 + 2k$,
$(-\infty, -1] \cup [1, \infty)$

$y = -\csc\left(\dfrac{\pi}{2}x + \dfrac{\pi}{2}\right)$

**55.** $y = \sec(x - \pi/2) + 1$    **57.** $y = -\csc(x + 1) + 4$
**59.** $x = \pi/2 + k\pi$    **61.** $x = k\pi/2$    **63.** $x = k\pi$
**65.** $x = k\pi/2$    **67.** $x = k\pi/2$    **69.** $x = 1/2 + k$
**71.** $(-\infty, -|A| + D] \cup [|A| + D, \infty)$    **73.** $y, x$

**75.** $5, \pi, \pi/2, [-2, 8]$    **77.** $0$    **79.** $y = \dfrac{1}{2} - \dfrac{1}{2}\cos(2x)$

## Section 2.4

*For Thought:*  **1.** T    **2.** T    **3.** F    **4.** T    **5.** F    **6.** F
**7.** T    **8.** F    **9.** T    **10.** T
*Exercises:*
**1.** tangent    **3.** domain    **5.** inflection    **7.** $\sqrt{3}$    **9.** Undefined
**11.** $0$    **13.** $1$    **15.** Undefined    **17.** $0$    **19.** $92.6$
**21.** $-108.6$    **23.** $500.0$    **25.** $-500.0$    **27.** $\pi/8$    **29.** $1$
**31.** $3$    **33.** $\pi/3$
**35.** $\pi/3, x = \pi/6 + k\pi/3$    **37.** $1, x = k + 1/2$

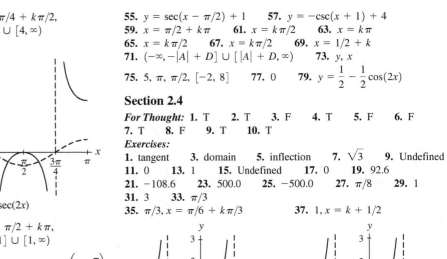

$y = \tan(3x)$    $y = \tan(\pi x)$

**39.** $\pi, x = \pi/2 + k\pi$    **41.** $\pi, x = k\pi$

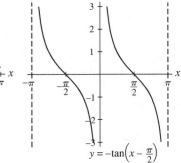

$y = -2\tan(x) + 1$    $y = -\tan\left(x - \dfrac{\pi}{2}\right)$

**43.** $2, x = 2k$    **45.** $\pi, x = 3\pi/4 + k\pi$

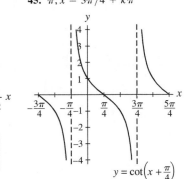

$y = \tan\left(\dfrac{\pi}{2}x - \dfrac{\pi}{2}\right)$    $y = \cot\left(x + \dfrac{\pi}{4}\right)$

**47.** $2\pi, x = 2k\pi$      **49.** $\pi, x = \pi/2 + k\pi$

**5.**

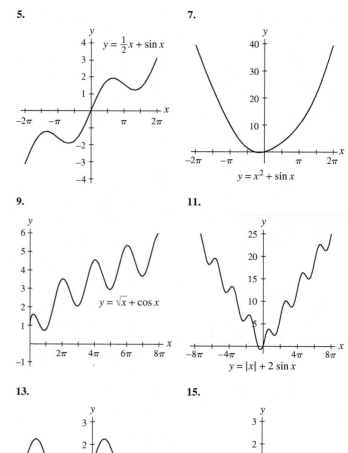

$y = \frac{1}{2}x + \sin x$

**7.**

$y = x^2 + \sin x$

**51.** $\pi/2, x = \pi/4 + k\pi/2$

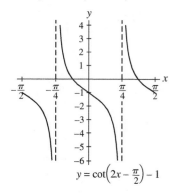

$y = \cot\left(2x - \frac{\pi}{2}\right) - 1$

**9.**

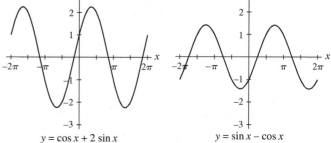

$y = \sqrt{x} + \cos x$

**11.**

$y = |x| + 2\sin x$

**53.** $y = 3\tan(x - \pi/4) + 2$      **55.** $y = -\cot(x + \pi/2) + 1$
**57.** $y = \tan[2(x - \pi/4)]$      **59.** $y = -\tan(\pi x/2)$
**61.** 0     **63.** Undefined     **65.** $\tan(2x + 3)$     **67.** $2\tan(x) + 3$
**69.** $y = x + 1$     **71.** $y = \sqrt{3}x - 3\sqrt{3} - 1$
**73. a.** 2.3 yr     **b.** Tangent     **75.** $1/y, 1/x, x/y$
**77.** $\pi, x = \pi/4 + k\pi/2$ for any integer $k$, $(-\infty, -3] \cup [3, \infty)$
**79.** 812 feet     **81. a.** Yes     **b.** No

## Section 2.5

*For Thought:* **1.** F    **2.** T    **3.** T    **4.** T    **5.** T    **6.** T
**7.** F    **8.** F    **9.** T    **10.** T
*Exercises:*
**1.**

$y = x + \cos x$

**3.**

$y = \frac{1}{x} - \sin x$

**13.**

$y = \cos x + 2\sin x$

**15.**

$y = \sin x - \cos x$

**17.**

$y = \sin x + \cos 2x$

**19.**

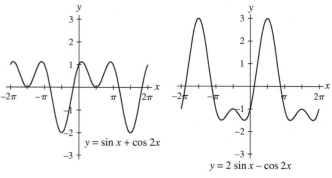

$y = 2\sin x - \cos 2x$

**21.**

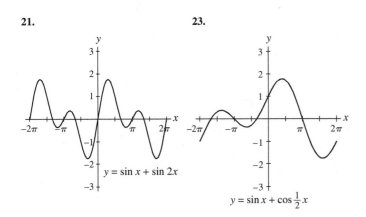

$y = \sin x + \sin 2x$

**23.**

$y = \sin x + \cos \frac{1}{2}x$

**45. a.**

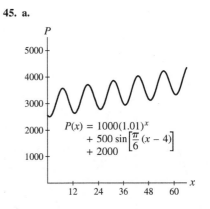

$$P(x) = 1000(1.01)^x$$
$$+ 500 \sin\left[\frac{\pi}{6}(x - 4)\right]$$
$$+ 2000$$

**b.** An exponential function
**47.** The graph looks like $y = x$.

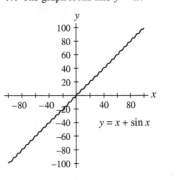

$y = x + \sin x$

**25.**

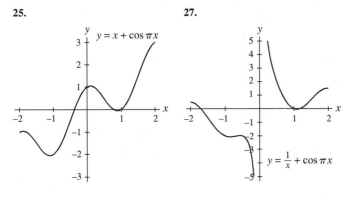

$y = x + \cos \pi x$

**27.**

$y = \frac{1}{x} + \cos \pi x$

**29.**

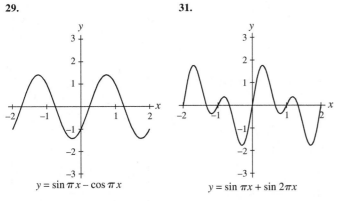

$y = \sin \pi x - \cos \pi x$

**31.**

$y = \sin \pi x + \sin 2\pi x$

**49.** The graph looks like $y = \tan x$ with increasing vertical translation on each cycle.

$y = x + \tan x$

**51.** $3, 4, 1, [4, 10]$    **53.** $2, (-\infty, -5] \cup [5, \infty)$
**55.** $1, (-\infty, \infty)$    **57.** 163 Elm Street

## Chapter 2 Review Exercises

**1.** 1    **3.** 1    **5.** $-1$    **7.** 0    **9.** $2\pi$    **11.** $\pi/2$
**13.** 2    **15.** $4\pi$    **17.** $(-\infty, \infty), [-2, 2]$
**19.** $\{x \mid x \neq \pi/4 + k\pi/2\}, (-\infty, \infty)$
**21.** $\{x \mid x \neq \pi/2 + k\pi\}, (-\infty, -3] \cup [-1, \infty)$
**23.** $\{x \mid x \neq k\}, (-\infty, \infty)$    **25.** $x = \pi/4 + k\pi/2$
**27.** $x = k$    **29.** $x = k\pi$    **31.** $x = k$

**33.**

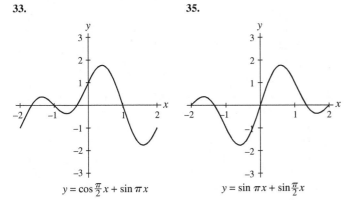

$y = \cos \frac{\pi}{2}x + \sin \pi x$

**35.**

$y = \sin \pi x + \sin \frac{\pi}{2}x$

**37.** (c)    **39.** (a)    **41.** (d)
**43.** $x = 4\sin(t) - 3\cos(t)$, 3.5, 6.3 sec, 5 cm

**33.** $3, \pi, 0, [-3, 3]$; points:
(0, 0), $(\pi/4, 3)$, $(\pi/2, 0)$,
$(3\pi/4, -3)$, $(\pi, 0)$

**35.** $1, 2\pi, \pi/6, [-1, 1]$; points:
$(\pi/6, 1)$, $(2\pi/3, 0)$,
$(7\pi/6, -1)$, $(5\pi/3, 0)$,
$(13\pi/6, 1)$

**49.** $y = 10 \sin\left[\dfrac{\pi}{4}(x + 2)\right]$

**51.** $y = 20 \sin\left[\dfrac{\pi}{2}(x - 1)\right] + 10$

**53.** $\pi/3, x = \pi/6 + k\pi/3$,
$(-\infty, \infty)$

**55.** $\pi/2, x = \pi/4 + k\pi/2$,
$(-\infty, \infty)$

**37.** $1, 2\pi, -\pi/4, [0, 2]$; points:
$(-\pi/4, 2)$, $(\pi/4, 1)$,
$(3\pi/4, 0)$, $(5\pi/4, 1)$,
$(7\pi/4, 2)$

**39.** $1, 4, 0, [-1, 1]$; points:
(0, 1), (1, 0), (2, -1),
(3, 0), (4, 1)

**57.** $4\pi, x = \pi + 2k\pi$,
$(-\infty, -1] \cup [1, \infty)$

**59.** $\pi/2, x = k\pi/2$,
$(-\infty, \infty)$

**41.** $1, \pi, -\pi/2, [-1, 1]$; points:
$(-\pi/2, 0)$, $(-\pi/4, 1)$, (0, 0),
$(\pi/4, -1)$, $(\pi/2, 0)$

**43.** $1/2, \pi, 0, [-1/2, 1/2]$; points:
$(0, -1/2)$, $(\pi/4, 0)$, $(\pi/2, 1/2)$,
$(3\pi/4, 0)$, $(\pi, -1/2)$

**61.** $\pi/2, x = \pi/3 + k\pi/2$,
$(-\infty, \infty)$

**63.** $\pi, x = k\pi/2$,
$(-\infty, -1/3] \cup [1/3, \infty)$

**45.** $2, \pi, -\pi/6, [-1, 3]$; points:
$(-\pi/6, 1)$, $(\pi/12, -1)$,
$(\pi/3, 1)$, $(7\pi/12, 3)$,
$(5\pi/6, 1)$

**47.** $1/3, 2, -1, [2/3, 4/3]$; points:
$(-1, 4/3)$, $(-1/2, 1)$, $(0, 2/3)$,
$(1/2, 1)$, $(1, 4/3)$

**65.** $2\pi, x = \pi/4 + k\pi$,
$(-\infty, -3] \cup [1, \infty)$

**67.**

**69.**

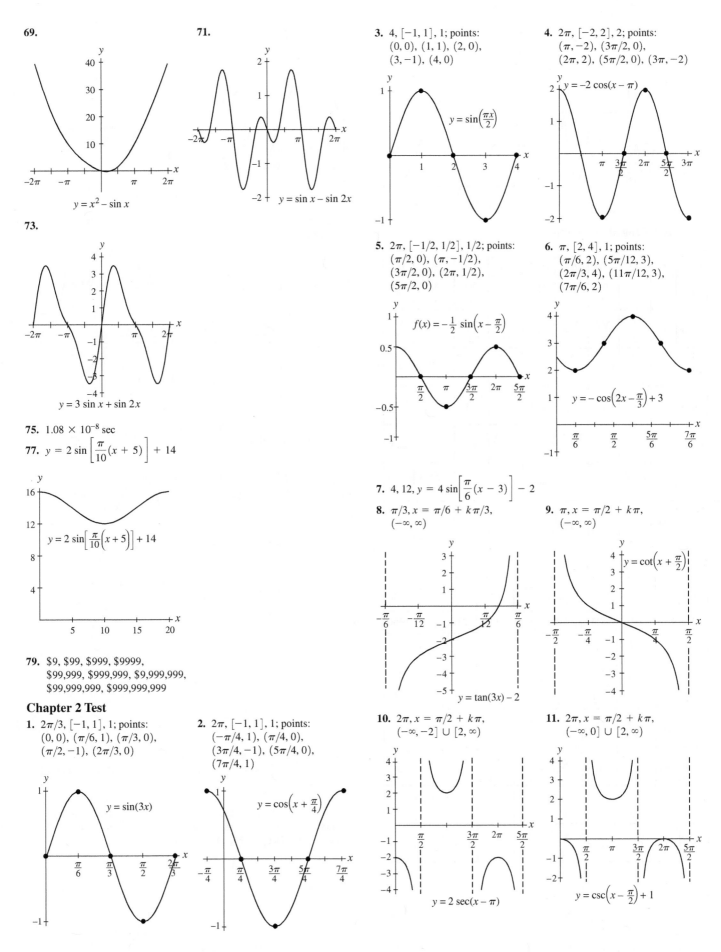

$y = x^2 - \sin x$

**71.**

$y = \sin x - \sin 2x$

**73.**

$y = 3 \sin x + \sin 2x$

**75.** $1.08 \times 10^{-8}$ sec

**77.** $y = 2 \sin \left[ \dfrac{\pi}{10}(x + 5) \right] + 14$

$y = 2 \sin \left[ \dfrac{\pi}{10}\left(x + 5\right) \right] + 14$

**79.** $9, $99, $999, $9999,
$99,999, $999,999, $9,999,999,
$99,999,999, $999,999,999

**Chapter 2 Test**

**1.** $2\pi/3$, $[-1, 1]$, 1; points:
$(0, 0)$, $(\pi/6, 1)$, $(\pi/3, 0)$,
$(\pi/2, -1)$, $(2\pi/3, 0)$

$y = \sin(3x)$

**2.** $2\pi$, $[-1, 1]$, 1; points:
$(-\pi/4, 1)$, $(\pi/4, 0)$,
$(3\pi/4, -1)$, $(5\pi/4, 0)$,
$(7\pi/4, 1)$

$y = \cos\left(x + \dfrac{\pi}{4}\right)$

**3.** 4, $[-1, 1]$, 1; points:
$(0, 0)$, $(1, 1)$, $(2, 0)$,
$(3, -1)$, $(4, 0)$

$y = \sin\left(\dfrac{\pi x}{2}\right)$

**4.** $2\pi$, $[-2, 2]$, 2; points:
$(\pi, -2)$, $(3\pi/2, 0)$,
$(2\pi, 2)$, $(5\pi/2, 0)$, $(3\pi, -2)$

$y = -2 \cos(x - \pi)$

**5.** $2\pi$, $[-1/2, 1/2]$, 1/2; points:
$(\pi/2, 0)$, $(\pi, -1/2)$,
$(3\pi/2, 0)$, $(2\pi, 1/2)$,
$(5\pi/2, 0)$

$f(x) = -\dfrac{1}{2} \sin\left(x - \dfrac{\pi}{2}\right)$

**6.** $\pi$, $[2, 4]$, 1; points:
$(\pi/6, 2)$, $(5\pi/12, 3)$,
$(2\pi/3, 4)$, $(11\pi/12, 3)$,
$(7\pi/6, 2)$

$y = -\cos\left(2x - \dfrac{\pi}{3}\right) + 3$

**7.** 4, 12, $y = 4 \sin\left[ \dfrac{\pi}{6}(x - 3) \right] - 2$

**8.** $\pi/3$, $x = \pi/6 + k\pi/3$,
$(-\infty, \infty)$

$y = \tan(3x) - 2$

**9.** $\pi$, $x = \pi/2 + k\pi$,
$(-\infty, \infty)$

$y = \cot\left(x + \dfrac{\pi}{2}\right)$

**10.** $2\pi$, $x = \pi/2 + k\pi$,
$(-\infty, -2] \cup [2, \infty)$

$y = 2 \sec(x - \pi)$

**11.** $2\pi$, $x = \pi/2 + k\pi$,
$(-\infty, 0] \cup [2, \infty)$

$y = \csc\left(x - \dfrac{\pi}{2}\right) + 1$

**12.**    **13.**

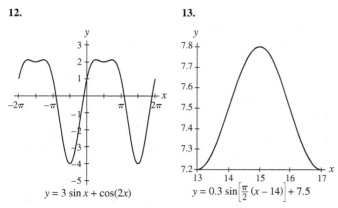

$y = 3 \sin x + \cos(2x)$     $y = 0.3 \sin\left[\dfrac{\pi}{2}(x-14)\right] + 7.5$

## Tying It All Together Chapters P–2

**1.**

| θ deg | 0 | 30 | 45 | 60 | 90 | 120 | 135 | 150 | 180 |
|---|---|---|---|---|---|---|---|---|---|
| θ rad | 0 | $\dfrac{\pi}{6}$ | $\dfrac{\pi}{4}$ | $\dfrac{\pi}{3}$ | $\dfrac{\pi}{2}$ | $\dfrac{2\pi}{3}$ | $\dfrac{3\pi}{4}$ | $\dfrac{5\pi}{6}$ | $\pi$ |
| sin θ | 0 | $\dfrac{1}{2}$ | $\dfrac{\sqrt{2}}{2}$ | $\dfrac{\sqrt{3}}{2}$ | 1 | $\dfrac{\sqrt{3}}{2}$ | $\dfrac{\sqrt{2}}{2}$ | $\dfrac{1}{2}$ | 0 |
| cos θ | 1 | $\dfrac{\sqrt{3}}{2}$ | $\dfrac{\sqrt{2}}{2}$ | $\dfrac{1}{2}$ | 0 | $-\dfrac{1}{2}$ | $-\dfrac{\sqrt{2}}{2}$ | $-\dfrac{\sqrt{3}}{2}$ | $-1$ |
| tan θ | 0 | $\dfrac{\sqrt{3}}{3}$ | 1 | $\sqrt{3}$ | und | $-\sqrt{3}$ | $-1$ | $-\dfrac{\sqrt{3}}{3}$ | 0 |
| csc θ | und | 2 | $\sqrt{2}$ | $\dfrac{2\sqrt{3}}{3}$ | 1 | $\dfrac{2\sqrt{3}}{3}$ | $\sqrt{2}$ | 2 | und |
| sec θ | 1 | $\dfrac{2\sqrt{3}}{3}$ | $\sqrt{2}$ | 2 | und | $-2$ | $-\sqrt{2}$ | $-\dfrac{2\sqrt{3}}{3}$ | $-1$ |
| cot θ | und | $\sqrt{3}$ | 1 | $\dfrac{\sqrt{3}}{3}$ | 0 | $-\dfrac{\sqrt{3}}{3}$ | $-1$ | $-\sqrt{3}$ | und |

**2.**

| θ rad | $\pi$ | $\dfrac{7\pi}{6}$ | $\dfrac{5\pi}{4}$ | $\dfrac{4\pi}{3}$ | $\dfrac{3\pi}{2}$ | $\dfrac{5\pi}{3}$ | $\dfrac{7\pi}{4}$ | $\dfrac{11\pi}{6}$ | $2\pi$ |
|---|---|---|---|---|---|---|---|---|---|
| θ deg | 180 | 210 | 225 | 240 | 270 | 300 | 315 | 330 | 360 |
| sin θ | 0 | $-\dfrac{1}{2}$ | $-\dfrac{\sqrt{2}}{2}$ | $-\dfrac{\sqrt{3}}{2}$ | $-1$ | $-\dfrac{\sqrt{3}}{2}$ | $-\dfrac{\sqrt{2}}{2}$ | $-\dfrac{1}{2}$ | 0 |
| cos θ | $-1$ | $-\dfrac{\sqrt{3}}{2}$ | $-\dfrac{\sqrt{2}}{2}$ | $-\dfrac{1}{2}$ | 0 | $\dfrac{1}{2}$ | $\dfrac{\sqrt{2}}{2}$ | $\dfrac{\sqrt{3}}{2}$ | 1 |
| tan θ | 0 | $\dfrac{\sqrt{3}}{3}$ | 1 | $\sqrt{3}$ | und | $-\sqrt{3}$ | $-1$ | $-\dfrac{\sqrt{3}}{3}$ | 0 |
| csc θ | und | $-2$ | $-\sqrt{2}$ | $-\dfrac{2\sqrt{3}}{3}$ | $-1$ | $-\dfrac{2\sqrt{3}}{3}$ | $-\sqrt{2}$ | $-2$ | und |
| sec θ | $-1$ | $-\dfrac{2\sqrt{3}}{3}$ | $-\sqrt{2}$ | $-2$ | und | 2 | $\sqrt{2}$ | $\dfrac{2\sqrt{3}}{3}$ | 1 |
| cot θ | und | $\sqrt{3}$ | 1 | $\dfrac{\sqrt{3}}{3}$ | 0 | $-\dfrac{\sqrt{3}}{3}$ | $-1$ | $-\sqrt{3}$ | und |

**3.** $(-\infty, \infty)$, $[-1, 1]$, $\pi$    **4.** $(-\infty, \infty)$, $[-1, 1]$, $\pi$

**5.** $\{x \mid x \neq \pi/4 + k\pi/2\}$, $(-\infty, \infty)$, $\pi/2$    **6.** $\{x \mid x \neq \pi/4 + k\pi/2\}$, $(-\infty, -1] \cup [1, \infty)$, $\pi$

**7.** $\{x \mid x \neq k\pi/2\}$, $(-\infty, -1] \cup [1, \infty)$, $\pi$    **8.** $(-\infty, \infty)$, $[-1, 1]$, 2

**9.** $(-\infty, \infty)$, $[-1, 1]$, 2    **10.** $\{x \mid x \neq 1/2 + k\}$, $(-\infty, \infty)$, 1

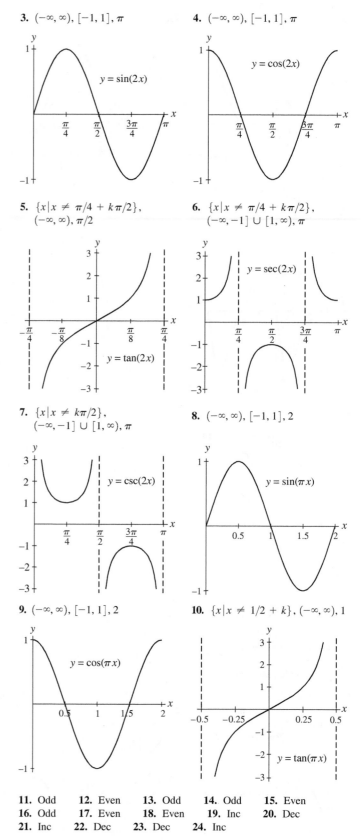

**11.** Odd    **12.** Even    **13.** Odd    **14.** Odd    **15.** Even
**16.** Odd    **17.** Even    **18.** Even    **19.** Inc    **20.** Dec
**21.** Inc    **22.** Dec    **23.** Dec    **24.** Inc

# Chapter 3

## Section 3.1

*For Thought:* **1.** F    **2.** T    **3.** T    **4.** F    **5.** F    **6.** F
**7.** F    **8.** T    **9.** F    **10.** T

*Exercises:*

**1.** Pythagorean    **3.** odd    **5.** odd    **7.** reciprocal    **9.** csc $x$
**11.** 1    **13.** $\cos^2 \alpha$    **15.** $-\cos^2 \beta$    **17.** 2 sin $\alpha$

**19.** $\cot(x) = \pm\sqrt{\csc^2(x) - 1}$    **21.** $\sin(x) = \dfrac{\pm 1}{\sqrt{1 + \cot^2(x)}}$

**23.** $\tan(x) = \dfrac{\pm 1}{\sqrt{\csc^2(x) - 1}}$

**25.** $\sin \alpha = \sqrt{5}/5$, $\cos \alpha = 2\sqrt{5}/5$, $\csc \alpha = \sqrt{5}$, $\sec \alpha = \sqrt{5}/2$, $\cot \alpha = 2$
**27.** $\sin \alpha = -\sqrt{22}/5$, $\tan \alpha = \sqrt{66}/3$, $\csc \alpha = -5\sqrt{22}/22$, $\sec \alpha = -5\sqrt{3}/3$, $\cot \alpha = \sqrt{66}/22$
**29.** $\sin \alpha = -3\sqrt{10}/10$, $\cos \alpha = \sqrt{10}/10$, $\tan \alpha = -3$, $\csc \alpha = -\sqrt{10}/3$, $\sec \alpha = \sqrt{10}$

**31.** $\cos x$    **33.** 0    **35.** 0    **37.** $\cos^2 \alpha$    **39.** $-\cos \beta$
**41.** Odd    **43.** Neither    **45.** Even    **47.** Even    **49.** Even
**51.** Odd    **63.** $-\tan^2 x$    **65.** $-1$    **67.** 1    **69.** $\tan^2 x$
**71.** csc $x$    **73.** $\sin^2 x - \cos^2 x$    **79.** 325°    **81.** $3\pi/2$ ft$^2$
**83. a.** 0    **b.** $\sqrt{2}/2$    **c.** Undefined    **d.** 0    **e.** $2\sqrt{3}/3$
**f.** 1    **g.** $\sqrt{3}/3$    **h.** $-1$    **85.** 4006

## Section 3.2

*For Thought:* **1.** T    **2.** F    **3.** T    **4.** T    **5.** T    **6.** F
**7.** T    **8.** T    **9.** F    **10.** F

*Exercises:*

**1.** complicated    **3.** numerator, denominator    **5.** D    **7.** A
**9.** B    **11.** H    **13.** G    **15.** $-\cos^2 \alpha$
**17.** $2\cos^2 \beta - \cos \beta - 1$    **19.** $\csc^2(x) + 2 + \sin^2(x)$
**21.** $4\sin^2(\theta) - 1$    **23.** $9\sin^2\theta + 12\sin\theta + 4$
**25.** $4\sin^4 y - 4 + \csc^4 y$    **27.** $(2\sin\gamma + 1)(\sin\gamma - 3)$
**29.** $(\tan\alpha - 4)(\tan\alpha - 2)$    **31.** $(2\sec\beta + 1)^2$
**33.** $(\tan\alpha - \sec\beta)(\tan\alpha + \sec\beta)$
**35.** $(\cos\beta)(\sin\beta + 2)(\sin\beta - 1)$    **37.** $(2\sec^2 x - 1)^2$
**39.** $(\cos\alpha + 1)(\sin\alpha + 1)$    **79.** Identity    **81.** Not identity
**83.** Identity    **85.** Not identity    **87.** Identity    **93.** 57.1 mph
**95.** 4, 3, 1/2
**97.** $\sin^2 x + \cos^2 x = 1$, $1 + \cot^2 x = \csc^2 x$, $\tan^2 x + 1 = \sec^2 x$
**99.** 201 workers; Chris female, Pat male

## Section 3.3

*For Thought:* **1.** F    **2.** F    **3.** F    **4.** T    **5.** T    **6.** T
**7.** F    **8.** F    **9.** T    **10.** T

*Exercises:*

**1.** cosine    **3.** cotangent    **5.** $\dfrac{7\pi}{12}$    **7.** $\dfrac{13\pi}{12}$    **9.** $-\dfrac{\pi}{12}$

**11.** $\dfrac{\pi}{12}$    **13.** 30° + 45°    **15.** 120° + 45°    **17.** $\dfrac{\pi}{3} - \dfrac{\pi}{4}$

**19.** $\dfrac{\pi}{3} + \dfrac{\pi}{4}$    **21.** $\dfrac{\sqrt{6} + \sqrt{2}}{4}$    **23.** $\dfrac{\sqrt{2} - \sqrt{6}}{4}$    **25.** $\dfrac{\sqrt{6} - \sqrt{2}}{4}$

**27.** $-\dfrac{\sqrt{6} + \sqrt{2}}{4}$    **29.** $\dfrac{\sqrt{6} + \sqrt{2}}{4}$    **31.** $-\dfrac{\sqrt{6} + \sqrt{2}}{4}$

**33.** $\cos 44°$    **35.** $\cos(11)$    **37.** $\cos(k)$    **39.** $\cos(3\pi/10)$
**41.** $\cos(2\pi/15)$    **43.** e    **45.** c    **47.** a    **49.** g    **51.** g
**53.** c    **55.** f    **57.** a    **59.** $\cos(15°)$    **61.** $\sqrt{3}/2$    **63.** 0
**65.** $\cos(2k)$    **67.** $-63/65$    **69.** $-304/425$    **71.** $\left(\sqrt{6} - \sqrt{2}\right)/4$

**73.** $\dfrac{2 - \sqrt{15}}{6}$    **75.** $-\sin\alpha$    **77.** $-\cos\alpha$    **79.** $-\sin\alpha$

**81.** $-\sin\alpha$    **95.** cot $x$    **97.** $\cos^2 \alpha$

**99.** $\pi/2$, $x = k\pi/2$ for any integer $k$    **101.** $\dfrac{a^2\left(8\pi + 3\sqrt{3}\right)}{6}$ ft$^2$

## Section 3.4

*For Thought:* **1.** T    **2.** F    **3.** T    **4.** F    **5.** T    **6.** F
**7.** F    **8.** T    **9.** T    **10.** T

*Exercises:*

**1.** $\dfrac{\pi}{3}$    **3.** $\dfrac{5\pi}{12}$    **5.** 60° + 45°    **7.** 30° − 45°    **9.** $\dfrac{\sqrt{6} + \sqrt{2}}{4}$

**11.** $2 + \sqrt{3}$    **13.** $\dfrac{\sqrt{2} - \sqrt{6}}{4}$    **15.** $-2 + \sqrt{3}$    **17.** 1

**19.** sin 21°    **21.** $-\sin(7\pi/10)$    **23.** sin 49°    **25.** $\tan(5\pi/18)$

**27.** $\tan(13\pi/42)$    **29.** 16/65    **31.** 87/425    **33.** $\left(\sqrt{6} + \sqrt{2}\right)/4$

**35.** $\dfrac{-2\sqrt{3} - \sqrt{5}}{6}$    **37.** $-\sin\alpha$    **39.** $-\sin\alpha$    **41.** $\dfrac{1 + \tan\alpha}{1 - \tan\alpha}$

**43.** $\tan\alpha$    **59. a.** $-\sin(x)$    **b.** $\cos(x)$    **c.** $-\tan(x)$
**d.** $-\csc(x)$    **e.** $\sec(x)$    **f.** $-\cot(x)$    **61.** $\cot(x)$

**63.** 16.9 in., 34.7 in.    **65.** $\dfrac{2\sqrt{3} - 3}{4}$

## Section 3.5

*For Thought:* **1.** T    **2.** T    **3.** F    **4.** T    **5.** F    **6.** T
**7.** F    **8.** F    **9.** T    **10.** T

*Exercises:*

**1.** 1    **3.** $\sqrt{3}$    **5.** $-1$    **7.** $\sqrt{3}$    **9.** $\dfrac{\sqrt{2 - \sqrt{3}}}{2}$

**11.** $2 - \sqrt{3}$    **13.** $\dfrac{\sqrt{2 + \sqrt{2}}}{2}$    **15.** $\dfrac{\sqrt{2 - \sqrt{2}}}{2}$

**17.** $-\sqrt{2 + \sqrt{2}}/2$    **19.** Positive    **21.** Negative    **23.** Negative
**25.** sin 26°    **27.** $\sqrt{2}/2$    **29.** $\sqrt{3}/6$    **31.** $\sqrt{3}/2$
**33.** $-\sin(7\pi/9)$    **35.** tan 6°
**51.** $\sqrt{5}/5, 2\sqrt{5}/5, 1/2, \sqrt{5}, \sqrt{5}/2, 2$
**53.** $\sqrt{26}/26, 5\sqrt{26}/26, 1/5, \sqrt{26}, \sqrt{26}/5, 5$
**55.** $-\sqrt{15}/8, -7/8, \sqrt{15}/7, -8\sqrt{15}/15, -8/7, 7\sqrt{15}/15$
**57.** $-24/25, -7/25, 24/7, -25/24, -25/7, 7/24$    **59.** $-24/25$

**61.** 161/289    **63.** $\dfrac{5\sqrt{34} - 25}{3}$    **65.** $A = \dfrac{d^2}{2}\sin(2\alpha)$

**67.** Not identity    **69.** Not identity    **71.** Not identity
**73.** Identity    **79. a.** $\sin x \cos y + \cos x \sin y$
**b.** $\sin x \cos y - \cos x \sin y$    **81.** $-\sqrt{5}/3$    **83.** 3/4
**85.** $4\pi/3 - 4\sqrt{3} + 4$ m$^2$

## Section 3.6

*For Thought:* **1.** T    **2.** F    **3.** T    **4.** F    **5.** T    **6.** F
**7.** T    **8.** T    **9.** T    **10.** T

*Exercises:*

**1.** $0.5(\cos 4° - \cos 22°)$    **3.** $0.5(\sin 36° - \sin 4°)$
**5.** $0.5(\cos(\pi/30) + \cos(11\pi/30))$
**7.** $0.5(\cos 2y^2 + \cos 12y^2)$    **9.** $0.5(\sin(3s) + \sin(s - 2))$

**11.** $\dfrac{\sqrt{2} - 1}{4}$    **13.** $\dfrac{\sqrt{2} + \sqrt{3}}{4}$    **15.** 2 cos 10° sin 2°

**17.** 2 sin 83.5° sin 3.5°    **19.** $-2\cos(4.2)\sin(0.6)$
**21.** $-2\sin(4y + 3)\sin(y - 6)$    **23.** $-2\cos(6.5\alpha)\sin(1.5\alpha)$
**25.** $-2\sin(4\pi/15)\sin(\pi/15)$    **27.** $\sqrt{6}/2$

**29.** $\dfrac{\sqrt{2 - \sqrt{2}}}{2}$    **31.** $\sqrt{2}\sin(x - \pi/4)$    **33.** $\sin(x + 2\pi/3)$

**35.** $\sin(x - \pi/6)$

**37.** $y = \sqrt{2} \sin(x + 3\pi/4)$, $\sqrt{2}, 2\pi, -3\pi/4$

**39.** $y = 2 \sin(x - \pi/4)$, $2, 2\pi, \pi/4$

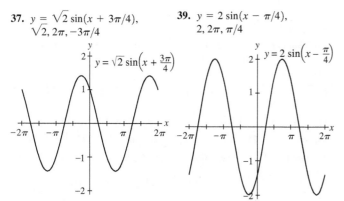

**41.** $y = 2 \sin(x + 7\pi/6)$, $2, 2\pi, -7\pi/6$

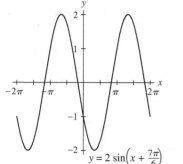

**43.** $5, -0.9$    **45.** $\sqrt{37}, -3.0$    **47.** $\sqrt{34}, -4.2$

**59.** $x = 2 \sin(t + \pi/6)$, 2 m    **65.** $-\sqrt{2}$    **67. a.** $\sin(5.6)$

**b.** $\sin(x)$    **c.** $\sin(9.6)$    **69.** $22.6°, 67.4°$    **71. a.** $m + n - 3$

**b.** $m + n - 2 - d$

### Chapter 3 Review Exercises

**1.** $\cos^2 \alpha$    **3.** $-\cot^2 x$    **5.** $\sec^2 \alpha$    **7.** $\tan 4s$

**9.** $-\sin 3\theta$    **11.** $\tan z$

**13.** $\sin \alpha = 12/13$, $\tan \alpha = -12/5$, $\csc \alpha = 13/12$, $\sec \alpha = -13/5$,
$\cot \alpha = -5/12$

**15.** $\sin \alpha = -4/5$, $\cos \alpha = -3/5$, $\tan \alpha = 4/3$, $\csc \alpha = -5/4$,
$\sec \alpha = -5/3$, $\cot \alpha = 3/4$

**17.** $\sin \alpha = -24/25$, $\cos \alpha = 7/25$, $\tan \alpha = -24/7$, $\csc \alpha = -25/24$,
$\sec \alpha = 25/7$, $\cot \alpha = -7/24$    **19.** Identity    **21.** Not identity

**23.** Odd    **25.** Neither    **27.** Even    **47.** $\sqrt{3} - 2$

**49.** $-\dfrac{\sqrt{2 + \sqrt{3}}}{2}$ or $-\dfrac{\sqrt{2 + \sqrt{6}}}{4}$

**51.** $y = 4\sqrt{2} \sin(x + \pi/4)$, $4\sqrt{2}, 2\pi, -\pi/4$

**53.** $y = \sqrt{5} \sin(x + 2.68)$, $\sqrt{5}, 2\pi, -2.68$

**55.** $2 \cos 17° \cos 2°$    **57.** $2 \cos \dfrac{\pi}{16} \sin \dfrac{3\pi}{16}$    **59.** $\sin 24° - \sin 2°$

**61.** $\cos \dfrac{x}{12} + \cos \dfrac{7x}{12}$    **63.** $y = \sqrt{5} \sin(x + 0.46)$, $\sqrt{5}$ in.

**65.** 600,001

### Chapter 3 Test

**1.** $2 \cos x$    **2.** $\sin 7t$    **3.** $2 \csc^2 y$    **4.** $\tan(3\pi/10)$

**9.** $y = 2 \sin(x + 5\pi/3)$, $2\pi, 2, -5\pi/3$

**10.** $\sin \alpha = 1/2$, $\cos \alpha = -\sqrt{3}/2$, $\tan \alpha = -\sqrt{3}/3$, $\sec \alpha = -2\sqrt{3}/3$, $\cot \alpha = -\sqrt{3}$

**11.** Even    **12.** $-\dfrac{\sqrt{2 - \sqrt{3}}}{2}$ or $\dfrac{\sqrt{2 - \sqrt{6}}}{4}$

### Tying It All Together Chapters P–3

**1.** $\sqrt{2}/2$    **2.** $\sqrt{2}/2$    **3.** $1/2$    **4.** $1/2$    **5.** $1/2$    **6.** $1/2$
**7.** $-1/2$    **8.** 1    **9.** $-1$    **10.** 0    **11.** Odd    **12.** Even
**13.** Even    **14.** Even    **15.** Even    **16.** Odd    **17.** Not identity
**18.** Not identity    **19.** Not identity    **20.** Not identity
**21.** $\beta = 60°, b = 4\sqrt{3}, c = 8$    **22.** $\alpha = 60°, \beta = 30°, c = 2$
**23.** $\alpha = 17.5°, \beta = 72.5°, a = 1.6, c = 5.2$
**24.** $\alpha = 36.9°, \beta = 53.1°, b = 2.7, c = 3.3$    **25.** angle
**26.** central    **27.** acute    **28.** quadrantal    **29.** coterminal
**30.** minutes    **31.** seconds    **32.** unit    **33.** $\pi$    **34.** $\alpha r$

## Chapter 4

### Section 4.1

*For Thought:* **1.** T    **2.** T    **3.** F    **4.** F    **5.** F    **6.** T
**7.** T    **8.** T    **9.** F    **10.** F

*Exercises:*
**1.** domain    **3.** domain    **5.** $\sin^{-1}(y)$    **7.** $\pi/4$    **9.** $\pi/3$
**11.** $-\pi/4$    **13.** 0    **15.** $-45°$    **17.** $30°$    **19.** $0°$    **21.** $-90°$
**23.** $33.8°$    **25.** $-19.5°$    **27.** $0.6°$    **29.** $\pi/4$    **31.** $\pi/6$
**33.** $3\pi/4$    **35.** $\pi/2$    **37.** $86.6°$    **39.** $101.5°$    **41.** $89.4°$
**43.** $-\pi/4$    **45.** $\pi/6$    **47.** $\pi/3$    **49.** 1.571    **51.** $\pi/3$
**53.** $\pi/3$    **55.** $\pi/3$    **57.** $5\pi/6$    **59.** 0    **61.** $\pi/2$
**63.** $3\pi/4$    **65.** 1.87    **67.** 0.15    **69.** 2.72    **71.** 3.06
**73.** 2.36    **75.** $\sqrt{3}$    **77.** $-\pi/6$    **79.** $\pi/6$    **81.** $\pi/4$    **83.** 1
**85.** $\pi/2$    **87.** $\pi/4$    **89.** $-\sqrt{3}/2$    **91.** 0    **93.** 0.8930
**95.** Undefined    **97.** $-0.9802$    **99.** $-0.4082$    **101.** 3.4583
**103.** 1.0183    **105.** $\sqrt{1 - x^2}$    **107.** $1/\sqrt{1 + x^2}$
**109.** $x/\sqrt{1 + x^2}$    **111.** $x/\sqrt{1 - x^2}$    **113.** $\sqrt{1 + x^2}$
**115.** $67.1°$    **117.** 39%

**119.** The domain of $y = \sin(\sin^{-1}(x))$ is $[-1, 1]$. So the graph is a line segment with endpoints $(-1, -1)$ and $(1, 1)$.

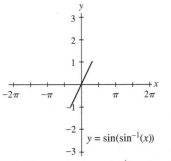

**121.** The graph of $y = \sin^{-1}(1/x)$ looks like the graph of $y = \csc^{-1}(x)$ because of the identity $\csc^{-1}(x) = \sin^{-1}(1/x)$.

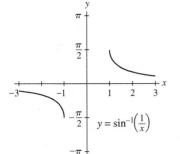

**123.** 16.0 cm, 14.4 cm    **125.** 5, $\pi/2$, $\pi/4$, $[-8, 2]$
**127.** $3\cos^2(x)$    **129.** $(4, 5), (49, 50), (499, 500), (4999, 5000)$

## Section 4.2

*For Thought:* **1.** F    **2.** T    **3.** F    **4.** T    **5.** T    **6.** T
**7.** F    **8.** F    **9.** T    **10.** T

*Exercises:*
**1.** identity    **3.** $\left\{x \,\middle|\, x = \dfrac{\pi}{2} + k\pi\right\}$    **5.** $\{x \mid x = 2k\pi\}$

**7.** No solution    **9.** $\left\{x \,\middle|\, x = \dfrac{3\pi}{2} + 2k\pi\right\}$    **11.** $\{x \mid x = k\pi\}$

**13.** $\left\{x \,\middle|\, x = \dfrac{3\pi}{4} + k\pi\right\}$    **15.** $\left\{x \,\middle|\, x = \dfrac{\pi}{3} + 2k\pi \text{ or } x = \dfrac{5\pi}{3} + 2k\pi\right\}$

**17.** $\left\{x \,\middle|\, x = \dfrac{\pi}{4} + 2k\pi \text{ or } x = \dfrac{3\pi}{4} + 2k\pi\right\}$

**19.** $\left\{x \,\middle|\, x = \dfrac{\pi}{6} + k\pi\right\}$    **21.** $\left\{x \,\middle|\, x = \dfrac{5\pi}{6} + 2k\pi \text{ or } x = \dfrac{7\pi}{6} + 2k\pi\right\}$

**23.** $\left\{x \,\middle|\, x = \dfrac{5\pi}{4} + 2k\pi \text{ or } x = \dfrac{7\pi}{4} + 2k\pi\right\}$    **25.** $\left\{x \,\middle|\, x = \dfrac{\pi}{3} + k\pi\right\}$

**27.** $\{\alpha \mid \alpha = 90° + k180°\}$    **29.** $\{\alpha \mid \alpha = 90° + k360°\}$
**31.** $\{\alpha \mid \alpha = k180°\}$    **33.** $\{\alpha \mid \alpha = 45° + k360° \text{ or } \alpha = 315° + k360°\}$
**35.** $\{\alpha \mid \alpha = 210° + k360° \text{ or } \alpha = 330° + k360°\}$
**37.** $\{\alpha \mid \alpha = 45° + k180°\}$    **39.** $29.2°, 330.8°$    **41.** $194.1°, 345.9°$
**43.** $79.5°, 259.5°$    **45.** No solution    **47.** $0.85, 5.43$
**49.** $3.39, 6.03$    **51.** $0.39, 3.53$    **53.** $2.68, 3.61$
**55.** $\{-360°, 0°, 360°\}$    **57.** $\{210°, 330°\}$    **59.** $\{56.4°\}$

**61.** $\{19.6°\}$    **63.** $y = \dfrac{1}{3}\cos^{-1}\left(\dfrac{x}{2}\right)$    **65.** $m = \sin^{-1}\left(\dfrac{2-t}{6}\right)$

**67.** $a = 3\tan^{-1}\left(\dfrac{b+d}{7}\right)$    **69.** $b = \dfrac{1}{\pi}\sin^{-1}\left(\dfrac{q}{3}\right) + 1$

**71.** $f^{-1}(x) = \dfrac{1}{2}\sin^{-1}(x), [-1, 1], \left[-\dfrac{\pi}{4}, \dfrac{\pi}{4}\right]$

**73.** $f^{-1}(x) = \dfrac{1}{3}\cos^{-1}(x/2), [-2, 2], \left[0, \dfrac{\pi}{3}\right]$

**75.** $f^{-1}(x) = \dfrac{1}{\pi}\tan^{-1}(x-3), (-\infty, \infty), \left(-\dfrac{1}{2}, \dfrac{1}{2}\right)$

**77.** $f^{-1}(x) = \dfrac{1}{\pi}\sin^{-1}(2-x) + 1, [1, 3], \left[\dfrac{1}{2}, \dfrac{3}{2}\right]$

**79.** $f^{-1}(x) = \dfrac{1}{3}\sin(x), \left[-\dfrac{\pi}{2}, \dfrac{\pi}{2}\right], \left[-\dfrac{1}{3}, \dfrac{1}{3}\right]$

**81.** $f^{-1}(x) = 2\sin(x-3), \left[3 - \dfrac{\pi}{2}, 3 + \dfrac{\pi}{2}\right], [-2, 2]$

**83.** $10.50$    **85.** $0.63 + 2k\pi$ or $2.52 + 2k\pi$ where $k$ is any integer
**87.** $1.37 + 2k\pi$ or $4.91 + 2k\pi$ where $k$ is any integer
**89.** $\pm 4.40$    **91.** $0.41 + 2k\pi$ or $2.73 + 2k\pi$ where $k$ is any integer
**93.** $0.88$    **95.** $3; 1$    **97. a.** $\pi/6$    **b.** $2\pi/3$    **c.** $-\pi/4$
**d.** $\sqrt{3}/2$    **e.** $0$    **f.** $-\pi/2$    **99.** $\sin^2(x)$    **101.** $3, 2\pi, \pi, [6, 12]$
**103.** 17th triangle

## Section 4.3

*For Thought:* **1.** F    **2.** T    **3.** F    **4.** T    **5.** F    **6.** T
**7.** T    **8.** F    **9.** T    **10.** F

*Exercises:*

**1.** $\left\{x \,\middle|\, x = \dfrac{2\pi}{3} + 4k\pi \text{ or } x = \dfrac{10\pi}{3} + 4k\pi\right\}$

**3.** $\left\{x \,\middle|\, x = \dfrac{2k\pi}{3}\right\}$    **5.** $\left\{x \,\middle|\, x = \dfrac{\pi}{3} + 4k\pi \text{ or } x = \dfrac{5\pi}{3} + 4k\pi\right\}$

**7.** $\left\{x \,\middle|\, x = \dfrac{5\pi}{8} + k\pi \text{ or } x = \dfrac{7\pi}{8} + k\pi\right\}$    **9.** $\left\{x \,\middle|\, x = \dfrac{\pi}{6} + \dfrac{k\pi}{2}\right\}$

**11.** $\left\{x \,\middle|\, x = \dfrac{k\pi}{4}\right\}$    **13.** $\left\{x \,\middle|\, x = \dfrac{1}{6} + 2k \text{ or } x = \dfrac{5}{6} + 2k\right\}$

**15.** $\left\{x \,\middle|\, x = \dfrac{1}{4} + \dfrac{k}{2}\right\}$    **17.** $\{7\pi/12, 23\pi/12\}$    **19.** $\{\pi/8\}$

**21.** $\{\pi/6, 5\pi/6\}$    **23.** $\{3\pi/4, 7\pi/4\}$    **25.** $\{240°, 300°\}$
**27.** $\{120°, 150°, 300°, 330°\}$    **29.** $\{60°, 120°, 240°, 300°\}$
**31.** $\{22.5°, 157.5°, 202.5°, 337.5°\}$
**33.** $\{45°, 75°, 165°, 195°, 285°, 315°\}$    **35.** $\{60°\}$
**37.** $\{11.25°, 33.75°, 101.25°, 123.75°, 191.25°, 213.75°, 281.25°, 303.75°\}$
**39.** $\pi/6, 7\pi/6, \pi/3, 4\pi/3$    **41.** $\pi/6, 5\pi/6, 7\pi/6, 11\pi/6$
**43.** $\pi/12, 5\pi/12, 9\pi/12, 13\pi/12, 17\pi/12, 21\pi/12$    **45.** $3\pi/4$
**47.** $\pi/3$    **49.** $\{\alpha \mid \alpha = 6.6° + k120° \text{ or } \alpha = 53.4° + k120°\}$
**51.** $\{\alpha \mid \alpha = 72.3° + k120° \text{ or } \alpha = 107.7° + k120°\}$
**53.** $\{\alpha \mid \alpha = 38.6° + k180° \text{ or } \alpha = 141.4° + k180°\}$
**55.** $\{\alpha \mid \alpha = 668.5° + k720° \text{ or } \alpha = 411.5° + k720°\}$

**57.** $\left\{x \,\middle|\, x = 0.07 + \dfrac{2k\pi}{5} \text{ or } x = 0.56 + \dfrac{2k\pi}{5}\right\}$

**59.** $\{x \mid x = 7.57 + 4k\pi \text{ or } x = 11.28 + 4k\pi\}$
**61.** $\{x \mid x = 0.43 + 2k \text{ or } x = 1.57 + 2k\}$    **63.** $\{x \mid x = 0.72 + k\}$
**65.** $\{\pi/2, 3\pi/2\}$    **67.** $\{\pi/6, 5\pi/6, \pi/2, 3\pi/2\}$
**69.** $\{0, \pi, 2\pi\}$    **71.** $\varnothing$    **73.** $\{\pi/2, 3\pi/2\}$
**75.** $\{0, \pi/2, \pi, 3\pi/2, 2\pi\}$    **77.** $\{\pi/6, 11\pi/6\}$
**79.** $\{\pi/12, 5\pi/12, 13\pi/12, 17\pi/12\}$
**81.** $\{0.4, 1.9, 2.2, 4.0, 4.4, 5.8\}$    **83.** $\{\pi/3\}$
**85.** $44.4°$ or $45.6°$    **87.** $12.5°$ or $77.5°, 6.3$ sec
**89.** $1 + 6k$ and $2 + 6k$ for $k$ a nonnegative integer

**91.** Infinitely many solutions    **93.** $\dfrac{y}{\sqrt{1 + y^2}}$

**95. a.** $-\pi/6$    **b.** $3\pi/4$    **c.** $\pi/4$    **d.** $\pi/4$
**97.** $\left(\sqrt{2 + \sqrt{3}}\right)/2$    **99.** $(30 + 25)^2 = 3025$

## Section 4.4

*For Thought:* **1.** F    **2.** F    **3.** T    **4.** T    **5.** T    **6.** F
**7.** F    **8.** T    **9.** F    **10.** F

*Exercises:*
**1.** $\{0, \pi\}$    **3.** $\{\pi/6, 5\pi/6, 7\pi/6, 11\pi/6\}$
**5.** $\{\pi/8, 3\pi/8, 5\pi/8, 7\pi/8, 9\pi/8, 11\pi/8, 13\pi/8, 15\pi/8\}$
**7.** $\{\pi/6, 5\pi/6, 7\pi/6, 11\pi/6\}$    **9.** $\{0, \pi/2, \pi\}$
**11.** $\{\pi/2, 3\pi/2, 2\pi/3, 4\pi/3\}$    **13.** $\{0, \pi, \pi/6, 5\pi/6, 7\pi/6, 11\pi/6\}$
**15.** $\{0, \pi, \pi/4, 3\pi/4, 5\pi/4, 7\pi/4\}$    **17.** $\{\pi/6, 5\pi/6, \pi/2\}$
**19.** $\{2\pi/3, 4\pi/3, \pi\}$    **21.** $\{3\pi/2, 0.3, 2.8\}$
**23.** $\{\pi/2, 3\pi/2, 0.6, 2.5, 3.8, 5.7\}$    **25.** $\{\pi/3, 5\pi/3, 0.7, 5.6\}$

**27.** {197.6°, 342.4°}  **29.** {32.5°, 126.4°, 233.6°, 327.5°}
**31.** {67.5°, 157.5°, 247.5°, 337.5°}  **33.** {330°}  **35.** {30°, 240°}
**37.** {79.4°, 190.6°}  **39.** {0, π}  **41.** {0, π}
**43.** {0, 0.3, 2.8, π}  **45.** {π/2, 3π/2}  **47.** {2π/3, π, 4π/3}
**49.** {11π/6}  **51.** {π/3, 4.9}  **53.** {0.7, 2.5, 3.4, 6.0}
**55.** {0, π}  **57.** {π/2, 3π/2}  **59.** {7π/12, 23π/12}
**61.** {7π/6, 11π/6}  **63.** {π/6, π/2, 5π/6}  **65.** {0°}
**67.** {26.6°, 206.6°}  **69.** {0°, 180°}
**71.** {30°, 90°, 150°, 210°, 270°, 330°}
**73.** {67.5°, 157.5°, 247.5°, 337.5°}  **75.** {221.8°, 318.2°}  **77.** ∅
**79.** {120°, 300°}  **81.** {30°, 45°, 135°, 150°, 210°, 225°, 315°, 330°}
**83.** {0°, 60°, 120°, 180°, 240°, 300°}
**85.** (0.905 + 2kπ, 0.786), (5.379 + 2kπ, −0.786)
**87.** $\dfrac{5\pi}{12} + \dfrac{k\pi}{2}$ for $k$ a nonnegative integer
**89.** {±2π, ±4π, ±6π}, 0.7246  **91.** π/2, 3π/2
**93. a.** D: [−1, 1], R: [−π/2, π/2]  **b.** D: [−1, 1], R: [0, π]
**c.** D: (−∞, ∞), R: (−π/2, π/2)  **95.** 2 sin y
**97.** 843/68 square units

## Chapter 4 Review Exercises

**1.** −π/6  **3.** −π/4  **5.** π/4  **7.** √3/2  **9.** 1  **11.** −π/4
**13.** π/4  **15.** π/6  **17.** π/6  **19.** 90°  **21.** 135°
**23.** 30°  **25.** 90°  **27.** {π}  **29.** {π/6, 5π/6}
**31.** {3π/4, 7π/4}  **33.** {30°, 150°}  **35.** {45°, 315°}
**37.** {150°, 330°}  **39.** {2.21, 4.07}  **41.** {0.11, 1.46, 3.25, 4.60}
**43.** {0.62, 3.76}  **45.** $t = \sin^{-1}\left(\dfrac{x}{3}\right)$  **47.** $y = \dfrac{1}{2}\sin^{-1}\left(\dfrac{a}{3}\right)$
**49.** $h = \cos^{-1}\left(\dfrac{1-q}{2}\right)$  **51.** $x = \dfrac{1}{\pi}\tan^{-1}\left(\dfrac{b+3}{5}\right)$
**53.** $y = \cos\left(\dfrac{x}{a}\right) + 2$  **55.** $f^{-1}(x) = \sin^{-1}(x)$, [−1, 1], $\left[-\dfrac{\pi}{2}, \dfrac{\pi}{2}\right]$
**57.** $f^{-1}(x) = \sin(x)$, $\left[-\dfrac{\pi}{2}, \dfrac{\pi}{2}\right]$, [−1, 1]
**59.** $f^{-1}(x) = \dfrac{1}{3}\sin^{-1}(x)$, [−1, 1], $\left[-\dfrac{\pi}{6}, \dfrac{\pi}{6}\right]$
**61.** $f^{-1}(x) = \dfrac{1}{4}\cos^{-1}\left(\dfrac{x}{6}\right)$, [−6, 6], $\left[0, \dfrac{\pi}{4}\right]$
**63.** $f^{-1}(x) = \dfrac{2}{\pi}\tan^{-1}(x - 4)$, (−∞, ∞), (−1, 1)
**65.** −π/6  **67.** None  **69.** −π/4  **71.** ±π/4  **73.** −π/6
**75.** $\left\{x \mid x = \dfrac{\pi}{3} + k\pi \text{ or } x = \dfrac{2\pi}{3} + k\pi\right\}$
**77.** $\left\{x \mid x = \dfrac{\pi}{6} \text{ or } \dfrac{\pi}{3} \text{ or } \dfrac{2\pi}{3} \text{ or } \dfrac{5\pi}{6} \text{ all plus } 2k\pi\right\}$
**79.** $\left\{x \mid x = \dfrac{\pi}{2} \text{ or } \dfrac{\pi}{6} \text{ or } \dfrac{5\pi}{6} \text{ all plus } 2k\pi\right\}$
**81.** $\left\{x \mid x = \dfrac{4\pi}{3} + 4k\pi \text{ or } x = \dfrac{2\pi}{3} + 4k\pi\right\}$
**83.** $\left\{x \mid x = \dfrac{\pi}{3} + 4k\pi \text{ or } x = \dfrac{5\pi}{3} + 4k\pi \text{ or } x = \pi + 2k\pi\right\}$
**85.** $\left\{x \mid x = \dfrac{\pi}{2} + k\pi\right\}$
**87.** $\left\{x \mid x = \dfrac{3\pi}{2} + 2k\pi \text{ or } x = \pi + 2k\pi\right\}$
**89.** {45°, 225°}  **91.** {90°, 180°}  **93.** No solution
**95.** {22.5°, 67.5°, 112.5°, 157.5°, 202.5°, 247.5°, 292.5°, 337.5°}
**97.** {0°, 180°}  **99.** {15°, 75°, 135°, 195°, 255°, 315°}
**101.** $\left(\pi/4 + 2k\pi, \sqrt{2}/2\right), \left(5\pi/4 + 2k\pi, -\sqrt{2}/2\right)$
**103.** 1.28 sec and 2.85 sec
**105.** 44 (20 on one side and 24 on the other)

## Chapter 4 Test

**1.** −π/6  **2.** 2π/3  **3.** −π/4  **4.** π/6  **5.** −π/4
**6.** 2√2/3  **7.** $\left\{\theta \mid \theta = \dfrac{3\pi}{2} + 2k\pi\right\}$

**8.** $\left\{s \mid s = \dfrac{\pi}{9} + \dfrac{2k\pi}{3} \text{ or } s = \dfrac{5\pi}{9} + \dfrac{2k\pi}{3}\right\}$  **9.** $\left\{t \mid t = \dfrac{\pi}{3} + \dfrac{k\pi}{2}\right\}$
**10.** $\left\{\theta \mid \theta = \dfrac{\pi}{6} + 2k\pi \text{ or } \theta = \dfrac{5\pi}{6} + 2k\pi \text{ or } \theta = \dfrac{\pi}{2} + k\pi\right\}$
**11.** {30°, 150°}  **12.** {270°}  **13.** {30°, 330°}
**14.** {0°, 90°, 180°, 270°}  **15.** {45°, 135°, 225°, 315°}
**16.** {19.5°, 90°, 160.5°}
**17.** {27°, 63°, 99°, 171°, 207°, 243°, 279°, 351°}
**18.** $t = \dfrac{1}{2}\sin^{-1}\left(\dfrac{a}{5}\right)$  **19.** $f^{-1}(x) = 4\cos^{-1}(5x - 1)$, [0, 2/5], [0, 4π]
**20.** $(2k\pi, 1), (\pi + 2k\pi, -1)$  **21.** 0.4 sec, 1.4 sec, 2.5 sec, 3.5 sec

## Tying It All Together Chapters P–4

**1.** √2/2  **2.** −1/2  **3.** 1  **4.** 1/2  **5.** −π/6  **6.** 2π/3
**7.** −π/4  **8.** 0  **9.** −√3/2  **10.** √3/2  **11.** 1  **12.** 1
**13.** 2π, 1, π/6, (π/6 + kπ, 0)  **14.** π, 1, 0, (kπ/2, 0)

$y = \sin\left(x - \dfrac{\pi}{6}\right)$

$y = \sin(2x)$

**15.** 2π/3, 1, 0, (π/6 + kπ/3, 0)  **16.** π, 1, π/2, (π/4 + kπ/2, 0)

$y = \cos(3x)$

$y = \cos(2x - \pi)$

**17.** 4π, 3, π, (π + 2kπ, 0)  **18.** 2π, 4, −π/3, (π/6 + kπ, 0)

$y = 3\sin\left(\dfrac{x}{2} - \dfrac{\pi}{2}\right)$

$y = 4\cos\left(x + \dfrac{\pi}{3}\right)$

**19.** 6, 4, 0, (3/2 + 3k, 0)  **20.** 2, 2, 1, (k, 0)

$y = -4\cos\left(\pi\dfrac{x}{3}\right)$

$y = -2\sin(\pi x - \pi)$

**21.** $\cos\alpha = -2\sqrt{2}/3$, $\tan\alpha = \sqrt{2}/4$, $\csc\alpha = -3$, $\sec\alpha = -3\sqrt{2}/4$, $\cot\alpha = 2\sqrt{2}$

**22.** $\sin\alpha = \sqrt{15}/4$, $\tan\alpha = -\sqrt{15}$, $\csc\alpha = 4\sqrt{15}/15$, $\sec\alpha = -4$, $\cot\alpha = -\sqrt{15}/15$

**23.** $\sin\alpha = 3\sqrt{34}/34$, $\cos\alpha = 5\sqrt{34}/34$, $\csc\alpha = \sqrt{34}/3$, $\sec\alpha = \sqrt{34}/5$, $\cot\alpha = 5/3$

**24.** $\cos\alpha = 3/5$, $\tan\alpha = -4/3$, $\csc\alpha = -5/4$, $\sec\alpha = 5/3$, $\cot\alpha = -3/4$

**25.** $r\omega$    **26.** $y/r$, $x/r$    **27.** $y$, $x$    **28.** fundamental
**29.** amplitude    **30.** phase shift    **31.** frequency
**32.** vertical asymptote    **33.** even    **34.** odd

# Chapter 5

## Section 5.1

*For Thought:* **1.** T    **2.** F    **3.** T    **4.** F    **5.** F    **6.** T
**7.** T    **8.** F    **9.** T    **10.** F
*Exercises:*
**1.** oblique    **3.** ambiguous    **5.** $\gamma = 44°$, $b = 14.4$, $c = 10.5$
**7.** $\beta = 134.2°$, $a = 5.2$, $c = 13.6$    **9.** $\beta = 26°$, $a = 14.6$, $b = 35.8$
**11.** $\alpha = 45.7°$, $b = 587.9$, $c = 160.8$    **13.** None
**15.** One: $\alpha = 30°$, $\beta = 90°$, $a = 10$
**17.** One: $\alpha = 26.3°$, $\gamma = 15.6°$, $a = 10.3$
**19.** Two: $\alpha_1 = 134.9°$, $\gamma_1 = 12.4°$, $c_1 = 11.4$; or $\alpha_2 = 45.1°$,
$\gamma_2 = 102.2°$, $c_2 = 51.7$    **21.** One: $\alpha = 25.4°$, $\beta = 55.0°$, $a = 5.4$
**23.** 18.4 mi    **25.** 1425 ft    **27.** 38.0 ft    **29.** 159.4 ft
**31.** 28.9 ft, 15.7 ft    **33.** 18.1 in.    **35.** 62.2°    **37.** 7.5 sec
**41. a.** 1    **b.** $3\pi/4$    **c.** $-\sqrt{3}$    **d.** $-2\sqrt{3}/3$    **e.** $-\sqrt{2}$
**f.** $-\pi/6$    **43. a.** 2    **b.** $2\pi/3$    **c.** 1/2    **d.** $\pi$
**45. a.** Odd    **b.** Even    **c.** Even    **d.** Even    **47.** $\sqrt{5} - 1$

## Section 5.2

*For Thought:* **1.** T    **2.** F    **3.** F    **4.** T    **5.** F    **6.** T
**7.** T    **8.** T    **9.** T    **10.** F
*Exercises:*
**1.** law of cosines    **3.** cosines    **5.** $\alpha = 30.4°$, $\beta = 28.3°$, $c = 5.2$
**7.** $\alpha = 26.4°$, $\beta = 131.3°$, $\gamma = 22.3°$
**9.** $\alpha = 163.9°$, $\gamma = 5.6°$, $b = 4.5$
**11.** $\alpha = 130.3°$, $\beta = 30.2°$, $\gamma = 19.5°$
**13.** $\beta = 48.3°$, $\gamma = 101.7°$, $a = 6.2$
**15.** $\alpha = 53.9°$, $\beta = 65.5°$, $\gamma = 60.6°$    **17.** $\alpha = 120°$, $b = 3.5$, $c = 4.8$
**19.** 0    **21.** 1    **23.** 0    **25.** 1    **27.** 0    **29.** 9.90 ft
**31.** 783.45 ft    **33.** 20.6 mi    **35.** 19.2°
**37.** $\angle A = 78.5°$, $\angle B = 57.1°$, $\angle C = 44.4°$    **39.** 3.8 mi, 111.6°, 40.4°
**41.** 11.76 m    **43.** $\theta_1 = 13.3°$, $\theta_2 = 90.6°$    **45. a.** 0.52° to 0.54°
**b.** 0.49° to 0.55°    **c.** yes    **47.** 603 ft
**51.** $\gamma = 53.3°$, $a = 33.9$, $b = 11.4$    **53.** $y = 2\sin(2(x - \pi/4)) + 3$
**55. a.** $3\pi/2$    **b.** $7\pi/4$    **c.** $-7\pi/6$    **d.** $2\pi/3$
**57.** $\dfrac{(3\sqrt{5} + 9)\pi}{2}$ ft

## Section 5.3

*For Thought:* **1.** F    **2.** T    **3.** F    **4.** T    **5.** T
*Exercises:*
**1.** $bh/2$    **3.** 9.8    **5.** 83.4    **7.** 66.3    **9.** 37.7    **11.** 11 ft$^2$
**13.** 44 mi$^2$    **15.** 40.9    **17.** 13.9    **19.** 40,471.9    **21.** 100
**23.** 20.8    **25.** 20.4    **27.** 277.7 in.$^2$    **29.** 35.5 in.$^2$    **31.** 2 ft$^2$
**33. a.** $A_T = (1/2)r^2\sin(\alpha)$    **b.** $A_S = r^2\alpha/2$
**c.** $A_L = (r^2/2)(\alpha - \sin\alpha)$    **35.** 97,534.8 ft$^2$    **37.** 37,536.3 ft$^2$
**39.** \$681    **43.** No such triangle    **45.** 23.7 ft$^2$
**49.** $\alpha = 46.7°$, $\gamma = 11.2°$, $c = 5.2$
**51.** $a = 6.3$, $\beta = 124.7°$, $\gamma = 22.1°$.
**53.** $\pi/9$, $2\pi/9$, $4\pi/9$, $5\pi/9$, $7\pi/9$, $8\pi/9$    **55.** 3.850 m$^2$

## Section 5.4

*For Thought:* **1.** T    **2.** F    **3.** T    **4.** T    **5.** F    **6.** F
**7.** T    **8.** T    **9.** T    **10.** T
*Exercises:*
**1.** vector    **3.** magnitude    **5.** parallelogram law    **7.** component
**9.**                                    **11.**

**13.**

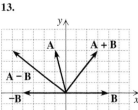

**15.** D    **17.** E    **19.** B    **21.** $|\mathbf{v_x}| = 1.9$, $|\mathbf{v_y}| = 4.1$
**23.** $|\mathbf{v_x}| = 7256.4$, $|\mathbf{v_y}| = 3368.3$    **25.** $|\mathbf{v_x}| = 87.7$, $|\mathbf{v_y}| = 217.0$
**27.** 2, 30°    **29.** 2, 135°    **31.** 16, 300°    **33.** 5, 0°
**35.** $\sqrt{13}$, 146.3°    **37.** $\sqrt{10}$, 341.6°    **39.** $\langle 4\sqrt{2}, 4\sqrt{2}\rangle$
**41.** $\langle -237.6, 166.3\rangle$    **43.** $\langle 17.5, -4.0\rangle$    **45.** $\langle 15, -10\rangle$
**47.** $\langle 18, -22\rangle$    **49.** $\langle 11, -13\rangle$    **51.** $\langle 0, -1\rangle$    **53.** $-13$
**55.** 32.5°    **57.** 27.3°    **59.** 90°    **61.** Perpendicular
**63.** Parallel    **65.** Neither    **67.** $2\mathbf{i} + \mathbf{j}$    **69.** $-3\mathbf{i} + \sqrt{2}\mathbf{j}$
**71.** $-9\mathbf{j}$    **73.** $-7\mathbf{i} - \mathbf{j}$    **75.** $\sqrt{17}$, 76.0°    **77.** $3\sqrt{10}$, 198.4°
**79.** $\sqrt{29}$, 158.2°    **81.** $\sqrt{65}/2$, 172.9°    **83.** $\sqrt{89}$, 32.0°
**85.** $2\sqrt{10}$, 108.4°    **87.** $\sqrt{5}$, 243.4°    **89.** 450.3 mph, 260 mph
**93.** $\sqrt{35}/4$    **95. a.** $\dfrac{\tan x + \tan y}{1 - \tan x\tan y}$    **b.** $\dfrac{\tan x - \tan y}{1 + \tan x\tan y}$
**97.** 57°, 35.9 ft, 55.4 ft    **99. a.** $64 + 16\sqrt{2} \approx 86.63$
**b.** $32 + 32\sqrt{2} \approx 77.25$
**c.** $208\sqrt{3}/9 + 40 \approx 80.03$, least volume is in (b).

## Section 5.5

*For Thought:* **1.** T    **2.** T    **3.** F    **4.** T    **5.** T    **6.** T
**7.** F    **8.** F    **9.** F    **10.** T
*Exercises:*
**1.** 8.5 lb, 69.4°, 20.6°    **3.** 8.3 newtons, 22.9°, 107.1°
**5.** 5.1 lb, 75.2°    **7.** 127.0 lb    **9.** 1026.1 lb    **11.** 236.6 lb
**13.** 11.5°    **15.** 97.1°, 241.9 mph    **17.** 66.5°, 38.5 mph
**19.** 108.2°, 451.0 mph    **21.** 198.4°, 6000 ft    **23. a.** $d = 0.2|\sec\beta|$
**b.** $r = 3\cos(\alpha)|\sec(\beta)|$    **25.** 1420.0 lb
**27.** $\alpha = 27.7°$ $\beta = 40.5°$, $\gamma = 111.8°$    **29.** 40.4°
**31.** 255.4 ft    **33.** 24

## Chapter 5 Review Exercises

**1.** $\alpha = 82.7°$, $\beta = 49.3°$, $c = 2.5$    **3.** $\gamma = 103°$, $a = 4.6$, $b = 18.4$
**5.** No triangle    **7.** $\alpha = 107.7°$, $\beta = 23.7°$, $\gamma = 48.6°$
**9.** $\alpha_1 = 110.8°$, $\gamma_1 = 47.2°$, $a_1 = 6.2$; $\alpha_2 = 25.2°$, $\gamma_2 = 132.8°$, $a_2 = 2.8$
**11.** 92.4 ft$^2$    **13.** 23.0 km$^2$    **15.** 5.5, 2.4    **17.** 2.0, 2.5
**19.** $\sqrt{13}$, 56.3°    **21.** 6.0, 237.9°    **23.** $\langle 1, 1\rangle$    **25.** $\langle -3.0, 8.6\rangle$
**27.** $\langle -6, 8\rangle$    **29.** $\langle 0, -17\rangle$    **31.** 6    **33.** $-4\mathbf{i} + 8\mathbf{j}$
**35.** $3.6\sqrt{3}\mathbf{i} + 3.6\mathbf{j}$    **37.** Parallel    **39.** Perpendicular
**41.** Parallel    **43.** 18.4 lb, 11.0°, 19.0°    **45.** 44.2°, 205.6 mph
**47.** Seth, 42,785 ft$^2$    **49.** \$2639    **51.** $\sqrt{15}/8$    **53.** 28%
**57.** 5 acres or 217,800 ft$^2$

## Chapter 5 Test

**1.** One: $\beta = 90°, \gamma = 60°, c = 2\sqrt{3}$
**2.** Two: $\beta_1 = 111.1°, \gamma_1 = 8.9°, c_1 = 0.7, \beta_2 = 68.9°, \gamma_2 = 51.1°,$
$c_2 = 3.5$
**3.** One: $\gamma = 145.6°, b = 2.5, c = 5.9$
**4.** One: $\alpha = 33.8°, \beta = 129.2°, c = 1.5$
**5.** One: $\alpha = 28.4°, \beta = 93.7°, \gamma = 57.9°$    **6.** $2\sqrt{10}, 108.4°$
**7.** $2\sqrt{5}, 206.6°$    **8.** $3\sqrt{17}, 76.0°$    **9.** $22.5 \text{ ft}^2$    **10.** $12.2 \text{ m}^2$
**11.** $a_1 = 3.66, a_2 = 2.78, \mathbf{v} = 3.66\mathbf{i} + 2.78\mathbf{j}$    **12.** Perpendicular
**13.** $642.8$ lb    **14.** $47.2°, 239.3$ mph

## Tying It All Together Chapters P–5

**1.** $1/2, \sqrt{3}/2, \sqrt{3}/3, 2, 2\sqrt{3}/3, \sqrt{3}$
**2.** $\sqrt{2}/2, \sqrt{2}/2, 1, \sqrt{2}, \sqrt{2}, 1$
**3.** $\sqrt{3}/2, 1/2, \sqrt{3}, 2\sqrt{3}/3, 2, \sqrt{3}/3$
**4.** $1, 0$, undefined, $1$, undefined, $0$    **5.** $\pi/2$    **6.** $-\pi/2$
**7.** $-\pi/6$    **8.** $\pi/6$    **9.** $\pi$    **10.** $0$    **11.** $5\pi/6$
**12.** $\pi/6$    **13.** $0$    **14.** $\pi/4$    **15.** $-\pi/4$    **16.** $\pi/6$
**17.** $\{x \mid x = k\pi \text{ where } k \text{ is any integer}\}$
**18.** $\{x \mid x = k\pi \text{ or } x = \pi/2 + 2k\pi\}$
**19.** $\{x \mid x = 3\pi/2 + 2k\pi\}$
**20.** $\{x \mid x = \pi/6, 5\pi/6, 2\pi/3, 4\pi/3 \text{ all plus } 2k\pi\}$
**21.** $\{x \mid x = -1/2 \text{ or } x = \pi/6 + 2k\pi \text{ or } x = 5\pi/6 + 2k\pi\}$
**22.** $\{x \mid x = \pi/12 + k\pi \text{ or } x = 5\pi/12 + k\pi\}$
**23.** $\{x \mid x = \pi/24 + k\pi/4\}$    **24.** All real numbers
**25.** $1, 2\pi/3, 0, (-\infty, \infty), [-1, 1]$    **26.** $3, \pi, 0, (-\infty, \infty), [-3, 3]$

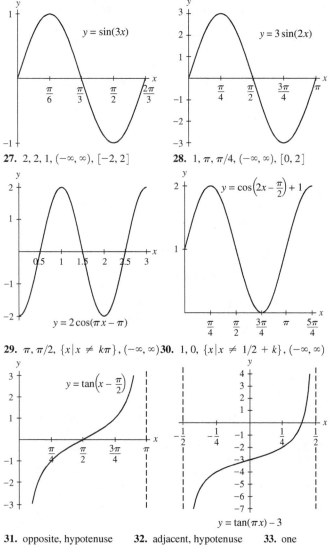

**27.** $2, 2, 1, (-\infty, \infty), [-2, 2]$    **28.** $1, \pi, \pi/4, (-\infty, \infty), [0, 2]$

**29.** $\pi, \pi/2, \{x \mid x \neq k\pi\}, (-\infty, \infty)$ **30.** $1, 0, \{x \mid x \neq 1/2 + k\}, (-\infty, \infty)$

**31.** opposite, hypotenuse    **32.** adjacent, hypotenuse    **33.** one
**34.** period    **35.** Pythagorean    **36.** even, odd    **37.** oblique
**38.** law of sines    **39.** law of cosines    **40.** triangle inequality

# Chapter 6

## Section 6.1

***For Thought:*** **1.** T   **2.** T   **3.** F   **4.** T   **5.** T   **6.** F
**7.** T   **8.** T   **9.** T   **10.** F
***Exercises:***
**1.** complex numbers    **3.** imaginary number    **5.** Imaginary, $0 + 6i$
**7.** Imaginary, $\frac{1}{3} + \frac{1}{3}i$    **9.** Real, $\sqrt{7} + 0i$    **11.** Real, $\frac{\pi}{2} + 0i$
**13.** $7 + 2i$    **15.** $-2 - 3i$    **17.** $-12 - 18i$    **19.** $26$
**21.** $29$    **23.** $4$    **25.** $-7 + 24i$    **27.** $1 - 4\sqrt{5}i$    **29.** $i$
**31.** $-1$    **33.** $1$    **35.** $-i$    **37.** $90$    **39.** $17/4$    **41.** $1$
**43.** $12$    **45.** $\frac{2}{5} + \frac{1}{5}i$    **47.** $\frac{3}{2} - \frac{3}{2}i$    **49.** $-1 + 3i$    **51.** $3 + 3i$
**53.** $\frac{1}{13} - \frac{5}{13}i$    **55.** $-i$    **57.** $-i$    **59.** $-4 + 2i$    **61.** $-6$
**63.** $-10$    **65.** $-1 + i\sqrt{5}$    **67.** $-3 + i\sqrt{11}$    **69.** $-4 + 8i$
**71.** $-6$    **73.** $0$    **75.** $0$    **77.** $9 + 6i$    **79.** $0$    **81.** $21$
**83.** Yes    **85.** No    **87.** Yes    **89.** Yes    **91.** Yes    **95.** Yes
**97.** $205.2$ pounds    **99.** $\alpha = 22.2°, \beta = 42.8°, \gamma = 115.0°$    **101.** $-2$
**103.** $6.9016$ ft or $\dfrac{\sqrt{47 - 10\sqrt{2} + 5\sqrt{2} + 1}}{2}$ ft

## Section 6.2

***For Thought:*** **1.** T   **2.** F   **3.** T   **4.** F   **5.** F   **6.** T
**7.** T   **8.** T   **9.** T   **10.** F
***Exercises:***
**1.** complex    **3.** absolute value    **5.** modulus, argument
**7.** quotient    **9.** $2\sqrt{10}$    **11.** $4$    **13.** $8$    **15.** $9$    **17.** $1$
**19.** $3\sqrt{2}$    **21.** $3\sqrt{2}(\cos 135° + i \sin 135°)$
**23.** $3(\cos 135° + i \sin 135°)$    **25.** $8(\cos 0° + i \sin 0°)$
**27.** $\sqrt{3}(\cos 90° + i \sin 90°)$    **29.** $2(\cos 150° + i \sin 150°)$
**31.** $5(\cos 53.1° + i \sin 53.1°)$    **33.** $\sqrt{34}(\cos 121.0° + i \sin 121.0°)$
**35.** $3\sqrt{5}(\cos 296.6° + i \sin 296.6°)$    **37.** $1 + i$    **39.** $-\frac{3}{4} + \frac{\sqrt{3}}{4}i$
**41.** $-0.42 - 0.26i$    **43.** $3i$    **45.** $-i\sqrt{3}$    **47.** $\frac{\sqrt{6}}{2} + \frac{3\sqrt{2}}{2}i$
**49.** $6i$    **51.** $\frac{3\sqrt{2}}{2} + \frac{\sqrt{6}}{2}i$    **53.** $9i$    **55.** $\sqrt{3} + i$
**57.** $0.34 - 0.37i$    **59.** $-40i, -0.8$    **61.** $8i, \frac{\sqrt{3}}{4} - \frac{1}{4}i$
**63.** $4\sqrt{2}, i\sqrt{2}$    **65.** $-7 - 26i, -\frac{23}{29} - \frac{14}{29}i$
**67.** $-18 + 14i, \frac{6}{13} + \frac{22}{13}i$    **69.** $-3 + 3i, 1.5 + 1.5i$    **71.** $9$
**73.** $4$    **75.** $-54 + 54i$
**79.** $6 \cos(9°) + 3 \cos(5°) + i(6 \sin(9°) + 3 \sin(5°))$,
$6 - 4i$, standard form
**81. a.** $-16 + 30i$   **b.** $\frac{3}{5} + \frac{1}{5}i$   **c.** $7i\sqrt{2}$
**83.** $y = \cos(1.5(x - \pi/3)) - 2$    **85. a.** $\pi$   **b.** $1$   **c.** $\pi/4$
**d.** $2\pi/3$    **87. a.** $A(n) = \dfrac{\pi(n^2 + n)}{8(n + \sqrt{3} - 1)^2}$
**b.** No. As $n \to \infty$, $A(n) \to \pi/8$, but the area of the triangle is $\frac{\sqrt{3}}{4}$, which is larger than $\pi/8$.

## Section 6.3

***For Thought:*** **1.** F   **2.** F   **3.** T   **4.** F   **5.** F   **6.** F
**7.** T   **8.** T   **9.** T   **10.** F
***Exercises:***
**1.** De Moivre's    **3.** $27i$    **5.** $-2 + 2i\sqrt{3}$    **7.** $-\frac{1}{2} + \frac{\sqrt{3}}{2}i$
**9.** $-18 + 18i\sqrt{3}$    **11.** $701.5 + 1291.9i$    **13.** $-16 + 16i$
**15.** $-8 - 8i\sqrt{3}$    **17.** $-3888 + 3888i\sqrt{3}$    **19.** $-119 - 120i$
**21.** $-7 - 24i$    **23.** $-44.928 - 31.104i$
**25.** $2(\cos 45° + i \sin 45°), 2(\cos 225° + i \sin 225°)$

**27.** $\cos \alpha + i \sin \alpha$ for $\alpha = 30°, 120°, 210°, 300°$

**29.** $2(\cos \alpha + i \sin \alpha)$ for $\alpha = \dfrac{\pi}{6}, \dfrac{\pi}{2}, \dfrac{5\pi}{6}, \dfrac{7\pi}{6}, \dfrac{3\pi}{2}, \dfrac{11\pi}{6}$

**31.** $1, -\dfrac{1}{2} \pm \dfrac{\sqrt{3}}{2}i$    **33.** $\pm 2, \pm 2i$    **35.** $\dfrac{\sqrt{2}}{2} \pm \dfrac{\sqrt{2}}{2}i, -\dfrac{\sqrt{2}}{2} \pm \dfrac{\sqrt{2}}{2}i$

**37.** $\pm \dfrac{\sqrt{3}}{2} + \dfrac{1}{2}i, -i$    **39.** $\pm\left(1 + i\sqrt{3}\right)$

**41.** $1.272 + 0.786i, -1.272 - 0.786i$    **43.** $\dfrac{1}{2} \pm \dfrac{\sqrt{3}}{2}i, -1$

**45.** $\pm 3, \pm 3i$    **47.** $-1 + i, 1 - i$    **49.** $0, \pm 2, 1 \pm i\sqrt{3}, -1 \pm i\sqrt{3}$

**51.** $2^{1/5}(\cos \alpha + i \sin \alpha)$ for $\alpha = 0°, 72°, 144°, 216°, 288°$

**53.** $10^{1/8}(\cos \alpha + i \sin \alpha)$ for $\alpha = 40.4°, 130.4°, 220.4°, 310.4°$

**55.** $-\dfrac{1}{4} + \dfrac{1}{4}i$    **57.** $1, -i$    **61.** $\sqrt{34}$    **63.** $6i$

**65.** $1/8, 3/8, 5/8, 7/8, 9/8, 11/8$    **67. a.** $a = 4, b = 6, c = 5$
**b.** If the sides are $a, b,$ and $c,$ and $\beta = 2\alpha,$ then $c = (b^2 - a^2)/a.$
The next two triangles are $a = 9, b = 12, c = 7,$ and
$a = 8, b = 12, c = 10.$

## Section 6.4

***For Thought:*** **1.** T    **2.** F    **3.** F    **4.** F    **5.** T    **6.** T
**7.** T    **8.** F    **9.** F    **10.** F

***Exercises:***
**1.** polar    **3.** four-leaf rose    **5.** $r \cos \theta, r \sin \theta$    **7.** $\left(3\sqrt{2}, \pi/4\right)$
**9.** $(3, \pi/2)$    **11.–21.** Odd:

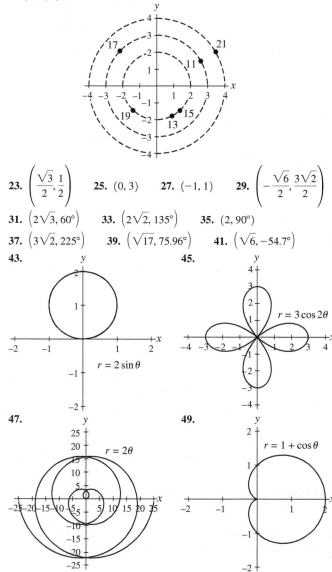

**23.** $\left(\dfrac{\sqrt{3}}{2}, \dfrac{1}{2}\right)$    **25.** $(0, 3)$    **27.** $(-1, 1)$    **29.** $\left(-\dfrac{\sqrt{6}}{2}, \dfrac{3\sqrt{2}}{2}\right)$

**31.** $\left(2\sqrt{3}, 60°\right)$    **33.** $\left(2\sqrt{2}, 135°\right)$    **35.** $(2, 90°)$

**37.** $\left(3\sqrt{2}, 225°\right)$    **39.** $\left(\sqrt{17}, 75.96°\right)$    **41.** $\left(\sqrt{6}, -54.7°\right)$

**43.**

$r = 2 \sin \theta$

**45.**

$r = 3 \cos 2\theta$

**47.**

$r = 2\theta$

**49.**

$r = 1 + \cos \theta$

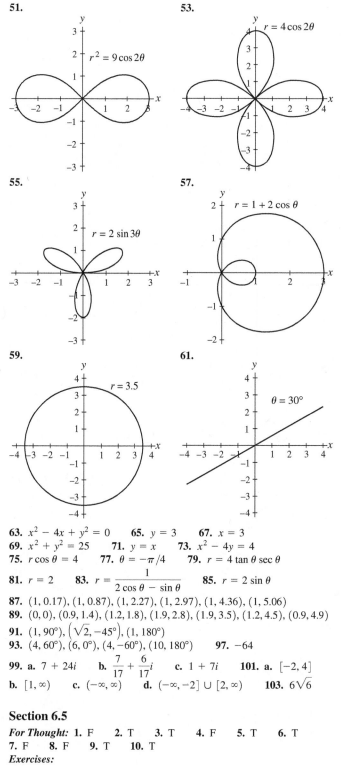

**51.**

$r^2 = 9 \cos 2\theta$

**53.**

$r = 4 \cos 2\theta$

**55.**

$r = 2 \sin 3\theta$

**57.**

$r = 1 + 2 \cos \theta$

**59.**

$r = 3.5$

**61.**

$\theta = 30°$

**63.** $x^2 - 4x + y^2 = 0$    **65.** $y = 3$    **67.** $x = 3$
**69.** $x^2 + y^2 = 25$    **71.** $y = x$    **73.** $x^2 - 4y = 4$
**75.** $r \cos \theta = 4$    **77.** $\theta = -\pi/4$    **79.** $r = 4 \tan \theta \sec \theta$

**81.** $r = 2$    **83.** $r = \dfrac{1}{2 \cos \theta - \sin \theta}$    **85.** $r = 2 \sin \theta$

**87.** $(1, 0.17), (1, 0.87), (1, 2.27), (1, 2.97), (1, 4.36), (1, 5.06)$
**89.** $(0, 0), (0.9, 1.4), (1.2, 1.8), (1.9, 2.8), (1.9, 3.5), (1.2, 4.5), (0.9, 4.9)$
**91.** $(1, 90°), \left(\sqrt{2}, -45°\right), (1, 180°)$
**93.** $(4, 60°), (6, 0°), (4, -60°), (10, 180°)$    **97.** $-64$

**99. a.** $7 + 24i$    **b.** $\dfrac{7}{17} + \dfrac{6}{17}i$    **c.** $1 + 7i$    **101. a.** $[-2, 4]$
**b.** $[1, \infty)$    **c.** $(-\infty, \infty)$    **d.** $(-\infty, -2] \cup [2, \infty)$    **103.** $6\sqrt{6}$

## Section 6.5

***For Thought:*** **1.** F    **2.** T    **3.** T    **4.** F    **5.** T    **6.** T
**7.** F    **8.** F    **9.** T    **10.** T

***Exercises:***
**1.** parametric
**3.**

| $t$ | $x$ | $y$ |
|-----|-----|-----|
| 0 | 1 | $-2$ |
| 1 | 5 | $-1$ |
| 1.5 | 7 | $-0.5$ |
| 3 | 13 | 1 |

**5.**

| $t$ | $x$ | $y$ |
|-----|-----|-----|
| 1 | 1 | 2 |
| 2.5 | 6.25 | 6.5 |
| $\sqrt{5}$ | 5 | $3\sqrt{5} - 1$ |
| 4 | 16 | 11 |
| 5 | 25 | 14 |

**7.**

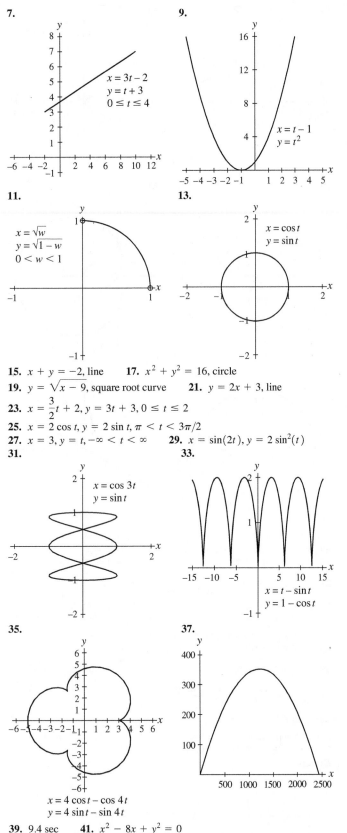

$x = 3t - 2$
$y = t + 3$
$0 \le t \le 4$

**9.**

$x = t - 1$
$y = t^2$

**11.**

$x = \sqrt{w}$
$y = \sqrt{1 - w}$
$0 < w < 1$

**13.**

$x = \cos t$
$y = \sin t$

**15.** $x + y = -2$, line    **17.** $x^2 + y^2 = 16$, circle
**19.** $y = \sqrt{x - 9}$, square root curve    **21.** $y = 2x + 3$, line
**23.** $x = \dfrac{3}{2}t + 2, y = 3t + 3, 0 \le t \le 2$
**25.** $x = 2\cos t, y = 2\sin t, \pi < t < 3\pi/2$
**27.** $x = 3, y = t, -\infty < t < \infty$    **29.** $x = \sin(2t), y = 2\sin^2(t)$

**31.**

$x = \cos 3t$
$y = \sin t$

**33.**

$x = t - \sin t$
$y = 1 - \cos t$

**35.**

$x = 4\cos t - \cos 4t$
$y = 4\sin t - \sin 4t$

**37.**

**39.** 9.4 sec    **41.** $x^2 - 8x + y^2 = 0$
**43.** $6(\cos(5\pi/3) + i\sin(5\pi/3))$    **45.** $\pi/3, \pi, 5\pi/3$    **47.** 4.0851 ft

## Chapter 6 Review Exercises

**1.** $-1 - i$    **3.** $-9 - 40i$    **5.** $-16 - 12i$    **7.** $-3 - 2i$
**9.** $-\dfrac{1}{13} + \dfrac{5}{13}i$    **11.** $3 + i\sqrt{2}$    **13.** $-1 + i$    **15.** $\sqrt{34}$
**17.** $2\sqrt{2}$    **19.** $5.94(\cos 135° + i\sin 135°)$
**21.** $7.6(\cos 252.3° + i\sin 252.3°)$    **23.** $-\dfrac{3}{2} + \dfrac{\sqrt{3}}{2}i$    **25.** $5.4 + 3.5i$
**27.** $-15i, -\dfrac{5}{6}$    **29.** $8 - i, 0.31 + 0.54i$    **31.** $-4\sqrt{2} + 4i\sqrt{2}$
**33.** $-128 + 128i$    **35.** $\dfrac{\sqrt{2}}{2} + \dfrac{\sqrt{2}}{2}i, -\dfrac{\sqrt{2}}{2} - \dfrac{\sqrt{2}}{2}i$
**37.** $2^{1/3}(\cos \alpha + i\sin \alpha)$ for $\alpha = 10°, 130°, 250°$
**39.** $5^{1/6}(\cos \alpha + i\sin \alpha)$ for $\alpha = 8.9°, 128.9°, 248.9°$
**41.** $5(\cos \alpha + i\sin \alpha)$ for $\alpha = 22.5°, 112.5°, 202.5°, 292.5°$
**43.** $\left(2.5, 5\sqrt{3}/2\right)$    **45.** $(-0.3, 1.7)$    **47.** $(4, 4\pi/3)$
**49.** $\left(\sqrt{13}, -0.98\right)$

**51.**

$r = -2\sin\theta$

**53.**

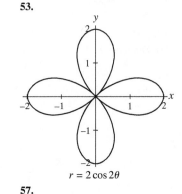

$r = 2\cos 2\theta$

**55.**

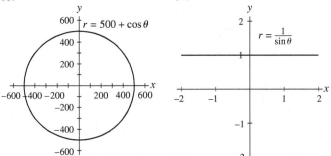

$r = 500 + \cos\theta$

**57.**

$r = \dfrac{1}{\sin\theta}$

**59.** $x + y = 1$    **61.** $x^2 + y^2 = 25$    **63.** $r = \dfrac{3}{\sin\theta}$    **65.** $r = 7$

**67.**

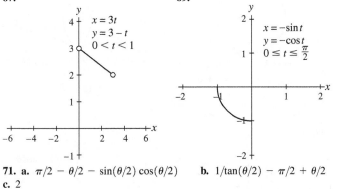

$x = 3t$
$y = 3 - t$
$0 < t < 1$

**69.**

$x = -\sin t$
$y = -\cos t$
$0 \le t \le \dfrac{\pi}{2}$

**71. a.** $\pi/2 - \theta/2 - \sin(\theta/2)\cos(\theta/2)$    **b.** $1/\tan(\theta/2) - \pi/2 + \theta/2$
**c.** 2

## Chapter 6 Test

**1.** $7 - 24i$    **2.** $\dfrac{1}{2} - \dfrac{1}{2}i$    **3.** $-1 + i$    **4.** $-4 + 4i\sqrt{3}$
**5.** $3\sqrt{2}(\cos 45° + i\sin 45°)$    **6.** $2(\cos 120° + i\sin 120°)$
**7.** $2\sqrt{5}(\cos 206.6° + i\sin 206.6°)$    **8.** $3\sqrt{2} + 3i\sqrt{2}$

**9.** $512i$   **10.** $\dfrac{3\sqrt{2}}{4} + \dfrac{3\sqrt{2}}{4}i$   **11.** $\left(\dfrac{5\sqrt{3}}{2}, \dfrac{5}{2}\right)$

**12.** $\left(-\dfrac{3\sqrt{2}}{2}, \dfrac{3\sqrt{2}}{2}\right)$   **13.** $(-26.4, -19.9)$

**14.**   **15.**

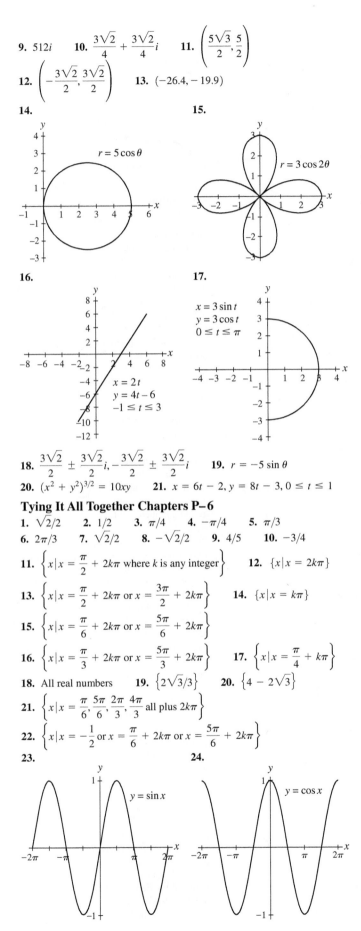

**16.**   **17.**

**18.** $\dfrac{3\sqrt{2}}{2} \pm \dfrac{3\sqrt{2}}{2}i, -\dfrac{3\sqrt{2}}{2} \pm \dfrac{3\sqrt{2}}{2}i$   **19.** $r = -5\sin\theta$

**20.** $(x^2 + y^2)^{3/2} = 10xy$   **21.** $x = 6t - 2, y = 8t - 3, 0 \le t \le 1$

## Tying It All Together Chapters P–6

**1.** $\sqrt{2}/2$   **2.** $1/2$   **3.** $\pi/4$   **4.** $-\pi/4$   **5.** $\pi/3$

**6.** $2\pi/3$   **7.** $\sqrt{2}/2$   **8.** $-\sqrt{2}/2$   **9.** $4/5$   **10.** $-3/4$

**11.** $\left\{x \,\middle|\, x = \dfrac{\pi}{2} + 2k\pi \text{ where } k \text{ is any integer}\right\}$   **12.** $\{x \mid x = 2k\pi\}$

**13.** $\left\{x \,\middle|\, x = \dfrac{\pi}{2} + 2k\pi \text{ or } x = \dfrac{3\pi}{2} + 2k\pi\right\}$   **14.** $\{x \mid x = k\pi\}$

**15.** $\left\{x \,\middle|\, x = \dfrac{\pi}{6} + 2k\pi \text{ or } x = \dfrac{5\pi}{6} + 2k\pi\right\}$

**16.** $\left\{x \,\middle|\, x = \dfrac{\pi}{3} + 2k\pi \text{ or } x = \dfrac{5\pi}{3} + 2k\pi\right\}$   **17.** $\left\{x \,\middle|\, x = \dfrac{\pi}{4} + k\pi\right\}$

**18.** All real numbers   **19.** $\left\{2\sqrt{3}/3\right\}$   **20.** $\left\{4 - 2\sqrt{3}\right\}$

**21.** $\left\{x \,\middle|\, x = \dfrac{\pi}{6}, \dfrac{5\pi}{6}, \dfrac{2\pi}{3}, \dfrac{4\pi}{3} \text{ all plus } 2k\pi\right\}$

**22.** $\left\{x \,\middle|\, x = -\dfrac{1}{2} \text{ or } x = \dfrac{\pi}{6} + 2k\pi \text{ or } x = \dfrac{5\pi}{6} + 2k\pi\right\}$

**23.**   **24.**

**25.**   **26.**

**27.**   **28.**

**29.**   **30.**

**31.**   **32.**

**33.** Heron's area formula   **34.** vector   **35.** scalar

**36.** magnitude   **37.** resultant   **38.** dot product   **39.** orthogonal

**40.** complex number   **41.** complex conjugates   **42.** De Moivre's

# INDEX OF APPLICATIONS

## Astronomy

achieving synchronous orbit, 69
catching a green flash, 93
communicating via satellite, 93
discovering planets, 131, 147
eclipse function, 224, 308
Eratosthenes measures the Earth, 62–63
filtering the sun, 266
first pulsar, 130
flat Earth, 267
Jupiter's satellites, 131
linear velocity of a point on the equator, 67
moon illumination, 132
parsecs, 92
partial eclipse, 308
photographing the Earth, 62
scattering angle, 275
sunspots, 130
total eclipse, 276, 307, 308
view from Landsat, 92

## Biology/Health/Life Sciences

aerobics, 19
army of ants, 169
blood velocity, 131
height of a tree, 90
lightest kid, 138
lung capacity, 131

## Business

across-the-board raise, 29
cost of a sundae, 19
cost-of-living raise, 29
counting coworkers, 176
maximizing the total profit, 207
maximum profit, 153
nines, 157
paying up, 12
periodic cost, 131
periodic revenue, 131
periodic sales, 244
pipeline detour, 306–307
price of a car, 38
total profit, 153
withholding, 38

## Construction

adjacent pipes, 275
angle of elevation, 92, 107, 266
angular window, 52
buckling bridge, 95
designing an addition, 266
eyebrow window, 177
four pipes, 169
hanging a pipe, 94
height of a skyscraper, 92
installing a guy wire, 90

installing an antenna, 275
laying pipe, 352
making a kite, 267, 282
making a shaft, 276
the Pentagon, 274
tiling a room, 148
triangular flashing, 282
triangular glazing, 282

## Consumer

costly computers, 268
cost of window cleaning, 19
cul-de-sac, 283

## Design

angle between spokes of a wheel, 52
angles on a bicycle, 52
designing an addition, 266
figuring the setback, 94
limiting the beam, 19
maximizing area, 94
planning a tunnel, 275
positioning a solar panel, 275
solving a lot, 95
spacing between teeth, 103
tank stand, 307

## Economics

depreciation rate, 39
expenditures versus income, 38
growth rate, 39
production function, 30
unemployment versus inflation, 29

## Entertainment

AM radio, 157
best view in a theater, 278
broadcasting the oldies, 157
Ferris wheels, 68, 69, 70
London Eye, 68
the survivor, 64
waveform of a guitar note, 139

## Environment

amount of daylight, 132
attack of the grizzly, 276
avoiding a swamp, 90
Bismarck temperature, 102
cloud height, 107
Devil's Tower, 107
giant redwood, 90
height of a rock, 92
large ocean waves, 132
ocean waves, 132
periodic temperature, 132

Phoenix temperature, 102
progress of a forest fire, 92
saving an endangered hawk, 267
tide in the Bay of Fundy, 101

## Geometry

aerial photography, 87, 89–90
angle of an airplane, 223
angle of depression, 90
angle of elevation, 52
arcs, chords, and angles, 283
area, area, area, 306
area of a parallelogram, 43
area of a polygon, 251
area of a sector, 63
area of a slice of pizza, 63
area of a triangle, 283–284
area of a wing, 282
area of irrigation, 59, 63, 106
areas, 282–283
central angle, 276
circle and square, 206
circles in conduits, 11
clock angle, 53
completely saturated, 197
concentric circles, 133
cutting cardboard, 138
diameter of a conduit, 11
dividing property, 306
equal areas, 284
filling a triangle, 326
finding the diagonal, 70
height of a balloon, 92
height of a crosswalk, 90
height of an antenna, 90
height of an object, 86–87, 264–265
height of a rock, 92
height of buildings, 107
a hexagon, 274
interesting angle, 189
intersecting lines, 108
lakefront property, 308
length of a chord, 271–272, 274
length of a pipe, 90
lot on a curve, 283
machine parts, 53
maximizing the volume, 64
maximum sector, 63
maximum triangle, 282
more circles, 308
overlapping circles, 184
the Pentagon, 274
perpendicular tangents, 11
pipe on an inclined plane, 94
property angles, 53
radius of a pulley, 88
radius of the Earth, 94, 95
regular hexagon, 95
regular polygon, 91

regular polygons, 326
related angles, 331
Robin and Marian, 90
shady triangle, 307
side of a pentagon, 276
smallest angle of a triangle, 223
solving a triangle, 184, 268
spiral of triangles, 237
steering geometry, 79–80
surveying a quadrilateral, 283
surveying property, 266
surveying triangular property, 283
tangent circles, 121
telling time, 70
Texas star, 108
Texas triangles, 108
three pulleys and a belt, 107
triangles and circles, 53
triangular flashing, 282
triangular glazing, 282
twin squares, 318
two common triangles, 189
two pulleys and a belt, 107
two squares and two triangles, 342
Venn watering diagram, 63
volume of a cube, 38
volume of a cup, 63
watering the lawn, 284
weather radar, 63
wing dimensions, 266

### Miscellaneous

acceleration of a missile, 294
angle bisector theorem, 267–268
arranging digits, 148
average rate of change, 147
Battle of Gettysburg, 255
Cellular One, 266
as the crow flies, 206
distance to the helper, 62
door prizes, 30
double boxed, 349
firing an M-16, 243–244
firing a torpedo, 275
head start, 267
large integer, 211
largest integer, 244
law of sines, 267
leaning ladder, 318
logs and tangents, 176
maximizing the sum, 39
metal star, 91
minimizing, 255
moving a refrigerator, 277
muzzle velocity, 244
natural numbers, 237
odd integer, 53
one in a million, 211
oscillating depth, 157
oscillating light, 211
oscillating temperature, 157
rabbit and fox, 267
renumbering Elm Street, 154
shipping pipes, 93
shipping stop signs, 103

shot down, 267
sine cubed, 251
sines and cosines, 197
sum and difference, 224
summing factorials, 342
tortoise and hare, 157
two wrongs make right, 133
watering the lawn, 93
wrong division, 43
in the wrong place, 307

### Physics

achieving synchronous orbit, 69
angular velocity of a bicycle wheel, 106
angular velocity of a helicopter
    blade, 64–65
angular velocity of a lawnmower blade, 65
angular velocity of a nail in a tire, 69
angular velocity of a radar antenna, 69
angular velocity of a record, 68
belt and pulleys, 69
bicycle velocity, 69
crooked man, 106
Ferris wheels, 68, 69, 70
force of draft horses pulling a stump, 287
force to move a tortoise, 297
ice force, 302
linear velocity of a lawnmower blade, 66, 68
linear velocity of a point on Earth, 67, 69
linear velocity of a propeller, 66
linear velocity of a record, 68
linear velocity of a router bit, 69
London Eye, 68
magnitude of a force, 302
mama's direction, 303
motion of a football, 198
motion of a projectile, 169–170
motion of a spring, 102–103,
    130, 151–152, 153, 204, 249, 255
moving a donkey, 302
muzzle velocity, 244
my three elephants, 303
path of a projectile, 241–242
positioning a human arm, 276
positioning a robotic arm, 272–273
pushing a shopping cart, 302
resultant force, 306
rock and roll, 302
Sisy's force, 302
speed of a table saw blade, 69
super force, 302
weight of a ball, 302
weight of a safe, 297–298
winch force, 302

### Production

making paper, 69
making radios, 38

### Sports

angle of completion, 267
baseball strategy, 252
choosing the right angle, 244
hiking, 274
kicking a field goal, 104

margin of error in a field goal, 62
packing billiard balls, 295
plastic rack, 91
rate of ascent, 93
shooting a target, 106–107
swim meet, 331
throwing a javelin, 103
wooden rack, 91

### Statistics/Demographics

age at first marriage, 12
counting votes, 80
disappearing dogs, 303
number of unmarried couples, 12
paper consumption, 19
whole lotta shakin,' 255

### Transportation

angle of inclination of a car, 298
area of a wing, 282
bicycle velocity, 69
boating, 274
cadence and speed, 63
capsize control, 19
clear sailing, 108
compression ratio, 19
course and ground speed of an
    airplane, 299–300
course of a Cessna, 306
course of a fighter plane, 266
course of a helicopter, 302
course of an airplane, 302
course of an ultralight, 306
crossing a river, 303
detecting a speeder, 91–92
Detroit pistons, 307
displacement of an engine, 19
distance and rate, 303
distance between Los Angeles
    and New York City, 58
distance from Fairbanks to
    Fort Yukon, 263–264
distance to North Pole, 62
due east, 302
due west, 302
finding the bore, 19
finding the head angle, 168
flop factor, 80
flying, 275
gaining altitude, 294
going with the flow, 302
minimizing the time, 303
motion of a car, 211
motorcycle trail, 90–91
observing traffic, 266
passing in the night, 92
price of a car, 38
pushing a cycle, 306
rake and trail, 91, 168
rebuilt engine, 19
rotating cogs, 63
sailing, 266
shortcut to Snyder, 90
smuggler's blues, 277
steering geometry, 79–80
ultralight, 302

# INDEX

## A

Absolute value of a complex number, 319, 320
Acute angles
  degree measure of, 46
  of right triangles, trigonometric functions of, 83–84
Addition. *See also* Sum(s)
  of complex numbers, 312–313
  of scalars, 287–288
Algebraic functions
  combining with trigonometric functions, 148–150
  converting compositions of functions to, 221–222
Ambiguous case (SSA), solving oblique triangles, 260–263
Amplitude of a sine wave, 113
Angles, 45–50
  acute. *See* Acute angles
  central. *See* Central angle
  coterminals. *See* Coterminal angles
  definition of, 45
  degree measure of, 45–48
  of depression, 85–88
  of elevation, 85–88
  finding, given the cosine, 216–217
  finding, given the sine, 215
  intercepted arc of, 45
  minutes and, 48–50
  obtuse, 46
  quadrantal, 46, 48
  radian measure of, 54–57
  reference, 98–100
  right, 46
  seconds and, 48–50
  sides of, 45
  in standard position, 45
  straight, 46
  between two vectors, 290–291
  vertex of, 45
Angular velocity, 64–65
  linear velocity in terms of, 67
Arc(s), intercepted, of an angle, 45
Arc length of a circle, 57–59
  constructing an eyebrow window, 177
Arc-length formula, 57
Area
  of a circle, 86
  Heron's area formula and, 280–281
  of a quadrilateral, 279
  of a sector of a circle, 59
  of a triangle, 278–285
Argument, 320
Arm motion, modeling, 272–273
Asymptotes
  of cosecant functions, 135
  of cotangent functions, 143
  of tangent functions, 140
  vertical, 133–134

Axes
  in complex plane, 319
  polar, 332

## B

Baseball throwing strategy, modeling, 252
Bearing, 263–264
Bifolium, 343
Binomials, multiplication of, 171

## C

Cartesian coordinate system, 2–12
  circles in, 5–6
  lines in, 6–7
  midpoint formula, 6
  ordered pairs and, 2
  plotting points on, 2
  polar-rectangular conversions and, 333–334
  Pythagorean theorem and, 3
  square roots and, 4–5
  *xy*-plane and, 2
Cartesian equations. *See* Rectangular equations
Center of a circle, 5
Central angle, 45
  finding, 58–59
Chords, length of, 271–272
Circles, 5–6
  arc length of, 57–59, 177
  area of, 86
  center of, 5
  circumference of, 285
  equation for, 5
  radius of, 5
  sectors of, area of, 59
  unit. *See* Unit circle
Circumference of a circle, 285
Cissoid, 343
Cofunction(s), of trigonometric functions, 180
Cofunction identities, 180–182
Combining trigonometric functions with algebraic functions, 148–150
Complex conjugates, 314, 323–324
Complex numbers, 311–330
  absolute value or modulus of, 319, 320
  addition of, 312–313
  complex conjugates and, 314, 323–324
  complex plane and, 319
  definition of, 311
  division of, 313–315
  equal, 311
  graphs of, 319
  imaginary and real parts of, 311
  multiplication of, 312–313
  powers of, 326–328
  powers of *i* and, 313

products and quotients in trigonometric form and, 321–323
  roots of, 326, 328–330
  roots of negative numbers and, 315–316
  as solutions to equations, 316
  standard form of, 311–312, 321
  subtraction of, 312–313
  trigonometric form of, 319–326
Complex plane, 319–326
Component form of a vector, 289–290
Composition of functions, 30–32, 220–222
  converting to algebraic functions, 221–222
  evaluating, 220–221
  finding inverses using, 35–37
Computations with degrees-minutes-seconds, 49–50
Conchoid of Nichomedes, 343
Conditional trigonometric equations, 213–256
  basic cosine equations, 225–228
  basic sine equations, 228–231
  basic tangent equations, 231–232
  inverse trigonometric functions and, 214–224
  inverses of general trigonometric functions and, 234–235
  involving compositions, 237–244
  isolating trigonometric functions and, 232–233
  of quadratic type, 245–252
  solving for a variable using inverse functions and, 233
Conjugates, complex, 314
Coordinate(s), in Cartesian coordinate system, 2
Coordinate plane, 2
Coordinate systems. *See* Cartesian coordinate system; Polar coordinate system
Cosecant function. *See also* Trigonometric functions
  asymptotes of, 135
  domain of, 135
  graphs of, 133, 134–137
  inverse, 218–220
Cosine(s), law of. *See* Law of cosines
Cosine equations
  basic, 225–228
  with a half angle, 239
Cosine function, 127. *See also* Trigonometric functions
  for common angles, 76–77
  of a difference, 179–180
  graphing, 114–115
  infinite series for, 80–81
  inverse, 82, 215–217, 218
  of a sum, 177–179, 180
  transformations of, 115–118

Cotangent function, 140. *See also*
    Trigonometric functions
    graphs of, 143–145
    identities of, 161
    inverse, 218–220
Coterminal angles, 46–48
    finding using radian measure, 57
Curve fitting, sinusoidal, 128–129

**D**

De Moivre, Abraham, 327
De Moivre's theorem, 327
Decimal degrees, converting to degrees-
    minutes-seconds, 49
Degrees
    conversion between radians and, 55–56
    decimal, converting to degrees-minutes-
        seconds, 49
    as measure of angles, 45–48
Denominators, rationalizing, 4
Dependent variable, 14
Depression, angle of, 85–88
Descartes, René, 2
Differences. *See also* Subtraction
    cosine of, 179–180
    expressing as a product, 201
    sine of, 184–185
    tangent of, 185–187
    of two trigonometric functions, 151
Directed length, 54
Directed line segments, 285
Direction of a vector, 286, 289–290
Division. *See also* Quotients
    of complex numbers, 313–315
Domains of functions, 15, 140, 143, 145
Dot product of two vectors, 290, 291
Double-angle identities, 189–190

**E**

Elevation, angle of, 85–88
Equal vectors, 286
Equations
    Cartesian. *See* Rectangular equations
    of a circle, 5
    complex number solutions to, 316
    conditional. *See* Conditional trigono-
        metric equations
    cosine. *See* Cosine equations
    linear, in two variables, 8
    parametric. *See* Parametric equations
    polar. *See* Polar equations
    proving equations are not identities, 166
    quadratic. *See* Quadratic equations,
        trigonometric
    rectangular (Cartesian). *See*
        Rectangular equations
    secant, 240
    sine. *See* Sine equations
    tangent. *See* Tangent equations
    trigonometric. *See* Conditional trigono-
        metric equations; Quadratic equa-
        tions, trigonometric; Trigonometric
        equations
Equilibrium position of a spring, 151
Euclid, 3

Even functions, 26–27, 164–166
Even identities, 165–166, 181–182

**F**

Factoring
    solving trigonometric equations using,
        245–247
    with trigonometric functions, 171
Families of functions, 20
First coordinate, 2
Folium of Descartes, 343
Force to move object, finding, 297
Formulas
    arc-length, 57
    for area of a triangle, 285
    for circumference of a circle, 285
    composition of functions given as, 31
    Heron's area formula, 280–281
    midpoint, 6
    quadratic, solving trigonometric equa-
        tions using, 247
    reduction, 202–204
45-45-90 triangles, 74
Fractions
    equal, trigonometric functions with,
        171–172
    in identities, subtracting, 172–173
    single, separating into two, 172
Frequency of a sine wave, 126–127
Function(s), 13–20
    algebraic. *See* Algebraic functions
    composition of. *See* Composition of
        functions
    concept of, 13–14
    constructing, 14
    cosecant. *See* Cosecant function
    cosine. *See* Cosine function
    cotangent. *See* Cotangent function
    definition of, 14
    domains of, 15, 140, 143, 145
    even, 26–27, 164–166
    families of, 20
    function notation for, 17–18
    graphs of, 15–17
    invertible. *See* Inverse trigonometric
        functions; Invertible functions
    odd, 27, 164–166
    one-to-one, 32–35
    period of, 112, 122–124,
        141–142, 145
    periodic, 112
    in polar coordinates, 337–338
    range of, 15, 145
    secant. *See* Secant function
    sine. *See* Sine function
    switch-and-solve method
        for, 33–34
    tangent. *See* Tangent function
    transformations of. *See* Transformations
        of functions
    trigonometric. *See* Cosecant function;
        Cosine function; Cotangent function;
        Secant function; Sine function;
        Tangent function; Trigonometric
        functions
Function notation, 17–18

Fundamental cycle of trigonometric func-
    tions, 112, 114, 145
Fundamental identity of trigonometry, 97

**G**

Garfield, James, 3
General sine wave, 122–133
    frequency of, 126–127
    graphing, 124–126
    period of, 122–124
    sinusoidal curve fitting and, 128–129
Graphs, 109–159
    in Cartesian coordinate system, 2
    of a circle, 6–7
    of combined trigonometric and alge-
        braic functions, 148–150
    of complex numbers, 319
    constructing sine functions for,
        117–118
    of functions, 15–17
    general sine wave and, 122–133
    inflection point on, 112
    of lines, 8
    of parametric equations, 344
    of polar equations, 334–337
    reduction formula and, 203–204
    reflection of, 22–23
    of secant and cosecant functions,
        133–139
    of sine function, 111–113
    stretching and shrinking and, 23–24
    symmetric, 26–27
    of tangent and cotangent functions,
        140–148
    unit circle and, 110–121
Guitar note, modeling waveform of, 139

**H**

Half-angle identities, 191–194
Heron's area formula, 280–281
Horizontal component of a vector,
    288–289
Horizontal lines, graphing, 8
Horizontal translation of functions, 20–22,
    117
Hypotenuse, 83

**I**

Identities
    reciprocal, 72. *See also* Cosecant func-
        tion; Cosine function; Cotangent
        function
    sum and difference. *See* Sum and dif-
        ference identities
    trigonometric. *See* Trigonometric
        identities
    trigonometric equations involving,
        240–241
Imaginary axis in complex plane, 319
Imaginary number *i*
    definition of, 311
    powers of, 313
Imaginary part of a complex number, 311
Inclined-plane problems, 296–299

Independent variable, 14
Inequalities, triangle, 270
Infinite series for sine and cosine, 80–81
Inflection point, 112
Initial point of a vector, 286
Initial side of an angle, 45
Intercepted arc of an angle, 45
Inverse trigonometric functions, 81–83,
    214–224
    compositions of functions and, 220–222
    cosecant, 218–220
    cosine, 82, 215–217, 218
    cotangent, 218–220
    evaluating, 219–220
    of general trigonometric functions,
        234–235
    secant, 218–220
    sine, 214–215, 218
    solving for variables using, 233
    tangent, 82, 217–218
Invertible functions, 33
    finding using composition, 35–37
    of one-to-one functions, 33–35
    switch-and-solve method for, 33–34

**K**

Key points
    for graph of sine function, 112
    for graph of tangent function, 141

**L**

Law of cosines, 268–277
    given three sides of a triangle (SSS),
        269–270
    given two sides and an included angle
        (SAS), 270
    length of a chord and, 271–272
    modeling arm motion and, 272–273
Law of sines, 258–268
    ambiguous case with (SSA), 260–263
    bearing and, 263–264
    finding the height of an object from a
        distance and, 264–265
    with two angles and an included side
        (ASA), 260
Length, directed, 54
Line(s), 6–7
    horizontal and vertical, graphing, 8
    linear equations in two variables and, 8
    ray on, 45
Line segments
    directed, 285
    parametric equations for, 345–346
Linear combinations of
    vectors, 291–292
Linear equations in two variables, 8
Linear velocity, 66–67
    in terms of angular velocity, 67
Lituus, 343

**M**

Magnitude of vectors, 286
    finding from component form, 289
Maximizing total profit, 207

Midpoint formula, 6
Minimizing total time, 304
Minutes (of degrees), 48–50
Modeling
    of baseball throwing strategy, 252
    of the best view, 278
    of the motion of a football, 198
    of the motion of a projectile, 169–170
    of the motion of a spring, 151–152,
        204, 249
    of the path of a projectile, 241–242
    of robotic arm motion, 272–273
    of tide, 101
    of the waveform of a guitar note, 139
Modulus of a complex number, 319, 320
Multiple angle tangent equations,
    238–239
Multiplication. *See also* Products
    of binomials, 171
    of complex numbers, 312–313
    dot product of vectors and, 290, 291
    of scalars, 286–287

**N**

Navigation problems, 299–301
Negative numbers, roots of, 315–316
Nephroid of Freeth, 343
Nonrigid transformations, 25
Numbers
    complex. *See* Complex numbers
    imaginary. *See* Imaginary number *i*
    negative, roots of, 315–316

**O**

Oblique triangles, 258
    ambiguous case (SSA), 260–263
    law of cosines for solving, 268–277
    law of sines for solving, 258–268
Obtuse angles, degree measure of, 46
Odd functions, 27, 164–166
Odd identities, 165–166, 181–182
One-to-one correspondence, 32
One-to-one functions, 32–35
    inverses of, 33–35
Opposites of vectors, 288
Ordered pairs, 2
    graphing by plotting, 15–17
Origin
    in Cartesian coordinate system, 2
    symmetry about, 27
Orthogonal vectors, 290

**P**

Parabolas, 15, 20
Parallel vectors, 290
Parallelogram law, 287
Parameters, 344
    eliminating, 344–345
Parametric equations, 343–349
    eliminating the parameter and, 344–345
    flight of a basketball and, 347
    graphs of, 344
    writing, 345–347
Parentheses, with ordered pairs, 2

Pentagon, finding length of a side of,
    271–272
Period of a function, 112, 122–124,
    141–142, 145
Periodic functions, 112
Perpendicular vectors, 290
Phase shift, 116
Plotting points
    in Cartesian coordinate system, 2
    in polar coordinate system, 332–333
Points
    in Cartesian coordinate system, 2
    inflection, 112
    initial, of a vector, 286
    in polar coordinate system, 332–333
    terminal, of a vector, 286
Polar axis, 332
Polar coordinate(s), functions
    in, 337–338
Polar coordinate system, 332
    plotting points in, 332–333
    polar-rectangular conversions and,
        333–334
    poles in, 332
Polar equations, 332–343
    applications of, 339–340
    converting between parametric
        equations and, 346–347
    converting between rectangular
        coordinates and, 333–334, 338–339
    polar coordinates and, 332–333
Poles, 332
Polynomials, Taylor, 121–122
Powers
    of complex numbers, 326–328
    of imaginary number *i*, 313
Principal square roots, 315
Products. *See also* Multiplication
    of complex conjugates, 314
    dot, of two vectors, 290, 291
    evaluating, 200
    expressing as sums, 199–200
    in trigonometric form, 321–323
Product-to-sum identities, 199–200
Profit, total, maximizing, 207
Projectiles
    motion of, 169–170
    path of, 241–242
Pythagorean identities, 161–162
Pythagorean theorem, 3, 85

**Q**

Quadrant(s), in Cartesian
    coordinate system, 2
Quadrantal angles, 46, 48
Quadratic equations, trigonometric,
    245–252
    solving using factoring, 245–247
    solving using quadratic formula, 247
    square-root property and, 245
    squaring each side of, 248–249
Quadratic formula, solving trigonometric
    equations using, 247
Quadrilaterals, area of, 279
Quotients. *See also* Division
    in trigonometric form, 321–323

## R

Radian(s), 55
  conversion between degrees and, 55–56
  finding coterminal angles using, 57
Radian measure of angles, 54–57
Radicand, 4
Radius of a circle, 5
Range of a function, 15, 145
Rationalizing the denominator, 4
Ratios, trigonometric, 71
Ray, 45
Real axis in complex plane, 319
Real part of a complex number, 311
Reciprocal identities, 72, 161–162. *See
  also* Cosecant function; Cosine func-
  tion; Cotangent function
Rectangular coordinate system. *See
  Cartesian coordinate system*
Rectangular equations, 334
  converting polar equations to, 338–339
  converting to polar equations, 339
Reduction formula, 202–204
Reference angles, 98–100
Reflection of graphs, 22–23
Resultants, 287, 295–296
Right angles, degree measure of, 46
Right triangles, 83–84
  hypotenuse of, 83
  Pythagorean theorem and, 3, 85
  sides of, 83
  solving, 84–85
  trigonometric functions in, 83–84
Rigid transformations, 25
Robotic arm motion, modeling, 272–273
Roots. *See also* Square roots
  of complex numbers, 326, 328–330
  of negative numbers, 315–316

## S

Scalar(s), 285
  multiplication and addition of, 286–288
Scalar quantities, 285
Secant equation, 240
Secant function. *See also* Trigonometric
  functions
  graphs of, 133–134
  inverse, 218–220
Second(s) (of degrees), 48–50
Second coordinate, 2
Sectors of circles, area of, 59
Shrinking functions, 23–24
Simplification
  with identities, 162–164
  of power of *i*, 313
  reduction formula and, 202–204
  of square roots, 4–5
  with sum and difference identities, 180,
    186
Sine(s), law of. *See* Law of sines
Sine equations
  basic, 228–231
  with double angle, 237–238
Sine function, 127. *See also*
  Trigonometric functions
  for common angles, 76–77
  constructing, 70–71

general, inverse of, 234–235
  graphing, 111–113
  infinite series for, 80–81
  inverse, 82, 214–215, 218
  modeling tide using, 101
  of a sum or difference, 184–185, 239
  transformations of, 115–118
Sine waves (sinusoids; sinusoid waves),
  112
  amplitude of, 113
  general. *See* General sine wave
  modeling a guitar note and, 139
Sinusoidal curve fitting, 128–129
Solving right triangles, 84–85
Speed
  finding, 299–300
  minimizing total time, 304
Spiral of Archimedes, 336–337
Springs
  equilibrium position of, 151
  modeling motion of, 151–152, 204, 249
Square roots
  of a negative number, 315–316
  principal, 315
  simplifying, 4–5
Square-root property, 245
Standard form
  of a complex number, 311–312, 321
  of the equation of a circle, 5
Standard position of an angle, 45
Straight angles, degree measure of, 46
Stretching functions, 23–24
Substitution, factoring using, 246–247
Subtraction. *See also* Differences
  of complex numbers, 312–313
  of fractions in an identity, 172–173
Sum(s). *See also* Addition
  cosine of, 177–179, 180
  evaluating, 201
  expressing as a product, 201
  expressing products as, 199–200
  of linear and trigonometric functions,
    149
  parallelogram law and, 287
  of rational and trigonometric functions,
    149–150
  sine of, 184–185, 239
  tangent of, 185–187
  of two trigonometric functions,
    150–151
  vectors as components of, 287
Sum and difference identities
  cosine of a difference and, 179–180
  simplification with, 180, 186
  sine of a sum or difference and,
    184–185
  tangent of a sum or difference and,
    185–187
Sum-to-product identities, 200–201
Switch-and-solve method, 33–34
Symmetric graphs, 26–27

## T

Tangent equations
  with angle multiples, 238–239
  basic, 231–232

Tangent function, 140. *See also*
  Trigonometric functions
  asymptotes of, 140
  domain of, 140
  graphs of, 140–143
  identities of, 161
  inverse, 82, 217–218
  of a sum or difference, 185–187
Taylor polynomials, 121–122
Terminal point of a vector, 286
Terminal side of an angle, 45
30-60-90 triangles, 75
Tide, modeling using sine function, 101
Time, total, minimizing, 304
Total profit, maximizing, 207
Transformations of functions, 20–26
  horizontal translation, 20–22
  multiple, 25–26
  reflection, 22–23
  rigid and nonrigid, 25
  of sine and cosine, 115–118
  stretching and shrinking, 23–24
  vertical translation, 24
Translation
  horizontal, of functions, 20–22, 117
  vertical, of functions, 24, 117
Triangle(s)
  angles of. *See* Angles
  area of, 278–285
  drawing with polar coordinates, 340
  45-45-90, 74
  oblique. *See* Oblique triangles
  right. *See* Right triangles
  solving using bearing, 263–264
  30-60-90, 75
Triangle inequality, 270
Trigonometric equations
  conditional. *See* Conditional trigono-
    metric equations
  involving identities, 240–241
Trigonometric form of a complex number,
  319–321
Trigonometric functions, 71–80. *See also*
  Cosecant function; Cosine function;
  Cotangent function; Secant function;
  Sine function; Tangent function
  approximate values for, 77–78
  cofunctions of, 180
  combining with algebraic functions,
    148–150
  evaluating, 72–73, 77–78
  evaluating using reference
    angles, 100
  factoring with, 171
  graphs of. *See* Graphs
  inverse, 81–83
  isolating, 232–233
  at multiples of 30°, 75–76
  at multiples of 45°, 73–74
  odd and even, 164–166, 181–182
  in right triangles, 83–84
  transformations of. *See* Transformations
    of functions
  trigonometric ratio definition of, 71–73
  unit circle definitions of, 110
  writing one function in terms of
    another, 163

Trigonometric identities, 160–212
  cofunction, 180–182
  cosine of a difference, 179–180
  cosine of a sum, 177–179, 180
  double-angle, 189–190
  with equal fractions, 171–172
  half-angle, 191–195
  identifying, 166
  product-to-sum, 199–200
  Pythagorean, 161–162
  reciprocal, 161–162
  reduction formula and, 202–204
  simplifying with, 162–164
  sine of a sum or difference, 184–185
  sum-to-product, 200–201
  tangent of a sum or difference, 185–187
  triple-angle, 191
  verifying, 170–176, 194
Trigonometric ratios, 71
Triple-angle identities, 191

## U

Unit circle, 5, 110–121
  graphing cosine functions and, 114–115
  graphing sine functions and, 111–113
  transformations of sine and cosine and,
    115–118
  trigonometric functions defined using,
    110–111
Unit vectors, 291–292

## V

Variables
  dependent, 14
  independent, 14
  solving for, using inverse functions,
    233
Vector(s), 285–295
  angle between, 290–291
  component form of, 289–290
  as components of sums, 287
  definitions related to, 285–286
  direction of, 286, 289–290
  dot product of, 290, 291
  equal, 286
  horizontal and vertical components of,
    288–289
  inclined-plane problems and, 296–299
  initial point of, 286
  navigation problems and, 299–301
  opposites of, 288
  orthogonal, 290
  parallel, 290
  perpendicular, 290
  resultants and, 295–296
  scalar multiplication and addition of,
    286–288
  terminal point of, 286
  unit, 291–292
Vector qualities, 285
  magnitude of, 286, 289–290

Velocity
  angular, 64–65
  linear, 66–67
Vertex of an angle, 45
Vertical asymptotes, 133–134
Vertical component of a
    vector, 288–289
Vertical lines, graphing, 8
Vertical translation of functions, 24, 117
Viewing angle, best, modeling, 278
Vinci, Leonardo da, 3

## W

Weight of an object,
    finding, 297–298
Witch of Agnesi, 343

## X

*x*-axis in Cartesian coordinate system, 2
*x*-intercept, 8
*xy*-plane, 2

## Y

*y*-axis, symmetry about, 26–27
*y*-axis in Cartesian coordinate
    system, 2
*y*-intercept, 8

## FUNCTION GALLERY

## THE SINE AND COSINE FUNCTIONS

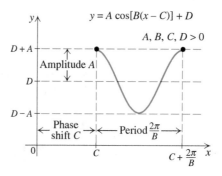

**FUNCTION GALLERY**

## PERIODS OF SINE, COSINE, AND TANGENT ($B > 1$)

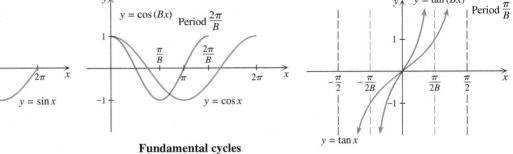

**Fundamental cycles**

**FUNCTION GALLERY**

## TRIGONOMETRIC FUNCTIONS

|  | $y = \sin(x)$ | $y = \cos(x)$ | $y = \tan(x)$ |
|---|---|---|---|
| Domain ($k$ any integer) | $(-\infty, \infty)$ | $(-\infty, \infty)$ | $x \neq \dfrac{\pi}{2} + k\pi$ |
| Range | $[-1, 1]$ | $[-1, 1]$ | $(-\infty, \infty)$ |
| Period | $2\pi$ | $2\pi$ | $\pi$ |
| Fundamental cycle | $[0, 2\pi]$ | $[0, 2\pi]$ | $\left[-\dfrac{\pi}{2}, \dfrac{\pi}{2}\right]$ |

|  | $y = \csc(x)$ | $y = \sec(x)$ | $y = \cot(x)$ |
|---|---|---|---|
| Domain ($k$ any integer) | $x \neq k\pi$ | $x \neq \dfrac{\pi}{2} + k\pi$ | $x \neq k\pi$ |
| Range | $(-\infty, -1] \cup [1, \infty)$ | $(-\infty, -1] \cup [1, \infty)$ | $(-\infty, \infty)$ |
| Period | $2\pi$ | $2\pi$ | $\pi$ |
| Fundamental cycle | $[0, 2\pi]$ | $[0, 2\pi]$ | $[0, \pi]$ |

**FUNCTION GALLERY**

# SOME BASIC FUNCTIONS OF TRIGONOMETRY

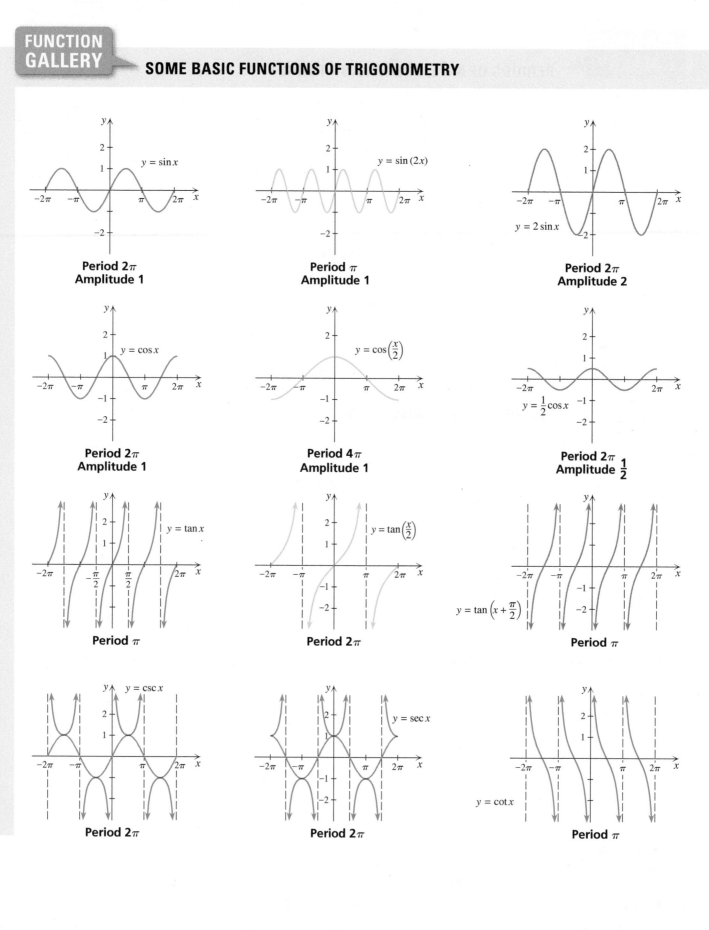

**Period $2\pi$**
**Amplitude 1**

**Period $\pi$**
**Amplitude 1**

**Period $2\pi$**
**Amplitude 2**

**Period $2\pi$**
**Amplitude 1**

**Period $4\pi$**
**Amplitude 1**

**Period $2\pi$**
**Amplitude $\frac{1}{2}$**

**Period $\pi$**

**Period $2\pi$**

**Period $\pi$**

**Period $2\pi$**

**Period $2\pi$**

**Period $\pi$**

**FUNCTION GALLERY**

**INVERSE TRIGONOMETRIC FUNCTIONS**

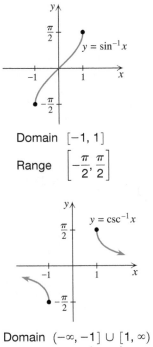

Domain  $[-1, 1]$

Range  $\left[-\dfrac{\pi}{2}, \dfrac{\pi}{2}\right]$

Domain  $[-1, 1]$
Range  $[0, \pi]$

Domain  $(-\infty, \infty)$

Range  $\left(-\dfrac{\pi}{2}, \dfrac{\pi}{2}\right)$

Domain  $(-\infty, -1] \cup [1, \infty)$

Range  $\left[-\dfrac{\pi}{2}, 0\right) \cup \left(0, \dfrac{\pi}{2}\right]$

Domain  $(-\infty, -1] \cup [1, \infty)$

Range  $\left[0, \dfrac{\pi}{2}\right) \cup \left(\dfrac{\pi}{2}, \pi\right]$

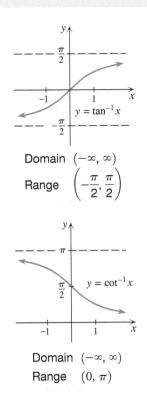

Domain  $(-\infty, \infty)$

Range  $(0, \pi)$

FUNCTION GALLERY

## FUNCTIONS IN POLAR COORDINATES

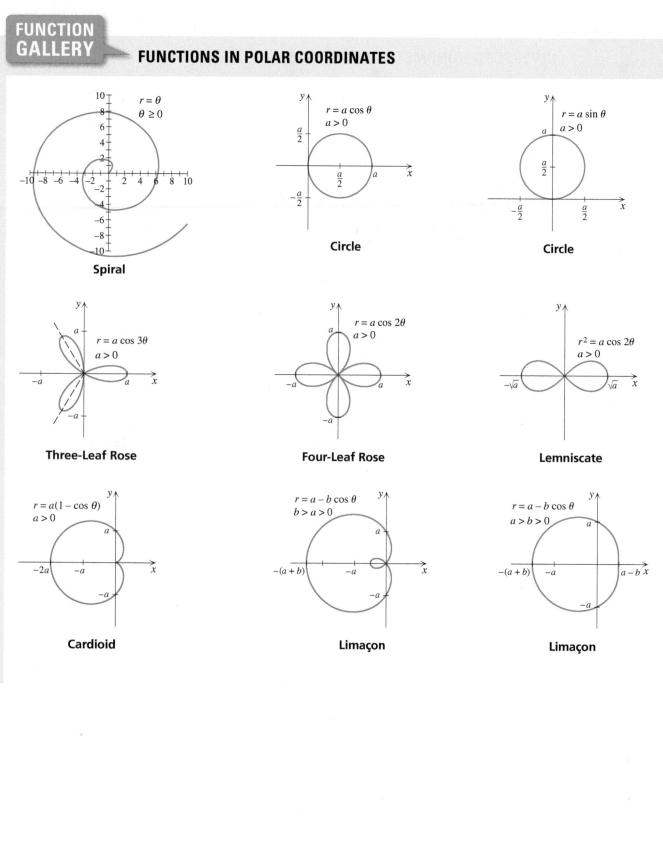

$r = \theta$
$\theta \geq 0$

**Spiral**

$r = a \cos \theta$
$a > 0$

**Circle**

$r = a \sin \theta$
$a > 0$

**Circle**

$r = a \cos 3\theta$
$a > 0$

**Three-Leaf Rose**

$r = a \cos 2\theta$
$a > 0$

**Four-Leaf Rose**

$r^2 = a \cos 2\theta$
$a > 0$

**Lemniscate**

$r = a(1 - \cos \theta)$
$a > 0$

**Cardioid**

$r = a - b \cos \theta$
$b > a > 0$

**Limaçon**

$r = a - b \cos \theta$
$a > b > 0$

**Limaçon**

 **Geometry**

### Rectangle

Area = $LW$

Perimeter = $2L + 2W$

### Square

Area = $s^2$

Perimeter = $4s$

### Triangle

Area = $\frac{1}{2}bh$

### Right Triangle

Area = $\frac{1}{2}ab$

Pythagorean theorem:
$c^2 = a^2 + b^2$

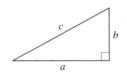

### Parallelogram

Area = $bh$

### Trapezoid

Area = $\frac{1}{2}h(b_1 + b_2)$

### Circle

Area = $\pi r^2$

Circumference = $2\pi r$

### Right Circular Cone

Volume = $\frac{1}{3}\pi r^2 h$

Lateral surface area = $\pi r\sqrt{r^2 + h^2}$

### Right Circular Cylinder

Volume = $\pi r^2 h$

Lateral surface area = $2\pi rh$

### Sphere

Volume = $\frac{4}{3}\pi r^3$

Surface area = $4\pi r^2$

### Metric Abbreviations

| Length | | Volume | | Weight | |
|---|---|---|---|---|---|
| mm | millimeter | mL | milliliter | mg | milligram |
| cm | centimeter | cL | centiliter | cg | centigram |
| dm | decimeter | dL | deciliter | dg | decigram |
| m | meter | L | liter | g | gram |
| dam | dekameter | daL | dekaliter | dag | dekagram |
| hm | hectometer | hL | hectoliter | hg | hectogram |
| km | kilometer | kL | kiloliter | kg | kilogram |

### English-Metric Conversion

| Length | Volume (U.S.) | Weight |
|---|---|---|
| 1 in. = 2.540 cm | 1 pt = 0.4732 L | 1 oz = 28.35 g |
| 1 ft = 30.48 cm | 1 qt = 0.9464 L | 1 lb = 453.6 g |
| 1 yd = 0.9144 m | 1 gal = 3.785 L | 1 lb = 0.4536 kg |
| 1 mi = 1.609 km | | |

| Length | Volume (U.S.) | Weight |
|---|---|---|
| 1 cm = 0.3937 in. | 1 L = 2.2233 pt | 1 g = 0.0353 oz |
| 1 cm = 0.03281 ft | 1 L = 1.0567 qt | 1 g = 0.002205 lb |
| 1 m = 1.0936 yd | 1 L = 0.2642 gal | 1 kg = 2.205 lb |
| 1 km = 0.6215 mi | | |

 # Trigonometry

## Trigonometric Functions

If the angle $\alpha$
(in standard position)
intersects the unit circle
at $(x, y)$, then

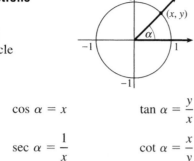

$$\sin \alpha = y \qquad \cos \alpha = x \qquad \tan \alpha = \frac{y}{x}$$

$$\csc \alpha = \frac{1}{y} \qquad \sec \alpha = \frac{1}{x} \qquad \cot \alpha = \frac{x}{y}$$

## Trigonometric Ratios

If $(x, y)$ is any point
other than the origin
on the terminal side of $\alpha$
and $r = \sqrt{x^2 + y^2}$, then

$$\sin \alpha = \frac{y}{r} \qquad \cos \alpha = \frac{x}{r} \qquad \tan \alpha = \frac{y}{x}$$

$$\csc \alpha = \frac{r}{y} \qquad \sec \alpha = \frac{r}{x} \qquad \cot \alpha = \frac{x}{y}$$

## Right Triangle Trigonometry

If $\alpha$ is an acute angle
of a right triangle, then

$$\sin \alpha = \frac{\text{opp}}{\text{hyp}} \qquad \cos \alpha = \frac{\text{adj}}{\text{hyp}} \qquad \tan \alpha = \frac{\text{opp}}{\text{adj}}$$

$$\csc \alpha = \frac{\text{hyp}}{\text{opp}} \qquad \sec \alpha = \frac{\text{hyp}}{\text{adj}} \qquad \cot \alpha = \frac{\text{adj}}{\text{opp}}$$

## Special Right Triangles

## Exact Values of Trigonometric Functions

| $x$ degrees | $x$ radians | $\sin x$ | $\cos x$ | $\tan x$ |
|---|---|---|---|---|
| $0°$ | $0$ | $0$ | $1$ | $0$ |
| $30°$ | $\frac{\pi}{6}$ | $\frac{1}{2}$ | $\frac{\sqrt{3}}{2}$ | $\frac{\sqrt{3}}{3}$ |
| $45°$ | $\frac{\pi}{4}$ | $\frac{\sqrt{2}}{2}$ | $\frac{\sqrt{2}}{2}$ | $1$ |
| $60°$ | $\frac{\pi}{3}$ | $\frac{\sqrt{3}}{2}$ | $\frac{1}{2}$ | $\sqrt{3}$ |
| $90°$ | $\frac{\pi}{2}$ | $1$ | $0$ | — |

## Basic Identities

$$\tan x = \frac{\sin x}{\cos x} = \frac{1}{\cot x} \qquad\qquad \cot x = \frac{\cos x}{\sin x} = \frac{1}{\tan x}$$

$$\sin x = \frac{1}{\csc x} \qquad\qquad\qquad\quad \csc x = \frac{1}{\sin x}$$

$$\cos x = \frac{1}{\sec x} \qquad\qquad\qquad\quad \sec x = \frac{1}{\cos x}$$

## Pythagorean Identities

$$\sin^2 x + \cos^2 x = 1 \qquad\qquad 1 + \cot^2 x = \csc^2 x$$

$$\tan^2 x + 1 = \sec^2 x$$

## Odd Identities

$$\sin(-x) = -\sin(x) \qquad\qquad \csc(-x) = -\csc(x)$$

$$\tan(-x) = -\tan(x) \qquad\qquad \cot(-x) = -\cot(x)$$

## Even Identities

$$\cos(-x) = \cos(x) \qquad\qquad \sec(-x) = \sec(x)$$